TERRAIN VEHICLE SYSTEMS

PROCEEDINGS OF THE

Second International Conference

OF THE INTERNATIONAL SOCIETY FOR

Terrain-Vehicle Systems

AUGUST 29 - SEPTEMBER 2, 1966

QUEBEC CITY, QUEBEC
ST. JOVITE, QUEBEC
HAMILTON, ONTARIO
CANADA

Editors:

J. N. SIDDALL

and

P. H. SOUTHWELL

UNIVERSITY OF TORONTO PRESS

Printed in Canada
Reprinted in 2018
ISBN 978-1-4875-7305-8 (paper)

Contents

HIGH SPEED TRANSPORTATION SYSTEMS

VEHICLE MOBILITY AND TERRAIN EVALUATION

ENGINEERING ASPECTS OF VEHICLES

VEHICLE MECHANICS

SOIL AND SNOW MEASUREMENTS AND INSTRUMENTATION

THEORY AND DEFORMATION OF SOILS

SOIL AND VEHICLE BEHAVIOUR

PERFORMANCE AND PREDICTION

HIGH SPEED TRANSPORTATION SYSTEMS

The Development of a Methodology for the Synthesis and Analysis of Transportation

PETER JOHN METTAM*

ABSTRACT

"Transportation" results from the synthesis of economic, sociologic, political, environmental, and technological factors into a means whereby people and goods can be moved with acceptable levels of safety, comfort, convenience, and cost. In defining what form transportation should take now or in the future, here or abroad, these factors must all be examined to the degree appropriate to each particular problem. The problems that are important to transportation planners are those of policy, research, and development, alternative approaches, and specific engineering developments. A methodology that may be used to structure each of these problem areas in any specific situation is discussed in terms of its philosophy, its logic, and its potential.

. . . for several hundred years the great evolutionary hormones of knowledge and technology have been pressing us . . . whether we like it or not, into a single coordinated humankind. The scattered and competing pasts are being bound together. Everywhere now we begin to see men and nations beginning the deliberate design of development with a growing confidence in the choice and creation of their own future.

John R. Platt, *The Step to Man*

TRANSPORTATION DEVELOPMENT PROBLEMS

THE PROBLEMS THAT are facing every country in the world today, engendered by mushrooming economic and population growth, impose an extraordinary burden on those responsible for transportation systems development. Without a seasoned, balanced, planned, and co-ordinated "blueprint" for the future transportation capabilities of a country, that country's future strength and economic well-being will be in jeopardy. This fact is recognized at the highest levels that forge political and technological progress in the United States.

*Booz-Allen Applied Research, Inc., Bethesda, Maryland.

Responsible as it has been for providing the wherewithal to enable some eighty per cent of all travel in this country to be accomplished, the Bureau of Public Roads has for some years been looking to the future in an endeavour to resolve just what its contribution to the nation's transportation systems of the future should be. Ideally, the goal of the future is to provide each individual with the means to travel for whatever purpose, to any destination, with whatever goods, and at whatever pace suits his purpose best. In the development of the nation's transportation systems, it is essential to keep such an ideal clearly in view while examining what can currently be accomplished, and what the future should be persuaded to produce.

The major problem areas to which the transportation planner must address himself are describable, in general terms, as follows:

Problems of Policy: should a particular road system be developed, or should it be deferred to other more sophisticated systems.

Problems of R & D Planning: R & D budgets reflect the future planning of highway technology and construction necessary to carry out planned programs. The question of which major R & D areas should be given priority in any 5 year planning period can only be resolved by examining the total needs for technology in the next five, ten or fifteen years, as required by particular policies.

Problems of System Evaluation: Engineers in the transportation industry, well aware of the transportation problem facing many areas in the United States at this time, have proposed and are proposing many different applications of present and developing technology. If these efforts are to be fairly encouraged and properly utilized they must be evaluated in the light of specific needs and these needs must be applied on a national rather than a parochial scale.

Problems of Technical and Engineering Alternatives: Within a transportation system, the particular functions performed have various practical means for fulfillment, either developed, under development, or proposed, such as in the highway safety area. The resolution of which of these means offer the most effective solutions is a necessary activity of transportation planners.

It is evident that these problem areas, while primarily concerned with highways, cannot be resolved without full awareness of the nature and capabilities of other modes of transportation, present or proposed, or without a thorough understanding of the interrelationship between society, economy, environment, and transportation. Since without people and their needs there would be no economy nor any demand for transportation, without society there would be no legal or political constraints on transportation, and without environment there would be no resistance to transportation, an adequate study of transportation can not be conducted without developing a clear understanding of the interactions between these elements and a transportation system.

Transportation as a subject has been studied for many years. When it first came under serious study, the problems involved were relatively few. The demand was definable in relatively simple terms, and the technology to fulfill the demands was limited and available. The demand was created in general by a need to trade. As society developed and became more sophisticated in its tastes, the economy grew, and technology became more diverse. The demand now arose from the need to trade, to supply industries with essential raw material, and to provide people with transportation. The transportation system of a region became an essential element of its economy and a good investment for wealthy entrepreneurs. Nowadays, the demand for transportation arises from all segments of the population and the economy, and transportation has moved from the realm of a purely private investment proposition, to that of a national asset.

In the early stages of transportation, the growth of society and technology was relatively slow, and there was no reason to feel that this would not continue to be so into the foreseeable future. On this basis, studies could project into the future with a high level of confidence. Our experience in this century, particularly in the last 25 years, and more particularly in the last 10 years, has shown us that this can no longer be accepted and that trends or predictions regarding future behaviour are highly suspect, since almost any month some new facet of the technological base for our future is uncovered that can have far reaching consequences. Consequently, we can no longer rely on projections into the future that are based on past performance or present trends, but we must determine for ourselves what we must achieve in the future to make the future acceptable. In the technological growth of transportation, this means that we must first examine, for probable economies and societies, what the character of transportation should be, then derive from this information what the most important developments in the future must be, and then, plan for them to occur on some practical time scale.

It is in these areas of the understanding of the many facets of transportation, their relationships, their significance to any situation, and their influence on transportation technology, that the Bureau of Public Roads AFT programme will be most significant.

THE METHODOLOGY

The methodology of the AFT Programme (The Analysis of the Functions of Transportation) is designed to develop a computer-oriented mathematical model that can be used in the transportation decision-making process. It is particularly oriented towards the four problem areas outlined earlier.

The procedure adopted in developing this model is that of Systems Analysis—a technique developed over the past 25 years that has its roots in the early operational analyses conducted during World War II, and in the parametric analyses common in aircraft design offices from that time to this. Since then it has grown in use in the aerospace industry in the development of multi-million dollar complex weapon systems, from aircraft to missiles to submarines. In the last five years, through the activities of Secretary MacNamara, and the Assistant Secretary for Defense, Systems Analysis, Dr. Alain Enthoven, it has become standard operating procedure in the Department of Defense: and within the last year, it has been recognized by President Johnson as a valuable tool for national budgetary activities. It would seem none too late for it to be applied to transportation.

THE ANALYTICAL PROCEDURE

Systems Analysis

Systems Analysis is a technique that endeavours to take a very complex subject, and devise a thorough understanding of that subject within the constraints or boundaries of a particular problem. The main steps in developing a systems analysis are as follows:

- State the objectives of the analysis
- Define the boundaries of the system
- Define the elements of the system
- Develop the logical functional structure of the system down to a level where analytical relationships are possible
- Determine all necessary analytical relationships
- Define a set of quantitative values appropriate to the problem of the analysis
- Conduct sample calculations to test and improve the structure
- Develop programming techniques to formalize and structure the model for computer use
- Programme and checkout on a computer and debug the programme
- Conduct a sensitivity analysis of the model for each of the problems to be considered

The results of such an analysis are then available to assist the decision-maker. Structured in this manner, it is possible to relate and evaluate the large number of complex relationships which bind the over-all United States Transportation System together. Tackling the problem can best be described by three major, consecutive areas of activity.

An exploratory activity aimed at providing a conceptual framework upon which to develop the succeeding parts of the programme, and which essentially accomplishes the first four steps and lightly touches on the next three.

An activity that converts the conceptual framework into a fully detailed mathematical model, ready for computer programming. This accomplishes steps five, six, seven, and eight, and touches on step nine.

An activity which programmes the mathematical model for the computer and then uses the programme to explore all of the interdependencies within the model. This defines the major analytical relationships among the transportation requirements, performance, and constraints that comprise the desired functions of transportation, and completes all the steps needed for a thorough systems analysis.

The objectives of the effort to date have been to:

Find the essential characteristics of transportation.

Develop a preliminary, analytical methodology systematizing these characteristics and their functional interaction.

Describe the interrelated characteristics of transportation (requirements, performance, and constraints) in detailed quantitative analytical fashion.

Conclude the development of a detailed analytical methodology interrelating the characteristics of transportation, such that the effect of changes in quantitative value can be assessed.

To meet these objectives, the four task areas have been:

Definition, refinement, and quantification of transportation requirement characteristics.

Definition, refinement, and quantification of transportation performance characteristics.

Definition, refinement, and quantification of transportation constraint characteristics.

Development, refinement, and finalization of a preliminary generalized mathematical model of transportation.

The Scope of the Work

The developed generalized model is to be capable of considering problems that vary in scale from national to local, that consider systems of *any* kind, and that can demand a state-of-the-art not yet developed.*

*The systems in the analysis are generalized and defined in terms of "1D, 2D, or 3D" systems, where:

1D systems are constrained to move along one path

2D systems can move by choice on the earth's surface

3D systems have no restriction

Figure (1) shows a categorization of possible systems.

Operating Medium \ Class of System	"One-Dimensional" Laterally and Vertically Constrained	"Two-Dimensional" Vertically Constrained	"Three- Dimensional" No Constraints
WATER	. Floating Pipeline . Chain Ferry	. Any Form of Ship . Air-Cushion Vehicle	. Submarines
		. Amphibians . Air-Cushion Vehicles	
LAND	. Railroads . Monorails . Cablecars . Subways	. Pedestrians . Cyclists Trucks. . Automobiles, Buses, . Tractors, Tanks . Walking Machines	. Mechanical Mole
AIR	. Pipelines Suspended By Balloons	?	. Balloons . Satellites . Airships . Aircraft . Guided & Ballistic Rockets

FIGURE 1. Classification of transportation systems: very wide range to be handled—to carry people, goods, or any combination

It is essential, for the model's eventual usefulness, that it be capable of examining transportation situation where mode mixes are used, particularly where intercity and intracity systems interact, so as to fully account for the problems of portal-to-portal movement. The four tasks,

The Requirements Task
The Constraints Task
The Performance Task
The Modelling Task

accomplish this work and are described in the rest of the paper.

The Requirements Task

The function of the Requirements Task is to develop the procedures for determining what is required of a transportation system by the society for which it is to be developed. The two major items that define the requirements are:

The Quantity of Movement of people and goods, by trip purpose, time of day, between all origins and destinations.

The Quality of Movement that becomes evident from the population willingness to move. This encompasses the level of safety, the degree of comfort, the measure of convenience, and the user cost that must be provided by an acceptable system.

The Requirements Task is broken down as shown in Figure 2 into several separate analyses:

The demand for people-movement

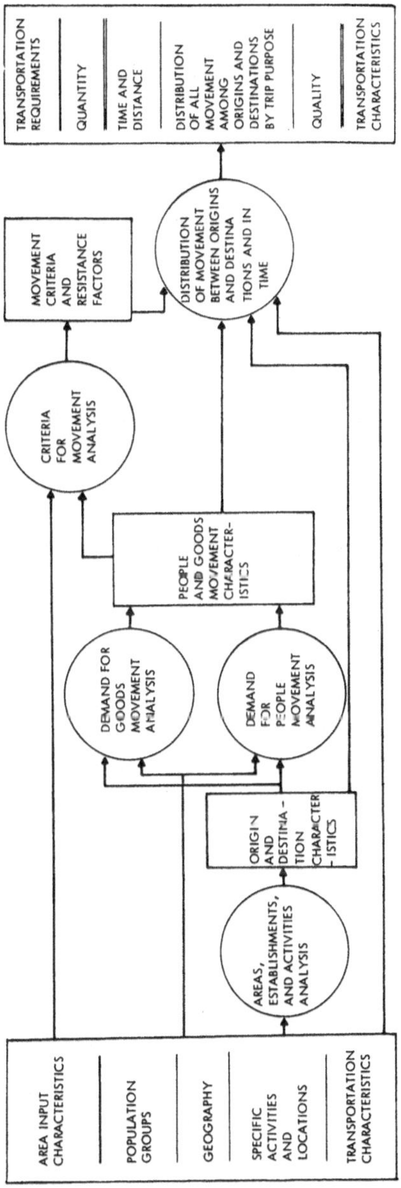

FIGURE 2. Second-level logic diagram of the requirements analysis

The demand for goods-movement
Area, establishment, and activity location
The criteria for movement
The distribution of the demand

The Demand for People-Movement identifies in what ways a family allocates its time and its money during any 24-hour period, and from this determines what trips to what activities family members would make if transportation and the activities were available.

The Demand for Goods-Movement develops similar information about goods-movement but in a more general fashion.

The Analysis of the Criteria for Movement determines both which activities would be chosen, and whether the movement will in fact take place to the chosen activities on the basis of criteria to be developed from the transportation characteristics of safety, comfort, cost, and convenience.

The Distribution of Demand seeks to develop the means for allocating the demand for movement to the activities on the basis of activity choice, and transportation characteristics.

The characteristics of safety, comfort, convenience and cost need some further discussion. In this study, they are considered to be the primary and sole descriptors of the qualities of transportation against which travelers will weigh their appreciation for the system.

Safety refers to the possibility of various degrees of damage or injury resulting from an accident.

Comfort relates to the physiological effects of the journey on the passenger and cargo, and their acceptance.

Convenience relates to the time/distance features of the journey—schedule (speed), frequency, meals, location relative to origin and destination.

Cost defines the dollar cost to the user, direct and indirect, and the non-dollar social cost.

These are the elements that form the basis for defining a resistance to movement between particular origins and destinations, and they constitute major inputs to the Performance Tasks, for some problem categories.

The Constraints Task

This task has as its purpose pulling together all those items about a system that are directly responsible for constraining or limiting its capabilities by restricting its performance. The two major divisions of constraint are:

Natural Environment
Operational Environment.

The *Natural Environment* study is aimed at defining, where possible in mathematical terms, those components of the natural environment, that affect transportation systems. The potential range of these components are listed in Figure 3. The possibility of developing adequate mathematical expressions for all characteristics of interest is not very high, mainly because of the random character of much of the data. However, by using existing classification schemes; wherever possible, accompanied by statistical analysis within each classification, the best possible compromise between effort and exactness will be achieved.

The *Operational Environment* study develops information about the human factors, legal and political constraints, and the effects of existing man-made systems on the operational transportation system, as indicated in Figure 4. Over-all, the Constraints Task affects almost all aspects of the Performance Analysis that follows.

The Performance Task

The operation of a total transportation system is a very complex, finely scheduled operation. There is no intent whatsoever in this analysis to attempt to directly simulate such interactions. The preferred approach is to synthesize the transportation system such that the results of dynamic processes, as functions of transportation system parameters, are utilized in the analysis, rather than the actual dynamic processes themselves. With this in mind, the following are the major subdivisions of the Performance Task, as shown in Figure 5.

Route Description and Characteristics in which the Requirements and Constraints information is identified for each route or leg of a network under examination, and the nature of the route or network is defined.

Schedule Requirements and Processing in which the system is sized and scheduled to meet the demand, incorporating any operational considerations as far as possible.

Vehicular Performance in which the performance characteristics necessary to meet the schedule, the requirements, and the constraints are synthesized.

Command and Control in which the necessary characteristics of command and control in order to provide an acceptable level of safety and schedule reliability are determined.

Operational Support Activities, that account for such necessary operations as loading and unloading, ticketing, assembly, passenger services, and system management operations.

Operational System Reliability and Maintenance in which the system reliability vs. system maintainability levels are explored, for the operational environments of concern. This provides a direct contribution to the analysis of system safety and cost.

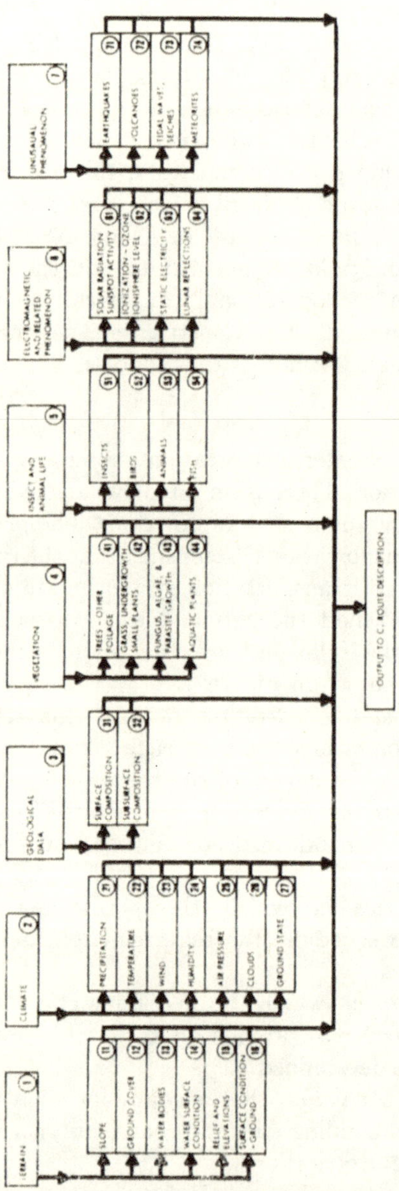

FIGURE 3. Natural environment

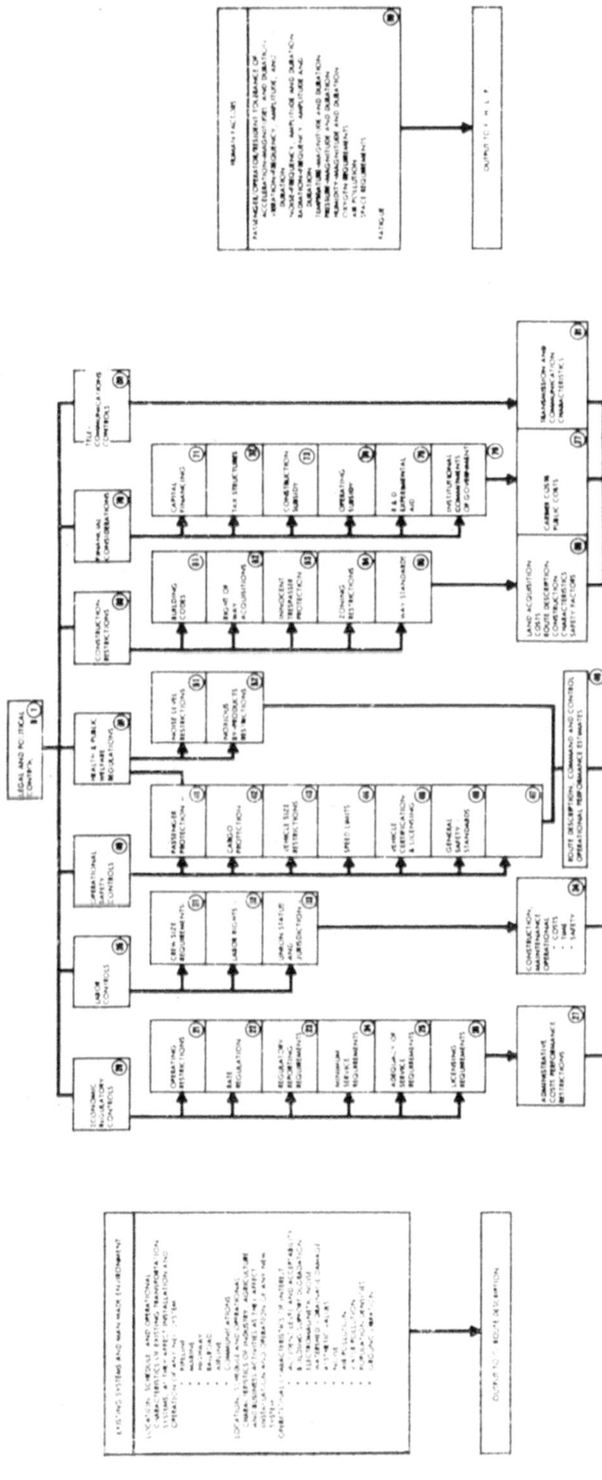

FIGURE 4. Operational environment

Construction, Installation, and Maintenance of Way and Terminal in which all necessary activities of this nature are defined in terms of manpower, equipment, and materials, so as to provide a firm base for system investment costs.

Safety Analysis in which the safety contributions of system operations, construction, and environment are determined and amalgamated to produce the overall safety figure.

Comfort and Convenience Analysis is a synthesis of the data from other sections that relate to comfort and convenience.

System Costs Analysis in which the total investment cost, the operating costs, and the indirect user costs are developed as direct functions of the preceding analysis. In this section also some of the intangible questions regarding subsidies, taxes, and other such quantities are examined.

The Development of the Generalized Model

An over-all structure for transportation problem formulation has been developed. The structure is subdivided into the three major task areas, in each of which much analysis work has been accomplished. The remaining structuring effort essential to effective completion of the study is that of assembling the results of the first three into the generalized model.

This is a multi-phase effort, that starts prior to the tasks just described. The logic structure begins to take shape, commencing with a top-level diagram (Figure 6) representing the total problem. This then breaks down into the three second-level diagrams, one for each major analysis task, and these three, at least in principle, can be subdivided into many more. However, continuing subdivisions become rather academic without some reference to analytical and functional background. It is at this point that the tasks begin, and they continue until the development of the detailed logic and the estimating relationships required are complete. The procedure adopted is to break down the logic diagrams successively as shown in Figure 7 to a final level, Figure 8, where analytical relationships can be provided as in Figure 9.

When this is done for all the tasks, the separate logic diagrams and analytical information are reassembled into an over-all model, for testing and programming. The preliminary model occupies a 15 ft $\times$ 8 ft panel, and is laid out as shown in Figure 10.

Each of the blocks shown in the layout represents a submodel within the main structure.

As an example, the Vehicle Performance area is subdivided into five submodels with the following functions:

Parametric Conversion (Figure 11) in which specific system data is converted to the general form of analysis utilized in the model.

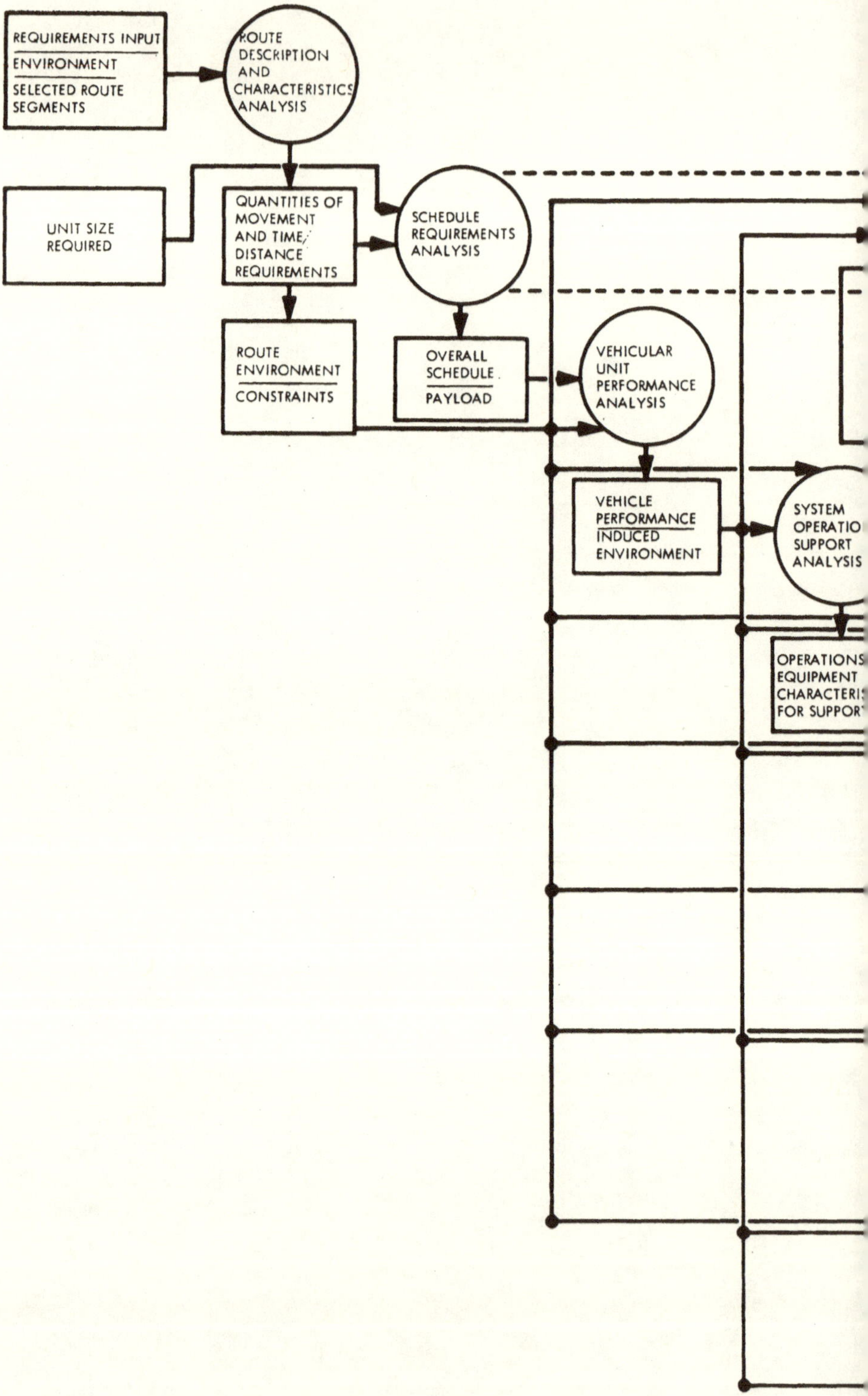

FIGURE 5. Second-level

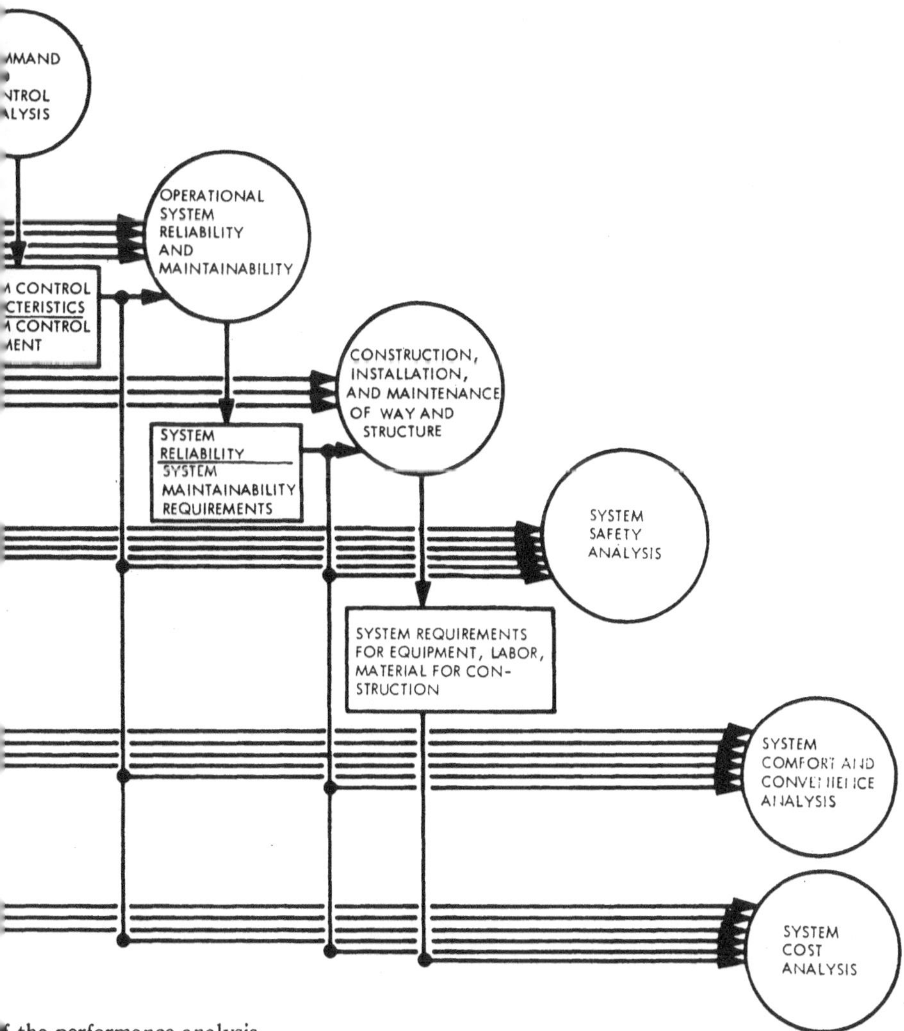

f the performance analysis

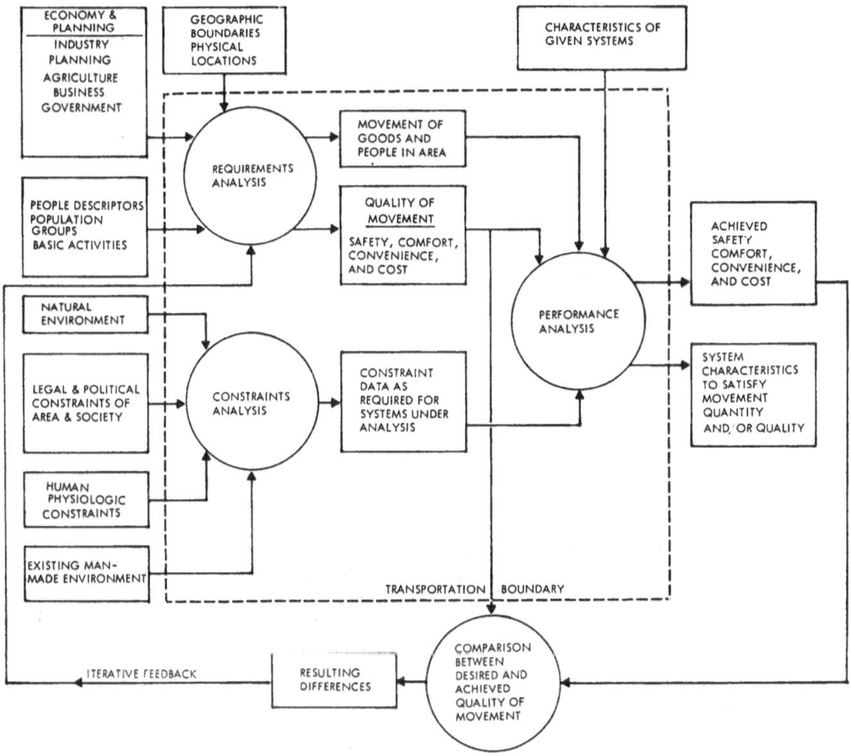

FIGURE 6. First-level logic diagram of transportation

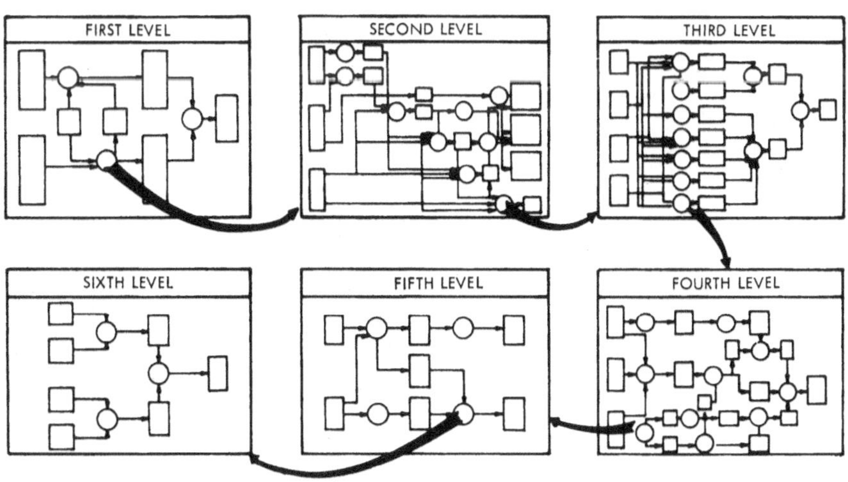

FIGURE 7. The procedure for developing the systems analysis structure

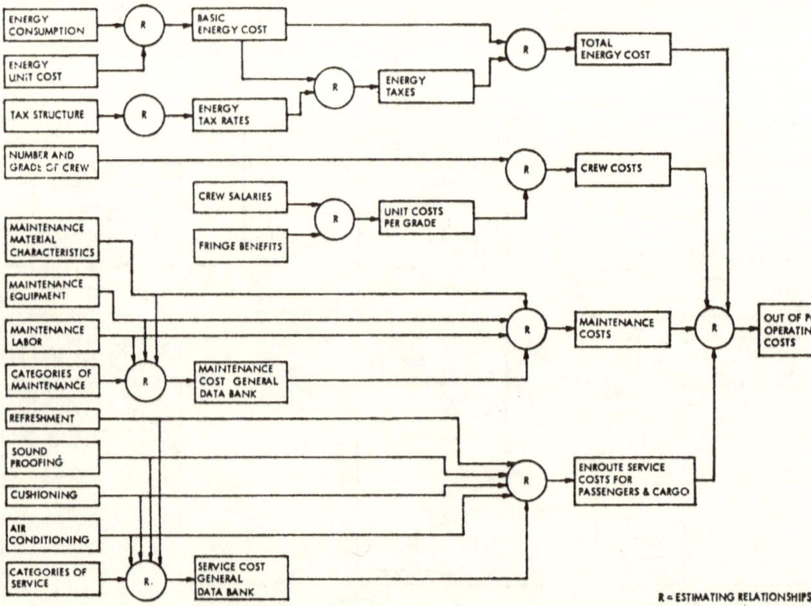

FIGURE 8. A typical fifth-level logic diagram (out-of-pocket operating costs)

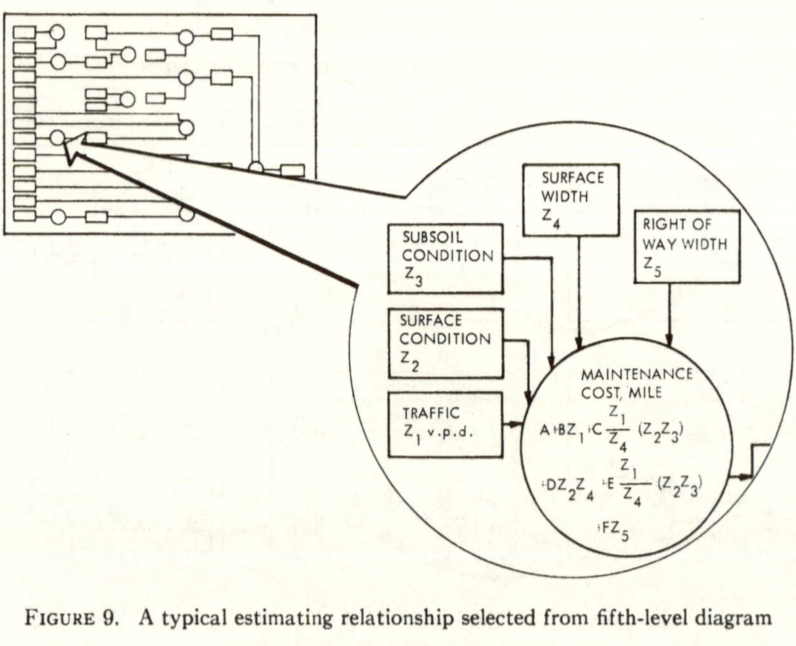

FIGURE 9. A typical estimating relationship selected from fifth-level diagram

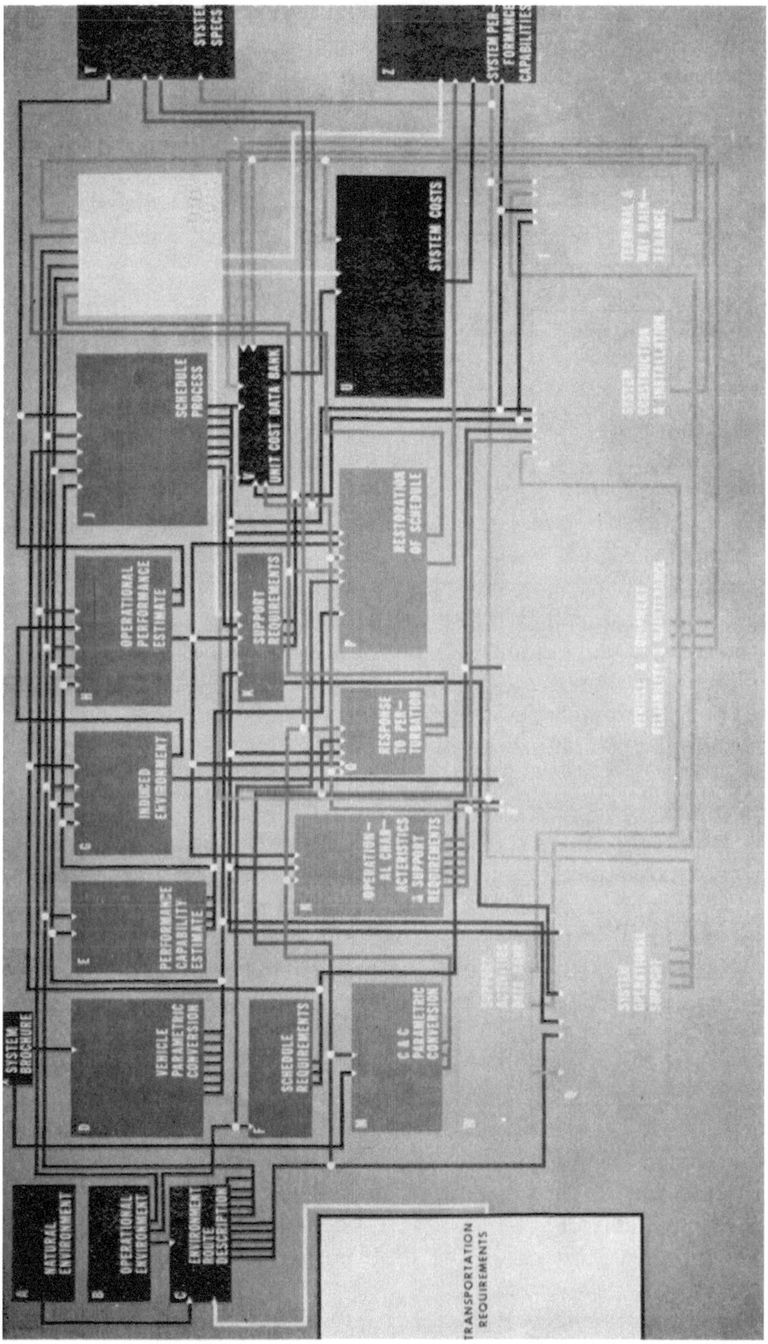

FIGURE 10. Analysis structure layout

Performance Capability Estimate (Figure 12) in which the over-all range of performance capability of the vehicular portion of the system is evaluated, in the absence of constraints.

Induced Environment (Figure 13) in which the non-translational characteristics of the system are defined—noise, vibration, pollution, and so on.

Operational Performance Estimate (Figure 14) in which the actual permissible performance, in the face of legal, social, political, and other environmental constraints is defined.

Support Requirements Analysis (Figure 15) in which all the support activities defined by the design and operation of the vehicular unit are identified.

The Utilization of the Model

The procedures that are planned for use of the model for various problem categories are as follows:

Define, from the problem statement, the known information in all areas of the model.

Identify what specific outputs are desired of the model.

State what parameters are still undefined; these must be provided by the user, or else assumed for the problem in hand.

Structure the sequence of analysis through the logic blocks of the model to provide the most effective computational procedure.

Describe the computational decision logic.

Define the ranges of values of interest.

Compute the desired answers.

The majority of these steps are carried out with the aid of a master file of punched cards that contains the complete library of estimating relationships for all types of problems. Given the desired outputs, the master file is exercised to select those series of relationships that are needed for each output. These series, when amalgamated in the correct logic sequence, form the model structure for the particular problem.

PRELIMINARY SAMPLE CALCULATIONS

To conclude, a brief mention of some early sample calculations is worthwhile. These were conducted towards the end of the first year of the programme, to test our procedures and to demonstrate the direction and depth of our analysis.

They represent three scales of problems in three areas of the United States.

Transportation among 10 major cities in the United States

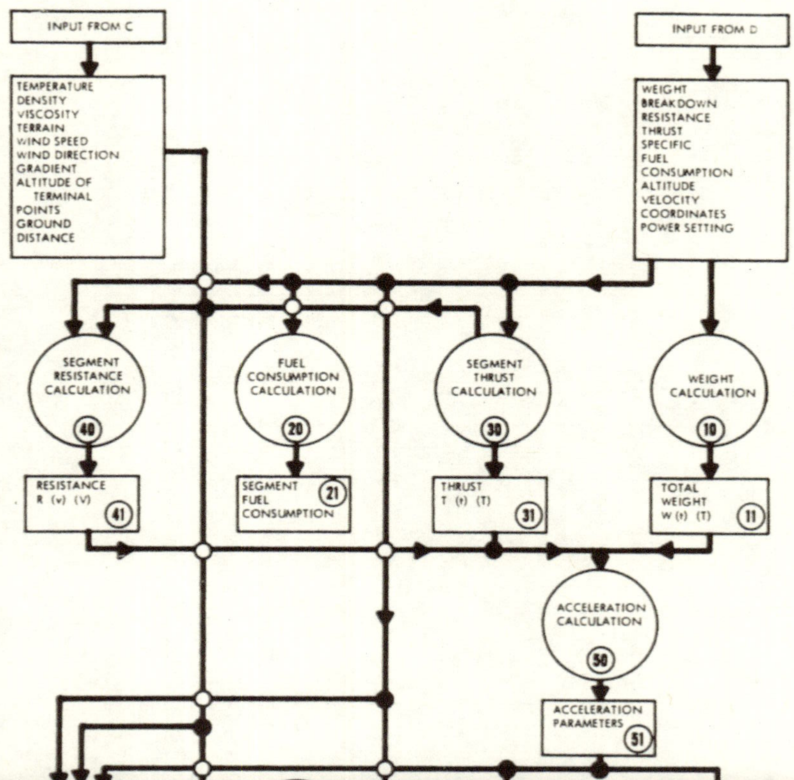
INPUT FROM C
TEMPERATURE
DENSITY
VISCOSITY
TERRAIN
WIND SPEED
WIND DIRECTION
GRADIENT
ALTITUDE OF TERMINAL POINTS
GROUND DISTANCE
INPUT FROM D
WEIGHT BREAKDOWN
RESISTANCE
THRUST
SPECIFIC FUEL CONSUMPTION
ALTITUDE
VELOCITY
COORDINATES
POWER SETTING
SEGMENT RESISTANCE CALCULATION 40
FUEL CONSUMPTION CALCULATION 20
SEGMENT THRUST CALCULATION 30
WEIGHT CALCULATION 10
RESISTANCE R (v) (V) 41
SEGMENT FUEL CONSUMPTION 21
THRUST T (t) (T) 31
TOTAL WEIGHT W (t) (T) 11
ACCELERATION CALCULATION 50
ACCELERATION PARAMETERS 51

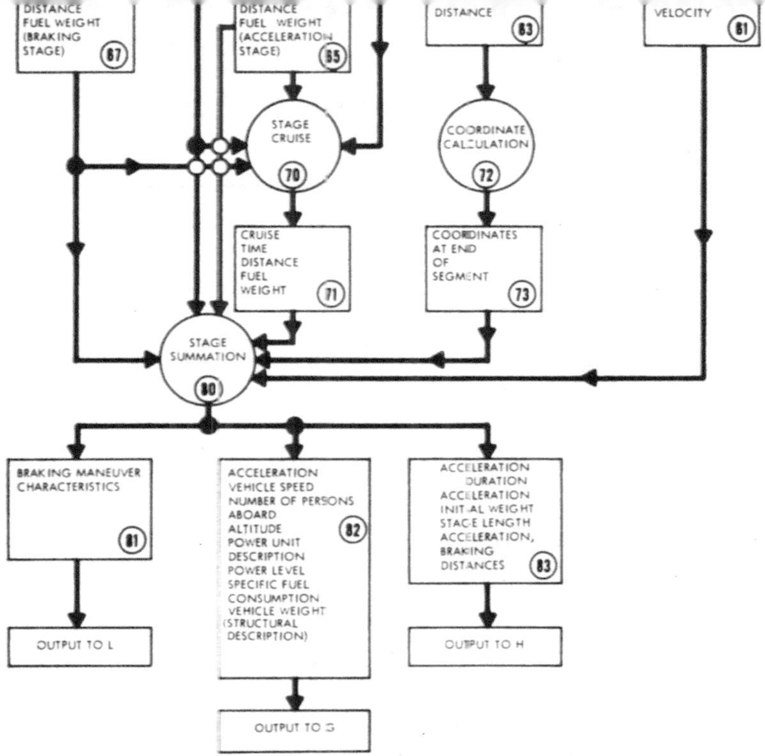

FIGURE 12. Performance capability estimate

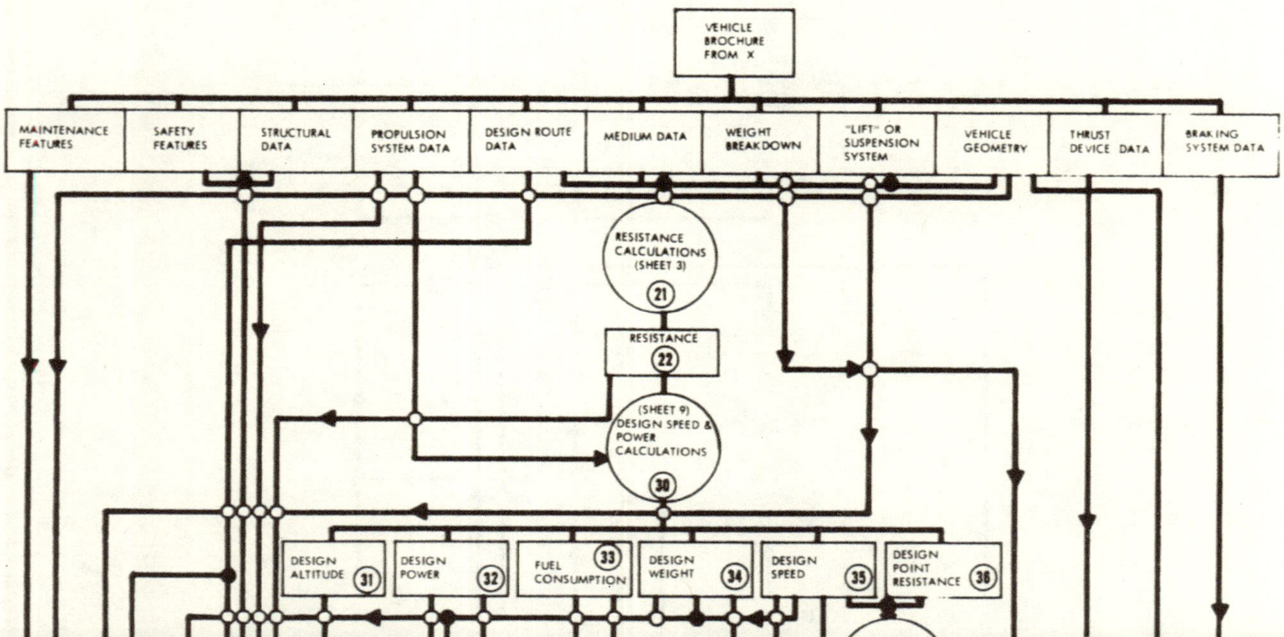

VEHICLE BROCHURE FROM X
MAINTENANCE FEATURES
SAFETY FEATURES
STRUCTURAL DATA
PROPULSION SYSTEM DATA
DESIGN ROUTE DATA
MEDIUM DATA
WEIGHT BREAKDOWN
"LIFT" OR SUSPENSION SYSTEM
VEHICLE GEOMETRY
THRUST DEVICE DATA
BRAKING SYSTEM DATA
RESISTANCE CALCULATIONS (SHEET 3) 21
RESISTANCE 22
(SHEET 9) DESIGN SPEED & POWER CALCULATIONS 30
DESIGN ALTITUDE 31
DESIGN POWER 32
FUEL CONSUMPTION 33
DESIGN WEIGHT 34
DESIGN SPEED 35
DESIGN POINT RESISTANCE 36

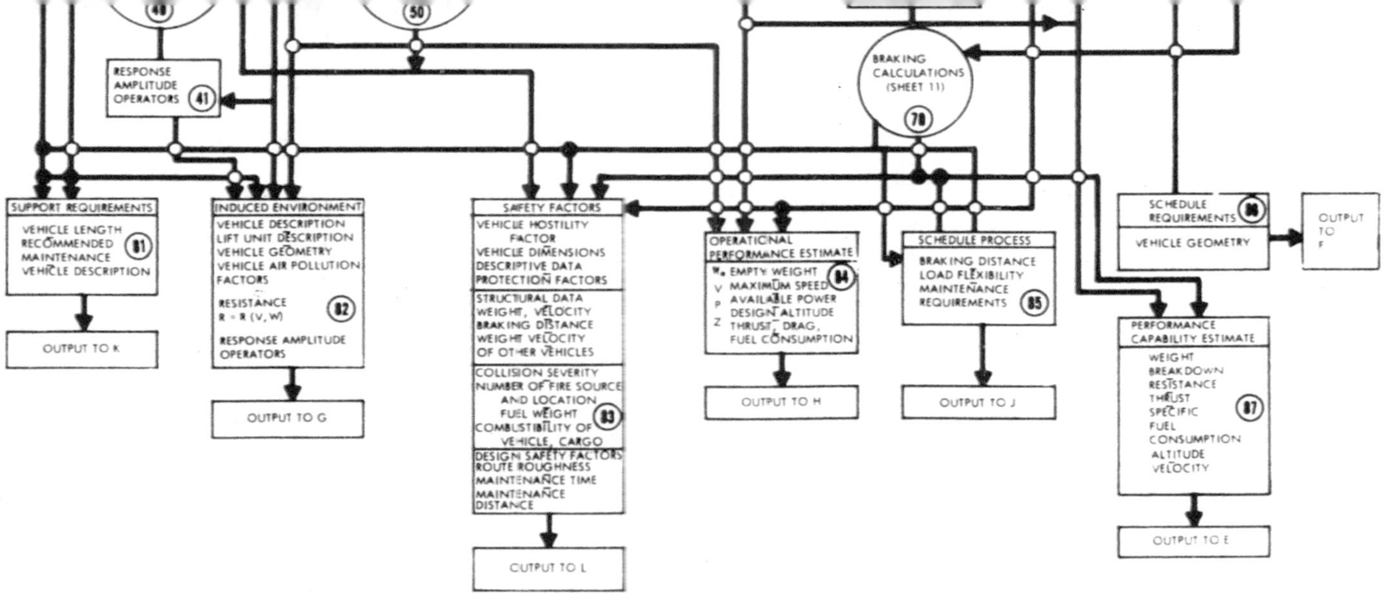

FIGURE 11. Vehicle parametric conversion

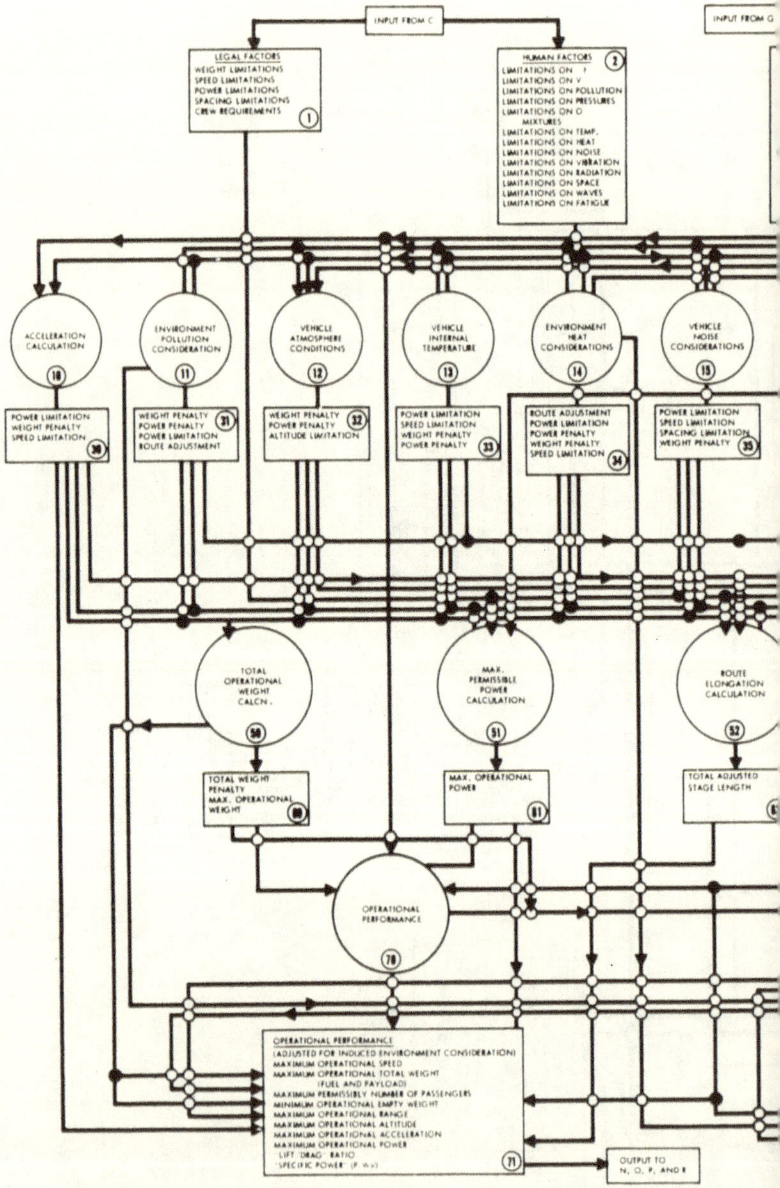

INPUT FROM C
INPUT FROM G
LEGAL FACTORS
WEIGHT LIMITATIONS
SPEED LIMITATIONS
POWER LIMITATIONS
SPACING LIMITATIONS
CREW REQUIREMENTS
1
HUMAN FACTORS
2
LIMITATIONS ON V
LIMITATIONS ON POLLUTION
LIMITATIONS ON PRESSURES
LIMITATIONS ON TEMP.
LIMITATIONS ON HEAT
LIMITATIONS ON NOISE
LIMITATIONS ON VIBRATION
LIMITATIONS ON RADIATION
LIMITATIONS ON SPACE
LIMITATIONS ON WAVES
LIMITATIONS ON FATIGUE
ACCELERATION CALCULATION
10
ENVIRONMENT POLLUTION CONSIDERATION
11
VEHICLE ATMOSPHERE CONDITIONS
12
VEHICLE INTERNAL TEMPERATURE
13
ENVIRONMENT HEAT CONSIDERATIONS
14
VEHICLE NOISE CONSIDERATIONS
15
POWER LIMITATION
WEIGHT PENALTY
SPEED LIMITATION
30
WEIGHT PENALTY
POWER PENALTY
POWER LIMITATION
ROUTE ADJUSTMENT
31
WEIGHT PENALTY
POWER PENALTY
ALTITUDE LIMITATION
32
POWER LIMITATION
SPEED LIMITATION
WEIGHT PENALTY
POWER PENALTY
33
ROUTE ADJUSTMENT
POWER LIMITATION
POWER PENALTY
WEIGHT PENALTY
SPEED LIMITATION
34
POWER LIMITATION
SPEED LIMITATION
SPACING LIMITATION
WEIGHT PENALTY
35
TOTAL OPERATIONAL WEIGHT CALCN.
50
MAX. PERMISSIBLE POWER CALCULATION
51
ROUTE ELONGATION CALCULATION
52
TOTAL WEIGHT PENALTY
MAX. OPERATIONAL WEIGHT
60
MAX. OPERATIONAL POWER
61
TOTAL ADJUSTED STAGE LENGTH
OPERATIONAL PERFORMANCE
70
OPERATIONAL PERFORMANCE
(ADJUSTED FOR INDUCED ENVIRONMENT CONSIDERATION)
MAXIMUM OPERATIONAL SPEED
MAXIMUM OPERATIONAL TOTAL WEIGHT
(FUEL AND PAYLOAD)
MAXIMUM PERMISSIBLY NUMBER OF PASSENGERS
MINIMUM OPERATIONAL EMPTY WEIGHT
MAXIMUM OPERATIONAL RANGE
MAXIMUM OPERATIONAL ALTITUDE
MAXIMUM OPERATIONAL ACCELERATION
MAXIMUM OPERATIONAL POWER
"LIFT/DRAG" RATIO
"SPECIFIC POWER"
71
OUTPUT TO N, O, P, AND R

FIGURE 14. Op

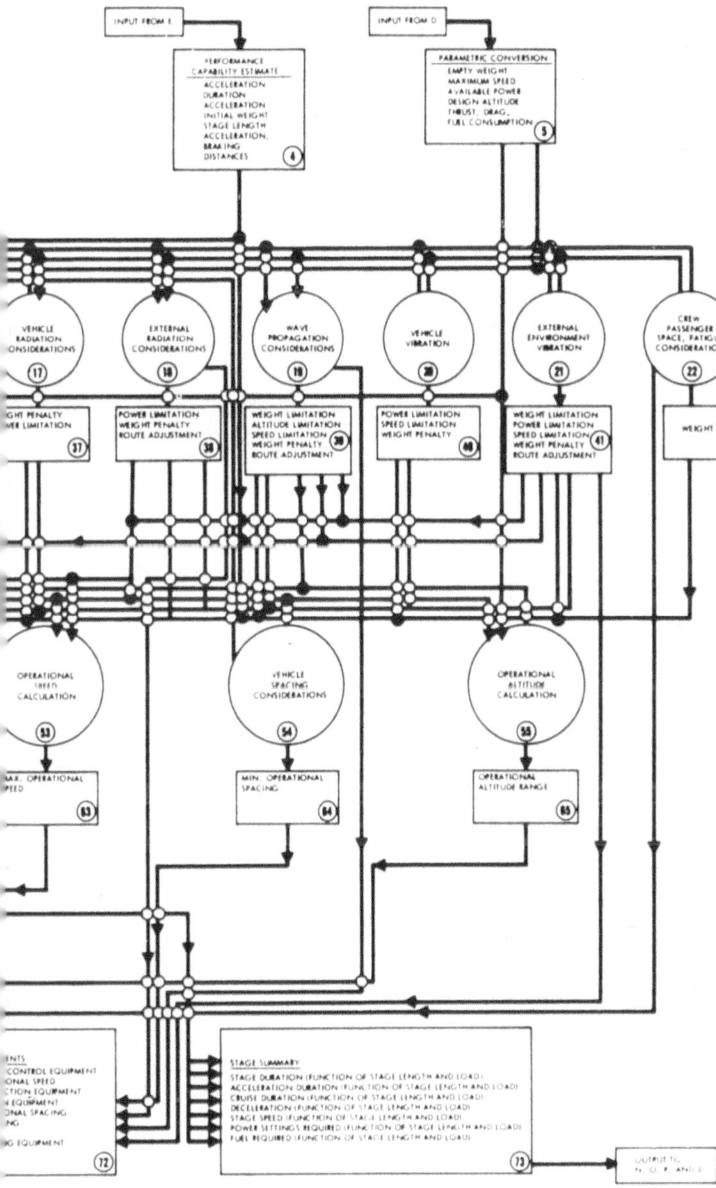

nance estimate

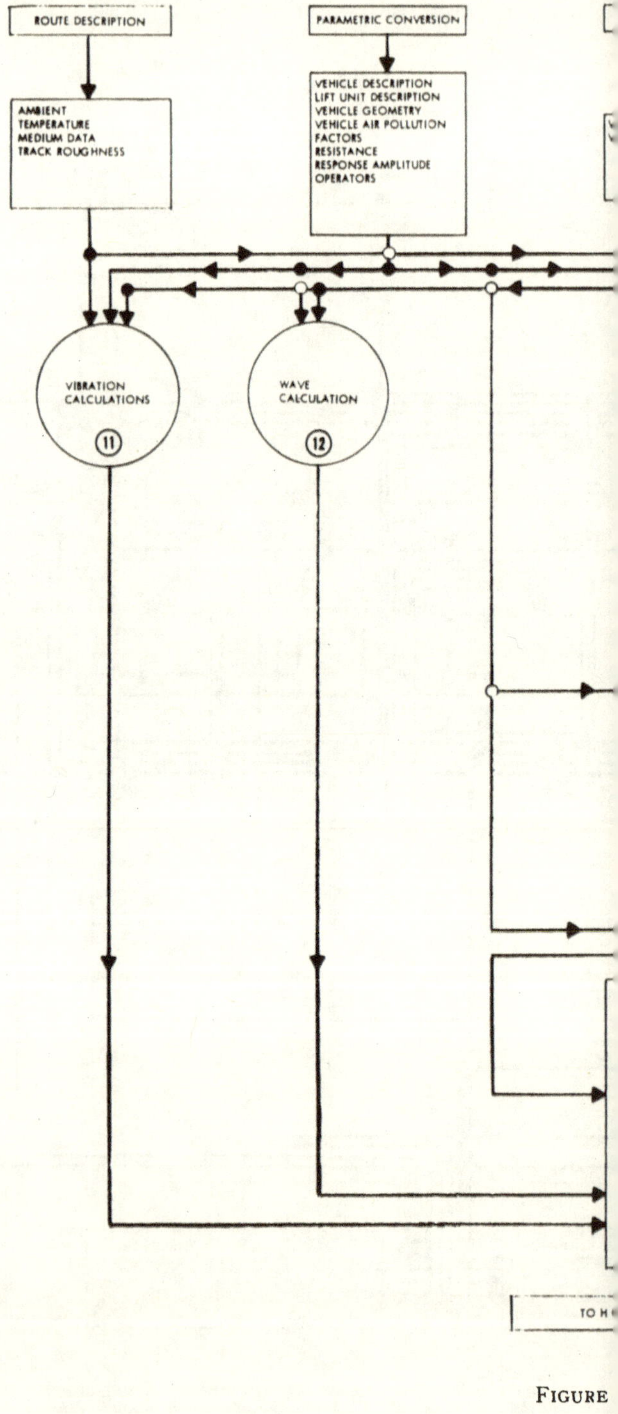
ROUTE DESCRIPTION
PARAMETRIC CONVERSION
AMBIENT
TEMPERATURE
MEDIUM DATA
TRACK ROUGHNESS
VEHICLE DESCRIPTION
LIFT UNIT DESCRIPTION
VEHICLE GEOMETRY
VEHICLE AIR POLLUTION
FACTORS
RESISTANCE
RESPONSE AMPLITUDE
OPERATORS
VIBRATION
CALCULATIONS
11
WAVE
CALCULATION
12

FIGURE

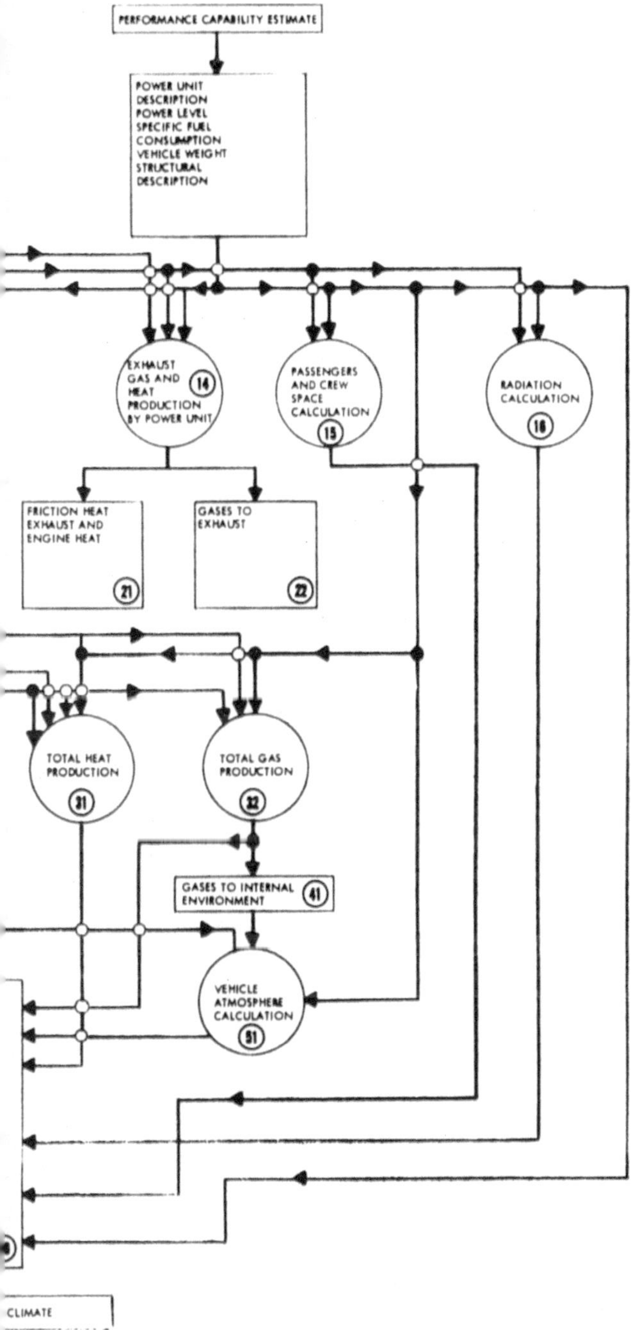

PERFORMANCE CAPABILITY ESTIMATE
POWER UNIT DESCRIPTION POWER LEVEL SPECIFIC FUEL CONSUMPTION VEHICLE WEIGHT STRUCTURAL DESCRIPTION
EXHAUST GAS AND HEAT PRODUCTION BY POWER UNIT 14
PASSENGERS AND CREW SPACE CALCULATION 15
RADIATION CALCULATION 16
FRICTION HEAT EXHAUST AND ENGINE HEAT 21
GASES TO EXHAUST 22
TOTAL HEAT PRODUCTION 31
TOTAL GAS PRODUCTION 32
GASES TO INTERNAL ENVIRONMENT 41
VEHICLE ATMOSPHERE CALCULATION 51
CLIMATE

vironment

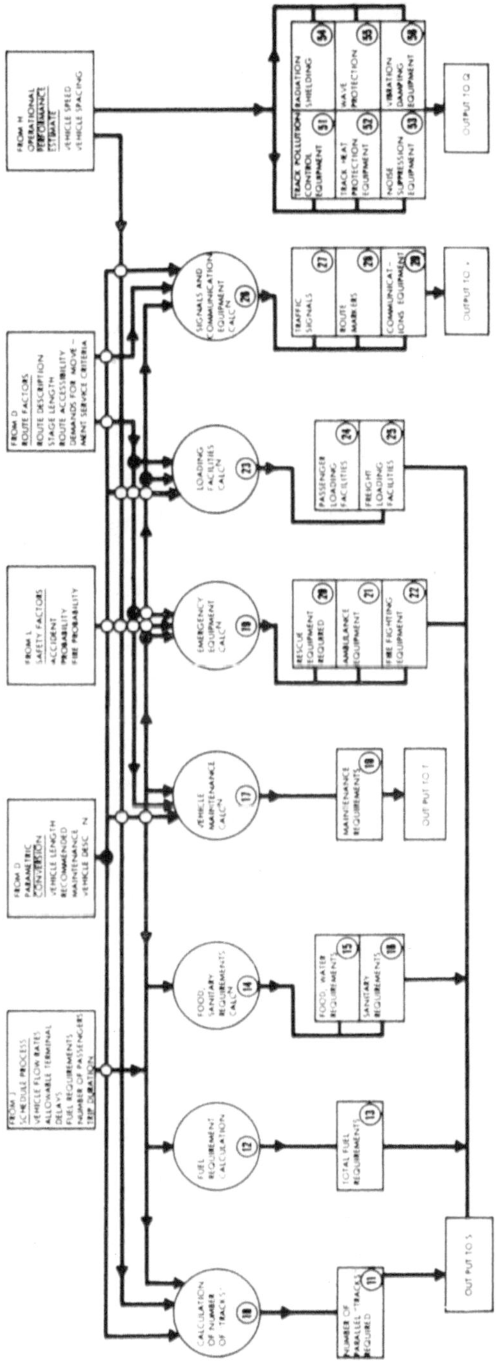

FIGURE 15. Support requirements

Transportation within a typical megalopolis

Transportation with a typical city within the megalopolis.

and they also represented three technological problems:

A comparison of present day road, rail, and air systems for a particular freight movement

Definition of specifications for a 1D rapid transit system for people and freight movement

Evaluation of an automated highway versus a conventional highway in a commuter situation.

The primary outcome of these sample calculations was:

(1) Evidence that the generalized model could handle a wide variety of problems

(2) Identification of the gaps in the analysis at that time

(3) Information to aid in defining the formulation of problem statement, for sufficient information to be provided to the model

(4) Development of a preliminary punched-card bank for further development

(5) Delineation of those areas in the model that were receiving emphasis out of proportion to their importance

The analytical procedure used is identified by the flow lines in Figure 16 which has been used over the last few months to develop the model in considerable detail to a point where it is capable of examining complex problems, within limiting assumptions that will be removed before the programme is completed. When that occurs, we shall have a working, effective methodology that will be of great value to decision-makers in transportation.

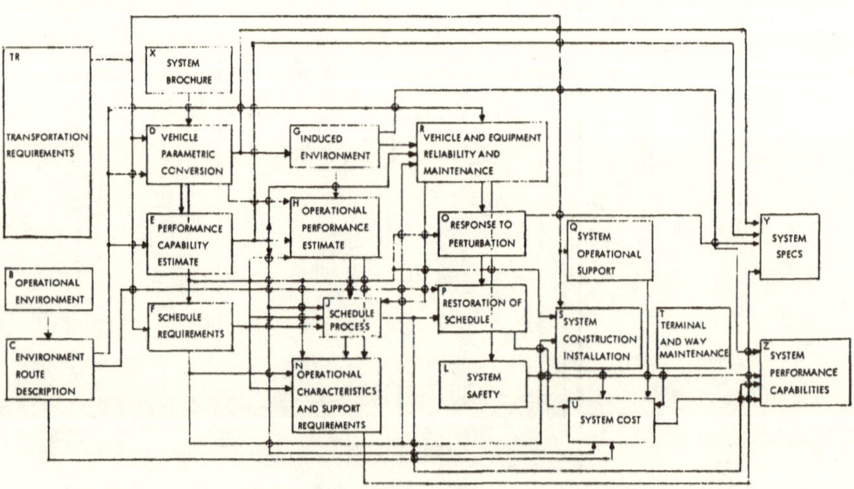

FIGURE 16. Model used in sample calculation

The New Tokaido Line

K. HATAGAWA*

ABSTRACT

A thirty-minute motion picture film which describes construction work, operation, and associated features of the new high-speed railway connecting Tokyo–Yokohama, the Nagoya, and the Osaka–Kobe industrial districts on the Tokaido route. Construction was begun in 1959 and regular operations started in October 1964.

*Director of the New York Office, Japanese National Railways.

Air Inlet Flow Characteristics For Pneumatic Tube Vehicle Systems

CHARLES M. HARMAN*

ABSTRACT

A combined vehicle and fluid dynamics study is made of a pneumatic tube vehicle system for a possible future ground transportation system. The paper is an attempt to develop a method for estimating the air inlet velocity when only the vehicle position, velocity and acceleration are known. Results make possible a reasonable estimate of the inlet velocity when the flow conditions for the air entering a pneumatic tube meet the following conditions:

(1) Reynolds number high for turbulent flow.
(2) Cauchy number much less than one for the velocity at the vehicle.
(3) Acceleration small compared to a free expansion into the tube.

THE PNEUMATIC TUBE vehicle system appears promising as a possible future ground transportation technique because when compared to alternatives it offers potential economics in both initial investment and operating cost. Before the extent of these advantages can be properly evaluated, studies, particularly concerning the combined vehicle and fluid dynamics of this system, need to be made. This is felt to be one such study.

This is a study of the air inlet flow to a pneumatic tube from the atmosphere when this air is propelling the vehicle along the tube. Specifically, this study is directed to the development of a method for estimating the inlet air velocity to a pneumatic tube when only the vehicle trajectory is known. Vehicle trajectory includes position in the tube, velocity, and acceleration of the vehicle during the time of its motion. The method is not dependent on knowledge of pressures or other air flow conditions in the tube and makes possible projection of inlet air flows based on vehicle trajectory parameters alone.

The theory describing the velocity variation of the air between inlet and vehicle is based on the use of quasi-steady state techniques and to superimpose on these, those effects which are associated entirely with

*Associate Professor, Department of Mechanical Engineering, Duke University.

the deviation from steady state. The steady (or quasi-steady) component is taken to be a one-dimensional, constant area, adiabatic flow of air with friction. This is also known as Fanno flow, All the pertinent relationships for this part of the flow problem can readily be found and are therefore not repeated here. The unsteady effects which are superimposed are developed in the following pages.

The mass of air entering the tube per unit time is proportional to the density, cross-sectional area and a velocity of the air, or:

$$(\Delta M/\Delta t) \sim \rho A V. \tag{1}$$

The instantaneous acceleration of the air is proportional to a change in velocity per unit time, or:

$$a \sim (dV/dt) \sim (\Delta V/\Delta t) \tag{2}$$

and the entering mass is proportional to the density, cross-sectional area, and characteristic length, or:

$$\Delta M \sim \rho A f(L). \tag{3}$$

Upon substitutions of Equation (2) and (3) into (1) and rearranging:

$$(\Delta V/V) \sim f(L)(dV/dt). \tag{4}$$

This is not represented to be other than a possible relationship among the variables affecting the variation from steady state.

The inlet velocity is then taken to be equal to the inlet velocity predicted by Fanno flow considerations based on the vehicle velocity and position, and upon a correction factor proportional to ΔV. This may be written

$$V_i = V_{\text{Fanno}} + \Delta V_{\text{unsteady}}. \tag{5}$$

The experimental phase of the study was carried out in the set-up shown diagrammatically in Figure 1. Stations, at which vehicle trajectory data is recorded, are shown. The photo-cell signal triggered by passage, was recorded at each of these stations, as shown on Figure 2. This information on vehicle velocity was converted to vehicle velocity versus time data for a number of trials and is shown in Figure 3. Vehicle velocities from this diagram were used in the calculations and this diagram was also used to construct a diagram of accelerations. The accelerations were determined by using the average slope over the last half of the trajectory to the point in question.

A constant temperature hot-wire anemometer was used to determine the instantaneous inlet air velocity. A typical record of the anemometer output is shown in Figure 2. The anemometer power is used to determine the air velocity at the probe through calibration equations. The

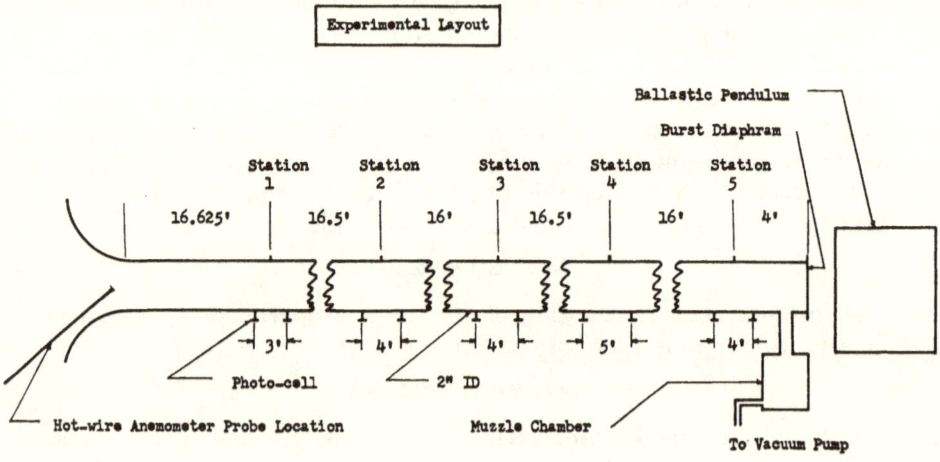

FIGURE 1

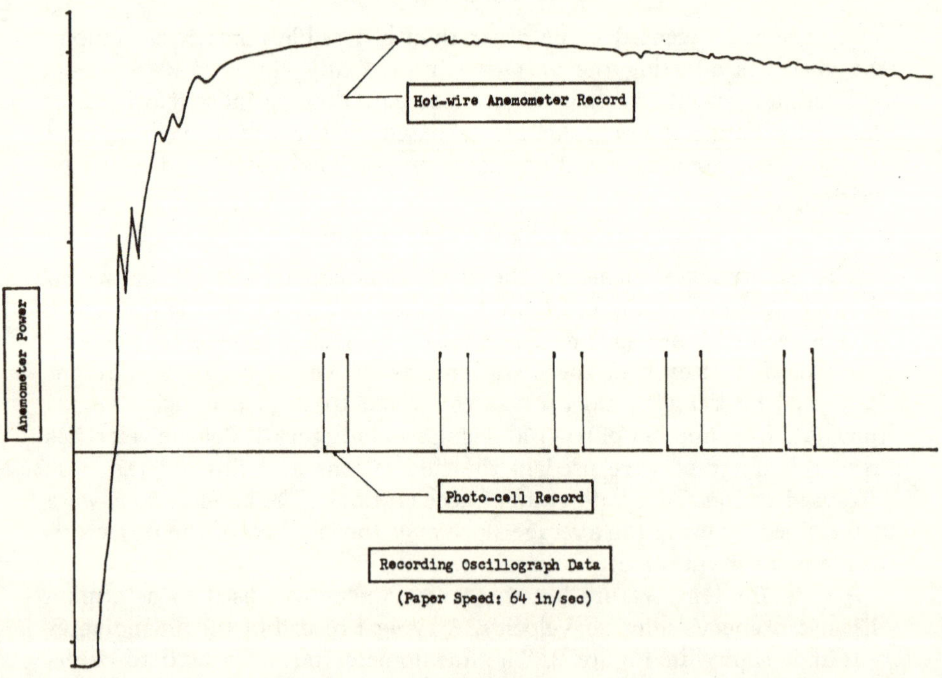

FIGURE 2

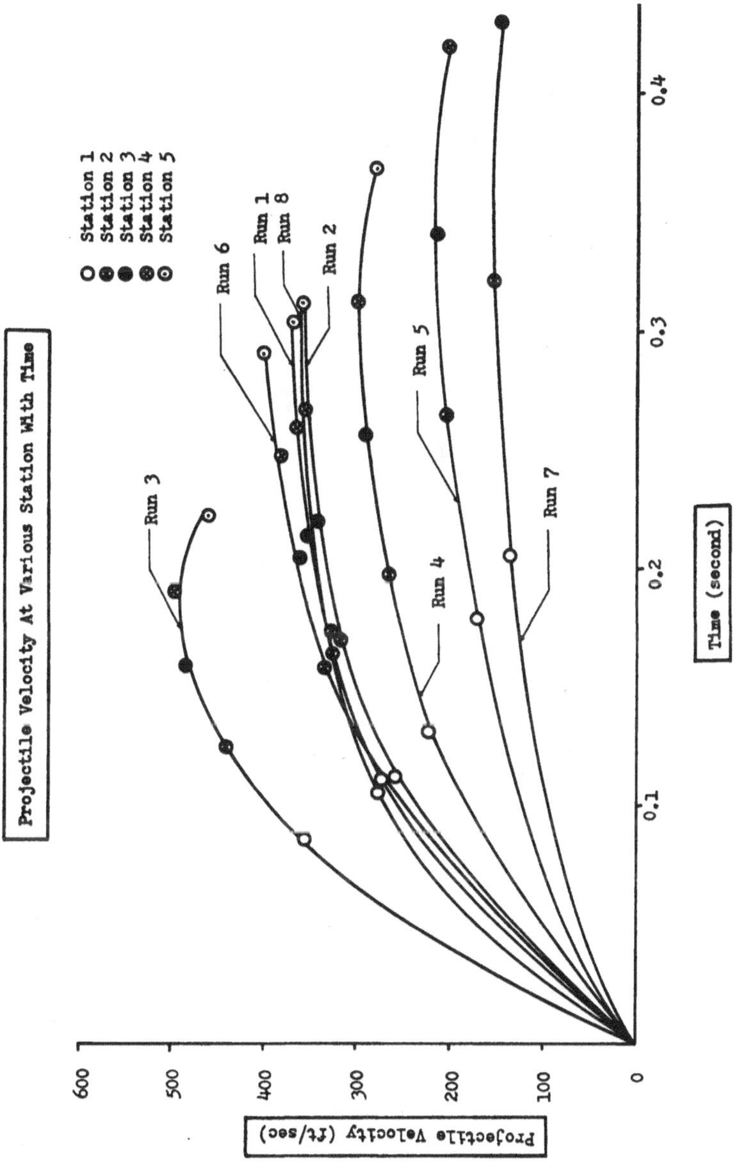

Projectile Velocity At Various Station With Time
Station 1
Station 2
Station 3
Station 4
Station 5
Run 3
Run 6
Run 1
Run 8
Run 2
Run 5
Run 4
Run 7
Projectile Velocity (ft/sec)
600
500
400
300
200
100
0
0.1
0.2
0.3
0.4
Time (second)

FIGURE 3

probe is located at the center of the air inlet as shown diagrammatically in Figure 1. The experimentally determined air inlet velocity is shown as a function of the vehicle acceleration in Figure 4 for the test runs carried out. The experimentally determined inlet velocity is also plotted versus the vehicle velocity at each of the stations and this is shown in Figures 5 through 9.

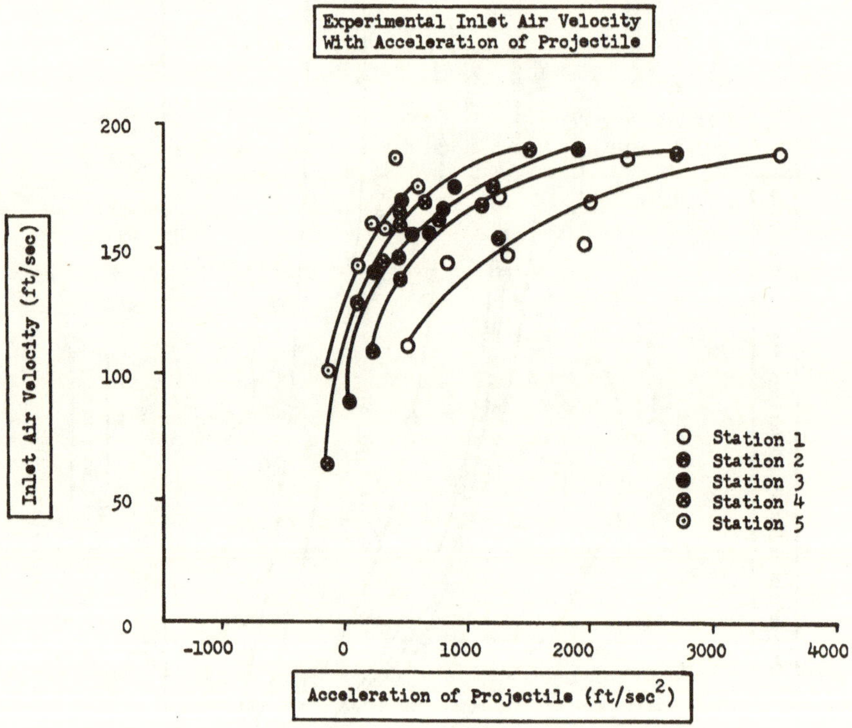

FIGURE 4

It was found that equation 5 predicts the inlet velocity reasonably well when:

$$\Delta V = (1/9V)(L/D)(dV/dt)_{\text{avg}}. \tag{6}$$

where V is the vehicle velocity and (L/D) is the length in duct diameters from entrance to vehicle. The solution of Equation 5 with Equation 6 substituted into it, was carried out on a computer for the experimental data obtained. These results were also plotted on Figures 5 through 9 and are signified as calculated points. A theoretical curve was drawn through the calculated points.

It can be seen that there was a reasonable amount of scatter in the

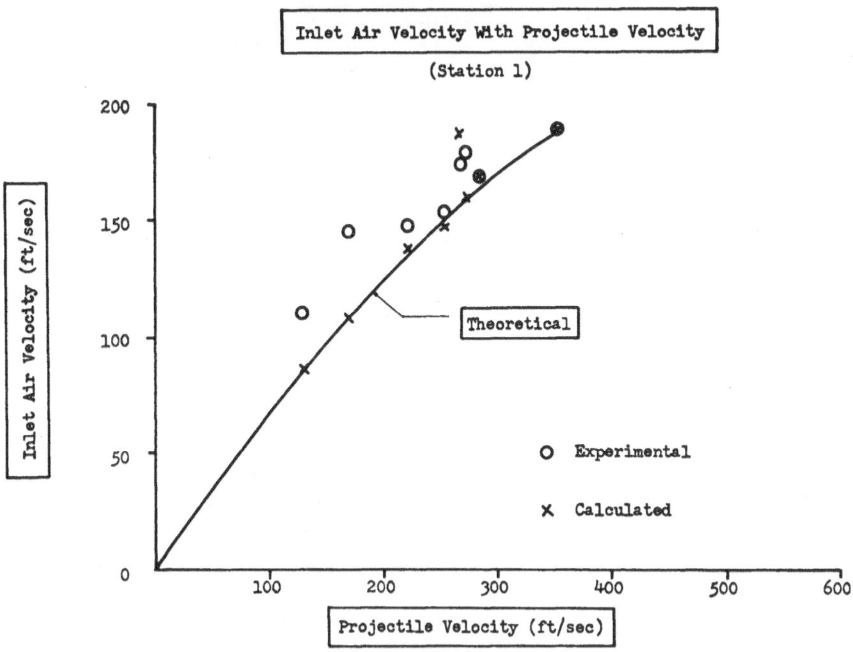

FIGURE 5

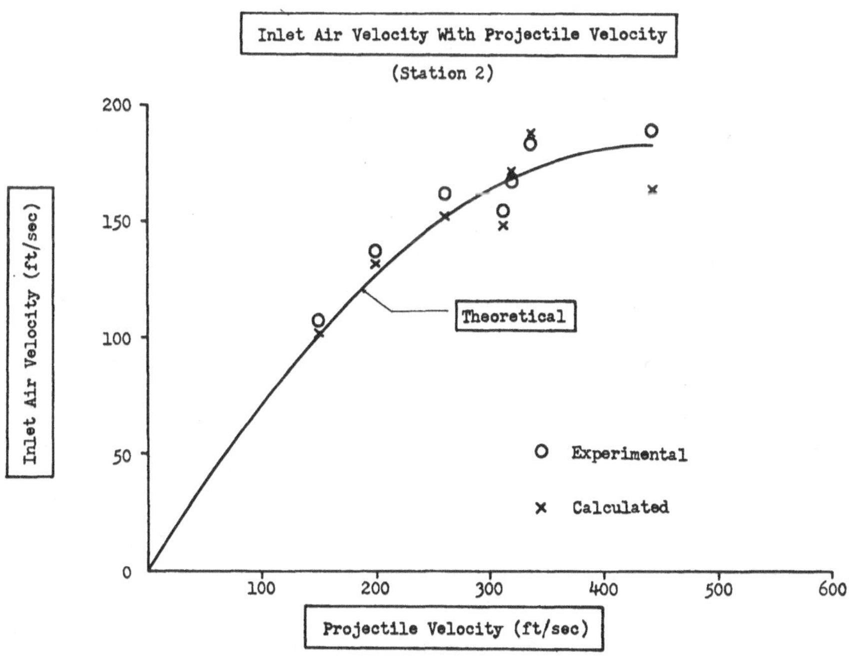

FIGURE 6

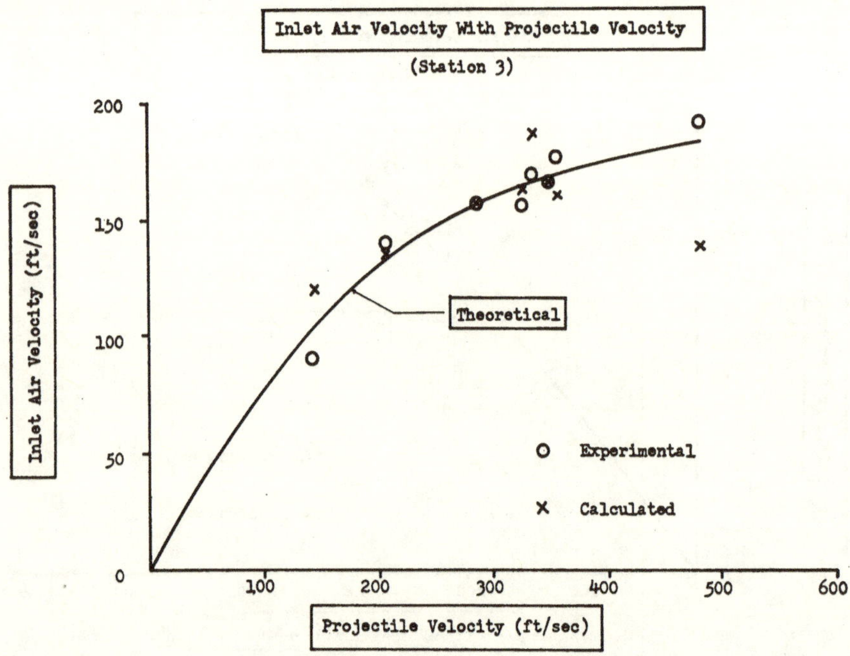

FIGURE 7

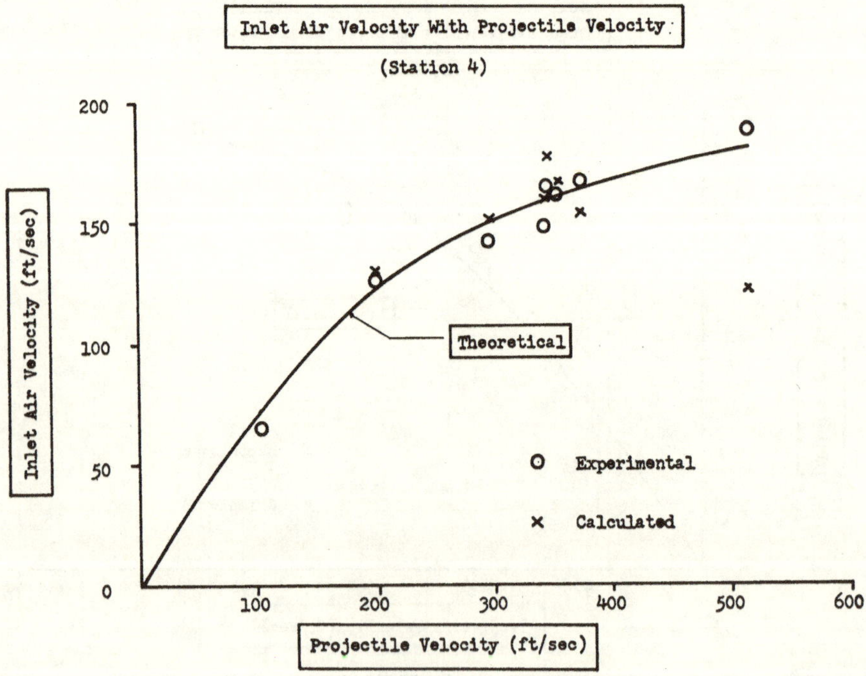

FIGURE 8

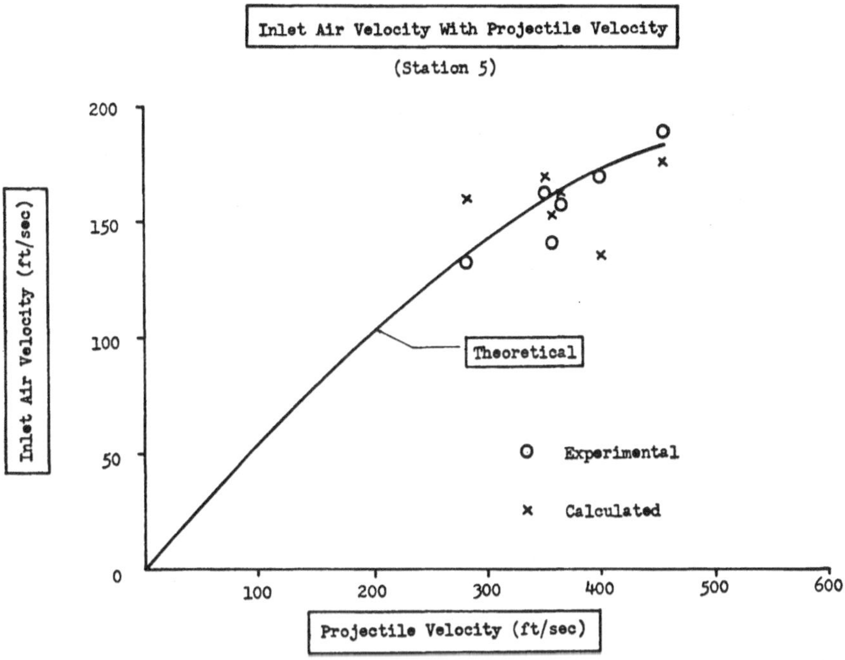

FIGURE 9

data and that the results, both theoretical and experimental, are somewhat lacking in precision. It is felt though, that a means is presented to make possible a reasonable estimate of the inlet velocity when the flow conditions for the air entering a pneumatic tube meet the following conditions:

(1) Reynolds number high for turbulent flow;
(2) Cauchy number much less than one for the velocity at the vehicle;
(3) acceleration small compared to a free expansion into the tube.

ACKNOWLEDGMENTS

The author is indebted to Mr. J. J. Murray and Dr. S. Kumar of the Army Research Office—Durham for their co-operation in the use of experimental facilities. The experimental phase was carried out in part by Kun Ha Kim, research assistant, Mechanical Engineering Department, Duke University, and forms a part of his M.S. thesis entitled "Air Inlet Flow Characteristics for Pneumatic Tube Vehicle Systems." The work in part relies on studies carried out by J. R. N. Rajan at Duke University and reported in his Doctoral dissertation entitled "Transient Viscous Subsonic Flow Studies in the Vacuum–Air Propulsion System."

French Aerotrain

J. BERTIN

This paper will be published in a special publication of the High Speed Transportation Committee of the ISTVS.

Urban Transport by Pneumatic Tube

JAMES J. MURRAY*

ABSTRACT

A historical survey of the development of the pneumatic tube is given, extending back to the first so propelled railway in Ireland, through that of the British and United States postal dispatch systems and the pioneer work of A. E. Beach of New York City in 1867. A pneumatic system is proposed, utilizing low vacuum ($\frac{1}{2}$ atmospheres), for mass tube transport in intra-city transport. Experimental results are presented, obtained from small vehicles studied in a two-inch diameter tube, 84 feet length under various low vacuum internal pressures. Vehicles were supported by slider bearing suspensions of solid lubricant materials. Dynamic pressures and velocities were determined throughout the runs along certain portions of the tube. A twelve-inch model results, in a tube over 2,000 feet in length, are shown as to velocity *vs.* pressure difference and distance of travel.

*Director, Engr., Sci. Div. Army Research Office, Durham, North Carolina, U.S.A.

VEHICLE MOBILITY AND TERRAIN EVALUATION

Soil Shear Test Device For Use in Bore Holes

NATHANIEL S. FOX

RICHARD L. HANDY*

ABSTRACT

Since July 1963, research has been carried out at Iowa State University Soil Engineering Department under the direction of Dr. Richard L. Handy, on a testing method and device for measuring soil shear strength within a bore hole. The original concept of this device was to run a direct shear test on the sides of a bore hole to determine soil cohesion and angle on internal friction. Several devices have been built and field-tested. The field data has been compared with laboratory data from samples taken with 3-inch Shelby tubes from the same holes which were subsequently field-tested. Both direct shear tests and triaxial tests were performed in the laboratory. Correlations for all soils tested were excellent. These soils include: dune sand, silty clay alluvium, loam, and loess.

The field shear test described differs from the vane shear, torsion shear, standard penetration and other in situ tests in several important ways. It is run at the sides of a hole, and not "blind" at the bottom; it was developed on a rational, not an empirical basis; most important, it allows for the separation of c and ϕ.

A complete test to determine c and ϕ requires 10 minutes to one hour depending on consolidation times (this does not include time to bore the hole). The test is run by two men, and the apparatus is completely portable. Tests have been performed as close as 3 inches from the ground surface, the only limit in depth is the height of the water table. Thus far it has not been possible to test below the level of the water table.

A patent application is being made; the patent will be the property of the Iowa State University Research Foundation. Commercial production is anticipated within one year.

*Soil Engineering Dept., Iowa State University, Ames, Iowa.

A Quantitative Method for Evaluating Terrain for Ground Mobility

J. H. SHAMBURGER*

ABSTRACT

Knowledge of the exact terrain conditions and vehicle characteristics is a prerequisite for predicting vehicle performance across specific terrain. Steps necessary to portray the terrain conditions within an area are: (*a*) identifying the pertinent environmental factors; (*b*) taking field measurements of these factors; (*c*) assigning significant ranges of values, that is, classes, of each factor; and (*d*) mapping the areal distribution of the factor classes.

Terrain factor-families that were considered significant from a mobility standpoint are surface composition, surface geometry, vegetation, and hydrologic geometry. Individual factors that describe each family were selected, and quantitative measurements were made in six areas in Thailand. These field data were reduced, and significant class ranges for each factor were established after thorough analysis of the reduced data. By utilizing the field data in conjunction with aerial photographs, a method was developed for identifying terrain conditions from the geometric, tonal, and textural characteristics of photographic patterns. Terrain characteristics were extrapolated from the sampled to unsampled areas by applying this technique, and factor–family maps at a scale of 1:50,000 were prepared in quantitative classes of the six study areas.

The studies reported herein were sponsored by the Advanced Research Projects Agency of the Office of the Secretary of Defense.

*Chief, Military Projects Section, Geology Branch, Soils Division, USAE Waterways Experiment Station, Vicksburg, Miss.

Terrain Reconnaissance with Electromagnetic Sensors

B. R. DAVIS*

ABSTRACT

Research studies are being made at the U.S. Army Engineer Waterways Experiment Station to determine which portions of the spectrum can be used to analyze terrain from a trafficability standpoint. These studies, for the most part, are being conducted in the laboratory and are concerned primarily with establishing correlations between changes in the quantity of energy reflected or emitted from laboratory-prepared soil samples and changes in soil parameters pertinent to trafficability. Regions of the spectrum investigated thus far include radar, infrared, radio, and gamma ray. The results of these investigations are presented, along with the conclusions and recommendations for future research needs.

The studies reported herein were conducted under the sponsorship and guidance of the Directorate of Research and Development, U.S. Army Materiel Command.

*Physicist, Terrain Analyzer Section, Army Mobility Research Branch, Mobility and Environmental Division, USAE Waterways Experiment Station, Vicksburg, Miss.

Mobility of Tracked Vehicles in Soft Soils

W. H. PERLOFF*

ABSTRACT

A study was conducted to develop an analysis which would predict the mobility characteristics of a tracked vehicle traversing soft soil. Available techniques were found to be inadequate to use vehicle design characteristics to predict behavior in cross-country travel. A computer programme vehicle which provided this prediction was developed, and successfully applied to compare mobility characteristics of existing tracked vehicles.

*Associate Professor of Soil Mechanics, Purdue University, Lafayette, Indiana.

Comparison of Engineering Properties of Selected Temperate and Tropical Surface Soils

M. P. MEYER*

ABSTRACT

Field and laboratory tests of trafficability and other engineering properties were conducted on 11 temperate fine-grained soils from the United States and 17 tropical fine-grained soils from Puerto Rico, Panamá Canal Zone, Hawaii, and Thailand to develop relations among the physical properties of the soils and to determine the degree of analogy that exists between soils of temperate and tropical climates. Soils were collected from the 6- to 12-in. layer. Parent materials consisted of basic and acidic igneous rock, volcanic ash, limestone, old water-deposited material, and recent alluvial clay and silt. Soils within each parent material group generally were of the same Unified Soil Classification System type and generally were located in similar topographic positions.

A comparison of the physical, mineralogical, and chemical properties, and relations of measurements obtained from standard and special engineering tests of soils grouped by parent materials indicates no significant differences by climatic zones.

A statistical study of the horizontal variation of soil properties, based on 20 observations in a 9- by 12-ft plot, shows similar mean standard deviations for temperate and tropical soils.

The equations that define relations between cone index and CBR and between cone index and unconfined compressive strength are approximately the same for temperate and tropical soils; 1 CBR unit equals about 50 cone index, and 1 unconfined compressive strength unit equals 8 cone index.

The magnitude of change in the Atterberg limits of a soil on drying and rewetting is correlated with the natural liquid limit of the soil; the greater the natural liquid limit the greater is the difference in limits between a natural and oven-dried sample.

It is concluded that temperate and tropical soils of similar parent materials and Atterberg limits generally have other engineering properties that are similar and behave similarly when subjected to standard engineering laboratory tests. Differences in behavior that exist among soils from each of the climates, therefore, can be attributed to differences that are reflected by the Atterberg limits.

The studies reported herein were sponsored by the Advanced Research Projects Agency of the Office of the Secretary of Defense.

*Engineer, Trafficability Section, Army Mobility Research Branch, Mobility and Environmental Division, USAE Waterways Experiment Station, Vicksburg, Miss.

Obstacle Performance of Tracklayer Vehicles

Z. J. JANOSI*

ABSTRACT

The maximum height of a vertical wall which can be negotiated by a tracked vehicle is first investigated. The limiting height is found on a purely geometric basis. The required slippages, both at the front and rear ends of the track, are found next. A simple analysis of the forces and moments involved while climbing the obstacle closes the discussion of the first problem.

The second problem discussed in this report involves two perpendicular corridors and a vehicle which attempts to complete a turn from one into the other. The relationship between the maximum width and length of the vehicle and the widths of the corridors is derived.

Finally, a vehicle is considered which crosses a field that is randomly strewn with trees. The length of the path necessary to cross the field is compared to the length of a path if the field were free of trees.

LIST OF SYMBOLS

l = length of vehicle or ground contact area (in.)
b = distance between C.G. and rear end (in.) and width of corridor
β = angle between vehicle and ground (°)
h = height of obstacle (in.) and distance between trees, projected to the distance of travel (miles)
h_m = maximum height of obstacle which can be negotiated by the vehicle (in.)
δ = height of C.G.
v_A = velocity at the leading end (in./sec.)
v_B = velocity at the rear end (in./sec.)
I = instantaneous centre of motion
i_A = slip, in the front
i_B = slip, in back
μ_A = coefficient of traction in front
μ_B = coefficient of traction in rear
H = tractive force in rear (lbs.)
W_B = load in rear (lbs.)

*Land Locomotion Laboratory, OTAC, Center Line, Michigan.

c = cohesion (lbs./in.^{2})
p = normal pressure (lbs./in.^{2})
α = angle of rod and corridor (°)
a = width of corridor and distance between front of vehicle and a tree when turning starts (in.)
w = width of vehicle (in. or ft.)
y = distance between inner edge of vehicle and corner (in.)
$\gamma = (90 - \alpha)°$
v_g = speed made good (mi./hr.)
v_a = average actual vehicle velocity (mi./hr.)
N = number of obstacles per square mile
L = length of trenches (miles)
θ = angle between direction of travel and trench and angle of turning
t = elasped time (hours)
S_s = straight line distance between the start point and the end point of the path (miles)
S_a = actual distance covered between start-point and end-point (miles)
P^1 = additional travel in case of parallel trenches (miles)
P = additional travel in case of randomly oriented trenches (miles)
$A = \sqrt{(W/2)^2 + a^2_{max}}$ (miles)
$B = \sqrt{(W/2)^2 + a^2_{min}}$ (miles)

OFF-THE-ROAD MOBILITY may be impaired either by the lack of sufficient soil strength or by rough terrain. The latter impediment is statistically more significant since very soft terrain occurs infrequently.

The roughness of the terrain may cause the driver to slow down because of possible damage to the passengers or to the vehicle. This problem area is investigated under research projects dealing with random terrain profiles, vehicle ride, and human tolerance.

It is remarkable, however, that very little work can be found in the literature about problems related to large obstacles. A few early reports by the Land Locomotion Laboratory and the results of a large scale experimental programme by the Waterways Experiment Station are the only available sources of information on this topic.

The Land Locomotion Laboratory undertook a project with the goal of establishing analytical methods for the evaluation of the obstacle performance of off-road vehicles. This report describes the first results of the theoretical work done under this project. More typical cases will be analyzed in the future, and experimental verification is also planned.

OBSTACLE PERFORMANCE OF TRACKED VEHICLES

Object

The object of this study was to establish theoretical methods for the analysis of the obstacle performance of off-the-road vehicles. Three separate cases are analyzed and presented in this report. Other situations will be investigated in the future.

Summary:

The first problem considered in the study was the establishment of a theoretical technique for the analysis of a vertical step-tracked vehicle system. The maximum possible step-height that a vehicle may negotiate is established from a geometrical viewpoint. The equilibrium of forces and moments, the available soil strength, and the required slip ratios are considered next.

The second problem involves two perpendicular corridors and a vehicle attempting to turn from one corridor into the other. At the outset the maximum allowable length of a "vehicle of zero width" is found. An equation is then derived which allows one to find the maximum allowable width of a vehicle whose length is less than the previously established maximum.

Finally, a field covered by randomly positioned large trees is considered. By using an analytical method first suggested by Brooks, a mathematical expression is derived for the establishment of the total excess travel caused by the presence of obstacles.

Conclusions:

Conclusions from the analyses are, of course, the same as those that could have been drawn from practical experience and simple logic. For example, it is evident that a vehicle cannot climb over a step whose height is greater than the distance between the centre of gravity and the rear end. It is also clear that a small vehicle moves more efficiently among obstacles than a large one. Therefore, the usefulness of this report is not new conclusions, but in the fact that the equations derived lend themselves to numerical analysis of the problems.

One conclusion, however, is not self-evident. It was found that when the approach to a vertical step is covered by soft soil, a vehicle is more likely to negotiate the obstacle than if the soil were firm.

Recommendations:

It is recommended that other typical cases, such as a vehicle crossing a ditch, be analyzed. Future efforts should also include non-conventional vehicles, such as articulated vehicles, walking machines, or the Airoll.

Furthermore, it is recommended that small scale and full scale tests be carried out since theoretical results are seldom convincing without experimental verification.

THEORETICAL ANALYSIS

1. *Vertical Step*:

The first type of obstacle considered is a vertical rigid step. To approach the problem in the simplest way possible, the "mechanics" of a rod is first described (Figure 1).

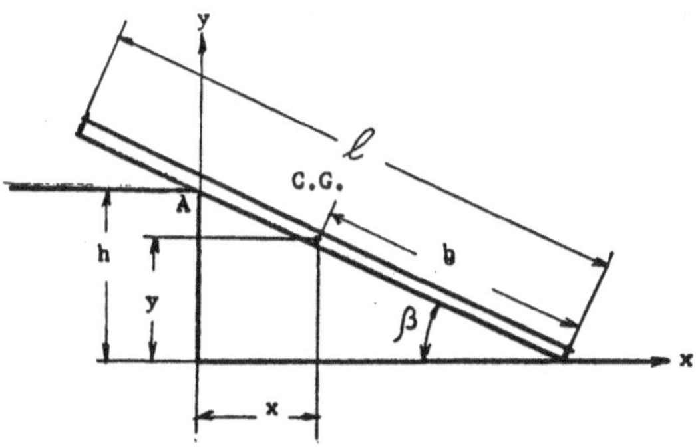

FIGURE 1. Thin rod over a vertical step

One end of the rod slides along the axis toward the left. Some point of the rod is constantly touching the top of the step denoted by A.

If the path of the centre of gravity (C.G.) is such that negative x values occur, the rod will negotiate the obstacle. If $x > 0$ for all β values, the rod cannot be pushed through its right end-point over the obstacle.

The equation of the path of the C.G. is

$$x = (h/\tan\beta) - b\cos\beta; \tag{1}$$

$$y = b\sin\beta. \tag{2}$$

It is clear that height h and distance b are the controlling parameters in this problem.

For the limiting case $x = 0$, Equation 1 becomes $h/\tan\beta = b\cos\beta$, or $h/b = b\sin\beta$, since:

$$|\sin\beta| \leqslant 1$$

$$h \leqslant b \tag{3}$$

must hold if the rod is to "negotiate" the obstacle, an obvious answer by itself. The same method is used for solving a problem much closer to the practical case (Figure 2). The dimensions δ and $(l - b)$ are the co-ordinates of the C.G. when the slab lies on the x axis, just touching the y axis.

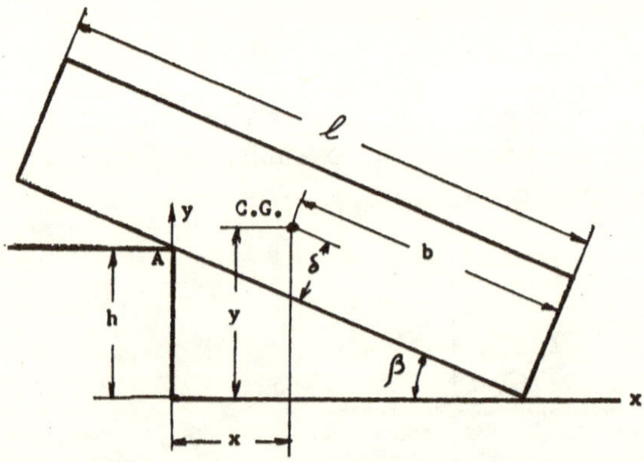

FIGURE 2. Thick slab over a vertical step

One is interested in the case in which a portion of the rod has already been pushed past point A.

The equation of the path of the C.G. is:

$$x = (h/\tan\beta) - b\cos\beta + \delta\sin\beta; \tag{4}$$

$$y = b\sin\beta + \delta\cos\beta. \tag{5}$$

The domain of β is:

$$\sin^{-1} h/l \leqslant \beta \leqslant \pi/2 \tag{6}$$

Again x must be negative for certain β values for the rod to "negotiate" the obstacle.

In the limiting case of $x = 0$, Equation 4 becomes:

$$(h/\sin\beta) + \delta\tan\beta = b. \tag{7}$$

From Equation 7:

$$h/b = \sin\beta - ([\delta/b][1/\cos\beta] - \cos\beta). \tag{8}$$

Equation 8 can be used to solve the problem graphically. Plotting the right hand side for the domain of β one can easily establish its maximum which is equal to the maximum h/b ratio that can be negotiated. It is more convenient, however, to use the results of the following derivation.

Differentiating Equation 8 with respect to β yields:

$$d/d\beta(h/b) = \cos\beta - (\delta/b)\sin\beta(2 + \tan^2\beta) \tag{9}$$

When the right hand side of Equation 9 is set equal to zero:

$$b/\delta = \tan\beta(2 + \tan^2\beta). \tag{10}$$

Equation 10 can be solved for β.

The roots of Equation 10 yield the numerical values of β at which h/b is maximum. To facilitate the solution, a graph has been plotted as shown in Figure 3. One has to find the β value at which

$$\tan\beta \times (2 + \tan^2\beta) = b/\delta.$$

Substituting this β into Equation 8 will yield (h/b) maximum.

Thus, the foregoing method allows one to establish the maximum height, h_m, of a vertical step which can possibly be negotiated by a track layer with given b and δ dimensions.

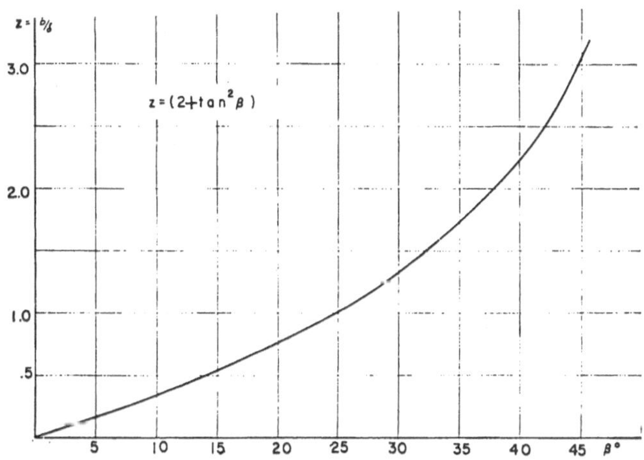

FIGURE 3. b/δ vs. β (see Equation 10).

If the height of a vertical obstacle is more than h_m, it is not worth while to conduct an analysis of the forces and moments involved because the vehicle cannot possibly climb such a "step."

SLIP RATIOS AT THE CONTACT POINTS

It will be shown that the slip ratios at the two end points of a track are of considerable importance. Therefore, a relationship between the necessary velocities will now be established.

Assuming that v_B is the velocity of the rear end of the "rod," v_A can be found as follows (Figure 4).

The instantaneous centre of motion is I. The instantaneous angular velocity is $\omega = (v_B/l \sin\beta)$, so that:

$$v_A = (v_B/l \sin\beta) l \cos\beta = v_B \cot\beta. \tag{11}$$

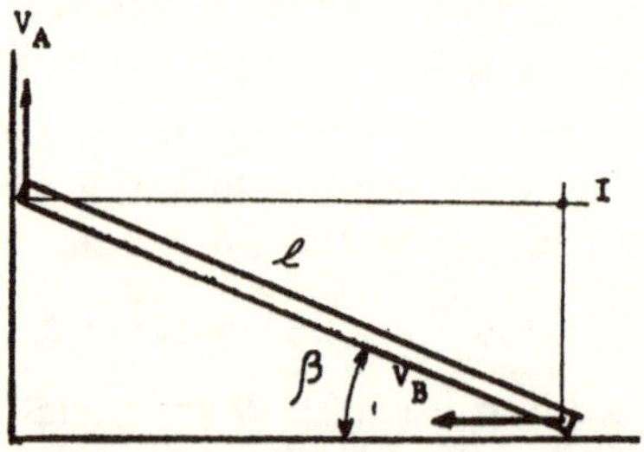

FIGURE 4. Velocities at the endpoints of the thin rod

An interesting geometric method is presented in Figure 5.

It is known from kinematics that the end points of the velocity vectors of each point located on a rigid rod form a line parallel to the rod when rotated by 90 degrees.

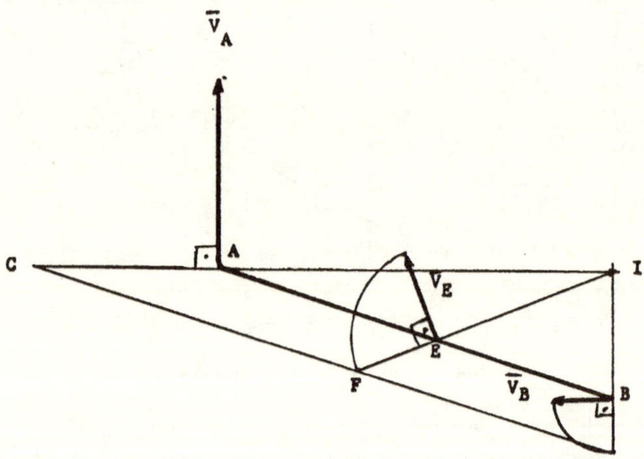

FIGURE 5. Geometric method for the establishment of the velocity distribution

Thus, rotate v_B to a vertical position and draw a line parallel to the rod. This line will intersect the horizontal drawn from A and $\overline{AC}$ will represent the magnitude of $\overline{v_A}$. Rotate $\overline{AC}$ back by 90° to obtain $\overline{v_A}$. (Any line drawn from I will yield an intercept between the two parallel lines ($\overline{EF}$). The length of the intercept ($\overline{EF}$) will represent the magnitude of the velocity vector at point E. The vector itself is perpendicular to ($\overline{EF}$).

Since $v_A = v_B \cot \beta$, $v_A \to \infty$ would be needed with $\beta = 0$. However, β is never zero in case of a vehicle (Figure 6).

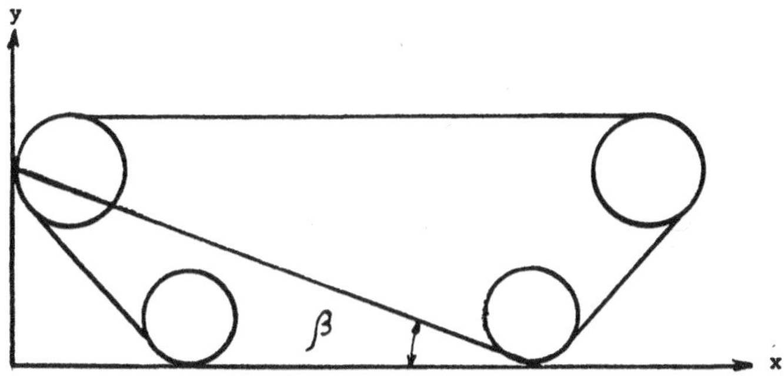

FIGURE 6. $\beta_{\min}$ for a tracked vehicle

Nevertheless, v_A must still be considerably higher than v_B as long as $\beta < 45°$. When $\beta > 45°$, the situation becomes the reverse.

For example at $\beta = 30°$, v_A must be 70 per cent higher than v_B. (cot 30° = 1.7).

The velocity of the track is constant relative to the vehicle. The track has to slip on the y and x axis.

Two different slip values are required:

$$i_B = (v_T - v_B)/v_T = 1 - (v_B/v_T); \qquad v_B = v_T(1 - i_B)$$

$$i_A = (v_T - v_A)/v_T = (v_T - v_B \cot \beta)/v_T = 1 - (v_B \cot \beta/v_T).$$

From here:

$$(1 - i_B) = (1 - i_A) \tan \beta,$$

$$(1 - i_B)/(1 - i_A) = \tan \beta; \tag{12}$$

and

$$i_B = 1 - (1 - i_A) \tan \beta, \tag{13}$$

or

$$i_A = 1 - (1 - i_B) \cot \beta. \tag{14}$$

Since in most practical cases, the vertical wall (y axis), is firmer than the soil covering the approach (x axis), it is likely that i_B will be greater than i_A. However, when β is more than 45°, i_B defined by the kinematics of the problem is less than i_A. Thus, one should expect difficulties when the height of the obstacle of the obstacle requires β angles greater than 45°.

Equilibrium of Forces

The case of a rigid wall and yielding approach will be investigated (Figure 7).

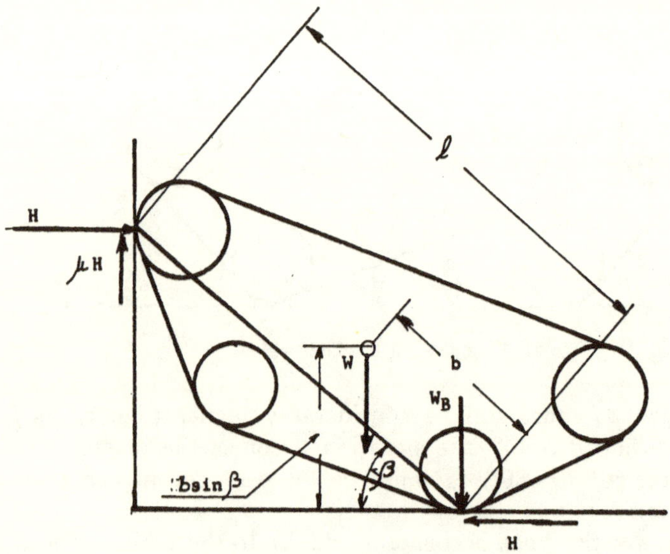

FIGURE 7. Equilibrium of forces

1. *Equilibrium of Moments*:

$$W(l - b) \cos \beta + Hl \sin \beta - W_B \, l \cos \beta = 0. \tag{15}$$

Here $H = \mu_B W_B$.

2. *Equilibrium of Vertical Forces*:

$$W - (\mu_A H + W_B) = 0, \tag{17}$$

or

$$W = \mu_A \mu_B W_B + W_B = (1 + \mu_A \mu_B) W_B; \tag{18}$$

from which $\mu_A = 1/\mu_B([W/W_B] - 1)$. Using Equation 15, one obtains $W/W_B = l/(l - b)(1 - \mu_B \tan \beta)$, so that:

$$\mu_A = 1/\mu_B[(l/[l - b])(1 - \mu_B \tan \beta)] - 1. \tag{19}$$

For the sake of simplicity assume that $l = 2b$. Then $\mu_A = 1/\mu_B - 2 \tan \beta$, or:

$$\mu_B = 1/(\mu_A + 2 \tan \beta) \tag{20}$$

Figure 8 shows the required μ_B values for various β values for $\mu_A = 0.5$ and $\mu_A = 1.0$; while Figure 9 depicts i_B as a function of β for $i_A = 10$ per cent and $i_A = 50$ per cent.

It is seen that when the slip and the "frictional" conditions are constant at the wall, the slip and the shear force may reduce at B with

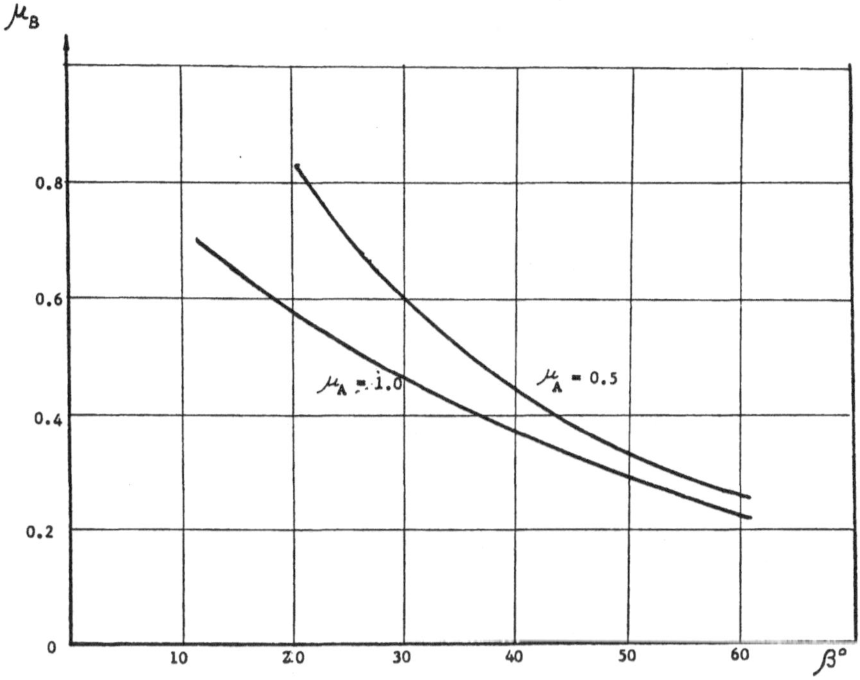

FIGURE 8. β VS. μ_B

increasing β angles. Thus the stage immediately after "lift-off" is the most critical when highest shear and highest slip are required.

When the soil values at the base of the step are known, one can establish the available μ_B by means of Coulomb's equation:

$$\mu_B = (c/p) + \tan\phi \tag{21}$$

where p is the average normal pressure under the tracks, l is the length of the ground contact area.

When the vehicle first touches the wall, the lifting force at the front is $\mu_A\mu_B W$. The moment created by this force around the rear end of the contact has to balance the moment of the weight. So that

$$l\cos\beta_{\min}(\mu_A\mu_B W) = b\cos\beta_{\min}W;\ \text{or}\ l\mu_A\mu_B = b;\ \text{and}$$

$$\mu_A = \mu_B b/l. \tag{22}$$

If the required μ_A is more than the available one, the vehicle will not be able to negotiate the obstacle. When the required μ_A is sufficiently small, the front of the vehicle will start to climb the wall.

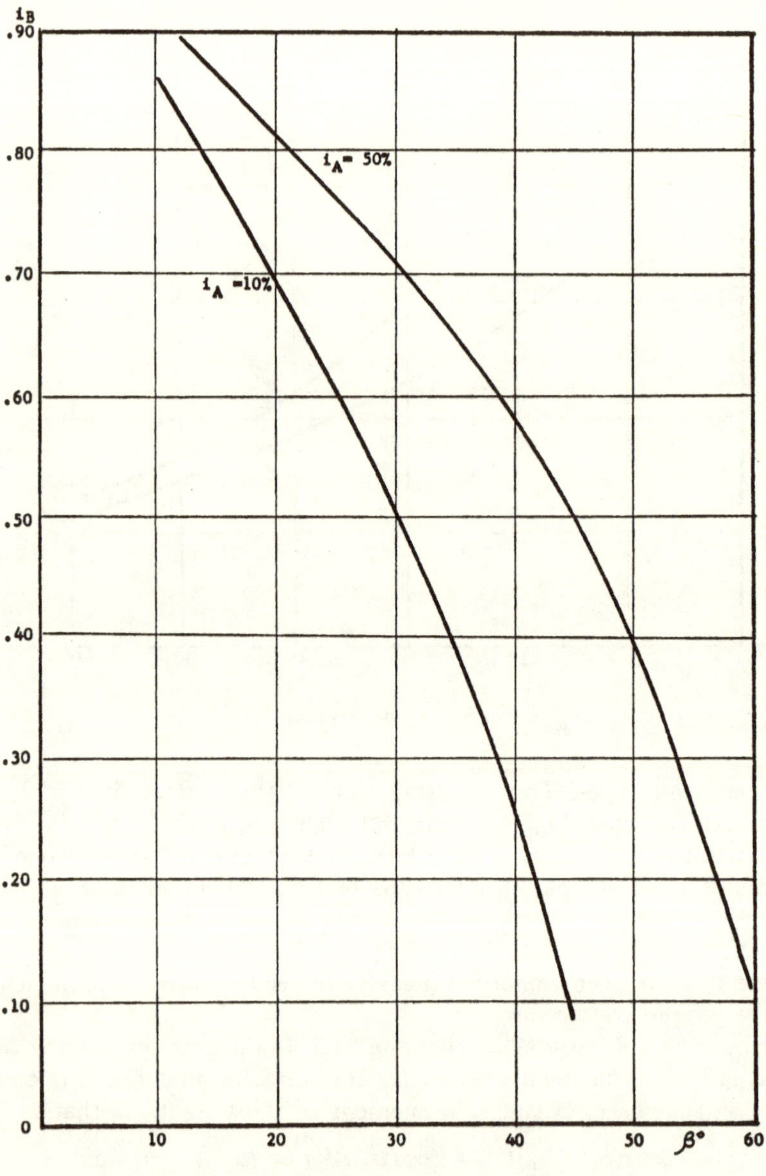

FIGURE 9. β vs. i_B

Next, one has to estimate the magnitude of the reduced ground contact area (A_B) after "lift-off" and find W_B from Equation 15. The required μ_A can be found from Equation 19. If it is found that the required μ_A is still smaller than the available value, the vehicle will continue climbing the wall. An example is presented in the following:

Example: The dimensions of a fictitious vehicle are shown in Figure 10. When the vehicle "stands on its rear sprocket," $b = 84$ in. First, one has to establish the highest possible step which this vehicle could climb.

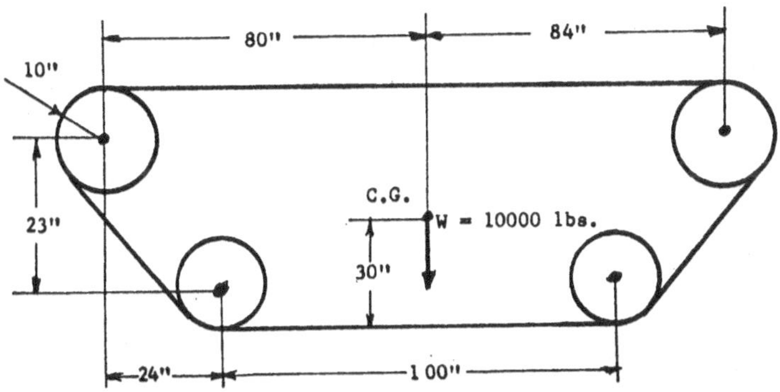

FIGURE 10. Dimensions of vehicle in Example 1

Solving Equation 10 by means of Figure 3 yields $\beta = 44°$ $(b/\delta = 2.8)$. Then from Equation 8, $h = 38$ in. Therefore, this vehicle can not possibly climb any step which is 38 in. or higher.

Assume that the step under investigation is 30 in. high and that the soil at the approach is characterized by $c = 1.0$, $\phi = 28°$, while $\mu_A = 0.8$. Since the weight is $W = 10{,}000$ pounds and the ground contact area is $2 \times 100 \times 20 = 4{,}000$ in.^{2}, the average pressure is $p = 2.5$ p.s.i. Therefore, $\mu_B = (c/p) + \tan\phi = (1.0/2.5) + 0.5317 = 0.9317$. So that, from Equation 22, $\mu_A = 0.0317(84/141) = 0.555$, which is less than 0.8; and, therefore, the vehicle will start climbing the step. Assume that the ground contact area reduces to 1,000 in.^{2} when the front of the vehicle reaches the top of the step. At this point, $\beta = 12.3°$.

Substituting into Equation 15 yields $10{,}000(141 - 45)0.9769 + (1{,}000 \times 0.1 + W_B 0.5317) \times 141 \times 0.2136 - W_B \times 141 \times 0.9769 = 0$. Here $l = 141$ in. and $b = 45$ in. was used.

From the above equation $W_B = 7{,}775$ lbs. and

$$\mu_B = H/W_B = (1000 \times 0.1 + 7775 \times 0.5317)/7775 = 0.5444.$$

The required μ_A is:

$$\mu_A = (1/\mu_B)([W/W_B] - 1) = (1/0.544)(10000/7775) - 1 = 0.53$$

Therefore, the soil covering the approach is firm enough to provide a large enough H force which in turn defines a small enough μ_A at the wall.

According to Equation 13:

$$i_B = 1 - \tan 9°40'(1 - i_A) = 1 - 0.17 + 0.17i_A = 0.83 + 0.17i_A$$

which shows that at the beginning of the climb at least 83 per cent slip is needed at the rear. In case of firm soil, the engine may not be powerful enough to provide the torque at the driving sprockets to create such a high slip.

2. *Cornering*:

The next problem considered is the establishment of a relationship between the dimensions of a tracked vehicle and the widths of two perpendicular "corridors" through which the vehicle can complete a 90° turn. (Figure 11)

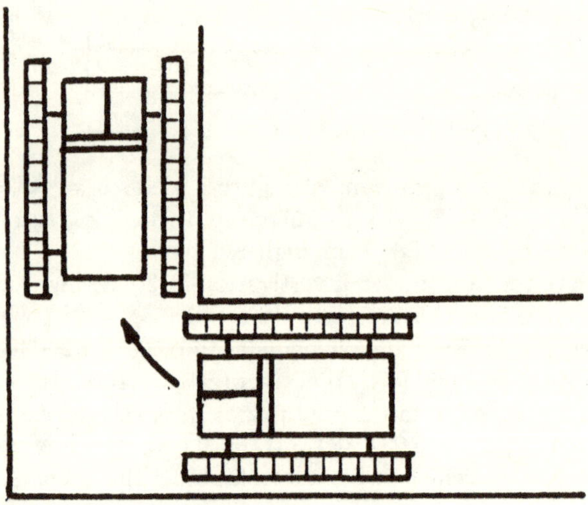

FIGURE 11. Vehicle negotiating corner

First the case of a thin rod is discussed. Figure 12 shows several positions of the rod while "turning around the corner."

It is seen that when corner-point A lies above or just on the envelope of the family of lines, the rod will have enough room to complete the turn.

Thus, one has to find the equation of the envelope and establish the co-ordinates of A so that the criterion expressed above can be satisfied.

The equation of an envelope is as follows:

Express the equation of the family of curves in the form

$$F(x, y, \alpha) = 0.$$

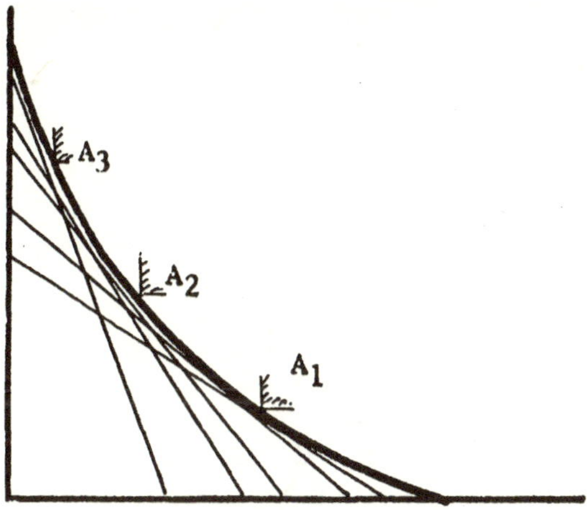

FIGURE 12. "Rod" negotiating corner

Find the equation $\partial F/\partial \alpha = 0$.
Eliminate α from the two equations.
For the present case, $F(x, y, \alpha) = 0$ is:

$$(x/\cos \alpha) + (y/\sin \alpha) - l = 0 \tag{23}$$

and:

$$\partial F/\partial \alpha = x \cos \alpha - y \sin \alpha + l \sin^2\alpha - l \cos^2\alpha = 0. \tag{24}$$

From Equations 23 and 24, one obtains: $\cos \alpha = (x/l)^{1/3}$, $\sin \alpha = (y/l)^{1/3}$. From which the equation of the envelope is

$$x^{2/3} + y^{2/3} = l^{2/3}. \tag{25}$$

The co-ordinates of the corner-point A are a and b, so that the criterion sought becomes

$$l \leqslant [a^{2/3} + b^{2/3}]^{3/2} \tag{26}$$

The case of a thick rod is more involved (Figure 13).

An approach similar to the previous method yields two equations:

$$x - l \cos^3\alpha - w \sin \alpha = 0 \tag{27}$$

$$y - l \sin^3\alpha - w \cos \alpha = 0 \tag{28}$$

It is not possible to eliminate α from Equations 27 and 28 by simple manipulations. Therefore, a different approach will be used.

Figures 14 and 15 represent the same two positions of the vehicle.

In Figure 15, however, the position of the co-ordinate axes are shown relative to the vehicle.

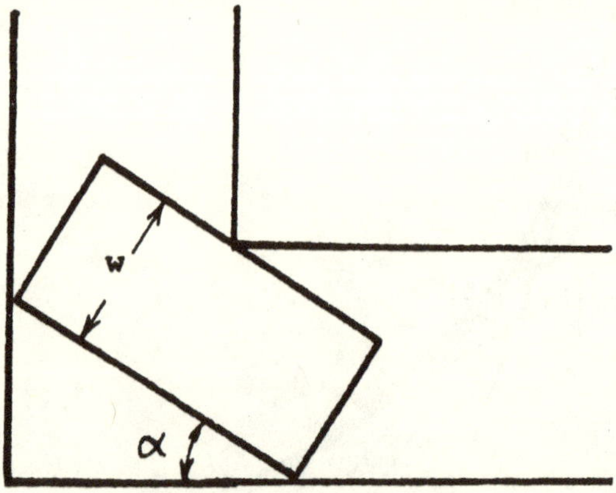

FIGURE 13. Definition of w and α

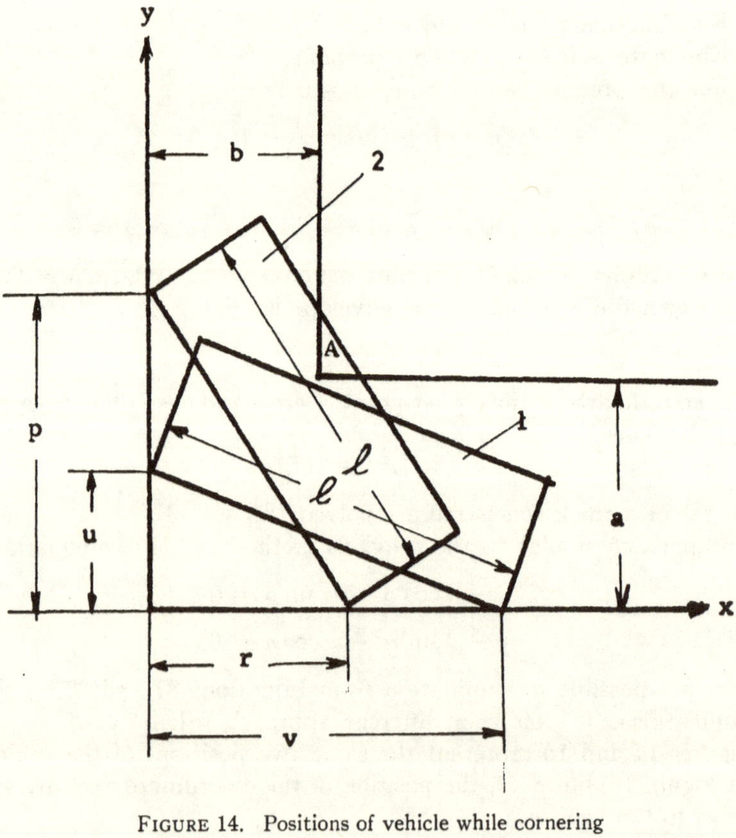

FIGURE 14. Positions of vehicle while cornering

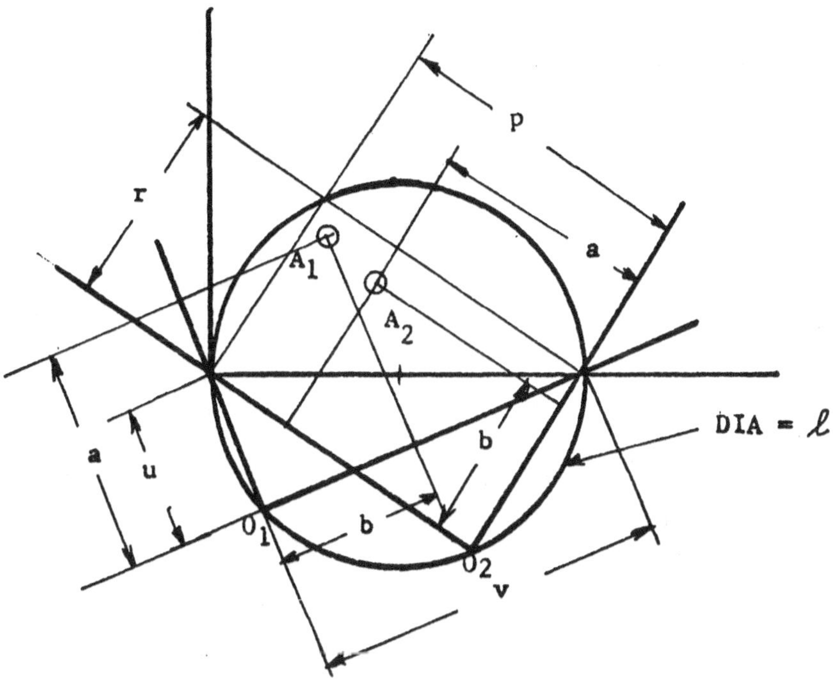

FIGURE 15. Positions of corner while vehicle turns

Imagine that point A is rigidly attached to the co-ordinate system. The problem can be easily solved graphically. By plotting an adequate number of points of the path of point A, one can find the smallest y value of the curve. If y_{min} is greater than w, the slab can turn around the corner.

According to Figure 16, the equation of the path of point A is

$$x = l\cos^2\gamma - a\cos\gamma + b\sin\gamma \tag{29}$$

$$y = l\sin\gamma\cos\gamma - a\sin\gamma + b\cos\gamma \tag{30}$$

Where $-\frac{1}{2}\pi \leqslant \gamma \leqslant 0$

(Note that when $b = 0$, a curve called Pascal's spiral is obtained.)

The criterion for negotiating the corner is: $y_{min} \geqslant w$.

To find y_{min}, the standard procedure is followed:

$$dy/\partial\gamma = l\cos^2\gamma - l\sin^2\gamma - a\cos\gamma - b\sin\gamma$$

$$l\cos^2\gamma - l + l\cos^2\gamma - a\cos\gamma - b\sin\gamma = 0$$

$$2l\cos^2\gamma - a\cos\gamma - b\sin\gamma - l = 0$$

$$2l\cos^2\gamma - \sqrt{a^2+b^2}\cos(\gamma - \tan^{-1}b/a) - l = 0. \tag{31}$$

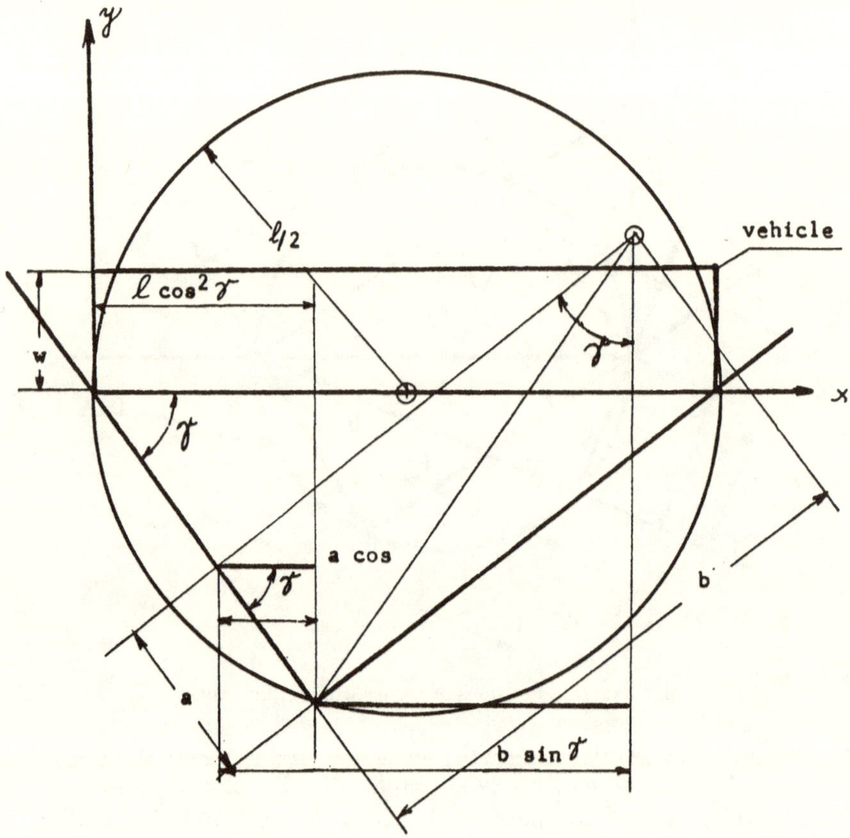

FIGURE 16. Dimensions for establishment of equation of "path of corner"

When $b/a = 1$, $\tan^{-1} b/a = \pi/4$. The solution of the equation is $\gamma = -\pi/4$. In this case Equation 30 becomes: $a\sqrt{2} - (l/2) \geqslant w$. (When $w = 0$, and $a = b$, $l \leqslant a2^{3/2}$, which is the same as $l^{2/3} \leqslant a^{2/3}+b^{2/3}$ for $a = b$).

Since $2l\cos^2\gamma = l + l\cos 2\gamma$, Equation 31 may be transformed to the following form: $l\cos 2\gamma = \sqrt{a^2+b^2}\cos(\gamma - \tan^{-1} b/a)$, or

$$l/\sqrt{a^2+b^2} = \cos(\gamma - \tan^{-1} b/a)/\cos 2\gamma. \quad (32)$$

It is not difficult to solve Equation 31 by trial and error.

Example:

Assume that $a = 4.0$, $b = 3.0$.

From $l^{2/3} \leqslant a^{2/3} + b^{2/3}$, $l \leqslant 9.8$.

(When $l = 9.8$, only an "infinitely thin" vehicle can negotiate the curve.) Suppose $l = 6.0$.

Then $l/\sqrt{a^2 + b^2} = 6.0/5.0 = 1.2$.

One has to find that γ value, by trial and error, to which an ordinate of 1.2 belongs.

Accordingly $\gamma_{min} = 39°$.

From Equation 29:

$y_{min} = -6 \times \sin 39x \cos 39 + 4 \sin 39 + 3 \cos 39 = 1.91$.

Answer: A 6 ft. long vehicle has to be less than 1.91 ft wide in order to negotiate a corner formed by a 4 ft and a 3 ft wide corridor.

3. *Random Obstacles*:

When a vehicle proceeds through a field strewn with impenetrable obstacles (large trees or deep and wide trenches), the straight-line distance between the starting point and the desired spot of arrival divided by the elapsed time yields a fictitious velocity called "speed made good" (v_g). The "speed made good" is always smaller than the average actual speed (v_a). The ratio of v_g and v_a is an important factor in the evaluation of the obstacle performance characteristics of a vehicle.

The Mobility Environmental Research Branch of the U.S. Army Engineers Waterways Experiment Station has done a great amount of experimental and analytical work in this area (1).

The Land Locomotion Laboratory has published a report on this topic which was written by P. F. Brooks in 1958 (2). The principles of Brooks' theoretical approach will be utilized in the following.

The problem is to find v_g/v_a in the case of a field in which there are N trenches per square mile. The trenches are randomly oriented and positioned. The length of the trenches is L.

It is clear that the number of trenches encountered by the vehicle is $L \times N$, if the trenches are perpendicular to the direction of travel. When all trenches are oriented at an angle θ with respect to the direction of travel, the number of trenches encountered is $L \times N \times \sin \theta$ (Figure 17).

The ratio of v_g and v_a may be rewritten as follows:

$$\eta = v_g/v_a = v_g t/v_a t = s_s/s_a \tag{33}$$

where t is the elapsed time, s_s is the straight-line distance between the starting point and the endpoint of the path, s_a is the actual path length. Since a square field of 1 square mile area is considered, $s_s = 1$ and $s_a = 1 + p$ may be assumed. Here p is the total additional distance to

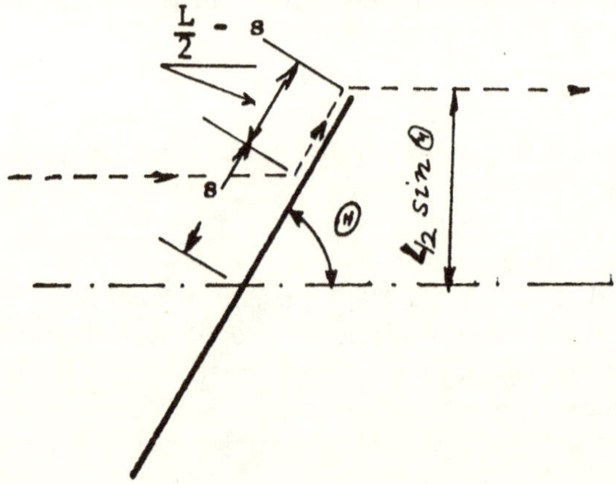

FIGURE 17. Randomly positioned trenches with constant angle θ

be covered. The total additional distance equals the product of the number of obstacles encountered and the average extra travel per obstacle.

When all the trenches are oriented at an angle θ, the additional travel is $L/2 - s$. However, the distance $(L/2 - s)\cos\theta$ is useful, so that the additional travel becomes $(L/2 - s)(1 - \cos\theta)$, Figure 17. When turning to the right, this distance becomes $(L/2 + s)(1 + \cos\theta)$. The two additional path lengths are equal when $s = -L/2\cos\theta$. Thus, when $s > (L/2)\cos\theta$, the vehicle must turn left; and when $s > -(L/2)\cos\theta$, the vehicle turns right.

The average additional distance (b') is

$$b' = \frac{1}{L}\left[(1 - \cos\theta)\int_{-L/2}^{L/2}\left(\frac{L}{2} - s\right)ds + (1 + \cos\theta) \times \int_{-L/2}^{-L/2\cos\theta}\left(\frac{L}{2} + s\right)ds\right] = \frac{L}{4}\sin^2\theta \quad (34)$$

The total additional distance is:

$$P' = (L \times N \times \sin\theta)(L/4)\sin^2\theta = (NL^2/4)\sin^3\theta.$$

Thus:

$$\eta' = 1/[1 + (L^2/4)N\sin^3\theta] \quad (35)$$

When the angle θ varies between 0 and π, the average additional distance becomes:

$$P = \frac{1}{\pi}\int_0^\pi P'\,d\theta = \frac{1}{\pi}\frac{NL^2}{4}\int_0^\pi \sin^3\theta\,d\theta = \frac{NL^2}{3}, \quad (36)$$

so that:

$$\eta = 1/(1 + [NL^2/3]) \quad (37)$$

The case of a field covered randomly by large trees is considered next. Again the number of trees per square mile is N, the number of trees encountered is $N \times w$. Here w is the width of the vehicle in miles. According to Figure 18, the additional travel is:

$$\sqrt{a^2 + (w/2 - s')^2} - a.$$

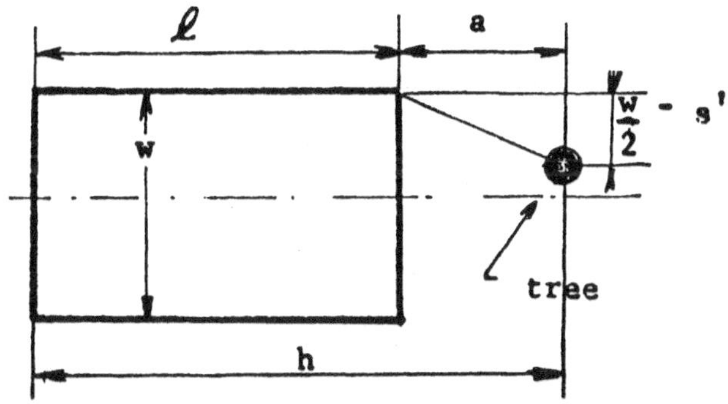

FIGURE 18. Vehicle approaching tree

This distance equals $w/2 - s'$ when $a = 0$; and it approaches zero when a approaches infinity. Thus, the adjustment in steering should take place as soon as possible. It is assumed that the driver will turn the vehicle by the necessary angle 0 as soon as the rear end of the vehicle passes the previous tree. Furthermore, it is assumed that the distance of the trees projected to the direction of travel is never less than the length of the vehicle.

If this distance is called h, then $h_{max} \geqslant h \geqslant h_{min} \geqslant l \geqslant w$.

Using the notation of Figure 17, $a = h - l$. So that $a_{max} = h_{max} - l$ and $a_{min} = h_{min} - l$. The additional distance will then be $(a/\cos\theta) - a$.

But: $$1/\cos\theta = \sqrt{a^2 + s^2}/a,$$

where $s = w/2 - s'$, so that one has to find the following average:

$$\frac{1}{a_{max} - a_{min}} \frac{2}{w} \int_{a_{min}}^{a_{max}} \int_0^{w/2} (\sqrt{a^2 + s^2} - a) ds\, da,$$

which is:

$$\frac{1}{3(a_{max} - a_{min})} \left\{ (Aa_{max} - Ba_{min}) + \frac{1}{8} W^2 \ln \frac{a_{max} + A}{a_{min} + B} + \frac{a^3_{max}}{W} \ln \frac{w/2 + A}{a_{max}} - \frac{a^3_{min}}{W} \ln \frac{w/2 + B}{a_{min}} \right\} - \frac{1}{2}(a_{max} + a_{min}). \qquad (39)$$

Multiply by wN, the number of trees encountered:

$$P = \frac{N}{3(a_{\max} - a_{\min})}\left\{w(Aa_{\max} - Ba_{\min}) + \left(\frac{w}{2}\right)^3 \ln\frac{a_{\max} + A}{a_{\min} + B}\right.$$

$$\left. + a^3_{\max} \ln\frac{w/2 + A}{a_{\max}} - a^3_{\min} \ln\frac{w/2 + B}{a_{\min}}\right\} - \frac{Nw}{2}(a_{\max} + a_{\min}).$$

Here $$A = \sqrt{(w/2)^2 + a^2_{\max}}; \quad B = \sqrt{(w/2)^2 + a^2_{\min}} \qquad (40)$$

and naturally, $\eta = 1/(1 + P)$.

REFERENCES

1. GRABAU, W. E., Terrain Evaluation for Mobility Purposes, *J. of Terramechanics*, Vol. 1, No. 2, (1964).

2. BROOKS, F. C., Effect of Impenetrable Obstacles on Vehicle Operational Speed, *Land Locomotion Laboratory Report, No. 28*, Warren, Michigan: (U.S. Army Tank-Automotive Center, 1958).

Trafficability Tests with Two Vehicles with Ten-Ton Wheel Loads

E. S. RUSH*

B. G. SCHREINER†

ABSTRACT

A programme of field tests was performed with two 80,000-lb. four-wheel-drive, construction-equipment-type vehicles to determine their performance on a range of natural soil conditions. Vehicle performance was evaluated on a self-propelled, go/no-go basis for a prescribed number of passes through each test course. Soil strength was measured before, during, and after traffic with the cone penetrometer and related trafficability equipment. Tests were performed on silt and clay. Vehicle performance was found to be related to soil strength and such vehicle factors as ground clearance and mechanical characteristics. These findings will extend existing empirical formulas for the prediction of vehicle performance on the basis of the vehicles' physical characteristics.

The studies reported herein were conducted under the sponsorship and guidance of the Directorate of Research and Development, U.S. Army Materiel Command.

*Engineer, Chief, Trafficability Section, Army Mobility Research Branch, Mobility and Environmental Division, USAE Waterways Experiment Station, Vicksburg, Miss.

†Engineer, Trafficability Section, Army Mobility Research Branch, Mobility and Environmental Division, USAE Waterways Experiment Station, Vicksburg, Miss.

Tropical Soil Study: Commonwealth of Puerto Rico

D. SLOSS*

ABSTRACT

A study to determine Bekker soil strength parameters for tropical soils was conducted in Puerto Rico during September, October, and November of 1965. Soils tested included laterites, alluvium, tidal flats, beach sands, and bog soils. Analysis of the data from the study indicates there are two major mobility problems that could be expected when operating on tropical soils. These are:

(1) Tidal flat and bog soils which exhibit decreasing bearing capacity with depth and cause immobilization due to excessive sinkage.

(2) Laterite type soils which cause immobilization due to slipperiness rather than due to excessive sinkage.

The study also demonstrated that a four-man field crew could transport all of the required soil-strength measuring equipment by commercial aircraft and conduct unsupported test operations for an extended period.

Subsequent field testing in Dade, Broward, Palm Beach, Orange, and Hillsborough counties of Florida provided a comparison of Bevameter shear curves with laboratory direct shear curves for medium sands. An average of eleven different sites comprised of this material showed an optimum water content versus direct shear strength relation at various normal loads. Further analysis of the direct shear strength versus normal load revealed that the friction angle of the medium sand obtained with the direct shear machine in the laboratory was quite close to the field Bevameter value, despite roots and gravel, provided there was no appreciable sinkage of the Bevameter shear head during the test. In cases of loose sand in the field which caused sinkage, the Bevameter friction angle was always less than the direct shear value.

*Land Locomotion Laboratory, Detroit Arsenal, Detroit, Michigan.

ENGINEERING ASPECTS OF VEHICLES

Vehicle System
Analytical Model Determination
and
Application to Lunar Surface Missions

TITUS ANDRISAN*

ABSTRACT

Scientific exploration of the moon carried out by manned and unmanned missions is soon to be accomplished by the National Aeronautics and Space Administration. Initial objectives of the lunar surface exploration will be to obtain scientific data on the characteristics of the lunar environment, surface, and subsurface conditions. A surface roving vehicle has been considered best capable of satisfying the functional and operational requirements to be met in obtaining the desired scientific data.

Realizing the difficulty that will be encountered in determining the vehicle system and subsystem requirements and limitations, a computerized programme of interrelated factors has been developed. Physical vehicle characteristics, mission requirements, and environmental factors are integrally related and correlated to various mission performance aspects. This will result in determining if the specific vehicle being considered can perform the desired mission successfully.

SCIENTIFIC EXPLORATION of the moon by manned and unmanned missions is soon to be accomplished by the National Aeronautics and Space Administration and the Soviet Academy of Science. The initial objective of the lunar surface exploration will be to obtain scientific data on the characteristics of the lunar environment, soil, and surface conditions. Such data may be acquired most effectively by manned and unmanned surface roving vehicles.

Numerous studies conducted by many aerospace and commercial organizations conclude that such a vehicle is potentially able to provide the functional and operational requirements to be met in obtaining the desired scientific data. The intent of this paper is to delineate the various vehicles considered appropriate for various missions and the major engineering problems relative to the vehicle system and subsystem design and validation.

*Senior Project Engineer, Brown Engineering Company, Inc., Huntsville, Alabama.

MISSION OBJECTIVES

The major objective of the Lunar Surface Vehicle (LSV) will be unmanned and manned operational support on the lunar surface which will contribute new scientific knowledge pertinent to the later manned landing programmes.

Specific objectives in order of occurrence are:

(*a*) Deployment from the softlander;

(*b*) Checkout of relay communication link from earth to the vehicle and back to earth;

(*c*) Determination of mobility capability over the lunar surface;

(*d*) Acquisition of data related to the lunar environmental, soil, and surface characteristics;

(*e*) Provide capability of performing additional mission support functions.

ASSUMPTIONS AND GUIDELINES

If the first manned lunar surface exploration is to be accomplished by the NASA before 1972, it is safe to assume that the following considerations will be valid:

(*a*) Utilization of Apollo Applications Program (AAP) systems and components;

(*b*) Availability of factual information about the lunar surface within close proximity of the LSV landing site;

(*c*) Availability of a softlander spacecraft, LEM/TRUCK, capable of being delivered to and safely landed on the surface of the moon;

(*d*) Availability and adaptability of a TV and Telemetry Communication system capable of transmitting satisfactorily;

(*e*) Earth based DSIF stations will have satisfactory reception and transmission capability.

Vehicle system design and successful performance will be dependent on numerous severe limitations. Allowable spacecraft payload volume, weight, c.g. limitations, and general launch environment, i.e., vibration, acceleration, etc., can be determined to provide those factors influencing packaging and transport considerations. Subsystem material, lubrication, sealing, and thermal requirements will be determined by evaluating recent information in addition to data collected from the Lunar Orbiter and Surveyor Softlander Spacecraft missions. Soil properties, surface characteristics, and lunar gravity will directly influence vehicle mobility conditions, tractive efforts, vehicle dynamics, payload capacity, and personnel transport.

Extreme lunar environmental factors that will affect vehicle operational performance, are shown below. These have been used as a basis for all initial design and operations analyses.

ATMOSPHERIC CONDITIONS

Vacuum: $\sim 10^{-13}$ mm Hg

Solar Flare: Flux density = 10^4 protons/cm^2sec. @ energy levels = 30–300 MEV with surface dose of 10^3 roetgens/flare

Solar Irradiance: Electromagnetic energy wavelengths = .3 − .4 millimicrons

Solar Wind: Flux density = 10^9 electrons/cm^2/sec. @ energy level of 1 KEV with surface dose of 10^6 roetgens/hours.

Meteoroid flux mass: $\text{Log}_{10} M = 1.34 \ \text{Log}_{10} M + 2.68 \ \text{Log}_{10} \ (.433/(p) - 14.48$

N = Number of impacts/sq. meter/sec.

M = Mass (grams)

$p = 0.5$ gm/cm^3

SURFACE AND SOIL CHARACTERISTICS

Surface Condition

Loose Dry Sand (Volcanic ash): 0—10%

Hard Compacted Sand: 10%—25%

Hard Surface: 25%—60%

Surface Temperature: ±270°F

Soil Characteristics

Angle of Friction (ϕ): 32°

Modulus of soil deformation (K_ϕ): 0.5 (0—10% Slope)
1.0 (10%—25% Slope)
3.0 (25%—60% Slope)

Stratification Factor of soil (n): 0.5 (0—10% Slope)
0.75 (10%—25% Slope)
1.00 (25%—60% Slope)

Modulus of Soil Deformation (K_c): 0

After considering the mission objectives, lunar environmental, and surface characteristics, guidelines, and development time limitations, the vehicle design requirements may be defined.

These will be:

1. Softlander Spacecraft/Lunar Surface Vehicle Interface:
   (*a*) To be independent of the spacecraft;
   (*b*) Must be compatible with the payload limitations, (i.e., size, c.g. location, launch, and landing loads, etc.);
   (*c*) Simple deployment mode.

2. Vehicle System Design Considerations:
   (*a*) Minimum size and weight;
   (*b*) Simplicity of design;
   (*c*) Maximum manoeuvrability;
   (*d*) Reliable mobility;
   (*c*) Reasonable obstacle negotiation.

Softlander Spacecraft/Lunar Surface Vehicle Interface

The softlander spacecraft presently being considered for delivering the first (LSV) to the moon is better known as the LEM Truck. Several varieties of payloads are being considered in attempting to provide the astronauts with the maximum surface explorational capability. The LSV and additional payload must be compatible with the equipment and operation of the LEM Truck, the Apollo Command module, and Service Module, all Ground Support Equipment, and the Apollo Mission Control Center.

Anticipated landing loads are in the range of 8–10 Earth g's and are considered to be the maximum inertial loading to be encountered. After the LEM Truck has landed safely on the lunar surface, the LSV will be automatically deployed by remote Earth operator control, Apollo Command Module, or from the landed LEM vehicle. A manual unloading method will be provided in the event of a failure of the other systems. Once the vehicle has been positioned on the surface, the relay communication link will be validated. The balance of the vehicle subsystem operational integrity will then be tested to determine operational readiness.

After the LEM has landed safely, and hopefully within 5 miles of the LSV landing site, the LSV will be remotely driven from its position to the LEM site. One or both of the LEM crew will then employ the LSV to assist them in conducting a preplanned scientific exploratory close traverse.

Mission Rational

In accordance with the NASA-Saturn-Apollo-Lunar Surface Exploration programme extensive considerations have been given to providing the astronauts with the proper equipment to assist them. A surface traversing vehicle is considered essential if an extensive surface exploration is to be accomplished. Numerous studies of lunar surface vehicles have been underway by aerospace and commercial organizations for some time. These studies have envolved vehicles of various shapes, sizes and weights; proposed to accomplish varying missions.

Vehicles that have been considered to date include:

	Weight Category
1. Surveyor Lunar Rover Vehicle (SLRV)	70—100
2. Prospector	1,000—1,500
3. Local Scientific Survey Module (LSSM)	1,200—1,800
4. Mobile Laboratory (MOLAB)	6,800—8,600
5. Mobile Lunar Excursion Module (MOLEM)	7,200—9,000
6. Mobile Command Module (MOCOM)	7,200—9,000
7. Lunar Exploration Systems for Apollo (LESA)	20,000—30,000

These vehicles reflect the logical considerations in the evolutionary sequence in vehicle size and mission capability. Effects of mission parameter variations, (i.e., objectives, associated systems, utilization of existing hardware (modified or unmodified), advancements in the state-of-the-art, spacecraft payload availability and limitations, vehicle mobility requirements, lunar environmental and surface/soil characteristics, etc.), on the total programme costs, schedule, scientific mission significance, etc. can now be ascertained more realistically.

Primary objectives of the LSV will be:

(*a*) Provide scientific data acquisition capability;
(*b*) Extend the surface exploration capability of the LEM crew;
(*c*) Locate and certify the desirability of future Apollo landing sites.

Due to certain vehicle limitations and lunar surface unknown factors, the first vehicle to actually be used will be operated at reduced speeds and limited range. Some of those limiting factors are:

(*a*) Total available power;
(*b*) Rate of power consumption;
(*c*) Surface/soil conditions encountered.

Accomplishment of all of the desired scientific objectives will be entirely dependent on the capability and safety of successfully completing the mission. Should the need arise to reduce the mission the astronaut will be required to evaluate the importance and feasibility of the desired experiments. An operations analysis will determine the desirable equipment that must be provided to enhance the mission objectives. To obtain flexibility in the scope of lunar surface operations, it would be desirable to provide two-manned ride capability on the vehicle.

Total mission rational may be classified into basic sequence of events:

1. *Delivery Sequence*: This phase will be effective during the transport of the vehicle from Earth to the Moon. During this period the power source will be isolated from all other subsystems with the exception of the Vehicle Command Control Receiver. This subsystem will be connected

in such a manner that a command sent from earth, the Command Module, or the LEM, will permit activation of all other subsystems to enable remote operation of the vehicle.

2. *Remote Operational Sequence*: After the vehicle has been successfully deployed to the lunar surface and all subsystems have been checked out for operational integrity, this mode will be activated. The vehicle will then be ready to be remotely driven to the location where the LEM has landed with the Apollo crew.

3. *Telecommunication and Television Sequence*: During remote operation of the vehicle, there is an obvious need for viewing the surface to be encountered. The telecommunication link will also provide a method of determining navigational data about the vehicle attitude, range, and speed and will prove valuable if the television system should prove inadequate.

4. *Manned Operational Sequence*: After the vehicle has reached the LEM landing site the vehicle will be converted from remote to manual operational control. Should the operator become incapable of maintaining operation of the vehicle, due to incurrence of an injury, the vehicle can be controlled remotely once more.

5. *Power Recharge Sequence*: Because of the weight restrictions imposed on the over-all vehicle system, the primary power supply will have limited capacity. To permit extended mission capability, the vehicle can replenish the primary power supply from the LEM Truck. When the vehicle power output has been reduced to 25 per cent efficient, a signal device will alert the operator of the condition thus preventing a possible power failure.

Vehicle System Design Approach

Achievement of operational integrity in the vehicle complex system requires the integration of all engineering disciplines involved. The consequence of failure of the system indicates that the reliability is the most serious problem to be faced. A vehicle having limited size and weight will also have limited component reliability which cannot be overcome by redundancy.

The vehicle system analysis will be considered from two basic viewpoints:

(*a*) Achieving a system representing the optimum in functional desirability and associated with no implications as to physical realization, development time or cost.

(*b*) Development of an optimum system within a set of economic, programmed, and physical constraints, but when subject to these given constraints, will achieve the ideal from a functional point.

The vehicle system analysis should first determine the entire system requirements abstractly. Interfaces between all subsystems must then be defined as well as those between each specific subsystem and the vehicle system.

Feasibility of attaining all requirements must be investigated, and all design problems that require extensions of the state-of-the-art should be identified.

A test plan must be established indicating tests to be run, determination of support operations essential for performing these tests, what data will be obtained and why. All quantities to be measured for vehicle system evaluation must be indicated, as well as what precision is required to yield meaningful results. These test results will be checked and will serve to improve the design. Procedures will be established to evaluate partial test results that will become available before the entire test programme is completed. It will be necessary to know what the test data will reveal about the system; otherwise it would be easy to produce a vehicle that, if tested, would yield meaningless data.

Optimization

The vehicle system analysis will encompass three basic considerations:

1. Optimum functional scheme identifying the relationship of the vehicle system design and operational decision.

2. Optimum plan of procedure pertaining to developing the vehicle system test programme and actual operation.

3. Optimum physical realization of the related subsystems and support operations to accomplish the programme.

Although it would be impossible at the present time to delineate every detail of the vehicle system and programme, it is important to have an explicit plan in sufficient detail before initiating the actual programme. Maintaining a current status throughout the various modifications that are to be anticipated, is also essential.

Design assurance can only be achieved through numerous evaluations and initiation of corrective action whenever necessary throughout the design period.

Providing a vehicle capable of satisfying the aforementioned requirements presents a challenge of great magnitude. Although finalization of the vehicle design cannot be achieved at the present time, because of insufficient valid information about the lunar surface/soil environment, the basic considerations developed by the various programmes will serve as a baseline for determining subsystem considerations and payload limitations. Proper evaluation of all subsystems is required to assure that the optimum system is achieved.

Variations in external surface temperatures and the over-all requirement to maintain a variable thermal equilibrium status, with the passenger and equipment, will perhaps be one of the most significant factors that will influence vehicle design. Due to the absence of an atmosphere, heat conduction to or away from the vehicle will be relatively insignificant, since the only path available is through the contact with the soil. This thermal path should be quite easy to control by proper design.

Radiation, on the other hand, during the lunar day will be coming from both the sun and the lunar surface. If the sky temperature is assumed to be absolute zero, thermal equilibrium temperature of the lunar surface will be approximately 270°F (730°R). The external surface of the vehicle will stabilize at the same approximate temperature.

Surface temperatures will drop to approximately −270°F (190°R) during the night from radiation losses to the sky. It would be safe to assume that the external surface of the vehicle will also approach this temperature since the heat rejected from this unit can be considered to be a small amount when compared to the radiation exchange.

The actual temperature variations we may expect a lunar surface transporter to experience can be determined only by a detailed thermodynamic study of the particular configuration under consideration. Onboard equipment exposed to the lunar night generates no heat, and is expected to reach a lower minimum temperature than that of the lunar soil nearby. Similarly, it is anticipated that equipment, operating and generating heat of its own during the lunar day, will probably have parts of the equipment stabilizing at considerably higher than the "ambient" temperature of +270°F. The upper limits of the lunar temperature is not foreign to our experience. However, the low temperatures anticipated, together with the resulting temperature range could, in as complex a mechanism as a lunar surface vehicle, present a materials problem of staggering proportions.

Solar radiation does not pose the problem for equipment that it does for the astronaut. For the expected operational life, approximately two years, the effect of solar radiation will probably not pose a major design consideration, if selection of material is judicious. However, the thermal shock from solar radiation and thermal stresses of a cyclic nature caused by vehicle operational movements could cause fatigue failure of structural members of the equipment unless precautions are taken in the vehicle design.

The meteoritic environment must be considered for any equipment that will be exposed for possibly 2 years or more. Meteorite shielding will be a major consideration in the structural design of life support equipment and containers, but in the selection of a means of locomotion, the meteoritic enviroment will be less of a governing criterion. Unlike a pressurized cab where penetration means puncture, with possible catastrophic effects on the occupants, the LSV and its tractive mechanism can be designed to sustain the statistical likelihood of certain energy impacts without severe shielding.

Another severe problem that must be faced is the evaporation of lubricants and adsorbed layers of gas on bearing surfaces and the resulting seizure, or "cold weld." The problem arises not only in bearings but also in electrical contacts such as commutator, brushes, slip rings, switches, and relays. Cold welding of metals in contact with each other is accomplished in a matter of days in the absence of air. It occurs after layers of adsorbed gases and impurities have been evaporated. Bare atom can then contact bare atom, and metals attach to each other as securely as a welded joint.

Operational lunar surface hardware will be designed with high safety factors, redundancy of critical mechanisms, and an over-all fail-safe approach that will allow for many contingencies.

Aside from the above environmental conditions, the basic requirements for lunar surface system mobility are derived from the functions that must be fulfilled and the surface(s) upon which the operation will take place.

Subsystem Design and Considerations

Subsystem selection will reflect extensive lunar surface roving vehicle considerations. Major emphasis is placed on the reliability of all subsystem selection to assure proper operation of the vehicle and accuracy of all accumulated data.

The vehicle design will utilize all of the advanced aerospace materials and manufacturing techniques to provide a unique vehicle designed to achieve the desired operation and obtainment of scientific data. Major subsystems that will be involved in the vehicle design will include:

Mobility Subsystem	Command and Control
Power Source	Telecommunication and T.V. Transmission
Structure	
Scientific Data Sensors	Human Factors

Mobility Subsystem

Selection of the mobility components reflect numerous analytical and parametric investigations conducted on lunar surface vehicles to determine the optimum mobility mode. The various components making up the mobility subsystem are: the wheel and hub assembly; the drive motor, transmission, and braking assembly, and the suspension system.

Lack of pertinent data on the composition and physical characteristics of the lunar soil and surface makes it necessary, at this time, to base a traction analysis only upon assumptions and current Ranger findings, with no clue pertaining to soil bearing capabilities, intraparticle friction, cohesion, or other soil mechanics performance criteria necessary to prepare logical estimates on traction device performance. Hence, traction analysis must also be hypothesized or synthesized.

Similar circumstances are responsible for the prevailing conservative approach which assumes that the terrain possesses very poor characteristics: poor bearing load capability, low cohesion etc. In general, operating on soils having these characteristics makes it necessary to distribute the vehicle weight over large surface areas for vertical support and to displace considerable amounts of soil to gain forward traction. Track laying vehicles have these capabilities.

Experience with vehicles propelled in this manner are generally unfavourable. Tracks consist of many moving parts which are difficult to lubricate properly. Low service life and reliability due principally to this factor are the results. Tracks also constitute relatively large masses that must be accelerated, hence, tracks are large consumers of power. Military experience with tanks and other tracked vehicles supports these conclusions.

After considering all other methods of surface negotiation, the wheel was selected as the optimum traction device for the LSV. A large wheel diameter is desirable to permit maximum obstacle negotiation. However, the diameter of the wheel will be limited to the maximum size that can be contained in the payload envelope. The consideration presented herein considered the minimum size payload envelopes. The first payload envelope considered for delivery of a surface roving vehicle was the Ranger Block V Spacecraft spherical payload container. The LEM Truck is the softlander craft presently considered for delivering the first vehicle to the moon.

One of the recommended wheel configurations presently being considered for the lunar surface vehicles (2) is a torus type "wire-mesh" construction. The wire-mesh wheel consists of an interwoven spring steel wire outer casing that will serve as the traction member and also as a primary suspension for normal shock loads. An inner or secondary spring system will serve to absorb secondary dynamic shock loads resulting from obstacles that may be encountered. The wheel will be enveloped by a lightweight, extremely abrasive resistant material. Centrally located in each wheel by means of tension struts is a cylindrical hub containing the drive, transmission, and braking assembly. A torsion bar suspension system will be employed to assure positive operator control and safe passenger and payload transport. To assure that proper correlation will be maintained between all related factors of the wheel design: deflection, ground-pressure, and traction, a computerized power profile programme was devised and will be discussed later in the paper.

Each drive wheel contains a D.C. electric drive motor-gear reduction unit. Torque output will be transmitted to the wheel through a torque tube that is hermetically sealed and pressurized to assure maximum life and reliability of the drive and transmission unit. Wheel speed control will be regulated by means of adjustable voltage regulation to each drive wheel.

The primary brake system will be an electrically activated disc system. In case of failure of this system, the backup braking mode that may be employed is a type referred to as dynamic braking. This unit is inherent in the wheel drive motor thus imposing no additional weight penalty. By reversing the power to the drive motor, the motor will function as a generator and provide a retarding force with the resultant electrical power dissipated as heat. Estimation of motor deceleration time and design of the braking circuit can be readily accomplished since this method has been employed extensively on large earth moving vehicles.

Power Source

Consideration must be given to incorporating the most favourable elements: size, weight, reliability, and cost, into the optimum primary power plant capable of satisfying the requirements of the over-all vehicle system. First requirement in selecting the optimum power source is the determination of total vehicle power requirement to accomplish every operational consideration. The fundamental laws of physics are invariant in space and time, therefore, the same dynamic relationships can be utilized to determine performance levels and power requirements on the moon as those employed to determine terrestrial requirements. Since the operations will take place in a strangely different inertial reference, there are some differences that can be anticipated from the way they would be performed on earth. These differences may be best illustrated by a series of analyses.

If the lunar vehicle under consideration weighs 1800 pounds on earth, and the surface gravity acceleration on the earth is 32.2 ft./sec.^{2} (980 cm/Sec.^{2}) compare to 5.32 ft./sec.^{2} (162 cm/Sec.^{2}) on the moon, then the vehicle weight on the moon would be: $Wm = We(\mathrm{gm/ge}) = 1800$ pounds $(162/980) = 295.6$ pounds.

Due to the lack of valid information concerning the lunar surface and soil characteristics, certain reasonable values must be established to permit preliminary analyses to be performed (3). A power profile computerized programme has been developed that will employ the assumed soil/surface characteristics in conjunction with any established vehicle physical characteristics to determine the desired vehicle performance limitations and power requirements (4). Data obtained from this programme includes wheel resistance, total vehicle resistance, and vehicle power requirements. Also obtainable is the total mobility energy required to negotiate various slopes and the power required to maintain steady state velocity.

The mobility power analyses indicates that approximately 2.5 Kw of power is required to permit the LSSM vehicle to negotiate the presently assumed lunar surface. Additional power will be required for other considerations: communications scientific experiments, television viewing, and other items, unless they provide their own power source.

The type of primary power to select for the LSSM vehicle will be an important consideration. Reliability, life expectancy, weight, rate of power consumption, and ability to perform in the hostile lunar environment are important factors that must be carefully examined. Realizing that there are several types of power systems capable of adequately satisfying all of the operational requirements, an extensive evaluation of each consideration will permit determination of the optimum system (5).

The primary power source presently considered most appropriate for the LSSM type vehicle is a silver-cadmium multi-celled rechargable battery providing an output of 28 volts. With a maximum anticipated power output requirement of approximately 3.0 Kw and a battery energy density of about 13.0 watt hr./lb., a silver-cadmium battery encased in an aluminum outer case and potted in epoxy would weigh approximately 230 lbs. Other equivalent power sources, although desirable, could not be provided within this weight limitation. The aforementioned power supply will, with careful utilization, permit an operational mode of 4 hours per sortie. A charge mode of 8 hours duration will be required to return the power source to full capacity again.

From the power profile computer programme it was determined that with the mobility power limited to 2.5 Kw, the maximum velocity that could be achieved on level unobstructed surface 7.4 m.p.h. Obviously it is not recommended that the vehicle be driven at top speed except in case of some emergency. With a prescribed speed of 5 mph and a 4 hour sortie, a maximum range of 20 miles can be accomplished per trip. Should an extended mission be desired, additional power may be substituted for some of the scientific equipment.

The power system will incorporate a state-of-charge regulator to monitor the stored net capacity of the battery. When the power output has decreased to 25 per cent—30 per cent efficient, a warning indication will be given of the approaching vehicle power charge mode. A combination voltage regulator-convertor will distribute various power levels to the various electronic subsystems. Blocking diodes in the systems will prevent undue battery discharge. Electrically coupled to the battery will be a unit to permit recharge when required.

Another factor that must be considered in attempting to predict the vehicle performance is the limitation imposed by having one or more subsystems inoperable. A generalized method for predicting the performance capacity of a vehicle may be programmed to determine the probability of continued operation and aid in considering the reliability requirements of the various subsystems involved (6). This technique may involve a family of vehicles in the analysis; those considered here include those which have a constant weight and have four, six, and eight wheels. These vehicles will be referred to as I, II, and III, respectively.

Realizing that a vehicle having one or more systems inoperable will incur restrictions of varied degrees depending on the surface being encountered, the various surfaces will also be grouped. Surfaces presently considered are grouped into two major categories. Annex A and G. Annex A[3] contains Tables 1 and 2 involving Maria and Upland surface considerations. These tables each contain five subsections which are subgrouped (a), (b), (c), (d), and (e). In this way surface references for Annex A may be indicated: A_{1a}, A_{1b}, A_{1c}, A_{1d}, A_{1e}, and A_{2a}, A_{2b}, A_{2c}, A_{2d}, A_{2e}.

Annex G may likewise be subdivided into two groups with the latter consisting of eight considerations. Surface references for Annex G may then be indicated: G_1 and G_{2a}, G_{2b}, G_{2c}, G_{2d}, G_{2e}, G_{2f}, G_{2g}, G_{2h}. Mobility factors that must be taken into consideration are: traction (t), drive motors (d), steering (h), and suspension (s).

Since the desired test data must reflect vehicle velocity capability while encountering each surface under consideration, each type of performance must be categorized. For example, vehicle I may develop velocity $(V_{A1})_{I}$ on surface A_1, and $(V_{G2})_{I}$ on surface G_2, while vehicle II maintains velocity $(V_{A1})_{II}$ and $(V_{G2})_{II}$. These values may then be tabulated into a velocity matrix as follows:

	Lunar Surface			
Vehicle	A_1	A_2	G_1	G_2
I	$(V_{A1})_{I}$	$(V_{A2})_{I}$	$(V_{G1})_{I}$	$(V_{G2})_{I}$
II	$(V_{A1})_{II}$	$(V_{A2})_{II}$	$(V_{G1})_{II}$	$(V_{G2})_{II}$
III	$(V_{A1})_{III}$	$(V_{A2})_{III}$	$(V_{G1})_{III}$	$(V_{G2})_{III}$

In a similar way, additional matrices for traction, drive motors, steering, and suspension can be established. All of these matrices combined together will establish a parameter providing an accurate evaluation of the mobility requirements of a vehicle necessary for a successful mission.

Structure

In selecting the materials to withstand lunar environmental conditions, the duration of exposure and anticipated dynamic conditions must be considered. Based on the presently known conditions, it is felt that conditions will be very severe on the Moon. The selection of suitable materials will obviously be required to contend such factors as friction, lubrication, abrasion, heat radiation, and possible sublimation effects.

Conventional tubular construction will be advocated for the frame and other appendages, employing aluminum alloy or other suitable materials. Thermal radiation and micrometeorite protection shall be provided where necessary, utilizing panels of light-weight materials having high reflectivity, energy-absorption, and self-sealing qualities. Excessive internally generated heat will be dissipated through adequate heat sinks and radiators.

Command and Control

Success of the mission will be dependent, to a great extent, upon the proper functioning of the command and control system of the vehicle. Performance requirements are as vital for a remote-controlled vehicle as they are for one that is manually controlled. Although the primary directional control of the vehicle during remote operation will employ television viewing, should this mode become inoperable a backup system must be available. Thus it is of utmost importance that the control system be highly reliable.

The task of achieving a reliable system will be placed on the design as well as technique standardization. Another factor that will increase the reliability is to adopt a command and control system having minimum complexity. A closed-loop-type system may be employed in controlling the vehicle. Output from various subsystems will utilize direct drive mechanisms employing electrical and/or mechanical means for activating the various functions. Each subsystem will be classified into its functional characteristics, such as: size, weight, distance, time of response, and accuracy. Output from the control system will then be classified into the usual characteristics, such as: force, pressure, velocity, acceleration, or position, thus enabling the output to be accurately defined.

The command and control system must be capable of regulating the heading of the vehicle during remote control operation.

Five major functions that the command and control system must perform to permit safe and accurate vehicle operation are:

1. Speed determination and control;
2. Distance traveled and control;
3. Attitude (pitch and roll);
4. Direction control;
5. Emergency override.

One of the wheels will function as a tracking wheel to measure vehicle speed and distance traveled. An azimuth gyro will provide directional reference to determine vehicle heading. Pitch and roll attitude will be measured by a vertical gyro reference.

An earth command and control console will be an integral part of this system and will contain amplifier networks and electromechanical components required to receive information from the vehicle. This information will then be processed to provide accurate and continual vehicle performance disclosure during remote control phase.

Sensors used include attitude reference, odometer, tachometer, and a directional reference azimuth gyro. Attitude reference will be achieved by a two-degree gyro. A simple revolution counter with suitable pickoffs, attached to the tracking wheel, will serve as an odometer.

Telecommunication and Television Transmission

The telecommunication subsystem will monitor the basic functions of surface operations and telemetry data transfer. To permit simplification, the television subsystem will be incorporated into the telecommunication system. The telemetry system will utilize existing tracking station capabilities.

The telecommunication system will have the sole responsibility of providing a two-way link between the vehicle and the softlander. Transmission of vehicle operation and scientific data will necessitate simultaneous multiplexing. Three methods of data transmission are considered adequate:

1. Transfer of engineering telemetry from the vehicle to the softlander will require a relatively narrow bandwidth and low power consumption.

2. Transmission of television pictures and soil bearing data, such as static forces, wheel drive, torque, and flotation information, will necessitate multiplexing.

3. Range measurements between the vehicle and the softlander will require use of a separate transmission band from the softlander to the vehicle and return.

Transmitters located on both the vehicle and the softlander must be capable of operating on low and high power output. Television transmission will be similar to that of the Surveyor spacecraft, incorporating advanced microminiature solid-state materials and techniques.

The telecommunication and television system will present a problem in that transmission to earth or the Command Module will require considerable power and directional beaming of the signal. On the other hand, low-power omnidirectional outputs will be limited by range and surface obstructions. The requirements would be considerably lower if vehicle control through real-time continuous television were not required. Permitting the vehicle to operate by multiple stop-start activation and restricting the vehicle speed to a maximum of 5 mph, non-real time pictures at one second intervals may be adequate for visual monitoring.

The television system in conjunction with the telecommunication system will be used for vehicle control, operation, and observational purposes. To provide control and observation, erection of the necessary equipment must be achieved first. Proper remote guidance of the vehicle is best achieved by a three-dimensional video or dual standard television system which will permit relatively accurate distance ascertainment.

Antenna requirements for lunar communications may vary from a high gain for the Lunar-Earth link, to an omnidirectional antenna for relay and lunar mobile operation. With the transmitting frequency for Earth-Lunar link regulated by the DSIF capability, it is desirable to use the largest antenna that can be tolerated and still provide the gain required to satisfy all communication requirements. Two antenna systems are considered adequate for the LSSM vehicle. A directional parabolic antenna for television and Telemetry transmission coupled through an electronic servo drive and control. The second antenna will be an omnidirectional antenna employed to provide a coherent detection of vehicle operational activity and transmission of all scientific data from the vehicle to the softlander.

Scientific Data Sensors

Scientific data considered most pertinent for the first mission will include soil bearing strength, surface topography, vehicle dynamic and mobility characteristics, and surface thermal limits. This collected data will permit charting of the actual mission from navigational plotting of the vehicle traverses. Geometrical relationship between terrain features and the softlander will be established to permit topographical referencing. A clinometer, located at the vehicle centre of gravity, will detect all attitude deviations during vehicle operation. This will indicate topographic deviations along the charted route.

Soil bearing strength will be determined employing instruments capable of obtaining all the pertinent measurements that can be correlated to acceptable laboratory readings and are compatible with the vehicle lunar weight. As the vehicle begins forward motion, the wheel drive motors impart a determinable torque to the drive wheels. By varying the wheel drive power, acceleration, and dynamic forces can be determined as well as traction effectiveness.

A dual-sensing thermo probe will be located at the wheel hub to determine ambient temperature within close proximity of the lunar surface and that of the drive motor compartment.

Should additional scientific experiments be desired, additional sorties could be scheduled until all possible experiments have been accomplished.

CONCLUSIONS AND RECOMMENDATIONS

The aspects of providing a surface roving vehicle capable of being transported to and landed on the Lunar surface by current spacecraft will enhance our Space exploration programme. Providing a vehicle capable of satisfying the above requirements in addition to performing specific operational and scientific missions involves numerous major engineering problems. The intent of this paper was to assess the various problem areas influencing the design of a Lunar Scientific Surface Roving Vehicle which was depicted herein as a baseline vehicle.

Advocating a relatively simple vehicle design, employing techniques and components of the present state-of-the-art, assures a greater margin of success not only on the mission but in savings of precious time and material.

Although it would be desirable that the mission operations follow a preplanned sequence of all scientific experiments, an uncontrolled random operation will yield ample valuable information and provide a greater degree of success.

Parametric information of the subsystems permit a weight determination and limitation. Predominant factors that became evident were:

(*a*) Having a limited volume and weight, the power source will be the major problem.

(*b*) Total desired scientific mission task would probably require numerous sorties.

(*c*) A small surface roving vehicle successfully completing a reduced mission, would yield pertinent and valuable data, sooner than with a larger vehicle and greater mission.

REFERENCES

1. Marshall Space Flight Center CD-501, *Saturn V Launch Vehicle System Specification*, July 1, 1963.
2. ANDRISAN, T. and J. CHALK, *Apollo Extension Systems—A Study of Local Scientific Survey Module*, Report A2-ABX-007, Brown Engineering Company, Inc., June 30, 1965.
3. MASON, MCCOMBS, and CRAMBLIT, *Engineering Lunar Model Surface* (*ELMS*), Kennedy Space Center, Sept. 4, 1964.
4. BLANKE, G. R. and D. R. JOHNSON, *Apollo Extension Systems Locomotion Energy Analysis for Lunar Surface Vehicles*, Report A2-ABX-002, Brown Engineering Company., Inc., June 30, 1965.
5. BELL, JR., J. H. and H. R. SCHMIDT, *Apollo Applications Program Power Systems Study*, Report A2-ABX-002, Brown Engineering Company, Inc., Dec. 31, 1965.
6. ANDRISAN, T. and E. R. MILES, *Report on Mobility Analysis and Vehicle Definition Study for Extreme Lunar Surface Conditions*, Report NSL E30-25, Northrop Space Laboratories, Sept., 1964.

Tires of Equivalent Soil Performance

ANTHONY J. RYMISZEWSKI*

ABSTRACT

This article is an extensive continuation of tire geometry–soil relationships initially studied under Dr. M. G. Bekker and R. A. Liston in 1958–1959. A graphical presentation of "equivalent" soil performance tires was formally reported by Bekker in *Machine Design,* February 1960. Liston also reported this possibility in Land Locomotion Research, Report Number 59.

Computer techniques for determining families of "equivalent" tires which produce the same DP/W in a hmoogeneous medium are discussed. Equations, tables, and performance curves are developed for various wheel loadings (250 to 8000 pounds), tire widths (6 to 60 inches), tire diameters (12 to 120 inches), and soil types. A novel computer-graphic approach utilizing present "state-of-the-art" mobility equations is presented wherein wheeled vehicle configurations (4 × 4, 6 × 6, 8 × 8, and 10 × 10) of "equivalent" soft soil performance can be selected.

This study will aid wheeled vehicle designers in selecting proper wheel forms for use in new vehicle configurations requiring optimum off-the-road soft soil mobility.

*Land Locomotion Laboratory, Detroit Arsenal, Detroit, Mich.

Application of Rubber Crawler "Ohtsu Mighty Pillar" for Engined Tillers

AKIRA OKAMURA*

ABSTRACT

Rubber tracks are described for application over pneumatic tires on small agricultural tractors. The performance is successful on snow, ice, sand, mud, paddy fields, and steep hills.

THE ADVANCE OF MECHANIZATION of agricultural equipment has replaced the use of livestock in agriculture. Now, on various modern agricultural jobs, the new Ohtsu Mighty Pillar is a unit which answers all requirements in farming. Transportation and plowing are generally regarded as the two main jobs in agriculture, and so rubber pneumatic tires are being introduced in this field. But there are some special places where the agricultural pneumatic tires do not run well, and therefore specially made wheels are used sometimes in accordance with the place and job.

It is pointed out that these specially made wheels are very troublesome when replacing to and from pneumatic tires.

We have released to the public the rubber crawler "Ohtsu Mighty Pillar" which is mounted outside conventional agricultural pneumatic tires. The attachment is very simple and can be done quickly. The purpose of the "Ohtsu Mighty Pillar" is to expand the use of tillers to snow, ice, sand, mud, paddy fields, and steep roads or field surfaces where the tillers with conventional tires are not able to work at the present time.

Japanese paddy fields are muddy and many traction problems are encountered in these special fields when using agricultural pneumatic tires. In this case, steel caterpillars are generally used on tillers, but this tends to cause damage to tiller and road when going on hard road surface. Thus it would be more profitable to use the rubber crawler. We have been doing research work on the application of rubber crawler to reaping and threshing machines and the combines used for harvesting.

*Ohtsu Tire and Rubber Co., Ltd., Japan.

This paper is to introduce the application of the crawler, now in regular production in our company, for the tillers. Figures 1 and 2 are "Ohtsu Mighty Pillar" with "HONDA TILLER TYPE F-190" attached.

FIGURE 1

FIGURE 2

CONSTRUCTION AND MOUNTING

The rubber crawlers are mounted on tires and idler wheels attached to either the front or rear parts of the tiller fitting frame. The crawlers and attachments may be attached to or taken off the tillers by very simple and quick procedures.

The name of parts and mounting procedures are as follows.

Mounting procedures:

(1) Mounting of hexagonal shaft;

a. Take off tires from tiller, and insert sub-wheel to the shaft of tiller. Then put back the tires.

b. Install the tires to end of shaft.

c. Insert hexagonal shaft to drive shaft of tiller, and align the pin holes of hub and shaft.

d. Then insert pin in the hole and put *R*-type pin into the pin (Figure 3).

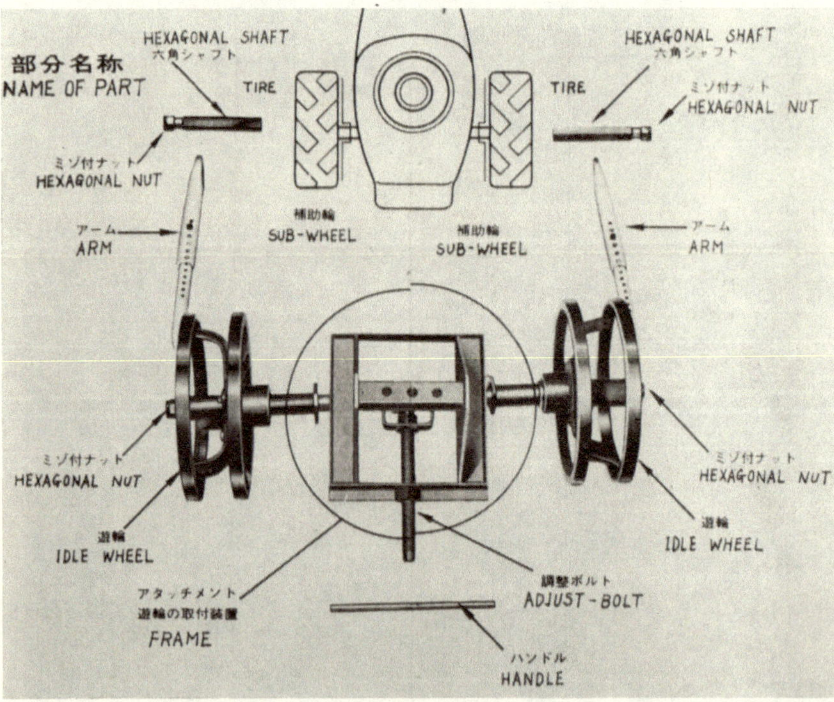

FIGURE 3A

(2) Mounting of Frame:

a. Take off hitch-pin from tiller hitch frame.

FIGURE 3B

b. Mount the frame to the tiller hitch frame to face adjust-bolt to the outside.

c. Align the pin holes of the frame and hitch frame. Insert hitch pin and put *R*-type pin in. (Figure 4.)

(3) Mounting of Arms:

a. Insert arms by putting big arm with bearing onto drive shaft, and putting small arm onto idle wheel shaft.

b. Put hexagonal nuts on the shafts, and insert split cotter. (Figure 5.)

(4) Mounting of Rubber Crawlers:

a. Insert rubber crawler while keeping idle wheels up.

b. Then insert crawler all around while keeping tire up. (Figure 6.)

(5) Adjustment of Rubber Crawler:

a. Insert handle to adjust-bolt.

b. Apply the tension to crawler by turning adjust-bolt, and controlling the tension of crawler so that a slight bend is encountered when pushing down by hand.

c. Insert arm pin into the hole after getting the big arm hole aligned to one of the small arm holes and put *R*-type pin in. (Figure 7.)

This size of rubber crawler is 20-245B used over AG 4.00–10 or 4.00–12 tires.

FIGURE 4

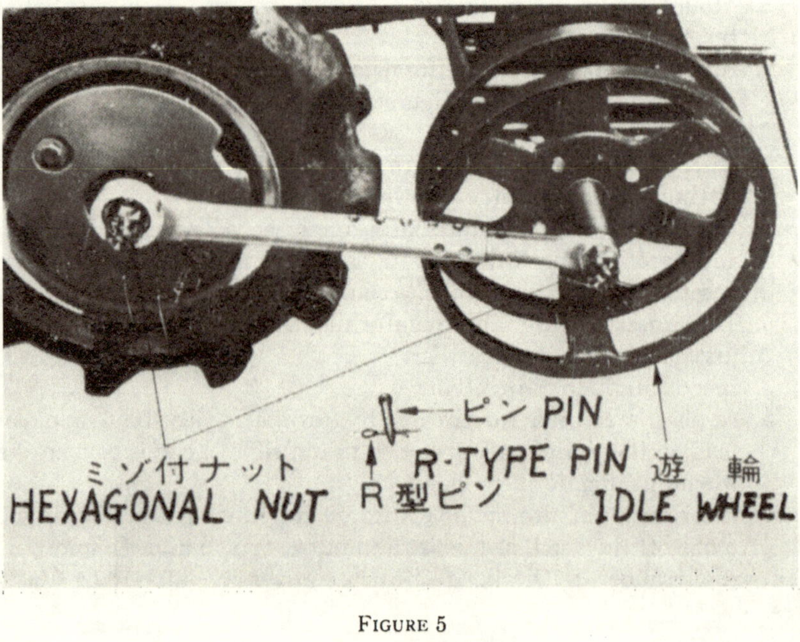

FIGURE 5

Figure 6

Figure 7

USE AND PERFORMANCE

The tillers with rubber crawler have excellent performance on snow and ice, mud, soft soil, sand, and steep slope roads. More satisfactory results have been achieved in application of the rubber crawler on the tillers for these fields than when only using conventional agricultural pneumatic tires in our field tests. These tillers with rubber crawlers have given very excellent performances with much less noise and thump than crawlers equipped with steel crawler for transportation jobs, and have less curb weight for production. The tillers with rubber crawlers are self-cleaning when snow and mud are encountered. The rubber crawlers have fewer maintenance troubles.

Several examples of rubber crawlers for special fields are as follows.

(*a*) Sand field:

We have had the most excellent performances on sand fields by these crawlers. Under these conditions they have longer life and more powerful pulling action and better flotation than other crawlers or special steel wheels. Thus they are preferred for transportation and pulling jobs on sand fields. (Figure 8.) Figures 8 and 9 show the tillers used to pull a root flower lifter.

FIGURE 8

FIGURE 9

(*b*) Paddy field:

The tillers with rubber crawlers are immediately able to do agricultural jobs without any replacement to steel wheels for paddy fields, and much smoother running and better stability from the tillers are achieved in plowing and cultivating. They give excellent flotation and strong traction in the paddy field with no variation in the pull. Figures 10 and 11 show the tillers used for rotary cultivating on paddy fields. Figure 12 shows how muddy the paddy field is.

(*c*) Steep slope road:

When the tillers with rubber crawlers are used for transportation on a steep slope road, they have given stronger traction, better stability and more satisfactory steering than the other steel caterpillars. They are able to go up as much as a 30° slope with a full load. Figure 13 shows the tiller with the rubber crawler going up a steep slope with heavy loads. Figure 14 shows the tiller with the rubber crawler going up a steep road pulling a loaded wagon trailer.

(*d*) Use for grading:

If the tillers with the rubber crawler are equipped with a grader plate, they are able to be used as a light grader for light public works and snow works. On a concrete road, rubber crawlers are preferable because they do not damage the road surface. Figures 15 and 16 show those that are used for public works.

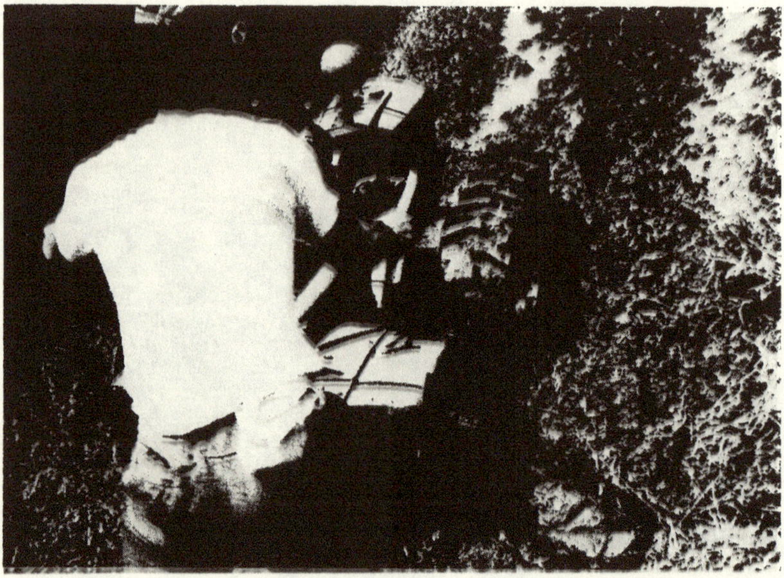

FIGURE 10

FIGURE 11

FIGURE 12

FIGURE 13

FIGURE 14

FIGURE 15

FIGURE 16

(*e*) Snow field:

The tillers with rubber crawlers give more satisfactory performance than others with a steel caterpillar. They are more successful in snow fields than steel crawlers, from the standpoint of speed, steering, and maintenance, etc. In the snow fields, the rubber crawlers remain self-cleaning and provide maximum pull while other types become clogged with snow and lose traction; snow builds up on the wheels, caterpillars, and other attachments, generating poor traffic conditions and sometimes causing mechanical troubles.

The self-cleaning action is provided as the rubber crawler is bent around the drive-wheel tires. The snow simply falls off.

If transport is needed on frozen fields, one is able to run with the rubber crawlers on which steel studs are embedded in the tread area. There are no slip troubles at all on the frozen field. Figures 17 and 18 show its use on skiing field. Figures 19 and 20 prove the difference between the snow built-up on crawlers of rubber and of steel. The left is rubber and right is steel. Figure 21 shows studs installed on the tread of rubber crawlers.

IV. Data:

(*a*) OHTSU MIGHTY PILLAR size

20-245B For Agricultural tires 4.00–10, 4.00–12, 4.50–10.

25-270B For Agricultural tires 5.00–12, 5.00–12.

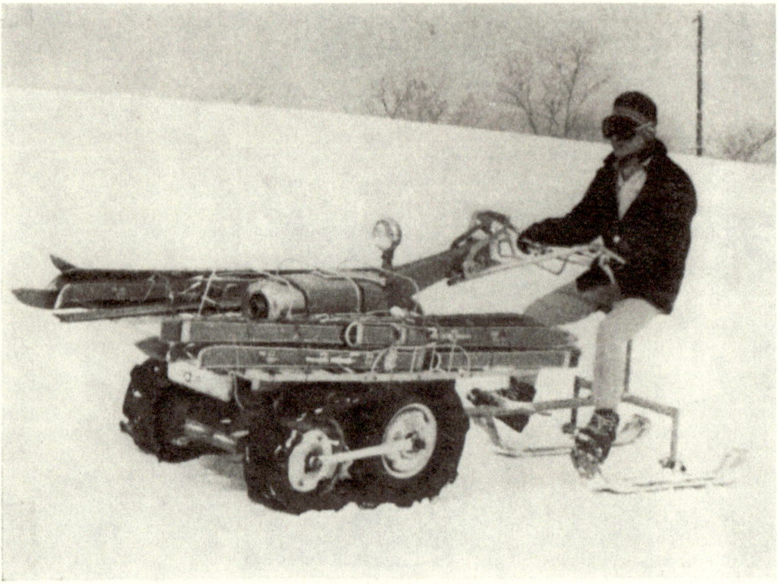

FIGURE 17

FIGURE 18

FIGURE 19

FIGURE 20

FIGURE 21

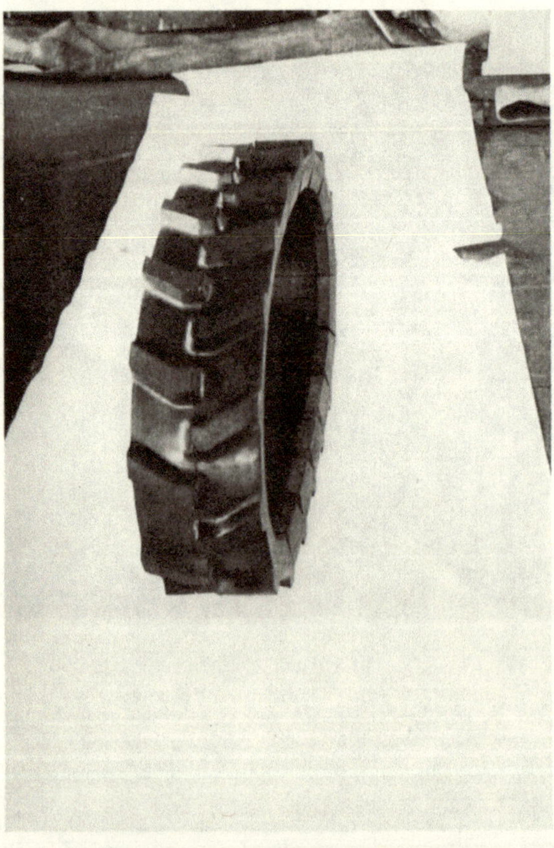

FIGURE 23

TRACTION

Traction curves have been obtained experimentally as follows. Figure 24 shows the traction curve for hard soil fields, Figure 25 for a sand field, Figure 26 for a very bad paddy field, and Figure 27 for a snow field.

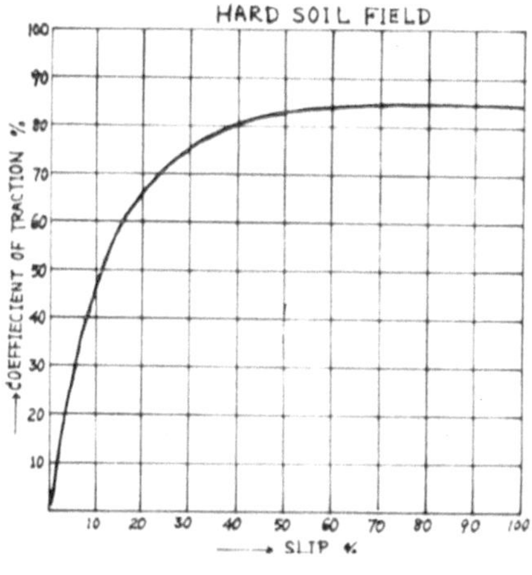

FIGURE 24

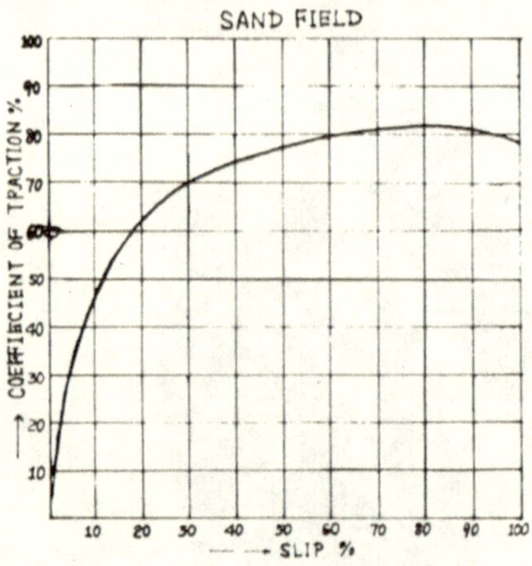
SAND FIELD
COEFFICIENT OF TRACTION %
SLIP %

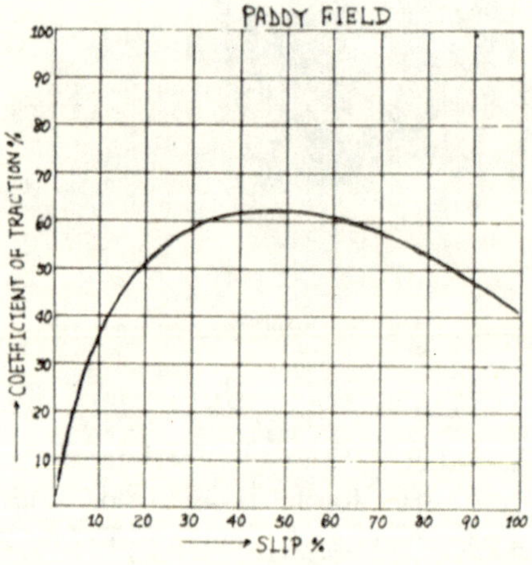
PADDY FIELD
COEFFICIENT OF TRACTION %
SLIP %

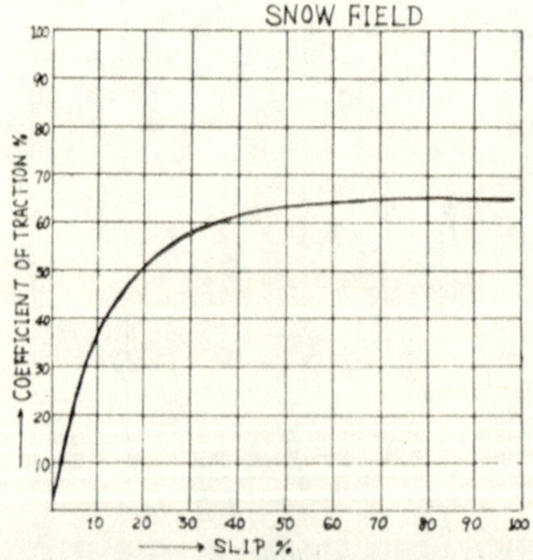
SNOW FIELD
COEFFICIENT OF TRACTION %
SLIP %

What Price Water Speed Improvements of Amphibians?

IRMIN O. KAMM*

ABSTRACT

Speed improvements of amphibious vehicles are almost entirely form dictated. The speed-drag characteristics of the various types of vehicles are discussed, ranging from the displacement type, through train concepts and planing hulls to hydrofoil vehicles. Emphasis shifts from basic hydrodynamics, as speeds are increased beyond the limits of displacement craft. A land vehicle marriage with higher forms of hydrodynamic shapes increase mechanical complexity, weight, and cost. Engineering and economic considerations become the deciding factors.

*Davidson Laboratory, Stevens Institute of Technology, Castle Point Station, Hoboken, N.J.

Transient Calibration of Vibration Transducers

JOHN N. MACDUFF*

FURMAN WYCHE BARTON†

ABSTRACT

The determination of the frequency response of vibration transducers and measurement systems from transient test data is outlined. The system used to produce the transient test signals in the field is described and the computation procedure is presented. Frequency response data obtained from normal harmonic test procedures are compared with the frequency response obtained from analysis of the transient data. Problems associated with the necessary numerical integration are considered and discussed.

CALIBRATION of the transducers and associated measurement system is one of the required phases of any programme of experimental dynamic investigation. For vibration and shock programmes, it is customary to excite the pickup with simple harmonic motion and obtain the output response as a function of the input frequency. Normally, this required calibration is carried out in the laboratory under carefully controlled conditions. Either the entire proposed measurement system may be used in this laboratory calibration, or individual calibration of the system components, including the transducers, amplifiers, filters, and recorders may be conducted. In this case, the system gain is then taken as the product of the individual gains, and the phase as the sum of the individual phases.

There is, of course, no real substitute for laboratory calibration under carefully controlled conditions. However, when measurements are to be made in the field, changes may be required in the calibrated measurement system and thus a modification of the laboratory calibration will result. The most common changes which occur between field and laboratory installations are changes in the lengths of connecting elements transmitting the information from the transducer to other components of the system. Cable length changes are common in field shock and vibration

*Department of Mechanical Engineering, Duke University, Durham, North Carolina.
†Department of Civil Engineering, Duke University, Durham, North Carolina.

work and tube length changes are the rule rather than the exception in dynamic measurement of blood pressure in medical research. Field conditions may also require changes in the amplifiers or filters used.

Static or single point testing is one common method used to check calibration in the field. For this type of check, the electrical components may be calibrated with either a direct current level change or a single frequency signal. The transducer may be inverted to give a one g signal or dropped in a blanket to give a transient of approximately known characteristics. While these and similar methods are useful, they do suffer from the common defect of indicating calibration at only one frequency, and may not provide information relative to attenuation or gain changes which are a function of frequency. It is desirable, therefore, to have available some simple direct method for field recalibration of a transducer or system over the entire frequency range of interest. It is the purpose of this paper to describe such a method. The procedure developed is not intended to be a substitute for careful laboratory frequency response testing. However, the proposed procedure does provide an approximate frequency response of the system as installed in the field where prolonged harmonic testing is not feasible.

OUTLINE OF METHOD

The basic procedure is rather simple and is based upon the use of the Fourier transform. The Fourier transform converts a function of time, $X(t)$, into a function of frequency, $X(j\omega)$, by means of the integral:

$$X(j\omega) = \int_0^\infty X(t)e^{-j\omega t}\,dt. \tag{1}$$

Hence, if we have a linear system with a transfer function G, an input $X(t)$, and an output $Y(t)$, as indicated in the block diagram of Figure 1,

FIGURE 1. Generalized block diagram

we have the symbolic equation: $X(t)xG = Y(t)$. If $X(t)$ and $Y(t)$ are known, they may be converted to functions of frequency by means of Equation 1, and solving for the frequency response gives:

$$G(j\omega) = Y(j\omega)/X(j\omega) \tag{2}$$

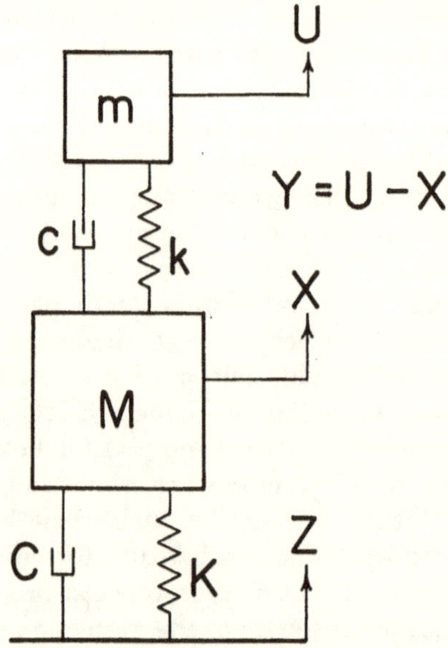

FIGURE 2. Idealized transient excitation system

To illustrate the application of this concept, the system of Figure 2 is used. Here the upper spring, dashpot, and mass is assumed to represent the transducer with an output proportional to Y, the relative motion between the transducer mass and its base. The lower system represents a spring-supported table with an absolute motion X, and an input motion Z.

For this system, the differential equations are:

$$(mD^2 + cD + k)Y + mD^2X = 0;$$
$$-(cD + k)Y + (MD^2 + CD + K)X = (CD + K)Z,$$

where D is the operator, d/dt.

Letting:

ω_1 = transducer natural frequency,
ζ_1 = transducer damping ratio,
ω_2 = base system natural frequency,
ζ_2 = base system damping ratio,

and assuming that the mass ratio, m/M, is very small, leads to the expression:

$$\begin{bmatrix} Y \\ X \end{bmatrix} = \frac{\begin{bmatrix} D^2 + 2\zeta_2\omega_2 D + \omega_2^2 & 0 \\ 0 & D^2 + 2\zeta_1\omega_1 D + \omega_1^2 \end{bmatrix}}{(D^2 + 2\zeta_1\omega_1 D + \omega_1^2)(D^2 + 2\zeta_2\omega_2 D + \omega_2^2)} \begin{bmatrix} 0 \\ (2\zeta_2\omega_2 D + \omega_2^2)Z \end{bmatrix}$$

From this, the ratio of the transducer relative motion Y to the table absolute motion X is:

$$Y/X = -D^2/D^2 + 2\zeta_1\omega_1 D + \omega_1^2. \tag{3}$$

If Y and X are known transient functions of time, they may be transformed to function of frequency by Equation 1. The frequency response transfer function, $G(j\omega)$, given by Equation 2, may then be expressed as:

$$Y(j\omega)/X(j\omega) = \omega^2/\omega_1^2 - \omega^2 + j2\zeta_1\omega\,\omega_1 \tag{4}$$

The same expression, Equation 4, is obtained if the operator D is replaced by $j\omega$ in Equation 3.

EXPERIMENTAL RESULTS

The test system used to obtain experimental results is shown in Figure 3. The table is supported by cantilever springs and auxiliary coil springs from a U-shaped member. The U-shaped member is guided on Teflon blocks as indicated. The table displacement, X, is measured relative to the base by means of a linear potentiometer.

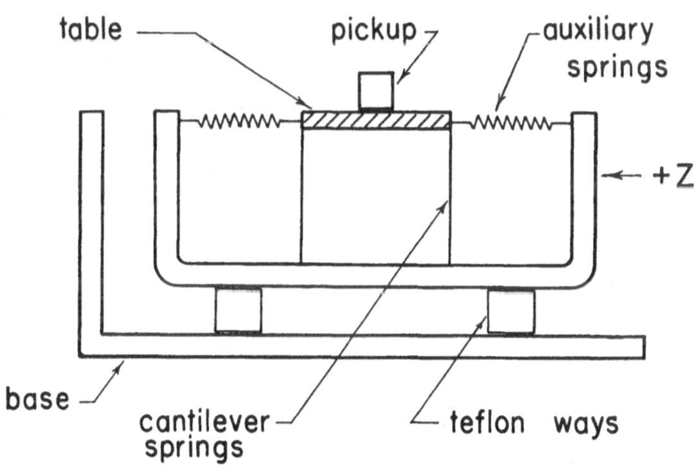

FIGURE 3. Transient calibrator

In use, the pickup is secured to the table, and the U-shaped frame is pushed manually along the ways until it contacts the base stop at some desired distance of travel. The pickup signal, amplified and filtered as in actual proposed field use, and the signal from the table displacement potentiometer, are recorded.

For this experimental investigation, an MB velocity transducer, Type 124, was used. This instrument has a natural frequency of about 4.75

c.p.s.; a listed damping ratio of 0.65; and an output of about 0.096 volts/(in./sec). The MB vibration meter used provides an approximate integration of the velocity signal. The low-pass filter, integral with the meter, has a time constant of approximately 0.03 seconds with an attenuation beyond 30 c.p.s. stated to be 20 db/octave.

For Test 1, the measurement system consisted of the pickup, meter, and filter with a Sanborn oscillograph used for recording. Figure 4 shows

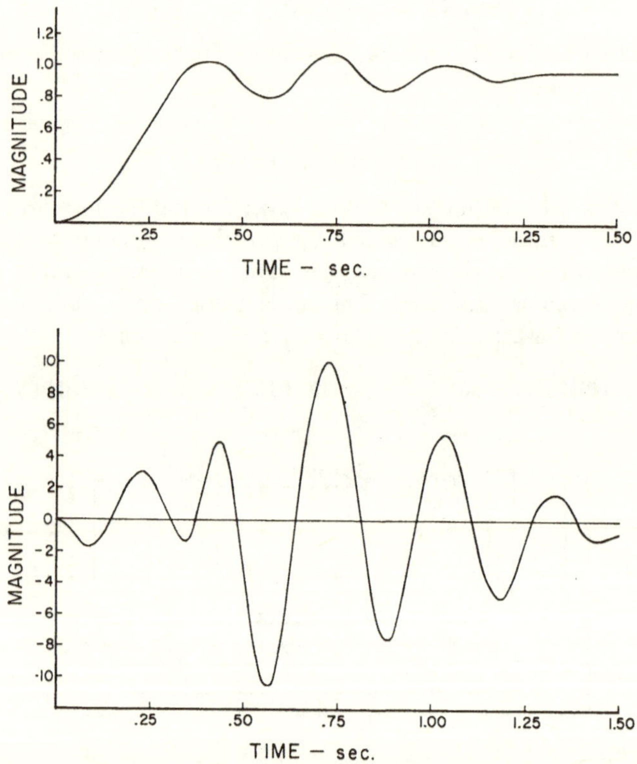

FIGURE 4. Test 1 oscillograph records

the input signal from the potentiometer, and the signal from the transducer as recorded on the Sanborn. In Figure 5, the frequency response obtained from this transient test data is compared with the ideal frequency response computed from the system characteristics listed by the manufacturer.

For Test 2, the pickup was connected directly to the Sanborn oscillograph without integration or filtering of the signal. Figure 6 compares the frequency response obtained from the transient test data with the ideal frequency response.

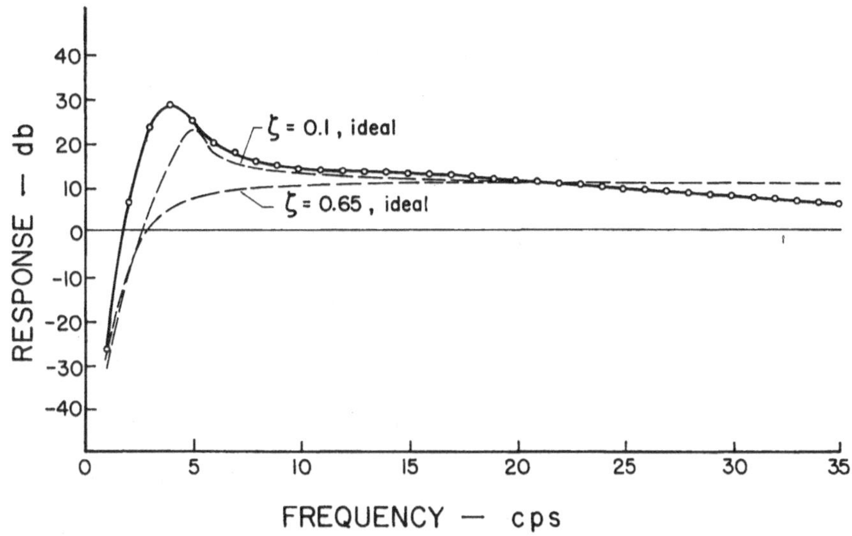

FIGURE 5. Frequency response for Test 1. Measurement system comprised of velocity transducer, amplifier, integrator, filter, and oscillograph

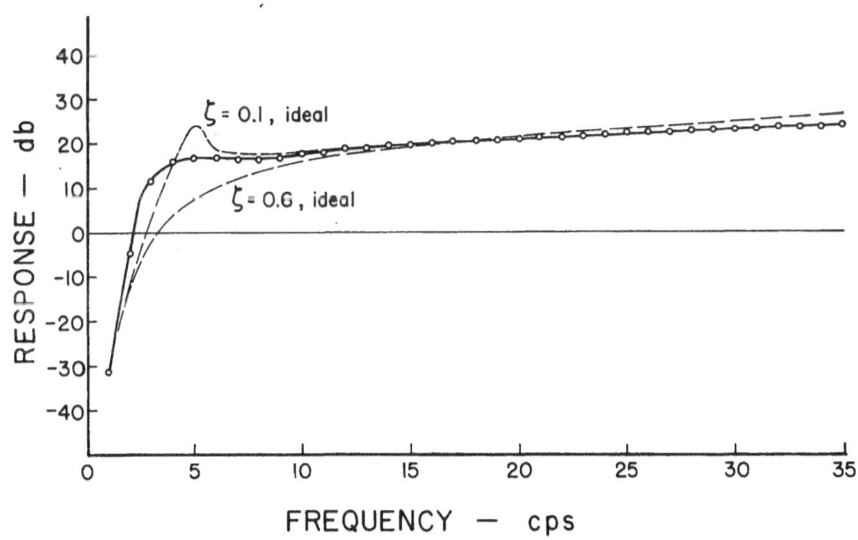

FIGURE 6. Frequency response for Test 2. Measurement system comprised of velocity transducer and oscillograph

NUMERICAL PROBLEMS

As with any form of data processing, there are inherent sources of error in the procedures used here. Errors occurring in the process of digitizing the transient records may be reduced by mathematically smoothing the data before attempting to calculate the frequency response. This tends to give a smoother output for the frequency response, but may destroy some of the higher frequency portion of the signal in the process.

The time increment, $h = t_n - t_{n-1}$, used in the numerical integration appears to be one of the variables affecting the accuracy of the frequency response obtained from the transient record. Huss and Donegan (1) suggest that this time interval should be on the order of 1/4th to 1/5th of the period of the highest frequency of interest. For simple transients such as the quarter sine wave pulse of Figure 7(a), this appears to be a

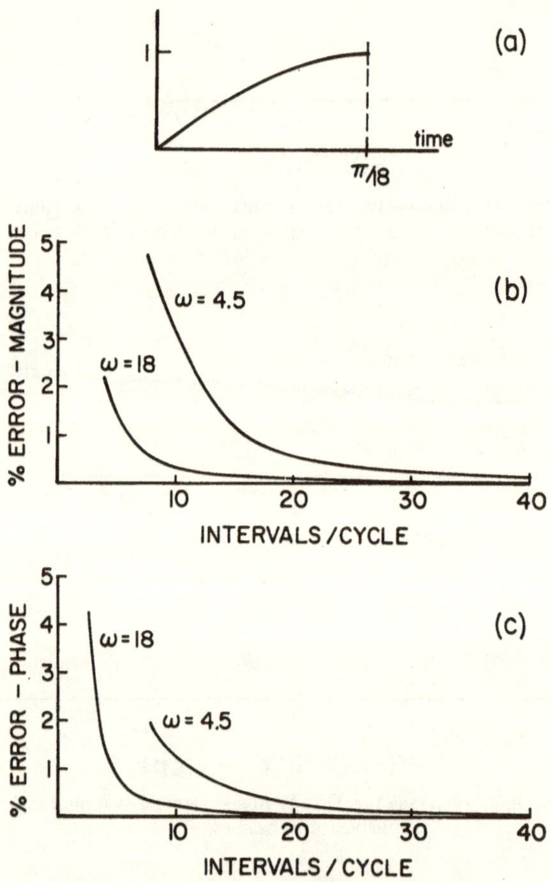

FIGURE 7. Effect of interval size for simple system. (a) assumed transient record; (b) magnitude of error in per cent vs intervals/cycle; (c) phase error in per cent vs intervals/cycle

good lower limit. Figure 7(*b*) shows the per cent error in magnitude as a function of the intervals/cycle of interest for $\omega = 4.5$ and $\omega = 18$ rad/sec. Figure 7(*c*) shows the error in phase for the same two frequencies. The error does not appear large, but it is believed that 10 to 12 intervals per cycle might be more conservative.

The record cutoff time appears to be a main source of error in the frequency response obtained from transient data. The cutoff time is the time at which integration of the record ceases instead of proceeding to infinity as theory dictates. Stopping the integration short of infinity tends to produce an oscillating error with a magnitude and frequency which is related to the ratio of the cutoff time to the time constants of the system.

The effect of cutoff time is illustrated by use of a simple first order system with the differential equation: $(D + 25)\,Y = f(t)$. If it is assumed that the transient excitation is a very short unit magnitude pulse, the Fourier transform is:

$$Y(j\omega) = \int_0^{t_0} e^{-(j\omega+25\,t)}dt$$

$$= \frac{1 - e^{-25\,t_0}(\cos \omega t_0 - j \sin \omega t_0)}{25 + j\omega}. \tag{5}$$

The magnitude of this frequency response is:

$$db = -10 \log(625 + \omega^2) + 10 \log[1 + (e^{-25\,t_0})^2 - 2e^{-25\,t_0} \cos \omega t_0] \tag{6}$$

The first term in Equation 6 is the ideal frequency response and the second term is the error created by stopping the integration at some cutoff time t_0. Figure 8(*a*) compares the ideal magnitude with the magnitude obtained by using a cutoff time of $t_0 = 0.04 = 1$ time constant; Figure 8(*b*) makes the same comparison with $t_0 = 0.08 = 2$ time constants; and Figure 8(*c*) gives the amplitude of the error as a function of t_0/T. Increasing the cutoff time increases the frequency of oscillation but decreases the amplitude. The negative amplitude of the error is somewhat greater than the positive amplitude until the ratio t_0/T is on the order of 5 or 6.

The same general phenomena is found in results obtained from higher order systems, such as the ideal transducer and low-pass filter of Figure 9. For this system, the ideal pulse response was computed and used as input data for the numerical integration of the frequency response on the computer. Figure 10 shows some of the results.

Figure 10(*a*) is the system frequency response with a damping ratio of 0.1; a cutoff time $t_0 = 0.095$; and a time interval of $h = 0.1$. As

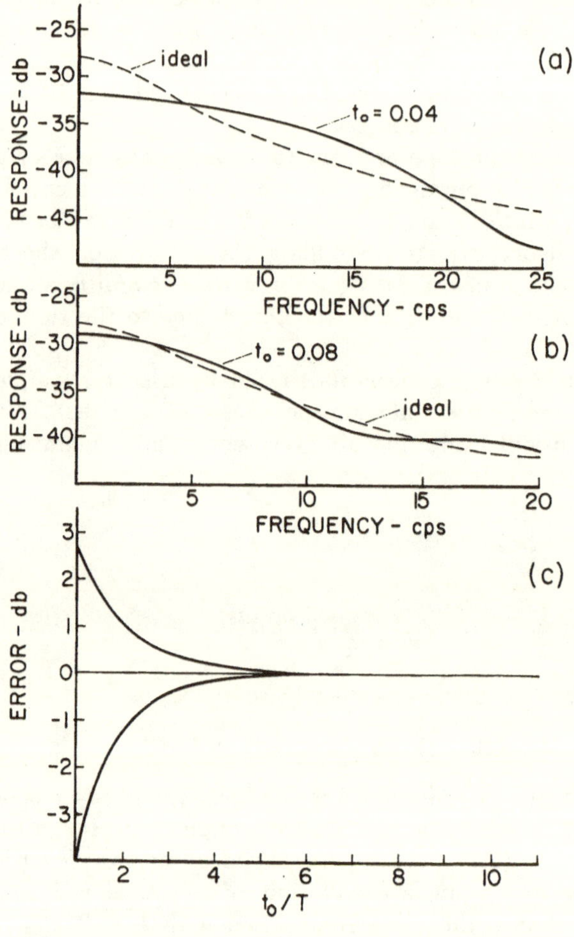

FIGURE 8. Effect of cutoff time for simple system. (*a*) Frequency response for $t_0 = 0.04 = 1$ time constant; (*b*) Frequency response for $t_0 = 0.08 = 2$ time constants; (*c*) amplitude of error in *db* vs t_0/T

Figure 10(*b*) indicates, with the same damping and cutoff time, increasing h to 0.01 makes a considerable improvement in the output data. These results tend to confirm the conclusions made from the study of the single time constant system that reducing the interval h reduces the amplitude of the error.

Figure 10(*c*) illustrates the effect of heavy system damping on the frequency response derived from a transient test. For this computation, a time interval of $h = 0.0005$ was used with a cutoff time of $t_0 = 0.095$.

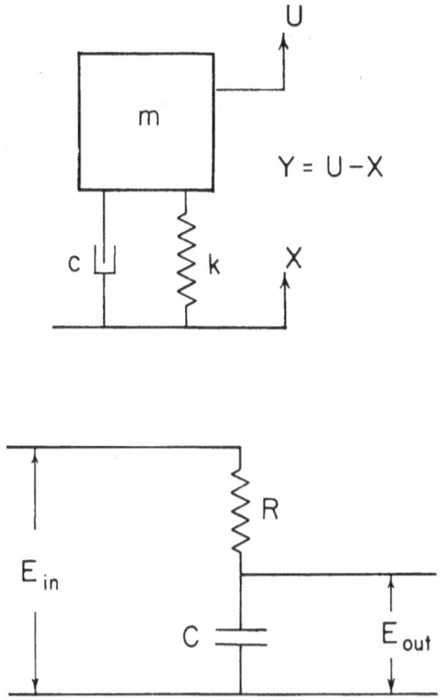

FIGURE 9. Ideal pickup and low-pass filter used for computer study of interval size and cutoff time effects

Increasing the damping ratio of the instrument from 0.1 to 0.6 improves the accuracy but the reduction in the amplitude of the error is not as drastic with increased damping as with changes in the time interval or cutoff time.

CONCLUSIONS

The proposed method does provide an adequate check calibration for instruments and systems with time constants of moderate magnitude ranging down to 0.03 seconds. For systems with extremely small time constants, such as accelerometers with very high natural frequencies, the presence of time constants on the order of 0.002 or smaller may present additional difficulties.

APPENDIX I

Numerical Integration for Fourier Transform

The numerical integration method used to obtain the frequency response from the transient record is based on the procedure developed

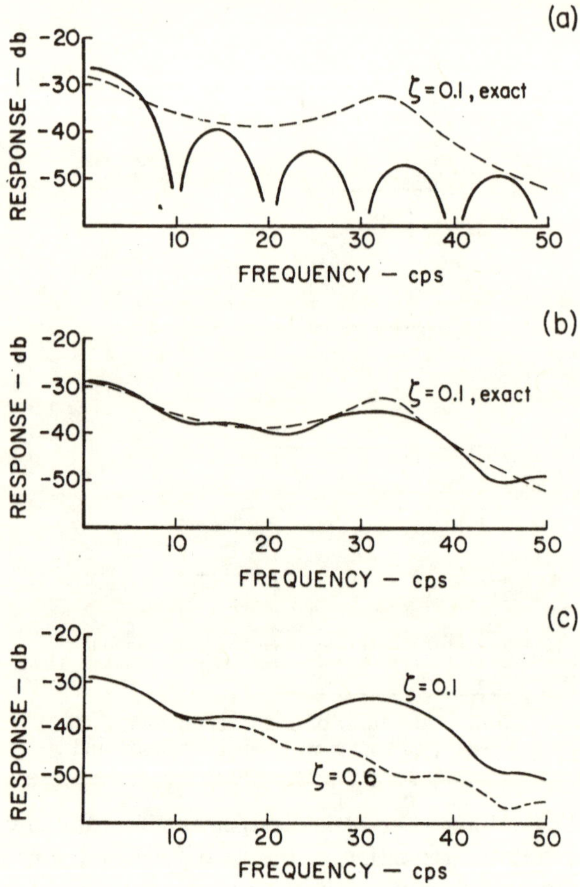

FIGURE 10. Frequency response of ideal system of Figure 9. (*a*) Frequency response with $\zeta = 0.1$; $h = 0.1$, $t_0 = 0.095$; (*b*) Frequency response with $\zeta = 0.1$; $h = 0.01$; $t_0 = 0.095$; (*c*) Frequency response with $h = 0.0005$; $t_0 = 0.095$; $\zeta = 0.1$ and 0.6

by Huss and Donegan (1). As shown in Figure 11, the continuous curve is replaced by equivalent area rectangles of time duration, $h = t_n - t_{n-1}$, and mean height r_n.

For the nth rectangle:

$$F_n(j\omega) = \int_{t_{n-1}}^{t_n} r_n e^{-j\omega t}\, dt$$

$$= \frac{r_n}{\omega}(\sin \omega t_n - \sin \omega t_{n-1}) + \frac{jr_n}{\omega}(\cos \omega t_n - \cos \omega t_{n-1})$$

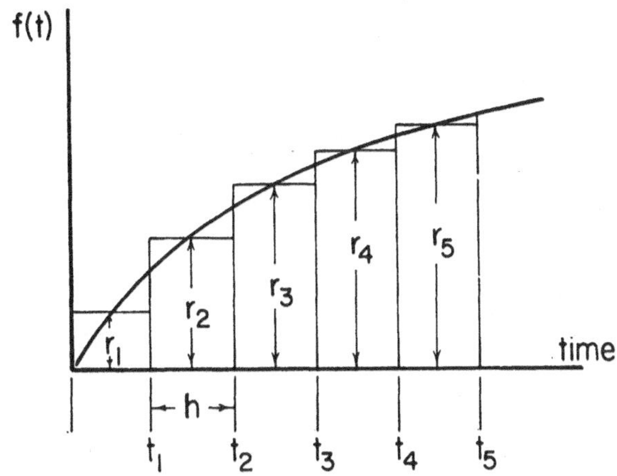

FIGURE 11. Representation of continuous curve by equivalent area rectangles

Letting: $Z = h\omega/2$, the contribution of the nth rectangle is:

$$F_n(j\omega) = hr_n(\sin Z \cos(2n - 1)Z/Z - jhr_n(\sin Z \sin(2n - 1)Z/Z).$$

The complete frequency response is the sum of the contribution of the individual rectangles, hence:

$$F(j\omega) = \sum_{n=1}^{\infty} F_n(j\omega)$$

APPENDIX II

Fourier Transform, Laplace Transform, Frequency Response, and Differential Equation

The relationship between the Fourier transform, the Laplace transform, frequency response, and the differential equation of a system may be seen by considering a linear system with a differential equation of the general form:

$$(D^n + a_1D^{n-1} + \ldots + a_{n-1}D + a_n)\,Y = f(t).$$

The Laplace transform is obtained by using the integral:

$$X(s) = \int_0^{\infty} X(t)e^{-st}\,dt$$

where s is some large number which may be complex. With zero initial conditions, the Laplace transform of the linear differential equation becomes:

$$[Y(s)]/[f(t)(s)] = 1/s^n + a_1s^{n-1} + \ldots + a_{n-1}s + a_n.$$

Since the Fourier transform is obtained by use of the same integral as the Laplace with s replaced by $j\omega$, the frequency response is of the form:

$$[Y(j\omega)]/[f(t)(j\omega)] = 1/(j\omega)^n + a_1(j\omega)^{n-1} + \ldots + a_{n-1}(j\omega) + a_n.$$

This is, of course, the result obtained if D is replaced by $j\omega$ in the original differential equation.

REFERENCES

1. Huss, Carl R. and James J. Donegan, Method and Table For Determining the Time Response to a Unit Impulse From Frequency Response Data and for Determining the Fourier Transform of a Function of Time, NACA TN 3598.
2. Harris, Cyril M. and Charles E. Crede, *Shock and Vibration Handbook*, McGraw-Hill Book Company, Inc., New York, N.Y., 1961.
3. Hamming, R. W., *Numerical Methods for Scientists and Engineers*, McGraw-Hill Book Company, Inc., New York, N.Y., 1962.

A Horsepower Measuring Device

C. E. GOERING

S. J. MARLEY

W. F. BUCHELE*

ABSTRACT

A Wheatstone bridge load transducer can, if excited by a tachometer generator, be used to measure horsepower directly. A "powerometer" was built at Iowa State University to utilize this principle. The powerometer serves as a coupling unit between a Wheatstone bridge, a tachometer generator, and a recorder, preferably an x–y plotter. The powerometer has been used in Agricultural Engineering classes for measuring response curves of a governed engine. The unit has also been used for measuring the drawbar performance of a tractor during a pulling demonstration. It is believed that the powerometer is a useful instrument for teaching, research, and extension. It should also be of value to industry engineers for testing farm equipment.

*Assistant Professor, Agricultural Engineering Department, University of Missouri, Columbia, Missouri; Assistant Professor, Agricultural Engineering Department, Iowa State University, Ames, Iowa; Professor, Agricultural Engineering Department, Iowa State University, Ames, Iowa.

VEHICLE MECHANICS

Systematic Test of Marsh Screw Rotors in Soil

I. ROBERT EHRLICH*

HOWARD DUGOFF†

ABSTRACT

A comprehensive test programme of a 1/6-scale buoyant screw rotor in dry sand and a very soft mud is described. Comparisons in performance are made between model speed, load, center of gravity location, rotor blade height, rotor blade angle, and rotor length/diameter ratio. Both drawbar pull and rotor torque are plotted *vs.* rotor slip as functions of these parameters.

A limited number of tests were also conducted with a $\frac{1}{3}$-scale model in an attempt to determine any model distortion. However, the results of these tests were inconclusive.

Recommended rotor designs for future vehicles are presented.

*Davidson Laboratories, Stevens Institute of Technology, Castle Point Station, Hoboken, N.J.

†710 Warburton Avenue, Yonkers, N.Y.

The Mechanics of An Unsprung Vehicle

CARROLL E. GOERING

WESLEY F. BUCHELE*

ABSTRACT

The behaviour of an instrumented prototype tractor was recorded while accelerating on a concrete strip under drawbar load and no-load conditions.

This behaviour of the prototype was compared with a mathematical model incorporating suitable parameters that was programmed on a computer and found to agree within 10 per cent.

*Assistant Professor of Agricultural Engineering, University of Missouri, Columbia, Missouri; Professor of Agricultural Engineering, Iowa State University, Ames, Iowa.

Zur Kinematik des Schreitschleppers

KURT HAIN*

VORSPRUCH

Schreitschlepper werden in der Hauptsache für Erdbewegungs- und Erd bearbeitungsgeräte verwendet. Durch die Anordnung von Schreitwerken an Stelle von Triebrädern lassen sich die spezifischen Bodendrücke zwischen einem solchen Gerät und dem Untergrund durch die grössere Berührungsfläche weitgehend herabsetzen. Für den Antrieb der Schreitschuhe sind schon mannigfaltige Vorschläge gemacht worden. In der vorliegenden Arbeit wird gezeigt, dass die Bewegung eines Schreitschuhes schon mit einem sechsgliedirgen Kurbelgetriebe ermöglicht werden kann. Mit Hilfe eines Getriebes, das nur relative Drehbewegungen aufweist, werden nicht nur Linearbewegungen eines Punktes, sondern einer gesamten Ebene erzeugt. Neben Hinweisen auf Einbau- und Verwendungsmöglichkeiten solcher Getriebe soll im folgenden vor allem auf deren Entwurf eingegangen werden.

Kinematics of Walking Tractor

ABSTRACT

Walking Machines are generally used in soil-working and soil-transporting machinery. Because of the walking shoes pressure between soil and tractor can highly be reduced. In this investigation it should be shown that the shoe motion is realized by a mechanism having six links only. There are six different mechanisms generating the same parallel motion of a coupler plane. The mechanism design will be explained in detail.

ZUR FORTBEWEGUNG von Fahrzeugen eignen sich am besten Räder, die um eine Achse drehen können. Zwischen Rad und Fahrbahn ist theoretisch nur eine Linienberührung, praktisch aber eine Berührung mit verhältnismässig kleiner Fläche gewährleistet. Dadurch ist die Zugfähigkeit solcher Geräte bzw. Fahrzeuge beschränkt; sie kann durch erhöhte Triebachsbelastung vergrössert werden, findet aber ihre Grenze im Triebrad-Schlupf. Die spezifischen Bodendrücke lassen sich, wie allgemein bekannt ist, durch eine grössere Berührungsfläche zwischen dem treibenden Glied und dem Untergrund wesentlich herabsetzen.

Bei Gleiskettenfahrzeugen wird dies durch eine Gliederkette erreicht, die durch Führungsräder in beliebiger Länge zur Auflage gebracht werden

*Mitteilung aus dem Institut für landtechnische Grundlagenforschung der Forschungsanstalt für Landwirtschaft Braunschweig-Völkenrode. Für die Vorarbeiten für diesen Aufsatz möchte Verfasser Herrn LYK den Dank aussprechen.

kann. Die zur Verbindung der Kettenglieder notwendigen Gelenke sind einem hohen Verschleiss ausgesetzt, da sie sich in Bodennähe bewegen und deshalb nur in unvollkommener Weise gegen das Eindringen von Staub und Schmutz zwischen Gelenkbolzen und Gelenkbohrung geschützt werden können. Der Schreitschlepper ist durch nacheinander sich in ihrer Bodenberührung ablösende Schreitschuhe gekennzeichnet, deren Berührungsfläche, wie bei der Gleiskette auch, beliebig gross vorgesehen werden kann. Es können also auch beim Schreitschlepper sehr niedrige spezifische Bodendrücke erreicht werden. Am Schreitschuh selbst brauchen keine gelenkigen Teile vorhanden zu sein; die Getriebe zur Bewegungserzeugung der Schreitschuhe können gut gekapselt und unter Umständen im Oelbad laufend, getrennt von der Auflagefläche angeordnet werden.

Bei der Erzeugung der Schreitbewegung müssen verhältnismässig schwierige Forderungen erfüllt werden. Zunächst ist zu verlangen, dass der geradlinig verlaufende Teil der Schreitschuhbewegung mit möglichst gleichförmiger Geschwindigkeit durchlaufen wird, vor allem aber, dass bei der Uebergabe der Fortbewegung einem von Gleitschuh zum anderen möglichst keine Geschwindigkeitsunterschiede zwischen beiden Gleitschuhen auftreten. Die wichtigste Forderung an den Gleitschuh selbst liegt aber in seiner geradlinigen Teilbewegung, damit bei der Fortbewegung des Fahrzeuges keine Vertikalbewegungen des Schwerpunktes zustande kommen. Da der Schreitschuh in einer grösseren Fläche aufliegen soll, muss verlangt werden, dass seine Bewegung relativ zur Fahrzeuggestellebene als Parallelbewegung erfolgt, d.h. sämtliche Punkte des Schreitschuhes müssen, zumindest während der Bewegung im Fortbewegungsbereich, zueinander parallele Gerade beschreiben. Vom Freiherrn von Bechtolsheim wurden bereits im Jahre 1930 Vorschläge für die Konstruktion einer schreitenden Zugmaschine gemacht (1). Unabhängig davon gibt es Entwicklungen für Schreitwerke bei Baggern (2), bei verrückbaren Bandförderern (3) und bei Abraumförderanlagen (4). Bei neueuren Untersuchungen wird besondere Rücksicht auf die Lenkbarkeit solcher Schreitwerke genommen (5). Es gibt auch wertvolle Hinweise auf spezielle getriebetechnische Probleme bei der Konstruktion von Schreitmechanismen (6).

FORDERUNGEN AN GETRIEBE FÜR SCHREITSCHLEPPER

Bild 1 zeigt einen Schlepper, bei dem die Hinterachstriebräder durch je einen Satz von Schreitschuhen ersetzt worden sind. Das Getriebe für jeden der vier Schreitschuhe hat eine doppelt gekröpfte Antriebskurbelwelle, die in A_0 gelagert ist und die Kurbelzapfen A and A' hat. Das Getriebe hat eine Koppelebene EE', mit der jeweils der zugehörige Schreitschuh fest verbunden ist.

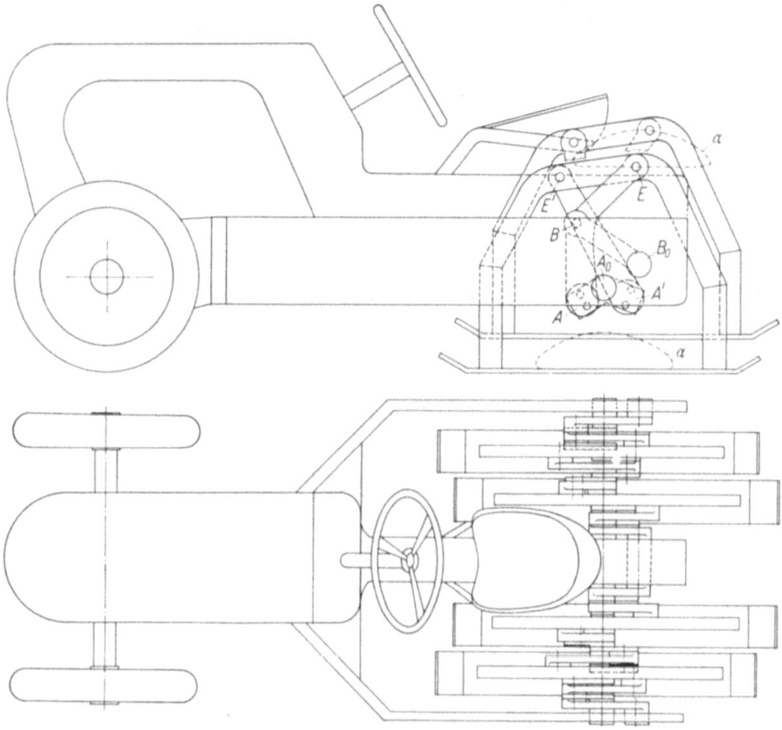

BILD 1. Schreitschlepper mit Vorderradlenkung

Diese Koppelebene wird so parallel geführt, dass jeder ihrer Punkte eine Koppelkurve α beschreibt, die aus einem geraden Teilstück und aus einer kurvenförmigen Erhebung besteht. Durch die paarweise Anordnung eines solchen Getriebes auf jeder Seite des Schleppers wird gewährleistet, dass zu jedem beliebigen Zeitpunkt einer der beiden Schreitschuhe auf dem Boden aufliegt.

Das Grund-Teilgetriebe für den Schreitschuhantrieb bildet ein einfaches Gelenkviereck nach Bild 2. Es hat eine gleichförmig umlaufende Antriebskurbel A_0A, ein hin- und herbewegtes Zwischenglied B_0B und eine diese beiden Glieder verbindende Koppel AB mit einem die Koppelkurve α beschreibenden Koppelpunkt E. Die Koppelkurve muss so gestaltet sein, dass der Koppelpunkt E eine möglichst genaue Gerade beschreibt, deren Länge einem Kurbelwinkel φ_0 entspricht, der unbedingt grösser als 180° sein muss. Dadurch ist gewährleistet, dass ein gleiches Gelenkviereck mit einer zum ersten um 180° versetzten Antriebskurbel seinen Koppelpunkt E so führt, dass er mit dem entsprechenden Punkt des ersten Gelenkviereckes sich auf dieser erwähnten Geraden ablöst, bzw.

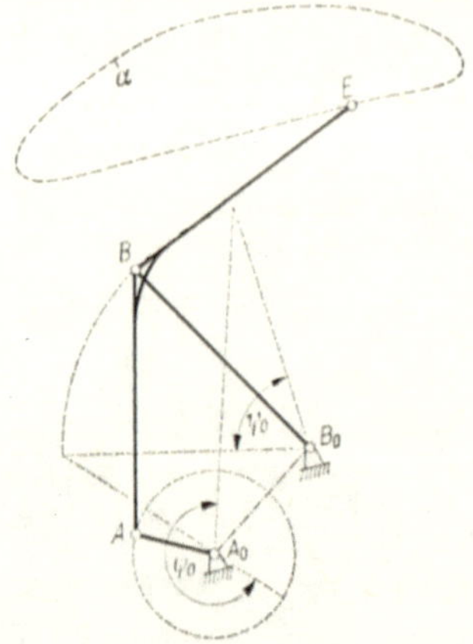

BILD 2. Gelenkviereck zur Erzeugung einer Koppelkurve mit geradlinigem Teilstück

dass beide sogar ein Stück gleichzeitig und mit möglichst gleicher Geschwindigkeit auf dieser Geraden laufen. Für den Rest einer vollen Kirbelumdrehung bewegt sich der Koppelpunkt E nach oben und kehrt in seine Anfangslage auf der Geraden zurück. Es ist klar, dass bei dieser Rückbewegung eine höhere Geschwindigkeit dieses Punktes vorhanden sein muss als bei seiner Vorwärtsbewegung.

Es genügt aber nicht, dass der Koppelpunkt E nur eine möglichst genaue Gerade beschreibt, er muss vielmehr diese Gerade mit möglichst gleichbleibender Geschwindigkeit durchlaufen, wenn die zugehörige Antriebskurbel A_0A ebenfalls gleichförmig umläuft. Das Gelenkviereck zur Erzeugung dieser Koppelkurbel muss also so entworfen werden, dass während der geradlinigen Bewegung gleichen Teilwinkeln der Antriebskurbel A_0A gleiche Wegstrecken des Punktes E auf seiner Teilgeraden entsprechen (7 bis 16).

DREIFACHE ERZEUGUNG DER KOPPELKURVE

Für die im folgenden zu erläuternde Theorie der Konstruktion eines sechsgliedrigen Getriebes mit mathematisch genauer Erzeugung von Parallelkoppelbewegungen ist der Roberts'sche Satz zur dreifachen Erzeugung der Koppelkurve mit Hilfe des Gelenkvierecks eine wichtige Voraussetzung (17). Danach gibt es für jede Koppelkurve, die von einem Koppelpunkt eines Gelenkviereckes erzeugt wird, noch weitere zwei Gelenkvierecke, die die gleiche Koppelkurve mit mathematisch genauer Uebereinstimmung ebenfalls erzeugen können.

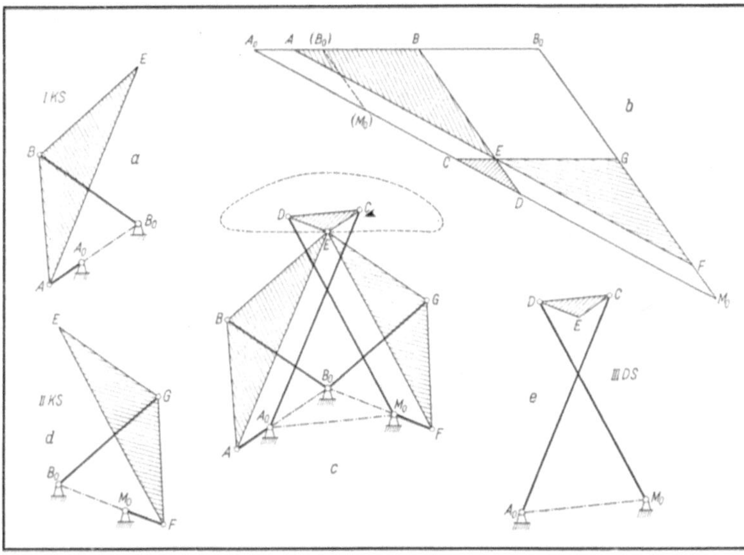

BILD 3. Gelenkvierecke nach Roberts für die dreifache Erzeugung einer Koppelkurve. (*a*) Grundgelenkviereck I KS für gegebene Koppelkurve als Kurbelschwinge; (*b*) Ableitung der beiden zusätzlichen Roberts'schen Gelenkvierecke aus der gestreckten Lage der Kurbelschwinge I KS: (*c*) Zusammenstellung der drei Roberts'schen Gelenkvierecke; (*d*) Zweites Roberts'sches Gelenkviereck II KS als Kurbelschwinge; (*e*) Drittes Roberts'sches Gelenkviereck III DS als Doppelschwinge

Im Bild 3*a* ist das Gelenkviereck des Bildes 2 nochmals aufgezeichnet, es ist eine sog. Kurbelschwinge, weil es eine unmlauffähige Antriebskurbel A_0A und ein zweites im Gestall gelagtertes, hin- und herschwingendes Glied B_0B besitzt. Zeichnet man Kurbel A_0A, Koppel AB und Schwinge B_0B in einer gestreckten Geraden auf, Bild 3*b* und zeichnet über der in Strecklage befindlichen Koppel AB das Koppeldreieck ABE, so erhält man durch einfaches Parallelenziehen die Abmessungen der beiden anderen erwähnten Gelenkvierecke. Man zeichnet durch A_0 die Parallele zur Koppelseite AE und durch B_0 die Parallele zur anderen Koppelseite BE; beide Parallele schneiden sich im Punkt M_0. Nun verlängert man die Koppelseiten AE bis zum Schnitt F mit B_0M_0 und die Koppelseite BE bis zum Schnitt D mit A_0M_0 und zieht die Parallele durch E zu AB bis zu den Schnittpunkten C und G mit A_0M_0 und B_0M_0.

Im Bild 3*b* trägt man von A_0 aus die Koppelgerade A_0B_0 des Ursprungsgetriebes, Bild 3*a*, als $A_0(B_0)$ auf und zeichnet über dieser Strecke das ähnliche Dreieck $A_0(B_0)(M_0)$ zum Koppeldreieck ABE. Dieses Dreieck ist im Bild 3*c* das Gestelldreieck $A_0B_0M_0$ der drei Gelenkvierecke, die mit ihrem gemeinsamen Koppelpunkt E mit mathematisch genauer Uebereinstimmung die gleiche Koppelkurve erzeugen. Die beiden neuen Gelenkvierecke sind getrennt hiervon in den Bildern 3*d* und 3*e* aufgezeichnet worden, wobie dei Abmessungen dieser beiden neuen Gelenkvierecke aus Bild 3*b* entnommen wurden.

Die drei Roberts'schen Gelenkvierecke zur Erzeugung der gleichen Koppelkurve stehen im engen Zusammenhang zueinander. So ist z.B. bekannt (18), dass diese drei Getriebe ent weder zwei Kurbelschwingen und eine Doppelschwinge oder nur drei Doppelkurbeln sein können. Die Kurbelschwinge ist, wie bereits erwähnt, dadurch gekennzeichnet, dass sie eine im Gestell gelagerte umlaufende Kurbel und ein zweites, im Gestell gelagertes hinund hergehendes Glied hat, die Doppelschwinge hat zwei im Gestell gelagerte hin- und herschwingende Glieder, und die Koppel dieses Gelenkvierecks läuft zu ihren benachbarten, im Gestell gelagerten Gliedern um. Die Doppelkurbel ist ein Gelenkviereck, bei dem beide im Gestell gelagerte Glieder volle Umlaufbewegungen ausführen können. Für den vorliegenden Fall findet man die beiden Kurbelschwingen I KS, Bild 3*a* und II KS, Bild 3*d* sowie die Doppelschwinge III DS, Bild 3*e*.

ERZEUGUNG VON PARALLELKOPPELBEWEGUNGEN

Die drei Roberts'schen Gelenkvierecke lassen sich auf sechsfache Weise zu sechsgliedrigen Getrieben so kombinieren, dass jeweils eine Koppelebene entsteht, die mathematisch genaue Parallelbewegungen ausführt, d.h. jeder beliebige Punkt dieser Ebene beschreibt die gleiche Koppelkurve, also z.B. die Koppelkurve der Gelenkvierecke im Bild 3 (19).

Diese sechs verschiedenen Getriebe sind im Bild 4 dargestellt. Jedes der drei Gelenkvierecke des Bildes 3 wird hierbei zweim al mit all seinen Abmessungen verwendet, und es wird in jedem Fall durch einen sog. Zweischlag eines der beiden anderen Gelenkvierecke ergänzt. Ein Zweischlag ist die gelenkige Verbindung zweier binärer Getriebeglieder.

Das Getriebe des Bildes 4*a* ergibt sich z.B. auf folgende Weise. Es besteht aus dem Gelenkviereck, Bild 3*a*, also aus dem Getriebe I KS und hat zusätzlich den Zweischlag $M_0F - FE$ des Gelenkvierecks II KS. Allgemein lässt sich hierfür unter Zugrundelegung des Gestelldreieckes $A_0B_0M_0$, Bild 3*c*, folgender Satz aufstellen:

Als Grundlage sechsgliedriger Getriebe zur Erzeugung mathematisch genauer Parallelbewegungen dienen die drei Roberts'schen Gelenkvierecke zur Erzeugung der gleichen Koppelkurve. Jedes der sechs möglichen sechsgliedrigen Getriebe setzt sich aus einem dieser drei Gelenkvierecke und einem Zweischlag aus je einem der anderen Gelenkvierecke zusammen. Der zusätzliche Zweischlag aus dem Zweitgelenkviereck ergibt sich

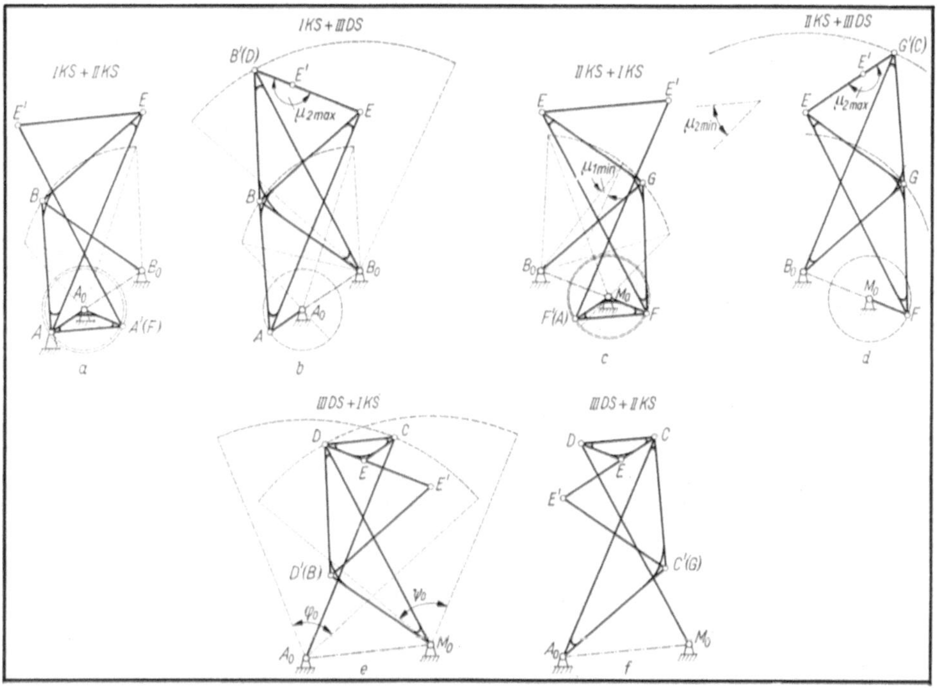

Bild 4. Sechs Gelenkvierecke aus den drei Roberts'schen Gelenkvierecken zur Erzeugung derselben Parallel-Koppelbewegung

durch den dritten, im Erstgelenkviereck nicht erscheinenden Gestellpunkt und dem Zweischlag von diesem Punkt zum Koppelpunkt. Der Gestellpunkt des Zweitgelenkviereckes wird mit demjenigen Gestellpunkt des Erstgelenkviereckes zusammengelegt, der im Zweitgelenkviereck nicht erscheint. Die beiden in diesem Doppelgestellpunkt zusammenfallenden Getriebeglieder des Erst- und des Zweitgelenkviereckes werden zu einem gemeinsamen ternären Glied vereinigt. Die gelenkige Verbindung des Koppelpunktes im Erstgelenkviereck mit dem übernommenen Koppelpunkt aus dem Zweitgelenkviereck ergibt die Koppelebene des sechsgliedrigen Getriebes, die die mathematisch genauen Parallelstellungen durchläuft.

Die sechs Getriebe des Bildes 4 sind also dadurch entstanden, dass man das Gelenkviereck I mit je einem Zweischlag aus den Getrieben II und III, das Gelenkviereck II mit je einem Zweischlag aus den Getrieben I und III und das Gelenkviereck III mit je einem Zweischlag aus den Getrieben I und II versieht.

Das Getriebe I KS + III KS, Bild 4*a*, hat z.B. das Gelenkviereck I

KS als Erstgetriebe und das Gelenkviereck II KS als Zweitgetriebe, aus dem der entsprechende Zweischlag entnommen wird. Im Erstgetriebe fehlt der Gestellpunkt M_0 (vergleiche Bild 3*c*). Von ihm aus wird also der Zweischlag $M_0F - FE$ dem Gelenkviereck I KS so beigefügt, dass der Gestellpunkt M_0 mit dem Gestellpunkt A_0 zusammenfällt, also mit dem Gestellpunkt A_0, der im Zweitgetriebe II KS nicht vorhanden ist. Auf diese Weise entsteht im Getriebe des Bildes 4 a das ternäre Glied A_0AA', und in dessen Gelenk $A'(F)$ ist ein binärer Hebel $A'E'$ angelenkt, der der Koppelseite FE des Getriebes 3*d* entspricht. Auf diese Weise entsteht die Koppelgerade $E'E$, die die Länge der Gestellgeraden A_0M_0 haben muss, weil der Gestellpunkt M_0 zum Zusammenfallen mit dem Gestellpunkt A_0 gezwungen wurde.

Das sechsgliedrige Getriebe des Bildes 4*b* hat als Erstgetriebe das Gelenkviereck I KS, dem nach der obengenannten Regel aus dem Getriebe III DS der Zweischlag $M_0D - DE$ entnommen wurde. In diesem zweiten sechsgliedrigen Getriebe wurde der Gestellpunkt M_0 mit B_0 zusammengebracht, so dass die Koppelebene EE' die Länge der ursprünglichen Gestellgeraden M_0B_0 haben muss.

In der gleichen Weise sind die restlichen vier Getriebe des Bildes 4 entstanden. Es war im Anfang also nur notwendig, mit den Hilfsmitteln der Getriebesynthese ein Gelenkviereck zu konstruieren, von dem ein Koppelpunkt eine vorgeschriebene Koppelkurve in vorgeschriebener Weise durchläuft. Zu diesem Gelenkviereck werden nach dem Satz von Roberts die beiden Aequivalenten, d.h. die gleiche Koppelkurve erzeugenden Gelenkvierecke ermittelt, und mit Hilfe dieser drei Gelenkvierecke findet man die sechs verschiedenen sechsgliedrigen Getriebe mit je einer Koppelebene EE'. die sämtlich die gleiche Koppel-Parallelbewegung verwirklichen.

KONSTRUKTIVE GESTALTUNG DER SCHREITSCHUH-GETRIEBE

Durch die Aufstellung der sechs möglichen sechsgliedrigen Getriebe zur Erzeugung einer parallelen Koppelbewegung besteht nunmehr eine willkommene Auswahlmöglichkeit, für den jeweils gegebenen Anwendungsfall das am besten geeignetste Getriebe herauszusuchen. Eine konstruktive Gestaltung des Getriebes I KS + II KS des Bildes 4*a* zeigt Bild 5. Die Antriebswelle A_0 ist als doppelt gekröpfte Kurbelwelle ausgebildet. In zwei dem Getriebeschema entsprechend zueinander versetzten Kurbelzapfen A und F sind die Hebel ABE und FE' so angeordnet dass zwischen ihnen in den beiden Gelenkpunkten EE' der Schreitschuh als Koppelebene EE' bewegt wird.

Das Getriebe I KS + III DS, Bild 4*b*, ist für den hier beschriebenen

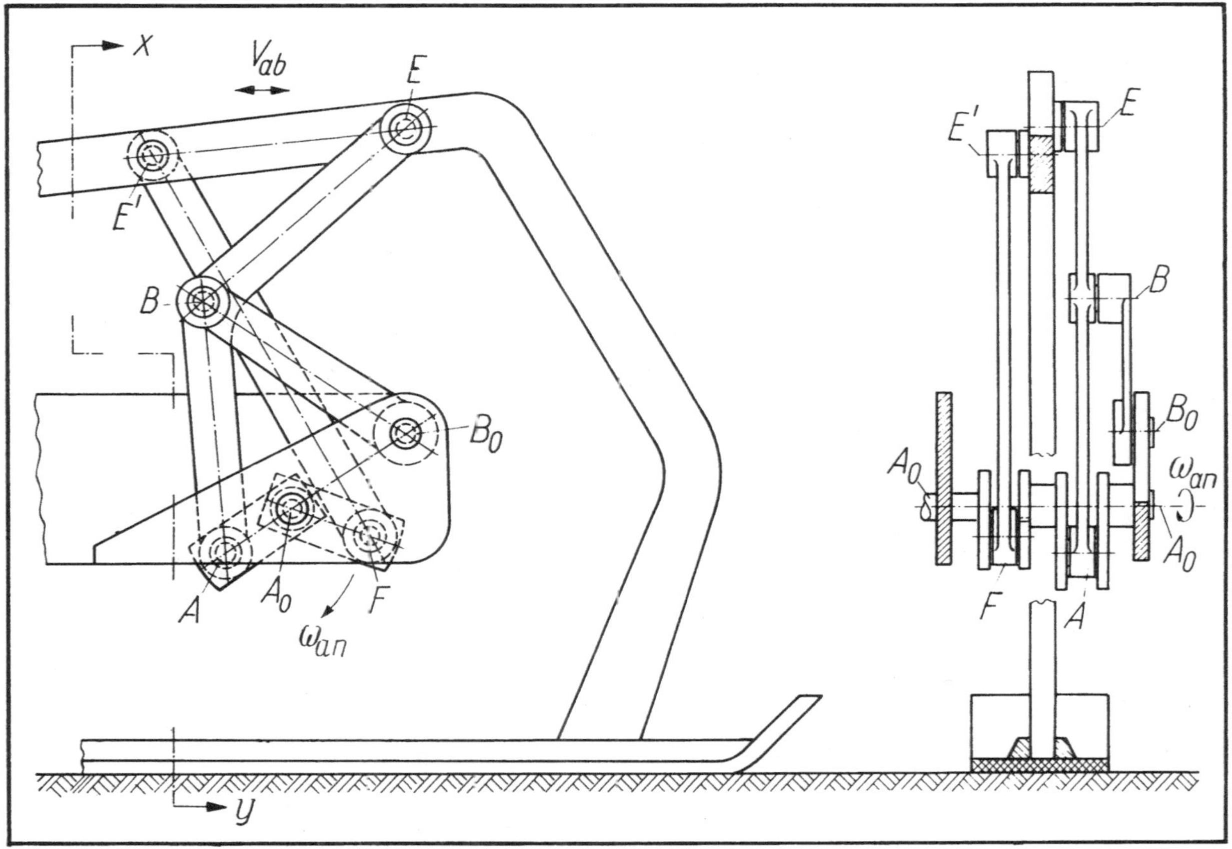

BILD 5. Sechsgliedriges Getriebe mit Kurbelschwinge als Grundgetriebe und mit doppelt gekröpfter Kurbelwelle zur Erzeugung einer Parallel-Koppelbewegung

Zweck nicht geeignet. Bei dieser Getriebeanordnung wird nämlich einer der beiden Uebertragungswinkel, im vorliegenden Falle der Winkel $\mu_{max} = 180°$. Wenn dieser Winkel zwischen dem Zweischlag $B'E'$–$E'E$ nämlich 0° oder 180° wird, ist es nicht möglich, das Gesamtgetriebe ohne besondere Hilfsmittel, wie z.B. durch Hilfsverzahnungen oder Federn, von der im Gestell gelagerten Antriebskurbel A_0A aus anzutreiben.

Das Getriebe II KS + I KS, Bild 4*c*, hat ungefähr den gleichen Aufbau wie das Getriebe 4*a*. Der Uebertragungswinkel $\mu_{1\,min}$ im Grundgelenkviereck als Winkel zwischen der Koppelgeraden *FG* und der Schwinge GB_0 ist zwar verhältnismässig klein, als Extremwert tritt er aber beim Rückgang des Schreitschuhes, also in der unbeanspruchten Zone auf, und deshalb ist das Getriebe durchaus noch verwendbar. Der zweite, im Getriebe des Bildes 4*c* zu berücksichtigende Uebertragungswinkel zwischen dem Zweischlag *EE'* und *E'F'* bleibt während der gesamten Bewegungsperiode zwischen 0° und 180°, sein kleinster Wert $\mu_{2\,min}$ ist neben dem Bild eingezeichnet worden.

Das Getriebe II KS + III DS des Bildes 4*d* ist wiederum nicht ohne weiteres verwendungsfähig. Es ist, wie das Getriebe des Bildes 4*b*, "nichtdurchlauffähig," weil der Uebertragungswinkel $\mu_{2\,max}$ in der eingezeichneten Getriebestellung 180° wird und deshalb das Getriebe nicht über die Antriebskurbel M_0F durchlaufend angetrieben werden kann.

Das sechsgliedrige Getriebe III DS + I KS, Bild 4*e*, hat ein Grundgelenkviereck, das als Doppelschwinge läuft, d.h. die beiden in den Gestellpunkten A_0 und M_0 gelagerten Glieder schwingen nur hin und her. Die Koppel *DC* dieses Gelenkviereckes vollführt aber relativ zu diesen ihren benachbarten Gliedern volle Umlaufbewegungen. Sie läuft auch relativ zum Gestell A_0M_0 um. Man kann sogar beweisen (20), dass beim gleichförmigen Antrieb dieser Koppel *DC* vom Gestell aus der zur gleichen Koppelebene gehörende Koppelpunkt nicht nur die gleiche Koppelkurve wie in den bisher beschriebenen Getrieben beschreibt, sondern dass die Koppelkurve dann uach in gleichen Zeitabschnitten vom Punkt *E* durchlaufen wird, genauso, als wenn z.B. im Getriebe I KS + II KS die Kurbel A_0A mit gleichförmiger Geschwindigkeit antreibt. Der gleichförmige Antrieb einer relativ zum Gestell umlaufenden, aber durch hin- und hergehende Hebel geführten Koppelebene kann z.B. mit Hilfe einer Oldham-Kupplung verwirklicht werden (20). Die Oldham-Kupplung ist bekanntlich in der Lage, zwischen versetzten parallelen Wellen winkeltreue Bewegungen zu übertragen, auch wenn diese beiden Wellen während des Laufes ihre Lage zueinander stetig oder in absatzweisen Bewegungen verändern.

Das Getriebe III DS + I KS ist im Bild 6 a nochmals aufgezeichnet worden. Die Punkte *E* und *E'* beschreiben die deckungsgleichen Koppel-

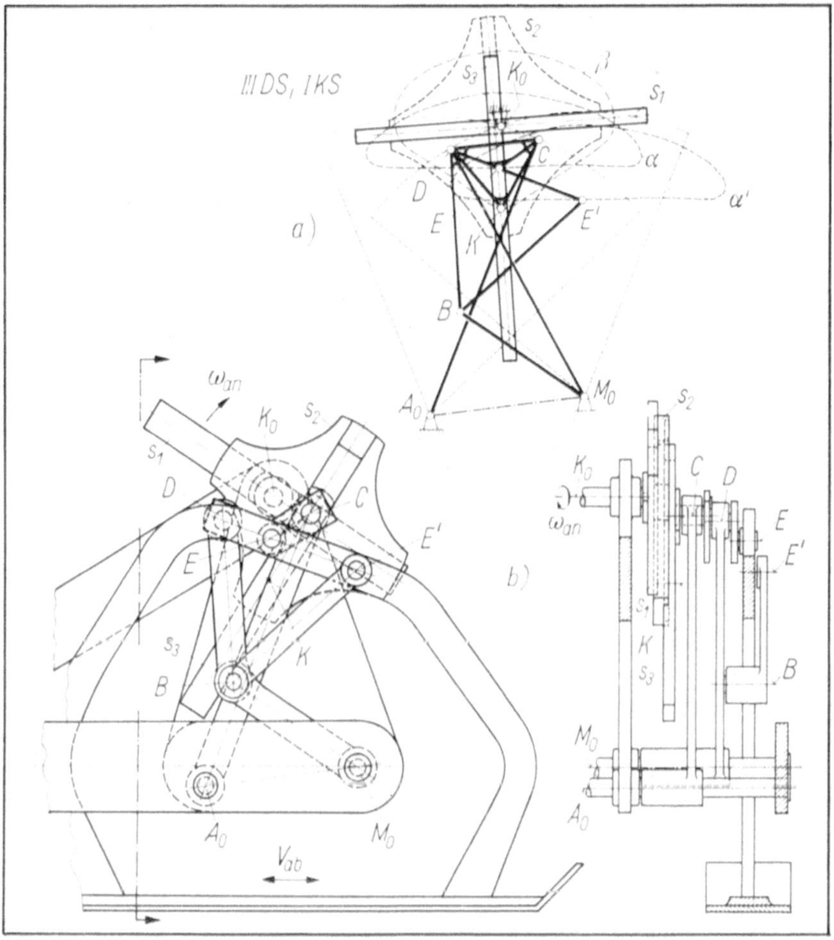

BILD 6. Sechsgliedriges Getriebe mit Doppelschwinge als Grundgetriebe und mit gekröpfter Koppel. Koppel-Antrieb über Oldham-Kupplung. (*a*) Getriebe-Schema. (*b*) Darstellung konstruktiver Einzelheiten

kurven α und α'. In der Koppelebene *DCE* wurde ein weiterer Koppelpunkt *K* ausgesucht, der eine Koppelkurve β mit möglichst geringer Ausdehnung beschreibt. In den Zentralpunkt K_0 der Kurve β wurde die Achse des Antriebsschleifenhebels s_3 gelegt. Durch die Wahl der Lage des Punktes K_0 in der Koppelkurve β mit möglichst geringer Ausdehnung ist auch eine geringe Relativbewegung in den Schleifenhebeln der Oldham-Kupplung gewährleistet, wenn der Koppelpunkt *K*, der in der Oldham-Kupplung selbst als besonderes Merkmal überhaupt nicht mehr erscheint, als Mitte der Schleifenführung des Hebels s_3 angenommen wird. Mit der Koppel DCE ist der zweite Schleifenhebel s_1 der Oldham-Kupplung fest verbunden; er macht also die umlaufende Bewegung dieser Koppel mit den überlagerten Hin- und Herbewegungen mit. Zwischen den beiden Schleifenhebeln s_1 und s_3 bewegt sich der Doppelschieber s_2 der Oldham-Kupplung.

Die konstruktive Gestaltung des Getriebes III DS + I KS des Bildes 6*a* ist im Bild 6*b* gezeigt. Es ist ein besonderer Vorteil, dass die in den Gestellpunkent A_0 und M_0 gelagerten Hebel nur hin- und herschwingenden Bewegungen vollführen; denn dadurch kann ihre Lagerung mit besonders guten Versteifungen ausgeführt werden. In der Seitenansicht des Bildes 6*b* ist zu erkennen, dass die Koppel CDE zwischen den beiden Hebeln A_0C und M_0BD angeordnet ist, dass sie aber durch die Bohrung D hindurchgehend auf der äusseren Seite des Hebels M_0BD eine Kröpfung mit der Welle E besitzt. Zwischen diesem Kurbelzapfen E und dem Gelenk E' des binären Hebels BE' liegt der eigentliche Gleitschuh EE'. Er beschreibt die gleichen Parallel-Koppelbewegungen wie der Gleitschuh des Getriebes nach Bild 5.

Die gelenkige Verbindung der Getriebeteile in vielgliedrigen Getrieben führt in vielen Fällen zu deren Anordnung in verschiedenen, nebeneinander liegenden Bewegungsebenen. Dadurch treten oft unangenehme Verkantungen in den Gelenklagerungen auf, die um so grösser werden, je weniger Platz in Achsrichtung der Gelenke für eine lange Lagerabstützung geschaffen werden kann. Wenn aber in der projizierten Ebene der Bewegungen aller Getriebeteile mit dem Platz nicht gespart zu werden braucht, kann man zu Zapfenerweiterungen übergehen, die dadurch gekennzeichnet sind, dass eine Lagerung die andere umschliesst.

Bild 7 zeigt zwei solcher Zapfenerweiterungen des Getriebes nach Bild 6 mit den gleichen Abmessungen. Das Getriebe des Bildes 7*a* hat also die gleichen Grundabmessungen, d.h. die Koppelebene EE' beschreibt mit jedem ihrer Punkte die deckungsgleiche Koppelkurve α'. Im Bild 7*a* wurde aber zur besseren Anpassung an die noch näher zu erläuternde Zapfenerweiterung ein anderer virtueller Koppelpunkt K gewählt, der die Länge der Schleifenhebel in der Oldham-Kupplung bestimmt. Dieser Koppelpunkt K beschreibt eine Koppelkurve β, und man kann aus deren Abmessungen die Führungslängen in den Schleifenhebel s_1 und s_3 festlegen. Im Bild 7*b* ist zu erkennen, dass der ternäre Hebel M_0BD einen Ring u_1 mit dem Mittelpunkt D besitzt. In diesem Ring läuft ein Stern mit den Rollen r_1. Eine dieser Rollen ist auf einem Bolzen gelagert, der gleichzeitig das Gelenk E für die Verbindung mit dem Koppelhebel EE' darstellt. Das dritte Gelenk C ist Mittelpunkt eines zweiten Ringes u_2,

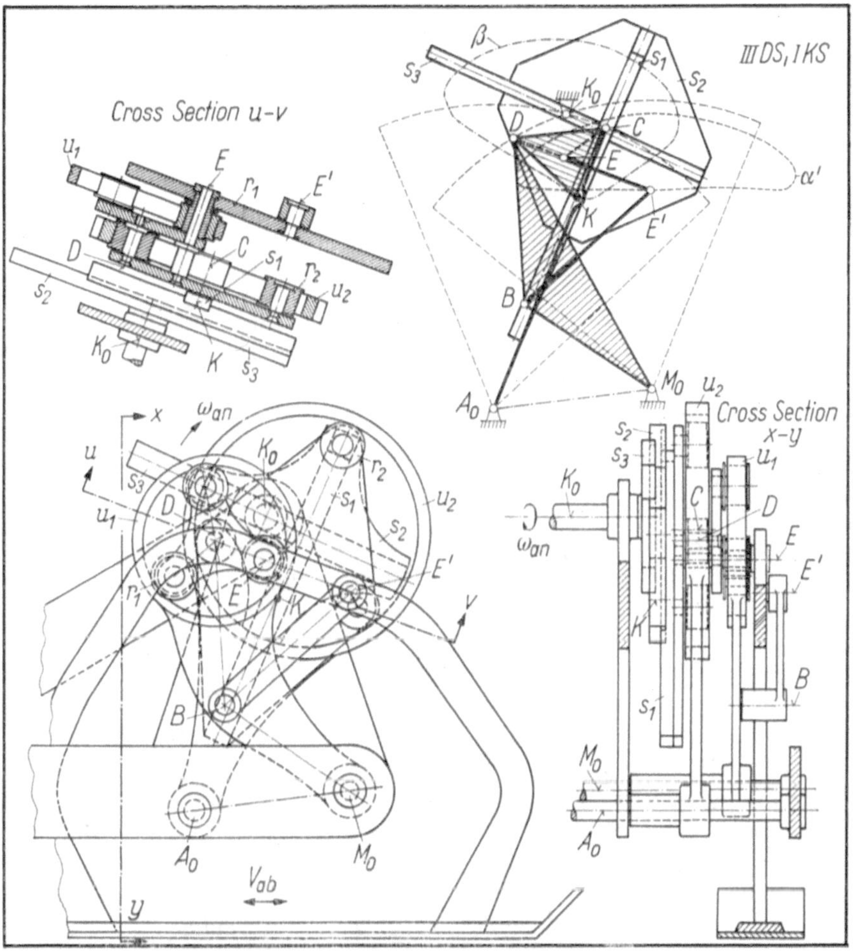

BILD 7. Sechsgliedriges Getriebe mit Doppelschwinge als Grundgetriebe und mit Zapfenerweiterungen der Koppel. Koppel-Antrieb über Oldham-Kupplung. (*a*) Getriebe-Schema; (*b*) Darstellung konstruktiver Einzlheten

der mit dem im Gestellpunkt A_0 gelagerten binären, hin- und herschwingenden Hebel fest verbunden ist. In diesem Ring u_2 läuft wiederum ein Stern mit drei Rollen r_2, und mit diesem Stern ist der Schleifenhebel s_1 der Oldham-Kupplung fest verbunden. Der Schleifenhebel s_3 der Oldham-Kupplung ist im Gestellpunkt K_0 gelagert, und dessen Lage ist der Form der Koppelkurve β so angepasst worden, dass die Führungslänge s_3 möglichst klein bleibt. Zwischen den Schleifenhebeln s_1 und s_3 bewegt sich auch in diesem Getriebe der Doppelschieber s_2. Der Antrieb des gesamten Getriebes erfolgt also von der Welle K_0, die, wie die übrigen Gelenkbolzen des Getriebes auch, nicht abgekröpft zu werden brauchen.

Das sechsgliedrige Getriebe III DS + II KS des Bildes 4*f* ist ebenfalls gut für den beabsichtigten Zweck verwendbar; es kann, wie das Getriebe nach Bild 4*e*, über die Koppelebene *EDC* mit Hilfe einer Oldham-Kupplung gleichförmig angetrieben werden.

LENKMÖGLICHKEITEN DES SCHREITSCHLEPPERS

Die Lenkmöglichkeiten des Schreitschleppers, d.h. also die Möglichkeiten des Kurvenfahrens, liegen im gleichen Bereich wie diejenigen der Gleiskettenfahrzeuge. Der im Bild 1 dargestellte Schreitschlepper hat noch die konventionelle Vorderrad-Achsschenkellenkung, bei der also die gelenkten Vorderräder allein die in der Kurve notwendigen Seitenkräfte aufzubringen haben.

Dem eigentlichen Verwendungszweck des Schreitschleppers entspricht aber der Wegfall auch der Vorderräder, und die Lenkimpulse müssen dann unmittelbar von den Schreitschuhen eingeleitet werden. Die einfachste Art des Lenkens besteht in der Anordnung von Freiläufen zwischen der Triebachse des Schleppers und den Schreitschuh-Aggregaten. Hierbei wird der kurvenäussere Antrieb durch die Ueberholmöglichkeit im Freilauf abgeschaltet. Diese Lenkungsart ist bei hohem Gleitwiderstand unbrauchbar, und es ist auch schwierig, vorgeschriebene Kurvenhalbmesser einzuhalten. Bei hohem Gleitwiderstand wird dagegen das Abschalten des kurveninneren Antriebes ausserordneltich wirkungsvoll. Wenn zwischen der Triebwelle und den Antriebswellen der Schreitschuhe Reibungskupplungen vorgesehen werden, ist nach deren Auskuppeln die Benutzung einer zusätzlichen Lenkbremse zweckmässig, und durch diese Kombination ist sogar ein Wenden um einen stillstehenden Schreitschuh möglich. Die Lenkbremse allein, d.h. also die Einleitung eines zusätzlichen Drehmomentes auf einer Seite des Doppelantriebes der Gleitschuhe, erfüllt am besten ihren Zweck bei den sog. Differentialantrieben, wenn also ein Differentialgetriebe für einen selbsttätigen Ausgleich der Drehmomentübertragung auf die beidseitigen Antriebsaggregate erfolgt. Mit Hilfe des Doppel-Differential-Gertriebes kommt man zur Anordnung von übersetzten Lenkbremsen, mit denen man dem Schreitschuh-Fahrzeug bestimmte vorgeschriebene Wendehalbmesser ohne Lenkbremsverluste aufzwingen kann (21).

Beim Kurvenfahren der Radfahrzeuge kann man durch Ausbildung der Lenkgestänge mit guter Annäherung zu einem einwandfreien Rollen sowohl des kurveninneren als auch des kurvenäusseren Rades kommen, was bei Gleisketten- oder Schreitschlepp-Fahrzeugen wegen der Flächenauflage der Ketten bzw. der Schreitschuhe niemals erreicht werden kann. Es geht hierbei also nur um die Frage, wie die Lenkung eines Schreitschuh-Fahrzeuges ausgelegt werden kann, damit die geringsten Reibungsverluste auftreten. Am günstigsten erscheint hier eine zwangläufige getriebetechnische Anpassung der Geschwindigkeiten des kurveninneren und kurvenäusseren Schreitschuhes an den angeforderten Kurvenhalbmesser. Hier bietet sich die Verwendung eines stufenlos verstellbaren Getriebes an, das nur eine der beiden Seiten der Schreitschuh-Getriebe relativ zur anderen Seite hinsichtlich der Geschwindigkeit zu vergrössern, bzw. zu verkleiner in der Lage ist, je nachden, ob eine Rechts- oder Linkskurve gefahren werden soll.

In der vorliegenden Arbeit wurde zum Antrieb der Gleitschuhe ein ungleichförmig übersetzendes Getriebe vorgeschlagen, das aber während der Schreitbewegung selbst mit möglichst gleichförmiger Uebersetzung arbeiten soll. Ungleichförmig übersetzende Getriebe lassen sich sehr leicht durch Gliedlängenänderungen verstellen. Die einfachste Verstellmöglichkeit besteht in der Lagenänderung eines oder mehrerer im Gestell angeordneter Lagerpunkte. Wenn es gelingt, ein solches Getriebe bei gleichbleibender Umlaufgeschwindigkeit der Antriebswelle so zu verstellen, dass die Schrittlänge sich verändert, so ist damit auch eine Veränderung der resultierenden Vorwärtsbewegung des gesamten Fahrzeuges verbunden. Bei dieser Schrittlängenverstellung müssen allerdings die Forderungen an die Form der Koppelkurve, wie sie im Vorangegangenen herausgestellt worden sind, im gleichen Masse erfüllt werden. Dies ist aber mit dem einfachen, beschriebenen sechsgliedrigen Getriebe nicht mehr möglich; vielmehr muss hier nach neuen masssynthetischen Verfahren gesucht werden, da solche im Schrifttum nur spärlich zu finden sind. Immerhin führt eine Schrittlängenverstellung auch zu einer sehr einfachen Möglichkeit des Kurvenfahrens, indem man jeweils nur die Schrittlänge des kurveninneren Rades verringert bzw. diejenige des kurvenäusseren Rades verlängert.

Eine weitere Möglichkeit für das Kurvenfahren des Schreitschleppers bestünde darin, den Schreitschuh mit einem Teil seines Getriebes um eine vertikale Achse relativ zur Fahrgestellebene zu verschwenken. Bei Anordnung einer reellen Achse wird schon bei geringen Schwenkbereichen ein verhältnismässig grosser Freiraum zwischen Schreitschuh und Fahrgestell notwendig. Unter Umständen ist hier eine Verbesserung durch die Verschwenkung der Schreitschuhe um ideele vertikale Achsen erzielbar. Wie weit hierbei die Lenkfähigkeit durch verschieden grosse Schwenkwinkel des kurveninneren und kurvenäusseren Gleitschuhes verbessert werden kann, müsste späteren Untersuchungen vorbehalten bleiben.

SCHLUSSBETRACHTUNGEN

In vorliegender Untersuchung sollte in der Hauptsache auf eine bestimmte kinematische Lösung für den Antrieb von Schreitschleppern eingegangen werden, wobei an diesem Beispiel insbesondere die Masssynthese zu erläutern war. Die Bewegungen der Gleitschuhe relativ zur Fahrgestellebene wurden hierbei in bestimmtem Masse als gegeben angenommen, und es ist beachtenswert, dass eine solche Parallel-Koppelbewegung schon mit einem sechsgliedrigen Getriebe, d.h. mit einem Mindestaufwand von fünf bewegten Getriebeteilen erreicht werden kann. Getriebe mit geringerer Gliederzahl können für diese Anforderungen niemals gefunden werden.

Zu jeder Schreitschuhgruppe gehören bei der im Vorangegangenen gestellten Aufgabe zwei mit Phasenverschiebung zueinander sich bewegende Getriebegruppen. Durch diese Zweifach-Anordnung ist zwar schon eine gewisse Verbesserung hinsichtlich des Massenausgleiches gegenüber einer Einfachanordnung zu verzeichnen; bei schnellaufenden Getrieben muss aber zusätzlich noch eine Verbesserung des Massenausgleiches angestrebt werden, was bei ungleichförmig übersetzenden Getrieben mit entsprechendem Aufwand mit mehr oder weniger guter Annäherung erreicht werden kann (22, 23, 24, 25).

Für die Fortbewegungsmöglichkeiten auf unebenem Untergrund bieten sich ausser der hier gezeigten getriebetechnischen Lösung noch andere Anordnungen an (26, 27, 28, 29), auf die aber in dem hier gesetzten Rahmen nicht eingegangen werden sollte.

SCHRIFTTUM

1. Stiebling, A., Die schreitende Zugmaschine des Freiherrn von Bechtolsheim. Diss. T. H. München, 1933.
2. Franke, W., Weiterentwicklung der amerikanischen Schreitbagger. VDI-Z. 91 (1949), Nr. 20, S. 525/527.
3. Hilldesheim, K., Rückbare Bandförderer, *Braunkohle, Wärme und Energie* 8 (1956), S. 134/142.
4. Kraus, H. W., Abraumförderanlagen mit Lenkschreitwerken. *Techn. Mitt. Schoppe u. Faeser*, Juli 1959, H. 2, S. 67/74.
5. Heuer, H., Bewegungsmöglichkeiten von Erdbaugeräten unter besonderer Berücksichtigung der Lenkschreitwerke. *Braunkohle* 13 (1961), Nr. 2, S. 52/61.
6. Shigley, J. E., The mechanics of walking vehicles. Mechanics of soil-vehicle systems. *Proceedings of the 1st International Conference on the mechanics of soil vehicle systems.* pp. 304/322 (Torino (Italy)–Saint Vincent, 1961).
7. Beyer, R., *Kinematische Getriebesynthese* (Berlin: Springer-Verlag, 1953).
8. Burmester, L., *Lehrbuch der Kinematik* (Leipzig, 1888).
9. Hain, K., *Angewandte Getriebelehre* (2. Auflage, Düsseldorf, 1961).
10. Harrisberger, L., *Mechanization of Motion* (New York, John Wiley and Sons, 1961).
11. Hartenberg, R. S. u. T. Denavit, *Kinematics Synthesis of Linkages* (New York: McGraw Hill, 1964).

12. KIPER, G., *Zur Synthese der ebenen Gelenkgetriebe.* VDI-Forschungsheft 433 (Düsseldorf, 1952).
13. KRAUS, R., *Getriebelehre, Band III: Massbestimmung* (Berlin: Verlag Technik, 1956).
14. LICHTENHELDT, W., *Konstruktionslehre der Getriebe* (Berlin: Akademie-Verlag, 1961).
15. MEYER ZUR CAPELLEN, W., Fünf- und sechspunktige Geradführung in Sonderlagen des ebenen Gelenkvierecks. *Forschungsbericht des Landes Nordrhein-Westfahlen,* Nr. 481 (1958), Köln-Opladen.
16. RAUH, W. K. u. L. HAGEDORN, *Praktische Getriebelehre. Bd. 1: Die Viergelenkkette.* (3 Aufl. Berlin: Springer-Verlag, 1964).
17. ROBERTS, S., On Three-Bar-Motion in Plane Space, *Proc. Lond. Math., Soc.* Bd. VII (1875), S 14./23.
18. LUCK, K., Zur Erzeugung von Koppelkurven viergliedriger Getriebe, *Masch. bautechn.* 8 (1959) H. 2, S. 97/104.
19. HAIN, K., Erzeugung von Parallel-Koppelbewegungen mit Anwendungen in der Landtechnik, *Grundlagen der Landtechnik,* H. 20 (Düsseldorf 1964), S. 58/68.
20. HAIN, K., Uebersicht über sämtliche Umlauf- und Schwingbewegungen in Getrieben aus Gelenkvierecken, *Grundl. Landtechn.* 15, (1965), Nr. 4 S. 97/106.
21. FRANKE, R., Beitrag "Ackerschlepper" im *Automobiltechnischen Handbuch.* (18. Auflage, Berlin, 1964).
22. MEWES, E., Massenkräfte in Landmaschinen und ihr Ausgleich, *Grundlagen der Landtechnik,* H. 6 (1955) Düsseldorf, S. 116/133.
23. MEWES, E., Massenausgleich an ebenen Kurbelgetrieben, *Zeitschr. f. Agnew. Math. und Mech.* 35 (1955), H. 9/10, S. 349/50, 36 (1956), H. 1/2, S. 80.
24. MEWES, E., Unbalanced Inertia Forces in Slider-Crank Mechanisms of Large Eccentricity. *ASME*-Paper Nr. 57-A-41; *Journal of Applied Mechanics,* Vol. 25 (Juni, 1958), No. 2, S.225/232.
25. DIZIOGLU, B., Dynamische Getriebesynthese der Kurbelausgleichsgetriebe, *Forschung auf dem Gebiete des Ingenieurwesens,* Bd. 26 (1960), Nr. 2, S.37/47.
26. BEKKER, M. G., *Theory of Land Locomotion* (Ann Arbor, Michigan: University of Michigan Press, 1957).
27. BEKKER, M. G., *Off-the-Road Locomotion* (Ann Arbor, Michigan: University of Michigan Press, 1960).
28. LISTON, R. A., Walking Machines, *Journal of Terramechanics,* Vol. 1 (1964), No. 3, pp. 18/31.
29. SHIGLEY, J. E., The Mechanics of Walking Vehicles, *Land Locomotion Laboratory, Report No. 71* (OTAC, Detroit, Michigan, 1957).

Schreitwerke mit dreidimensionaler Schreitplattenbewegung

W. F. G. KAMM*

ZUSAMMENFASSUNG

Konstruktion, Aufbau, und Bewegungsablauf von Lenkschreitwerken mit dreidimensionalen Schreitplattenbewegungen; Steuerung durch Modellnachahmung; Vergleich mit richtungsgebundenen Schienen- und Raupenfahrwerken; Anwendungsbeispiele und Einsatzmöglichkeiten für Fahrbrecher und ähnliche Geräte.

Walking Mechanisms with Tridimensional Shoe Movement

ABSTRACT

Design, arrangement, and operating sequence of steered walking mechanisms with tridimensional shoe movement; preassigned control from model, comparison with equipment with fixed direction of movement, rail, and crawler mountings; examples of application; employment possibilities for excavators, mobile crushing plants, and similar equipment.

MEIN REFERAT BEFASST SICH mit Schreitwerken, d.h. einem Bewegungsmittel, welches zum gelegentlichen Transport, im wesentlichen aber zur Ortsveränderung von Lasten und grossen Geräten Anwendung findet, wobei die kennzeichnenden Merkmale für die Ortsveränderung grosse Gewichte, vorwiegend unebenes Gelände, nicht zu häufig vorkommende Bewegungen und relativ geringe Geschwindigkeit sind.

Es handelt sich bei diesen grossen zu bewegenden Geräten in den allermeisten Fällen um schwere Eimerketten- oder Schaufelradbagger, Absetzer, Bandschleifenwagen, Bandantriebsstationen, und ähnliche Geräte, wie sie in Tagebaubetrieben für Kohle, Braunkohle, Bauxit, und Erz zur Anwendung gelangen.

Man hat solche Geräte, die ein Gewicht bis zu einigen tausend Mp besitzen, ursprünglich auf Rädern laufen lassen. Bild 1 zeigt einen Eimerkettenabsetzer aus dem Jahre 1946, der drei Schienenfahrwerke mit insgesamt 256 Rädern besitzt.

Um von den Schienen und allen damit in Zusammenhang stehenden Schwierigkeiten abzukommen, hat man derartig grosse Geräte sehr bald auf Raupen gesetzt. (Bild 2)

*Vorstandsmitglied der Eisenwerk Weserhütte AG Geschäftsfürhrer der Weserhütte Otto Wolff GmbH; Bad Oeynhausen.

BILD 1

BILD 2

Bild 3 zeigt Anordnungen von Raupen für Tagebaugeräte. Die Verbindungslinien der Stützpunkte bilden die Stützfigur, deren Grösse für die Standsicherheit von Bedeutung ist.

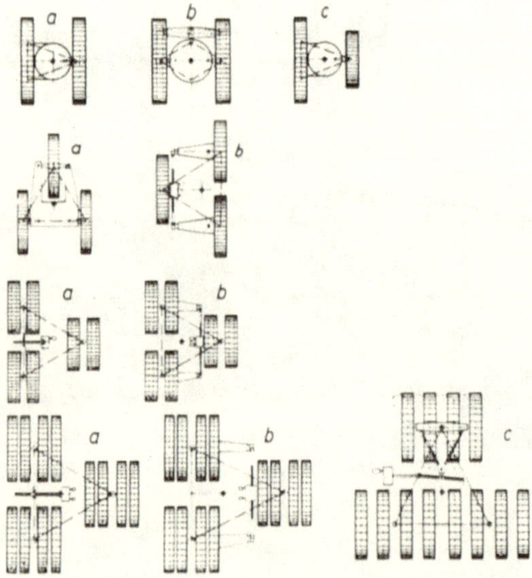

BILD 3

Die Verwendung von Raupen für die Ortsveränderung von grossen Geräten bringt gegenüber den schienengebundenen Rädern schon erhebliche Vorteile, weil nicht so hohe Anforderungen an die Bodenbeschaffenheit gestellt werden. Die Fahrbahn braucht weit weniger vorbereitet zu werden und die spezifische Belastung ist geringer. Eine gewisse Manövrierfähigkeit lässt auch das Verfahren in verschiedenen Richtungen bedingt zu.

Zur Entwicklung der Weserhütte-Schreitwerke ist es durch die rückbaren Bandstrassen in Braunkohlen-Tagebauen gekommen. In den Jahren 1955–58 wurden in Deutschland erstmalig rückbare Bandstrassen mit einer Gurtbreite von 2.200 mm erstellt, die entsprechend dem Abbaufortschritt immer weiter gerückt werden. Bisher waren die Bandstrassen maximal mit 1.400 mm Gurtbreite gebaut worden. Die Antriebe, die schwersten Teile der Bänder, waren auf einem Ponton verlagert und konnten, wenn auch z.T. mit grossen Schwierigkeiten, mittels Zugmaschinen und schweren Windwerken auf dem Erdboden gleitend bewegt werden. Der Entwicklungssprung auf die neu zu erstellende Bandbreite war ausserordentlich gross. Das Gewicht einer Antriebs station von 2.200 mm Bandbreite gegenüber 1.600 mm wurde doppelt so gross. Bild 4 zeigt eine Antriebsstation für ein rückbares Band.

BILD 4

Zunächst liegt es natürlich nahe, eine Antriebsstation auf Schienenfahrwerk zu setzen und sie mit vier Radsätzen mit je vier oder mehr Laufrädern auszurüsten. Jeder Laufrollensatz steht auf einem Schwellenrost, der in die künftige Fahrtrichtung ausgelegt ist. Der Antrieb hat beim Rücken nur wenige Meter zurückzulegen. Die ausgelegten Schienen bestimmen seine Richtung und seinen zukünftigen Bestimmungsort derart, dass man, um eine genaue Materialübergabe zu erhalten, immer genau vorher seinen Einsatzort festlegen muss. Die zu dem Antrieb gehörige Bandstrasse muss demnach ebenfalls genau auf eine vorher abgesteckte Linie ausgerichtet werden. Dieses Ausrichten des Antriebs, um einen guten Gurtlauf zu gewährleisten, war nicht einfach. Das

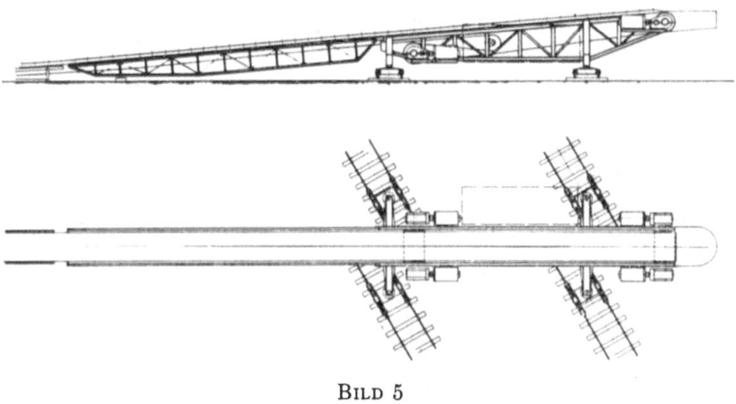

BILD 5

Vorstrecken der Roste war zeitraubend, weil bei einer Richtungsänderung sich auch der Abstand der einzelnen Roste veränderte. (Bild 5)

BANDANTRIEBE MIT RAUPENFAHRWERK

Bild 6

Zur gleichen Zeit entstand ein Antrieb, der mit einem Raupenunterwagen mit Drehkranz ausgerüstet ist. Der Raupenwagen sitzt unter dem Hauptrahmen der Antriebsstation, der normalerweise bei kleineren Antrieben, ausgebildet als Ponton, direkt auf dem Boden aufliegt. Durch

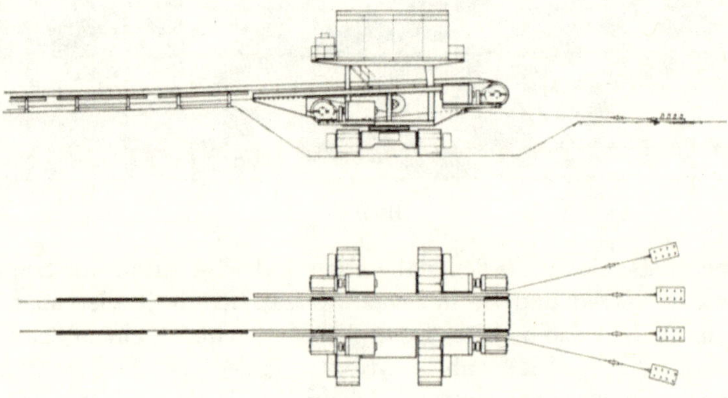

BILD 6

das Raupenfahrwerk baut der Antrieb hoch. Bis zum Drehkranz wurden mindestens 1.8 m benötigt. Diese zusätzliche Höhe ist insofern unangenehm, weil die Hubhöhe einen grossen Einlaufradius des Bandes benötigt und Antriebsleistung verbraucht. Man half sich dadurch, dass man den Antrieb in einem jeweils auszuhebenden Einschnitt stellte.

Ein Unterwagen mit Raupen wiegt etwa immer 30–40 Prozent des Gesamtgewichtes eines Gerätes. Wenn die Bandantriebsstation nur selten versetzt wird, steht der Raupenunterwagen u.U. 6 Monate still und ist damit totes Kapital. Die Fahrtrichtung des Antriebs müsste vorher in etwa festgelegt sein, da ein Drehen der Raupen auf der Stelle nur bedingt möglich ist. Es wurde daher von einer Braunkohlengrube eine Antriebsstation in Auftrag gegeben, bei der das Raupen fahrwerk mit einer zusätzlichen Hubvorrichtung ausgerüstet ist.

Bild 7 stellt diesen Antrieb dar, der genau wie schon gezeigt, auf einen Raupenunterwagen aufgesetzt ist. Zusätzlich ist an jeder Ecke des Hauptrahmens ein Hydraulikstempel mit Bodenplatte eingesetzt worden, um damit die Möglichkeit zu schaffen, die gesamte Station aufzuheben.

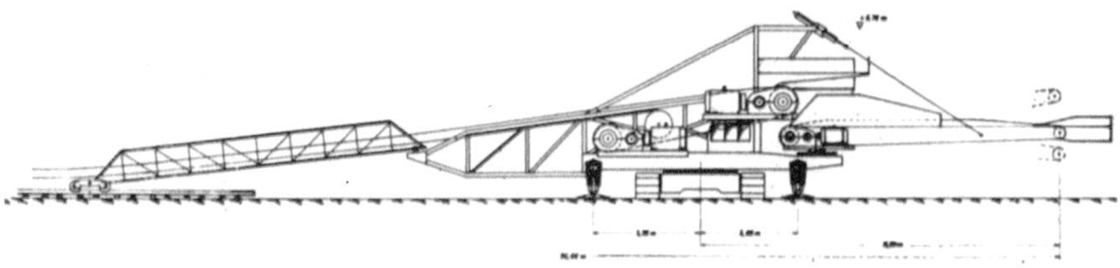

BILD 7

Im angehobenen Zustand kann dann das Raupenfahrwerk, ohne den Boden aufzuwühlen und ohne grosse Kräfte, in seine neue Bewegungsrichtung eingeschwenkt werden. Nach dem Einschwenken setzt das Hydraulikhubwerk den Antrieb wieder auf das Raupenfahrwerk. Die Hydraulikstempel (Bild 8) bringen dabei nicht nur den Vorteil mit, das

BILD 8

Raupenfahrwerk leicht einschwenken zu können, sondern erweisen sich auch sonst als sehr nützlich Mit ihnen kann die Station beim Einsinken in weichen Boden leicht herausgehoben werden. Sie sind ausserdem so konstruiert, dass sie mit Hilfe einer Seitenführung erhebliche Horizontalkräfte aufnehmen können. Die zusätzliche Bauhöhe, bedingt durch ein Raupenfahrwerk, musste allerdings auch bei dieser Konstruktion in Kauf genommen werden.

Der o.g. Antrieb entspricht bezüglich seiner Beweglichkeit wohl den Forderungen, die an einen rückbaren Antrieb gestellt werden. Er ist allerdings durch sein fest eingebautes Fahrwerk und durch die hydraulische Hubeinrichtung sehr aufwendig. Man suchte nach besseren Lösungen. Es lag der Gedanke nahe, die Raupen fortfallen zu lassen und die hydraulischen Stempel allseitig beweglich zu machen, um sie als Schreitfüsse zu verwenden. So kam es zu der Entwicklung des Lenkschreitwerkes der WESERHÜTTE.

Schreitwerke selbst sind seit vielen Jahren bekannt. Ich möchte Ihnen zunächst einige Bilder von verschiedenen ausgeführten Schreitwerksbauarten zeigen, um Ihnen dadurch den Unterschied zu den später zu behandelnden Lenkschreitwerken anschaulich zu machen.

EXZENTERSCHREITWERK

Bild 9 zeigt einen englischen Schürfkübelbagger auf Exzenterschreitwerken. Dieser Bagger, der sich nur wenig fortbewegen muss, da er durch seinen langen Ausleger ein grosses Feld bestreichen kann, trägt an seinem Oberbau Exzenter mit beweglichen Schreitkufen. Jede Kufe kann in einer Vertikalebene bewegt werden.

BILD 9

Der Bagger steht auf einem Unterbau mit Kugeldrehkranz, mit welchem vor dem Schreiten die Bewegungsrichtung eingestellt wird. Der Drehkranz ist sehr aufwendig, da er das gesamte Gewicht des Oberbaues tragen muss. Das Schreitwerk ist richtungsgebunden, da das Schreitwerk am Oberbau befestigt ist und der Bagger daher nur in der eingestellten

Richtung des Auslegers schreiten kann, jedoch nicht quer oder unter einem beliebigen Winkel zur Auslegerachse. Beim Drehen der zwei Exzenter, die durch eine Welle verbunden sind, drücken beide Kufen das Gerät hoch und ziehen es weiter. Der Bagger bewegt sich dabei mit seinem Ponton nachschleifend vorwärts. Jede Umdrehung des Exzenters bedeutet 1 Schritt.

Eine Variante zu dem Schreitwerk mit starrem Exzenter ist das Schreitwerk mit rollendem Exzenterrad, wie es ebenfalls bei den grossen englischen und amerikanischen Schürfkübelbaggern ausgeführt wird. Hierbei wird das Verhältnis Hub zu Weg verbessert.

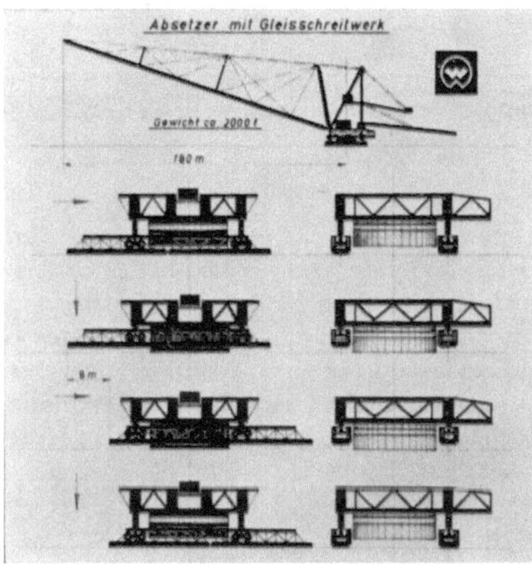

BILD 10

Bild 10

Dieses Bild zeigt einen russischen Absetzer mit 180 m Auslegerlänge und einem richtungsgebundenem *Gleisschreitwerk*, bei dem das Verhältnis Weg zu Hub wesentlich vergrössert ist. Die Fahrbahn, die hier immer unter 90° zur Auslegerrichtung steht, ist bis zu 12 m lang. Das Gewicht des Absetzers beträgt ca. 2.000 Mp. Durch Anwendung des Gleisschreitwerkes gegenüber einem Raupenunterwagen konnte etwa 15–20 Prozent des Absetzergewichtes eingespart werden.

Bild 11 zeigt einen russischen Schürfkübelbagger, der mit einem richtungsgebundenem hydraulischen Kufenschreitwerk ausgerüstet ist. Dieses Schreitwerk unterscheidet sich von dem Exzenterschreitwerk durch die hydraulische Betätigung der Schreitwerke. Hier sind die Schreitkufen gelenkig an zwei Hydraulikzylinder angesetzt, die sich

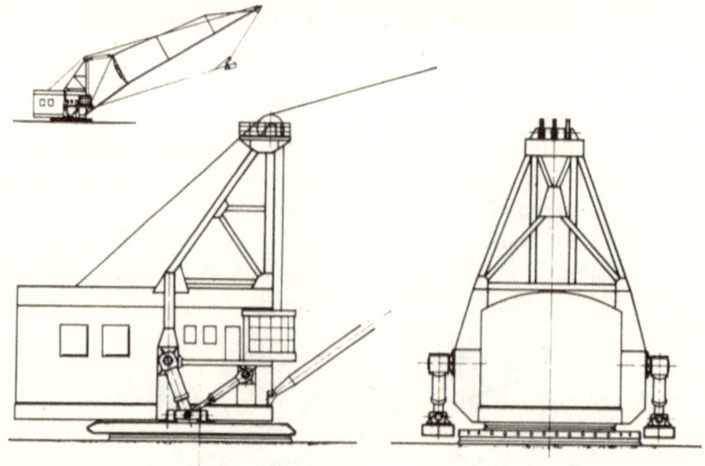

Bild 11

ebenfalls nur in einer Vertikalebene bewegen können. Um ein gleichmässiges Arbeiten der Zylinder der rechten und linken Seite zu bewirken und um ein Voreilen einer Seite zy verhindern, ist eine hydraulische Regeleinrichtung eingebaut. Die Schrittweite bei diesen Geräten beträgt 2 m, die Schreitgeschwindigkeit wird mit max. 2 m/min angegeben.

Das Topfschreitwerk auf Bild 12 ist eine tschechische Entwicklung. Hier ruht ein 600 Mp schwerer Absetzer auf einem Unterbau von etwa 16 m Durchmesser, In diesem Unterbau befindet sich unterhalb des

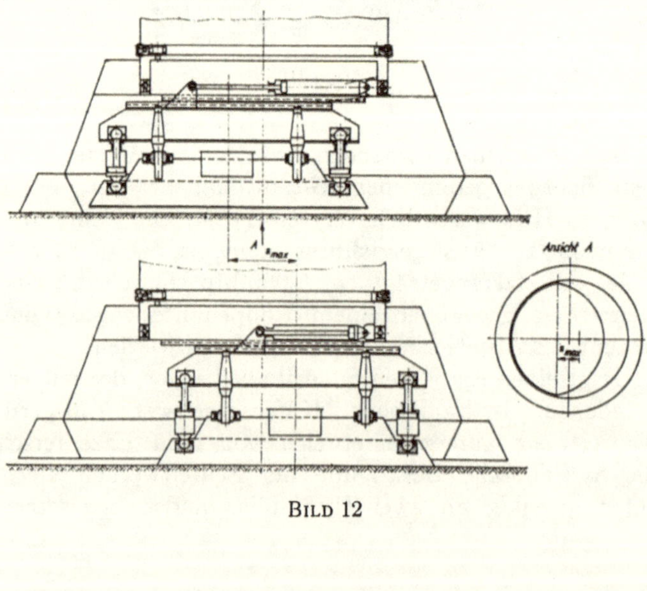

Bild 12

Kugeldrehkranzes die Schreitvorrichtung, bestehend aus einer Fahrbahn und drei Hubzylindern. Bei diesem Gerät ist die Schreitvorrichtung nicht am Oberbau, sondern unterhalb des Kugeldrehkranzes angebracht. Daher kann—wenn der Absetzer auf dem Aussenring steht—die Schrittrichtung beliebig eingestellt werden. Dieses Schreitwerk ist nicht richtungsgebunden. Der Absetzer kann längs, quer oder unter jedem Winkel zur Auslegerachse schreiten und ist daher sehr beweglich. Ein Nachteil liegt in der leichten Verschmutzung zwischen Innen- und Aussentopf.

Auf dem Bild 13 sind die Schreitwerksarten noch einmal schematisch gegenüber gestellt. In der rechten Spalte sind Hub und Weg in Kurvenform dargestellt.

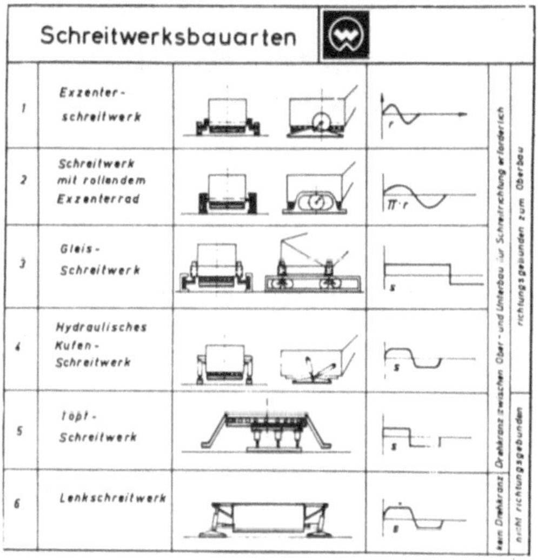

Bild 13

DAS WESERHUETTEN-LENKSCHREITWERK

WESERHUETTE-Lenkschreitwerke sind sum Transport und zur Ortsveränderung schwerster Lasten geeignet. Ein Schreitwerk besteht aus 2, 3, oder 4 Schreitfüssen, welche von einem zentralen Steuergerät aus ferngesteuert werden.

Auf Bild 14 sehen Sie einen Bandwagen mit vier Schreitfüssen. Die Schreitwerke sind nicht richtungsgebunden, wie die oben erwähnten Schreitwerke wie Schienen- oder Raupenfahrwerke, sondern können ihre Last sowohl in jeder Richtung bewegen als auch auf der Stelle drehen.

Die Anzahl der Schreitfüsse ergibt sich aus der zu bewegenden Last.

Bild 14

So werden z.B. für ein 400 Mp schweres Gerät vier Schreitfüsse mit je 100 Mp Tragkraft benötigt. Die Anschlusspunkte für die Schreitfüsse müssen am Gerät entsprechend ausgebildet sein. Durch einfache Verbindungsbolzen sind die Schreitfüsse (Eigengewicht 18 Mp) schnell an- und abschraubbar und können daher für die Bewegung verschiedener Geräte eingesetzt werden.

Bild 15

Bild 15: Der Transport und Anbau der Schreitfüsse an die zu bewegende Last erfolgt mit Hilfe eines Autokranes oder mit einem *Schreitfusstransportwagen.*

Auf Bild 16 ist der Bewegungsablauf der Lenkschreitwerke veranschaulicht. Es unterscheidet sich von den herkömmlichen Schreitgeräten

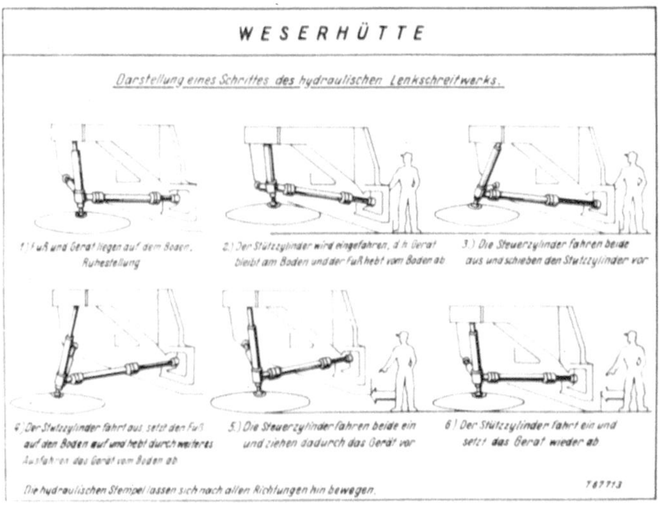

BILD 16

dadurch, dass hier die Schreitplatte nicht nur in vertikaler Ebene, sondern räumlich bewegt werden kann. Um das zu erreichen, werden drei Hydraulikzylinder, ein Stützzylinder, und zwei Steuerzylinder benötigt.

Alle Zylinder sind kugelgelenkig verlagert, um eine Raumbewegung zuzulassen. Die beiden Steuerzylinder greifen seitlich in einer gewissen Höhe ebenfalls über Kugelgelenke den Stützzylinder an. Sie sind um ca. 90° versetzt angeordnet. Werden die Druckoelströme aller drei Zylinder durch eine entsprechende Steuerung koordiniert, so kann eine vorgeschriebene Bewegung der Fussplatte ausgeführt werden. Dieses Lenkschreitwerk kann durch die durchführbare Raumbewegung jede beliebige Richtung einschlagen.

Bild 17

Bei Anordnung von mehreren Füssen an einem Gerät können durch entsprechende Steuerung der einzelnen Füsse sämtliche Bewegungen, die beispielsweise ein Vierfüssler durchführen kann, ermöglicht werden. Neben der Bewegung einer Last in jeder Richtung kann auch eine überlagerte Drehbewegung, eine Drehbewegung um einen zu wählenden Punkt durchgeführt werden. Ein Drehkranz, der bei den herkömmlichen Schreitwerken erforderlich ist, entfällt hier.

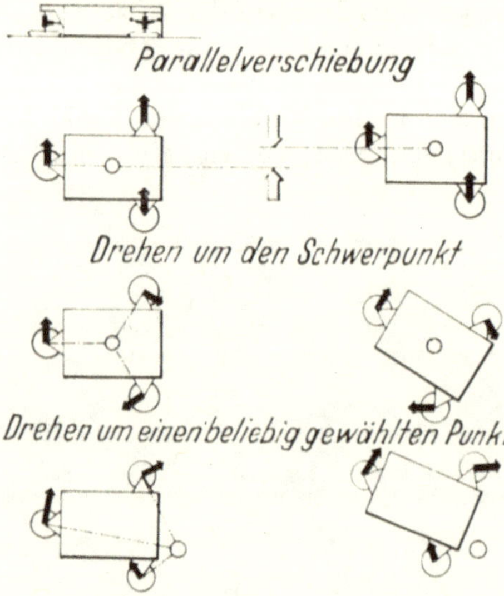

Bild 17

Bild 18

Aus Zweckmässigkeitsgründen wurden die Zylinder nicht direkt am Gerät gelagert, sondern es wurde als Zwischenglied eine Stahlkonstruktion, das Schreitwerksgerüst, das gleichzeitig als Ölbehälter dient,

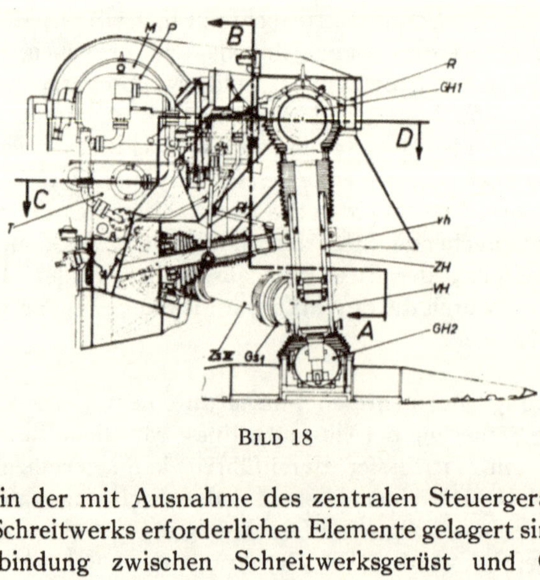

Bild 18

vorgesehen, in der mit Ausnahme des zentralen Steuergerätes alle zum Betrieb des Schreitwerks erforderlichen Elemente gelagert sind. Durch die lösbare Verbindung zwischen Schreitwerksgerüst und Gerät wurde erreicht, dass das Schreitwerk als geschlossene Baueinheit auch an anderen mit entsprechenden Anschlussmöglichkeiten versehenen Geräten angeschlossen werden kann, d.h. eine aus mindestens zwei Schreitwerken bestehende Schreitwerksgruppe kann zum Fortbewegen mehrerer Geräte benutzt werden. Neben den Zylindern sind in dem umbauten Raum der Stahlkonstruktion noch der Motor mit Hochdruck- und Steuerpumpe, die Rohrleitungen mit Sicherheitsund Rückschlagventilen, Filtern, und Steuerschieberr und die Steuerungselemente gelagert.

Die durch die Schreitbewegung auftretenden räumlichen Bewegungen der Zylinder erforderten kugelige Gelenkverbindungen, wobei die an den Zylindern befestigten Stützkugeln in mit Kunststoff ausgekleideten Kugelschalen gelagert sind. Auf die herkömmlichen Lagerwerkstoffe konnte nicht zurückgegriffen werden, da ihre Lebensdauer wegen der geringen Gleitgeschwindigkeiten in den Lagern und der damit verbundenen mangelhaften Ölpolsterausbildung von vornherein infrage gestellt war. Der gewählte Kunststoff erträgt bei kleinen Geschwindigkeiten nicht nur hohe Flächenpressungen—ein Punkt, der bei der Dimensionierung der Gelenke besonders beachtet werden muss, sondern er ist auch durch besondere Zusätze unempfindlich gegen Schmiermittelmangel. Das Problem der Druckölzuführung zu den einzelnen Zylinderräumen wurde mit Gelenkanschlüssen gelöst.

Die Bodenplatte hat mit 3.8 m ϕ bei 100 Mp Tragkraft nur eine Bodenpressung von 1 kp/cm². Daher können Schreitwerke auch auf weichem Boden und unwegsamem Gelände eingesetzt werden.

Steuerung

Schreitfüsse werden durch eine elektrohydraulische Folgeregelung von einem Zentralsteuergerät aus ferngesteuert.

Auf Bild 19 ist der Regelkreis für einen Zylinder dargestellt. Am Modell wird der Sollwert vorgegeben. Dieser wird verglichen mit dem 1st-Wert des Zylinders.

Ein Stellungsfehler erzeugt eine Fehlerspannung, die über einen Transistorverstärker ein elektro-hydraulisches Regelventil betätigt und den Stellungsfehler sofort durch die Hydraulik ausgleicht. Jeder Schreitfuss hat drei Zylinder, also auch drei solche Regelkreise. Das Modell ist eine kinematisch-ähnliche Nachbildung eines Schreitfusses. Wird also am Modell die untere Fusskugel in irgendeiner Richtung räumlich bewegt, folgt die Fusskugel am Schreitfuss genau nach.

Um die Genauigkeit der ausgeführten Bewegung festzustellen, wurde folgender Versuch unternommen (Bild 20): An dem Stützzylinder wurde eine Reissnadel befestigt und mit dem Schreitwerk die von einem handgesteuerten Modell vorgegebene Linien führung aufgeschrieben. Das

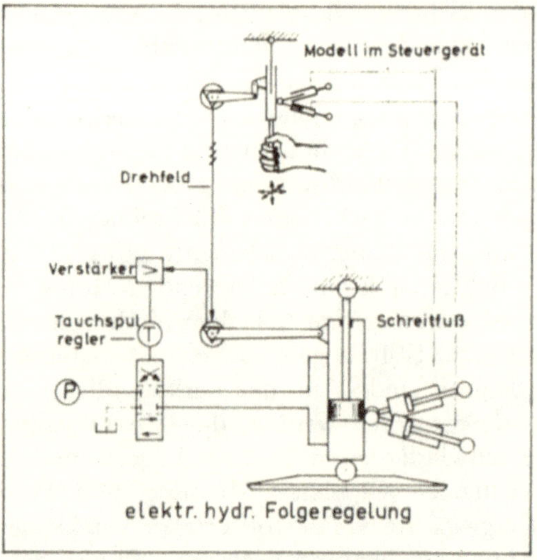

BILD 19

BILD 20

Ergebnis (Bild 21) war gut. Es zeigte sich nur ein Fehler von ca. 1 cm, eine Abweichung, die für den Schrittablauf unwesentlich ist.

Bild 22

Je nach Anzahl der einzelnen Schreitfüsse an einem Gerät befinden sich auch in einem Steuergerät 2, 3, oder 4 Modelle. Die Fusspunkte der

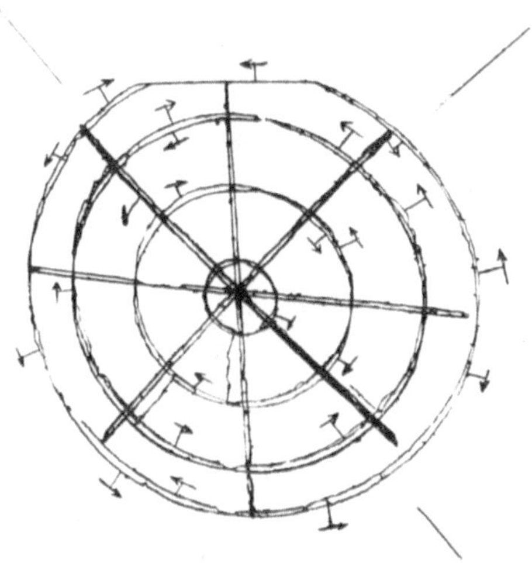

BILD 21

BILD 22

Modelle sind an einer Platte befestigt. Verschiebt man diese Platte, die man als Modellebene des Erdbodens bezeichnen kann, so werden alle Modellfüsse der Vorgabe entsprechend bewegt und koordiniert. Automatisch werden bie Kurvenfahrten unterschiedliche Schrittweiten und Schrittrichtungen diktiert. Bewegt wird die Modellebene durch Handhebel (Bild 23), die entsprechend den Schrittkomponenten Hub, Schritt,

BILD 23

Techn. Daten ausgeführter Weserhütte-Lenkschreitfüße				
		Type 30/800	Type 100/500	Type 140/500
Stützkraft	t	30	100	140
Durchmesser der Bodenplatte	m	2,8	3,8	3,8
Stützfläche	m^2	6,1	11,3	11,3
spez. Bodendruck bei Nennlast	kg/cm^2	0,63	1,03	1,4
Öldruck	atü	180	180	180
Stützzylinder-Durchmesser	mm	150/120	265/190	320/240
Schreitzylinder-Durchmesser	mm	130/90	210/150	210/150
Schreitzylinderkraft	t	12	31	31
Schrittweite	mm	0÷800	0÷500	0÷500
Schrittgeschwindigkeit	m/min	0÷5	0÷3	0÷2
Antriebsleistung	kW	55	70	70
Pumpen-Fördermenge	l/min	450	270	270
Dienstgewicht	t	ca. 8,5	ca. 16,5	ca. 18

BILD 24

Richtung, und Schwenken zugeordnet sind, oder durch motorisch betriebene Kurvenscheiben, die dem Schrittrhythmus angepasst sind. Das Gerät kann sich so automatisch entsprechend den eingestellten Werten fortbewegen. Lediglich eine Änderung der Werte muss eingeleitet werden. Das kann durch eine Beeinflussung mit der Hand an den Einstellknöpfen oder durch den Impuls eines Tastgerätes erfolgen.

Bild 24 zeigt die technischen Daten der ausgeführten Lenkschreitwerke mit 30, 100, und 140 Mp Tragkraft bei 180 atü Oeldruck.

VERBUNDSCHREITWERK

Bild 25

Eine Variante des eben genannten Schreitwerkes ist das ebenfalls von der WESERHÜTTE entwickelte, statisch bestimmte Verbundschreitwerk. Hier werden nicht Einzelschreitfüsse an die zu bewegende Last angeschlossen, sondern es werden drei Schreitfüsse unter einer Last zusammengefasst. Die drei Bodenplatten werden durch ein Gestänge verbunden. Dadurch sind nur drei vertikale und drei horizontale Zylinder erforderlich. Die Last ruht auf drei vertikalen Zylindern und ist also in vertikaler Richtung statisch bestimmt.

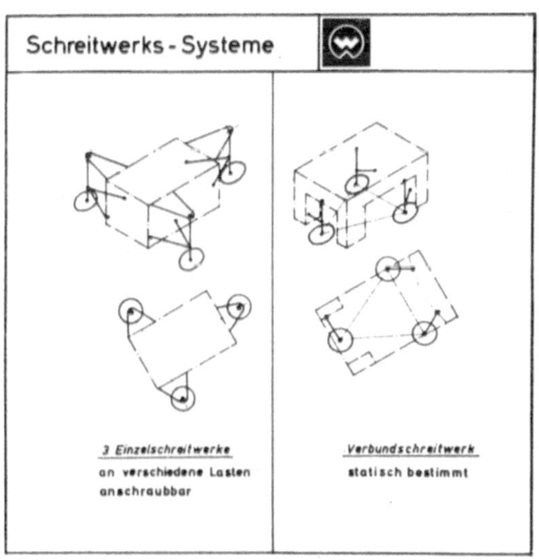

BILD 25

Auch in horizontaler Richtung können keine Zwangskräfte durch Fehlsteuerung auftreten. Da alle Zylinder kugelig gelagert sind, können alle Zylinder eine Pendelbewegung machen, wenn auch nur ein horizontaler oder ein vertikaler Zylinder verstellt wird.

Die Steuerung ist verhältnismässig einfach. Es wird hierbei auf die elektro-hydraulische Folgeregelung mit Modellvorgabe verzichtet: die elektrisch betätigten Ventile für die Hydraulik-Zylinder werden einfach von einer Programmwalze je nach gewünschter x-y- oder Drehrichtung beaufschlagt. Dieses Schreitwerk kann ebenfalls sowohl in alle Richtungen schreiten als auch auf der Stelle drehen. Es braucht keinen Kugeldrehkranz und ist nicht richtungsgebunden zum Oberbau.

Allerdings ist das Verbundschreitwerk nicht so universell an verschiedene Lasten anzusetzen, weil die Abstände der drei Fusspunkte infolge der mechanischen Verbindung nicht ohne weiteres verändert werden können. Daher gehört dieses Schreitwerk immer zu einer bestimmten Maschine. Ich zeige Ihnen nun einige Bilder über die Bandantriebe für rückbare Bandstrassen, welche mit Lenkschreitwerken ausgerüstet wurden.

Bild 26

Die erste Antriebsstation, die 1956 mit Lenkschreitwerken gebaut wurde, ist mit drei Füssen ausgestattet. Zwei sind seitlich angebracht und

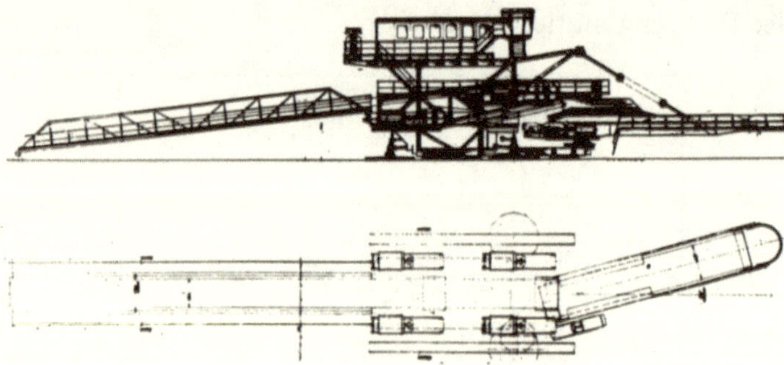

BILD 26

einer hinten in der Mitte. Der Antrieb ruht mit dem Hauptträger, der pontonartig ausgebildet ist, auf dem Boden. Die Schreitfüsse sind seitlich angeordnet. Der Nachteil der zusätzlichen Bauhöhe durch das Fahrwerk entfällt also hier.

Bild 27: die drei Schreitfüsse werden oben vom Steuergerät aus automatisch ferngesteuert.

Bild 28: das Dienstgewicht beträgt ohne Schreitwerke ca. 280 Mp, mit Schreitwerken und Steuerungselementen 330 Mp. Wäre er mit einem Raupenfahrwerk ausgerüstet, so betrüge das Dienstgewicht ca. 380 Mp. Hinzu käme noch das Gewicht einer längeren Einlaufbrücke bedingt durch die grössere Bauhöhe.

Bild 27

Bild 28

Bild 29

Die installierte elektrische Leistung beträgt

4 × 430 kW	=	1720 kW	Bandantrieb
1 × 200 kW	=	200 kW	Schwenkantrieb
4 × 30 kW	=	120 kW	Schreitwerkantrieb
2 × 10 kW	=	20 kW	Hub- u. Schwenkwerkantrieb
Summe	=	2060 kW	

Das Elektrohaus hat eine Grundfläche von ca. 64 qm. Die 3 Schreitfüsse haben eine Tragfähigkeit von je 120 Mp. Die auf dem Boden ruhende Fläche beträgt

Standfläche (Ponton)	40 m^2
Schreitfläche (Bodenplatte)	34 m^2
Bodendruck der Standfläche und der Schreitfläche	0, 85 kp/cm^2

Die Schreitgeschwindigkeit beträgt 1 m/Min.

Bild 29

Bild 30

Inzwischen ist die Entwicklung der Lenkschreitwerke so weit fortgeschritten, dass man sich in verschiedenen Tagebaubetrieben das Rücken der 200–400 Mp schweren Bandantreibsstationen ohne Schreitfüsse gar nicht mehr vorstellen kann. Es ist zur Selbstverständlichkeit

BILD 30

geworden, dass bei jedem Umbau des Bandsystems (Bild 31) die Schreitfüsse an die zu bewegende Station angesetzt werden. Hier ist der Einsatz von Lenkschreitwerken besonders günstig, weil die Antriebsstationen längere Zeit an der gleichen Stelle stehen und daher mit einem Satz Schreitfüsse mehrere Stationen nacheinander gerückt werden können.

BILD 31

Die Schreitfüsse arbeiten also dauernd, was die Investitionskosten günstig beeinflusst.

Bild 32

Diese Antriebsstation hat eine Antriebsleistung von 4 × 630 kW. Sie zieht ein Förderband von 2.200 mm Breite und bis zu 2.000 m Länge und fördert bis zu 11.000 m^3/h Abraum. Das Dienstgewicht beträgt ca. 280 Mp und wird mit 2 verstärkten Schreitfüssen mit je 140 Mp Tragkraft gehoben.

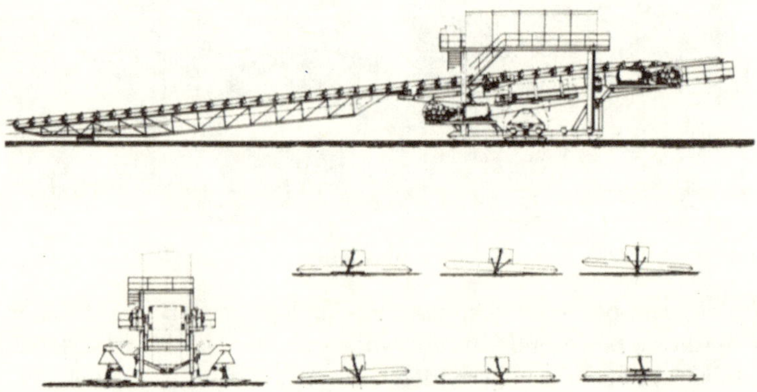

Bild 32

Bild 33

Bie dieser schweren Station wünschte der Kunde nur 2 Schreitfussanschlüsse, um Anbauzeit, und Betriebskosten zu sparen. Beim Rücken mit 2 Schreitfüssen werden die Schreitfussanschlüsse an dem pontonförmigen Hauptgerüst kurz vor der Schwerachse angebracht, Der rüchseitige Stützpunkt der Bodenwanne wird dann einfach mitgeschleift.

Bild 33 zeigt die Station während der Montage. Die Horizontalkraft eines Schreitfusses beträgt 30 Mp. Diese reicht aus, um die Einlaufbrücke mitzuschleifen und ca. 2.000 m Gummigurt u.U. aus einem Tal mitzuziehen oder um eine Böschung von 8–14 Prozent hochzuschreiten.

In der Praxis hat sich gezeigt, dass der Ponton der Einlaufbrücke sich nach kurzem Weg in den Boden einfrass und Erdmassen vor sich herschob. Daher wird bei grösseren Wegstrecken die Einlaufbrücke mit gummibereiften Radsätzen angehoben.

Bild 34

Es kommt vor, dass diese Stationen auch manchmal zu entfernten Einsatzorten wandern müssen. Dabei schafft das Schreitwerk bei 3 m/Min. Schreitgeschwindigkeit etwa 1.000 m an einem Tag. Hierbei muss

Bild 34

auch erwähnt werden, dass eben falls ein Dieselaggregat zur Erzeugung der elektrischen Energie für die Pumpenmotoren mitgezogen wurde.

Bild 35

Bei noch schwereren Stationen werden auch 3 oder 4 Schreitfüsse eingesetzt. Bei dieser Station ist das Elektrohaus nicht hoch, sondern seitlich neben das Band gesetzt.

BILD 35

Bild 36 zeigt eine Umkehrstation mit Lenkschreitwerk im Braunkohlentagebau Arjuzanx in Frankreich.

BILD 36

Bild 37 zeigt nochmals eine Bandantriebsstation mit 2 Schreitfüssen.

Neben den Antriebsstationen werden Lenkschreitwerke auch an Bandschleifenwagen verwendet.

BILD 37

Bild 38

Um das Fördergut von jedem Punkt einer Bandstrasse aus auf einen anderen Förderer oder Absetzer zu übergeben, fahren in der Bandstrasse Bandschleifenwagen. Biese werden teils ausgeführt mit Schienen- oder

BILD 38

Raupenfahrwerk. Aber auch der Bandschleifenwagen mit Schreitfüssen hat seine Vorzüge. Er ist beweglich und kann beim Rücken der Bandstrasse direkt zur Seite schreiten.

Bild 39 zeigt einen Bandschleifenwagen mit Lenkschreitwerk, der gerade aus der Bandstrasse ausgeschert ist. Er ist mir vier Lenkschreitfüssen ausgerüstet und liegt während des Stillstandes auf vier pontonartigen Standfüssen auf. Die Schreitfüsse werden mit der bereits beschriebenen automatischen Steuerung koordiniert und bewegt. Mit dem

Bild 39

Steuergerät kann die Schrittweite und damit die Fahrgeschwindigkeit vorgegeben werden. Zwei Tastvorrichtungen, die an der Rückschiene entlang gleiten, beeinflussen die Richtungslenkung des Steuergerätes. Das Gerät wird bei auftretendem Fahrfehler immer zur Schiene hin korrigiert und zwar am Kopf- und am Schwanzende. Die Korrektur erfolgt schon während des Ablaufes eines Schrittes.

Wenn der Bandschleifenwagen von einem Band in das andere umgesetzt wird, kann, nachdem die Tastvorrichtung entriegelt ist, durch das Steuergerät jede gewünschte Bewegung des Gerätes durchgeführt werden.

Bild 40

Beim Ausscheren müssen 3 oder 4 Gerüststösse aus der Bandstrasse herausgenommen werden. Der Gurt wird entspannt, die Abwurfschurre wird geöffnet, der Untergurt wird auf den Boden gelegt, der am Schwanzende auf der Bandstrasse laufende Kufenwagen wird durch ein Wind-

Bild 40

werk, welches im Gerüstträger verlagert ist, angehoben. Der Obergurt wird mit einem Kran angehoben. Das Gerät kann nun seitlich zwischen Ober- und Untergurt aus der Bandstrasse herausschreiten. Der von einem Kran gehobene Gurt wird nach dem Wiedereinbau der entnommenen Gerüststösse in das Bandstrassenbett zurückgelegt. Das Wiedere-

Bild 41

infahren in das nächste Kippenband erfolgt sinngemäss in umgekehrter Reihenfolge.

Bild 41

Hier sehen Sie, wie zwischen dem Bandschleifenwagen und dem nachgeschalteten Absetzer noch ein Zwischenförderer eingesetzt ist. Dieser Zwischenförderer, der neben der Bandstrasse steht, ist ebenfalls mit Lenkschreitwerken ausgerüstet. Vom Führerstand aus kann der Aufgabepunkt vom Bandschleifenwagen und der Abwurfpunkt vom Absetzer genau beobachtet und der Zwischenförderer entsprechend eingestellt werden.

FAHRBARE BRECHANLAGEN

Bild 42

In den meisten Steinbrüchen wird wie folgt abgebaut: An der Felswand wird gesprengt. Das abgesprengte Gestein wird vom Bagger auf LKWs

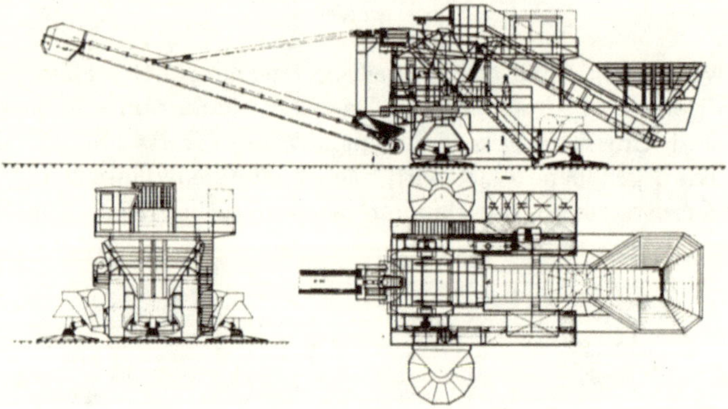

BILD 42

geladen und zum stationären Brecher gefahren, und wird dort verarbeitet. Mit laufendem Abbaufortschritt wird der Abstand vom Bagger zum Brecher immer grösser. Es müssen mehr LKWs eingesetzt werden. Der Verschleiss und die Personalkosten, teils im zweischichtigen Betrieb, steigen beträchtlich. U.U. muss die Brechanlage abgerissen und an anderer Stelle wieder aufgebaut werden.

Deshalb geht man auch hier zu fahrbaren Brechern über, die möglichst ohne Einschaltung von LKWs die Steine direkt vom Bagger in einen Aufgabetrichter übernehmen, in einem Brecher vorbrechen und über ein bewegliches Schwenkband das Brechgut kontinuierlich auf ein Förderbandsystem abgeben.

Die Betriebskosten eines fahrbaren Brechers mit Bandbetrieb gegenüber einem stationären Brecher mit LKW-Betrieb sind soviel geringer, dass etwaige höhere Anschaffungskosten in kürzester Zeit wieder hereingewirtschaftet sind. Man hat zunächst die leichteren Fahrbrecher auf gummibereifte Fahrzeuge gesetzt. Ein schwerer Fahrbrecher ähnlich dem gezeigten Bild mit ca. 400 Mp Dienstgewicht wurde von uns mehrfach auf Raupenfahrwerken gebaut. Schliesslich erkannte man, welche Vorteile auch hier Schreitwerke gegenüber Raupenfahrwerken bringen:

(1) Die Massenkräfte vom Brecher und vom Aufgabetrichter her gehen nicht über die Raupen, sondern über die 3 pontonförmigen Standfüsse direkt in den Boden. Die Schreitwerke werden nur eingeschaltet, wenn der Brecher stillsteht oder leer läuft.

(2) Die 3 Schreitplatten sind kugelig gelagert und können sich dem Boden anpassen. Raupenplatten können insbesondere auf felsigem Untergrund leichter zerstört werden.

(3) Das Schreitwerk ist beweglicher. So kann sich die Brechanlage auf engstem Raum drehen oder mit dem Aufgabetrichter immer in den günstigsten Schwenkbereich des Baggers stellen und dabei den Abwurfpunkt am Ausleger halten. Bei einem Raupenfahrwerk würde es durch Zick-Zack-Fahren Verlustzeiten geben.

Bild 43

Steht in einem Steinbruchbetrieb nur 1 Fahrbrecher bzw. ist die Austauschbarkeit der Schreitfüsse nicht gefordert, wird man den Brecher mit einem Verbundschreitwerk ausrüsten. Hier sind die Kräfte beim Schreiten eindeutig zu rechnen. Das Dienstgewicht—und damit der

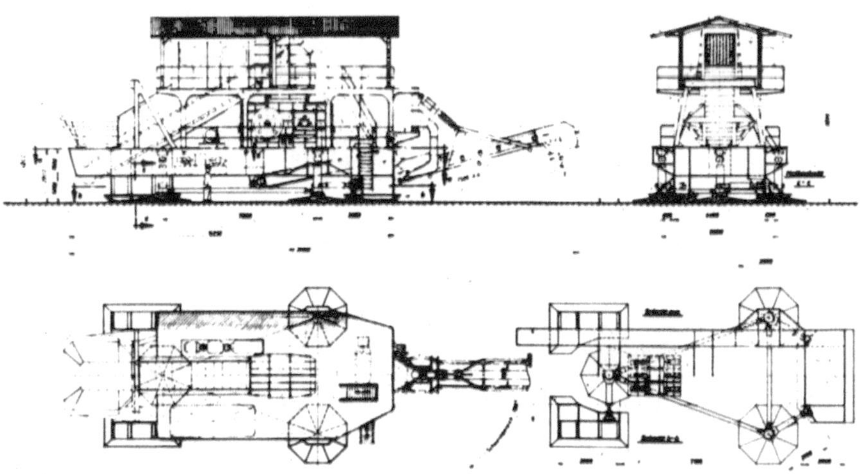

BILD 43

Preis—geht bei diesem Brecher mit Verbundschreitwerk gegenüber Raupenfahrwerk auf 75 Prozent, also von 400 Mp auf ca. 300, herunter. Rechnet man nur den Unterwagen ohne Aufbauten, so wiegt der Unterwagen mit Verbundschreitwerk gegenüber einem Unterwagen mit Raupen nur 55 Prozent.

Bild 44

Weitere Einsatzgebiete neben Erdbaugeräten in Tagebaubetrieben und Brechern in Steinbrüchen sind z.B. Bohrtürme auf Bohrfeldern oder Schiffskörper in Werften. Es ist auch schon einmal untersucht worden, Konverterbirnen in einem Hüttenwerk zur schnellen Ausmauerung abzutransportieren oder z.B. schwere Lasten in einem strahlungsgefährdeten Raum ferngesteuert zu bewegen.

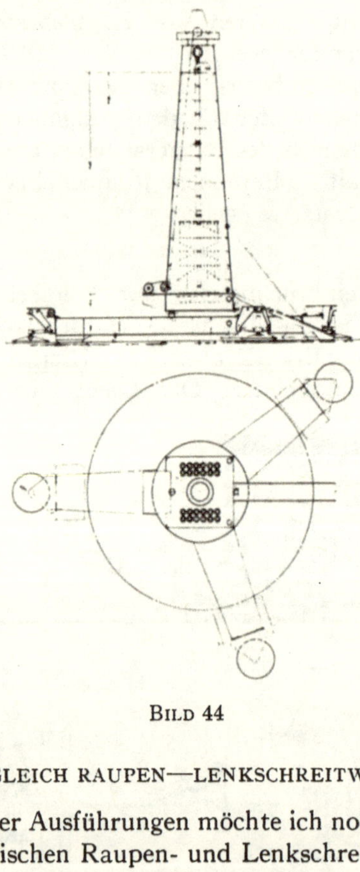

Bild 44

VERGLEICH RAUPEN—LENKSCHREITWERKE

Zum Schluss meiner Ausführungen möchte ich noch einmal die wesentlichen Merkmale zwischen Raupen- und Lenkschreitwerken vergleichen.

Der Einsatz von Lenkschreitwerken ist immer dann zweckmässig, wenn das Fahrwerk nicht für Dauerbetrieb ausgelegt werden muss, sondern viel Stillstandzeiten hat. Ist das Verhältnis Stillstandzeit zu Fahrzeit so günstig wie z.B. bei Bandantriebsstationen, dass mit einem Schreitwerk mehrere Geräte nacheinander transportiert werden können, braucht für verschiedene Lasten nur ein Fahrwerk angeschafft zu werden.

Als Nachteile sind bei einem Schreitwerk anzusehen:

(*a*) *Bild 45*: *Die diskontinuierliche Bewegung*, die ein Schreitwerk in Schrittrichtung und in Hubrichtung macht, kann sich bei manchen Geräten nachteilig auswirken, besonders, wenn sie mit kontinuierlich fahrenden Raupengeräten zusammenarbeiten müssen.

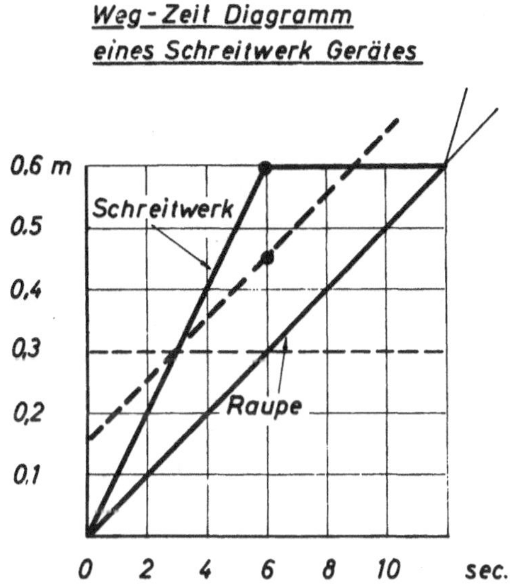

BILD 45

(*b*) Schreitwerke benötigen eine höhere *Antriebsleistung*, weil sie nicht nur den Fahrwiderstand überwinden, sondern auch Hubarbeit aufbringen müssen. So braucht ein Raupenfahrwerk, um 100 Mp mit v = 3 m/min zu bewegen, etwa 30 kW, während ein Schreitwerk etwa 75 kW benötigt. Bei kurzen Wegen spielt diese höhere Leistungsaufnahme jedoch meistens keine grosse Rolle.

(*c*) Die *Schreitgeschwindigkeit* der Schreitwerksgeräte liegt z.Zt. bei 3–4 m/min, bedingt durch die zusätzliche Hubarbeit. Da die Hubarbeit bei Raupen entfällt, können diese vergleichsweise etwas schneller fahren.

Als Vorteile sind bei einem Lenkschreitwerk folgende Punkte hervorzuheben:

(*a*) Lenkschreitwerke benötigen keine befestigte *Fahrbahn*, wie z.B. Schienen mit Schwellen und Schotterunterbau.

(*b*) Lenkschreitwerke sind nicht richtungsgebunden und daher *beweglicher* als Raupen. Sie können von Hand oder automatisch ferngesteuert werden. Fahrfehler oder Richtungsänderungen können direkt ausgesteuert werden. Sie benötigen keinen Kugeldrehkranz zwischen Ober- und Unterbau und können trotzdem den Oberbau auf der Stelle drehen.

(*c*) Die universelle Verwendbarkeit der *Einzelschreitfüsse*. Sie können leicht *abgeschraubt* werden und für den Transport verschiedener Lasten eingesetzt werden.

(*d*) Das *Verbundschreitwerk* ist *statisch bestimmt*, so dass horizontal und vertikal keine unvorhersehbaren Kräfte auftreten können.

(*e*) Wenn das Schreitwerksgerät mit dem Ponton auf dem Boden aufliegt, gehen die *Massenkräfte* der Maschinen—Brecher, Förderer, usw.—direkt in den Boden. Die Schreitwerke sind entlastet; sie können trotz laufender Maschinen *gewartet* oder *repariert* werden. Raupen müssten z.B. beim Auswechseln einer Bodenplatte aufgebockt werden, d.h. die ganze Anlage müsste den Betrieb unterbrechen.

(*f*) Die Bodenplatten sind *kugelig* gelagert und können sich daher besser dem Boden anpassen.

(*g*) Schreitwerksgeräte sind *leichter* und können daher mit geringerem *Bodendruck* gebaut werden.

Auf Bild 46 ist der Gewichtsanteil des Fortbewegungsmittels, Schreitwerk oder Raupe, in Abhängigkeit vom Bodendruck gebracht worden. Man kann beispielsweise ein Raupengerät mit einem bestimmten Gewicht nicht so bauen, dass ein bestimmter Bodendruck unterschritten wird. Schon das Raupenfahrwerk allein hat bei ganz schweren Geräten

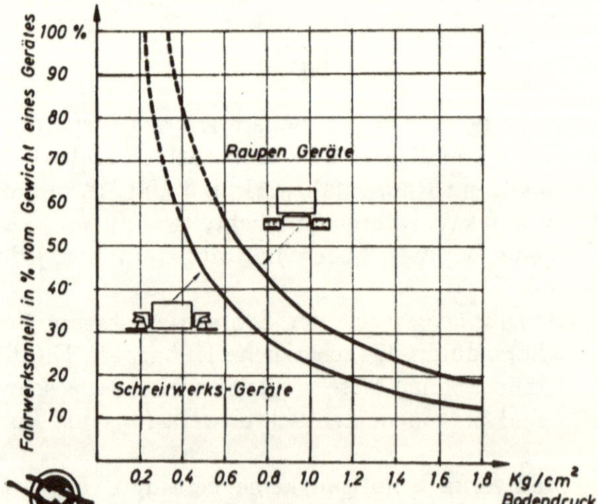

BILD 46

durch sein Eigengewicht einen Bodendruck von etwa 0.4 p/cm². Der Anteil des Raupengewichtes wird gegenüber zum Nutzgewicht bei schweren Geräten immer grösser, Schreitwerksgeräte liegen, wie die linke Hyperbel zeigt, bedeutend günstiger. Es können bei manchen Geräten durch Anwendung von Lenkschreitwerken 15–25 Prozent vom Dienstgewicht und damit vom Anschaffungspreis eingespart werden.

A Review of The State of Knowledge of The Directional Behaviour of Automotive Vehicles

LEONARD SEGEL*

ABSTRACT

The technical literature dealing with the directional behaviour of automotive vehicles is reviewed for purposes of (*a*) demonstrating the origins of this subject as a specialized field of interest, (*b*) following the historical development of the manner in which additions to knowledge have been made, and (*c*) appraising the state of the art for predicting the directional behavior of automotive vehicles. It appears that a substantial body of knowledge has been developed within the last decade facilitating the analytic description of a large number of the directional phenomena exhibited by pneumatic-tired vehicles. The conclusion is drawn that many investigations must be similarly instituted and completed to produce an equivalent understanding of (*a*) the behavior of the man-vehicle system and (*b*) the relationship of man-vehicle performance to accident reduction and the operational reliability of the man–vehicle–highway system.

THE OBJECTIVES of this paper are (*a*) to perform a general appraisal of the state of knowledge relative to the directional behaviour of automotive vehicles, and (*b*) to review those portions of the literature in which this knowledge is documented. Prior to performing these tasks, an attempt is made to define the scope of the vehicle-handling phenomenon. This definition serves to establish a perspective from which to view the significance of handling theory to vehicle and/or highway design/operations.

THE SCOPE OF THE VEHICLE-HANDLING PHENOMENON

The term "handling" implies a concern with the behaviour of the man-vehicle combination wherein the subjective reaction of the human operator plays a major role. It is possible to stress the role of the vehicle by asking these questions:

1. Does the automobile respond to steering control in such a way that the operator can change his course or path quickly, easily, and precisely?

*Staff Scientist, Cornell Aeronautical Laboratory, Inc., Buffalo, New York.

2. Does the automobile tend to follow the curved or straight path selected by the operator irrespective of disturbing forces deriving from the road contour or from cross winds?

3. Do the wheels and tires tend to stay on the road, thus providing stability of path irrespective of the ride disturbance produced by the vertical profile of the road?

4. Does the vehicle have a response to steering and steering-wheel displacement and steering-force levels that are conducive to safe control under emergency conditions?

Implicitly contained in all these questions is the factor which the professional driver often calls (*a*) "the feel of the road," or (*b*) the ability to sense what the vehicle is doing. For the purposes of this discussion, it shall be assumed that all of the above questions deal with those facets of directional behaviour which collectively, enter into an assessment of the handling qualities of an automotive vehicle.

Experience indicates that the automobile is a vehicle possessing directional stability and control characteristics that appear to be reasonably well matched to the requirements of the human operator. On the other hand, it must be pointed out that the process of driving a car requires considerable precision in course keeping and course changing, and it is not an overstatement to say that all of us put cars through manoeuvres which can be considered hazardous or safe, depending on one's point of view. They are hazardous, for example, if we consider the margin for error, in feet, allowed between two vehicles moving at high speed in opposite directions, or they are safe from the standpoint that cars have sufficiently good handling qualities such that the average driver can steer his vehicle along the roadway with the precision demanded by the geometry of the road. Obviously, the validity of the above statement holds only for present-day design practice and for present-day operating speeds; consequently, extensions of this view into the future must be made with caution.

THE DIRECTIONAL BEHAVIOUR OF AUTOMOTIVE VEHICLES—A REVIEW

In the early period of the history of the automobile, efforts to improve or understand stability and manoeuvrability were largely empirical. As reviewed elsewhere (1), theories of directional behaviour could not be advanced until various investigators first demonstrated the process whereby a pneumatic tire can sideslip and thus produce side forces proportional to the lateral distortion of the tire contact print. The early studies of tire mechanics were made in France and Germany, culminating in a German text (2) on the subject in 1931. At that time, cars in the

United States had a solid front axle and the centre of gravity was located considerably aft of the midpoint of the wheel base. Consequently, they often exhibited oversteer behaviour, and beyond a certain speed, became dynamically unstable. This experience provided the original motivation for investigating the mechanics of the rolling tire in the United States, and for relating this mechanical behaviour to the directional properties of automobiles. The literature (3), (4), (5), (6) shows that one United States car manufacturer took the lead in this endeavour, resulting in the development of a semi-empirical understanding of so-called "oversteer" and "understeer" behaviour. (Undoubtedly, other United States car manufacturers were also busy looking into these matters, but chose not to disclose their efforts.)

During this same period, many investigators (primarily in France, Germany, and Russia) began to examine the "whys and wherefores" of vehicle stability and manoeuvrability. An examination of the literature shows that during this period, the language barrier was extremely effective in blocking the interchange of knowledge. The most productive research was performed in Germany prior to the beginning of World War II, resulting in the publishing (in 1940) of what is acknowledged to be the first theoretical analysis of the directional dynamics of a pneumatic-tired vehicle (7). It should be noted that this analysis was restricted to a vehicle possessing only two degrees of freedom, namely, yaw and side-slip, and no experimental data were obtained to bear out the predictions made by the theory.

In the United States, the demands of the war effort forced the temporary cessation of the experimental measurements reported by Stonex (5). Subsequently, a sustained interest in gaining an understanding of the transient behaviour of automotive vehicles led to the initiation (in 1953) of a vehicle dynamics research programme at the Cornell Aeronautical Laboratory. Initial efforts were directed toward the development of a mathematical model for the directional behaviour of an automobile without a steering system and its subsequent verification by full-scale response tests (1), (8). Subsequent efforts dealt with improvements and expansion of the originally developed theory, such as (*a*) including the dynamics of the steering system, (9) (*b*) analyzing the path response to a rolling road surface, (*c*) removing the restrictions of linearity, both for tire and suspension characteristics (10), (*d*) developing variable-steering and variable-stability vehicles for handling-qualities research (11), (12), and (*e*) investigating those aspects of tire mechanics having a bearing on the directional behaviour of automobiles (13).

Starting around 1960, other U.S. investigators began to publish results of research dealing with the directional stability and manoeuvrability of automotive vehicles. These publications (for example (14–20)) have indicated that investigators have been concerned with (*a*) vehicle behaviour at or near the skid, (*b*) aerodynamic characteristics of vehicles and cross-wind stability, and (*c*) influence of suspension characteristics on directional behaviour. During this same period, a body of literature began to develop wherein the subject of interest was the man-vehicle interaction process. Attention has been directed to the steering process (21–23) and efforts have been made, for example, to select optimum steering-wheel displacement and force levels (24).*

Some of the vehicle stability and manoeuvrability research performed on the European scene was reviewed in 1961 by Bergman, Fox, and Saibel (25) who comment on the work of Mitschke (26), Gebelein and Schlick (27), Paslay and Slibar (28), (Enke 29), Mitterlehner (30), (31), and Fiala (32). These papers range over a broad spectrum of topics in vehicle stability, such as (*a*) investigating the effects of front-, rear- and four-wheel steering, (*b*) confirming the aerodynamic stability requirements previously formulated by Kamm (33), (*c*) writing the equations of motion for a four-wheeled vehicle with six degrees of freedom, (*d*) considering the influence of suspension kinematics on vehicle behaviour using experimental tire data instead of equations, (*e*) demonstrating how free-control stability (the steering wheel is free) improves with an increase in the natural frequency of the steering system, and (*f*) indicating the importance of steering-system flexibility and body roll in influencing vehicle stability. Segel (9) has reviewed the work of additional European investigators (35), (36), who have been concerned with the dynamics of the free-control vehicle, this work (9) constituting one of the more recent treatments of this particular subject. The major result of this latter investigation was the finding that the directional stability and control of the free-control automobile can only be characterized satisfactorily by adding a quasi-linear representation of a steering system† to the dynamic model of the fixed-control vehicle as represented by a set of three-degrees-of-freedom, fourth-order equations of motion.

Within the last five years, there has been a significant upsurge of interest within the technical community on matters relating to the directional behaviour of automotive vehicles. Certain portions of the literature appearing during this period have already been cited. In addition, there is the work being pursued by investigators in the United Kingdom (for example (36–40)), which research has grown to a level

*Note that the scope of this paper precludes mention of studies that deal with the driving process in general.

†A mechanism with two degrees of freedom with Coulomb friction incorporated as the single nonlinear element.

warranting a full symposium devoted to the "Control of Vehicles During Braking and Cornering"* in June of 1963. Further, the influence of aerodynamic forces on the directional behaviour of vehicles is again receiving considerable attention (41), (42), and it should be noted that a careful examination has been made (in Germany) of the role of cross winds in causing single-vehicle accidents on the highway (43). Articles concerned with vehicle stability (such as (44–50)) are appearing with increasing regularity in automotive engineering journals, resulting, for example, in a recent issue of the *Automobiltechnische Zeitschrift* (Vol. 67, 7, July 1965) being devoted to "Road Stability," Sometimes suitable automotive journals are not available. For example, we find the results of vehicle research (51) being reported in *Australian Road Research.*

It should be noted that the foregoing citations have dealt with single vehicles only. Because of the interests of the military user, vehicle researchers have studied the directional behaviour of the multielement vehicle (see, for example, (52–56)).

CONCLUDING REMARKS

No attempt has been made in the foregoing review to list every single document bearing on the subject at hand. Many items have been omitted, either because they have been outdated by a subsequent document or because they belonged to a time period that fell outside the historical format of this presentation. Fortunately, a few efforts have been made to consolidate this growing body of literature by means of books and monographs (57–60).

The foregoing review indicates that the time is not too distant when the directional dynamics and statics of automotive vehicles will be covered by adequate theory, particularly for vehicles operating on smooth, level surfaces. For those operational situations wherein ride motions influence the lateral distortion of tires and thereby influence the directional behaviour of the vehicle, additional research appears to be indicated in order to adequately analyze the "road-holding problem." The work of Kurz (61) was a beginning step in this regard, culminating in the more recent investigation performed by Metcalf (13).

Although we have an understanding of the physical aspects of directional behaviour that is reasonably substantial, many investigations must be instituted and completed to produce a similar understanding of (*a*) the overall behaviour of the man-vehicle system and (*b*) the relationship of man-vehicle performance to accident reduction and the

*A symposium held under the auspices of the Automobile Division, The Institution of Mechanical Engineers, in collaboration with the Road Research Laboratory, D.S.I.R.

operational reliability of the man-vehicle-highway system. It is to this area that the attention of the automotive research community will be increasingly turned.

REFERENCES

1. SEGEL, L., Research in the Fundamentals of Automobile Control and Stability, *Trans.*, Soc. of Automotive Eng., Vol. 65, 1957.
2. BECKER, G., H. FROMM, and H. MARUHN, *Schwingungen in der Automobillenkung*, Krayn Verlag, Berlin, 1931.
3. FOX, M. L., Relation Between Curvature and Speed, *Proc.* Seventeenth Annual Meeting of Highway Research Board, 1937, p. 202.
4. OLLEY, M., National Influence on American Passenger Cars, *Proc.*, Inst. of Automobile Eng., Vol. 32, 1938, p. 509.
5. STONEX, K. A., Car Control Factors and Their Measurement, *Trans.* Soc. of Automotive Eng., Vol. 48, No. 3, 1941, p. 81.
6. OLLEY, M., Road Manners of the Modern Car, *Proc.*, Inst. of Automobile Eng., Vol. 51, p. 147 (1946–47).
7. RIECKERT, P., T. E. SCHUNK, The Mechanics of Rubber-Tired Motor Vehicles (Zur Farhmechanik des Gummibereiften Kraftfahrzeugs), *Ingenieur Archiv* 11, 1940, p. 210.
8. SEGEL, L., Theoretical Prediction and Experimental Substantiation of the Response of the Automobile to Steering Control, *Proc.* of the Automobile Div., The Inst. of Mech. Eng., No. 7, 1956–57, p. 310.
9. SEGEL, L., On the Lateral Stability and Control of the Automobile as Influenced by the Dynamics of the Steering System. To be published in the *J. of Engineering for Industry*.
10. RADT, H. S., H. B. PACEJKA, Analysis of the Steady-State Turning Behavior of an Automobile, *Proc.* of the Symposium on Control of Vehicles During Braking and Cornering, The Inst. of Mech. Eng., London, England, 1963.
11. SEGEL, L., An Investigation of Automobile Handling as Implemented by a Variable-Steering Automobile, *Human Factors*, August, 1964, p. 333.
12. SEGEL, L., The Variable Stability Automobile; Part I: Concept and Design, Soc. of Automotive Eng., Special Publication No. 275.
13. METCALF, W. H., Effect of a Time-Varying Load on Side Force Generated by a Tire Operating at Constant Slip Angle, Soc. of Automotive Eng. Preprint No. 713C, June 1963.
14. BEAUVAIS, F. N., C. GARELIS, and D. H. IACOVONI, An Improved Analog for Vehicle Stability Analysis. Paper presented at Soc. of Automotive Eng. Int. Congress, January 1961.
15. KELLY, K. B., P. R., KYROPOULOS, and W. F. TANNER, Automobile Aerodynamics, Soc. of Automotive Eng. Preprint No. 148, March 1960.
16. GOLAND, M. and F. JINDRA, Car Handling Characteristics, *Automobile Engineer*, Vol. 51, 1961, p. 296.
17. F. JINDRA, The Influence of the Wheel Suspension on Directional Control and Stability when Cornering (Der Einfluss der Radaufhangung auf die Richtungshaltung und Stabilitat bei Kurvenfahrt), *Automobiltechnische Zeitschrift*, Vol. 63, No. 5, 1961, p. 133.
18. WINKELMANN, O. J., Requirements on the Road Stability of Automobiles (Anforderangen an das Fahrverhalten von Kraftfahrzeugen), *Automobiltechnische Zeitschrift*, Vol. 63, No. 5, 1961, p. 121.
19. SETHNA, P. R., Dynamics of a Four-Wheeled Vehicle and the Effects of Suspension Geometry, *Proc.* of the Automobile Div., Inst. of Mech. Eng., 1962–63, p. 201.
20. BUNDORF, R. T., D. E. POLLOCK, and M. C. HARDIN, Vehicle Handling Response

to Aerodynamic Inputs. Paper presented to Soc. of Automotive Eng., Summer Meeting, 1963.

21. LEARNER, D. B., A Minimum Analog Driving Simulator, *IRE Transactions* on Human Factors in Electronics, September 1960.
22. OLSON, P. L., R. A. WACHSLER, Relative Controllability of Dissimilar Cars, *Human Factors*, Vol. 4, 1962, p. 375.
23. SHERIDAN, T. B., Vehicle Handling: Mathematical Characteristics of the Driver. Paper presented at Soc. of Automotive Eng. Annual Meeting, 1963.
24. HAYNES, A. L., H. S. MIKA, and L. M. FORBES, Simulation Techniques in Automotive Human Factors Research, *Proc.* of the National Conference on Driving Simulation, Santa Monica, California, 1961.
25. BERGMAN, W., E. A. FOX, and E. SAIBEL, Dynamics of an Automobile in a Cornering Maneuver On and Off the Highway, *Proc.* 1st Int. Conf. Soil Vehicle Systems, Torino, Italy, June 1961.
26. MITSCHKE, M., The Directional Behavior and Stability of Four-Wheeled Motor Vehicles (Fahrtrichtungshaltung und Fahrstabilitat von vierradrigen Kraftfahrzeugen), *Deutsche Kraftfahrtforschung und Strassenverkehrstechnik*, Heft 135, 1960.
27. GEBELEIN, H., G. SCHLICK, Investigations to Improve the Safety of the Motor Vehicle Through Improvement of the Stability and Directional Behavior (Untersuchungen zur Erhöhung der Sicherheit der Krafthfarzeuge durch Verbesserung der Stabilität und der Fahrtrichtungshaltung), *Deutsche Kraftfahrtforschung und Strassenverkehrstechnik*, Heft 122, 1959.
28. PASLAY, P. R., A. SLIBAR, The Motion of Automobiles in Unbanked Curves, *Ingenieur Archiv*, Vol. 24, 1956, p. 412.
29. ENKE, K., A Method for Calculating the Bank Angle of Vehicles (Ein Verfahren zur Berechnung der Kurvenlage des Kraftwagens), Dissertation, Technische Hochschule Karlsruhe, 1957.
30. MITTERLEHNER, G., The Steering Stability of Pneumatic-Tired Vehicles (Die Lenkstabilität des luftbereiften Kraftwagens gegenuber Kleinen Storungen), *Ingenieur Archiv*, Vol. 27, 1959, p. 88.
31. MITTERLEHNER, G., The Influence of Vehicle Suspension on Steering Stability (Der Einfluss der Kraftwagenfederung auf die Lenkstabilität), *Ingenieur Archiv*, Vol. 29, 1960, p. 100.
32. FIALA, E., The Vehicle Dynamics of Road Vehicles with Consideration of Steering Stability (Zur Fahrdynamik des Strassenfahrzeuges unter Berucksichtigung der Lenkstabilität), *Automobiltechnische Zeitschrift*, Vol. 62, No. 3, 1960, p. 71.
33. KAMM, W. I. E., Directional Stability of Motor Vehicles, Stevens Institute of Technology, Experimental Towing Tank Technical Memorandum No. 103, 1953.
34. SLIBAR, A., P. R. PASLAY, Behavior of Vehicles Subjected to Wind Gusts, *Ingenieur Archiv*, Vol. 28, 1959, p. 213.
35. BOHM, F., On the Directional Stability and Sensitivity to Cross-Wind of a Vehicle Traveling Straight Ahead (Ueber die Fahrtrichtungs stabilität und die Seitenwind emp findlichkeit des Kraftwagens bei Geradeaus-Fahrt), *Automobiltechnische Zeitschrift*, Vol. 63, Nos. 5 and 10, 1961, pp. 128 and 339.
36. JONES, G., The Skidding Behavior of Motor Vehicles, *Proc.* of the Automobile Div., Inst. of Mech. Eng., No. 1, 1962–63, p. 65.
37. ELLIS, J. R., Understeer and Oversteer, *Automobile Engineer*, Vol. 53, No. 5, 1963, p. 178.
38. ELLIS, J. R., Tyre Mechanics and Vehicle Handling, *Automotive Design Engineering*, September 1963, p. 61.
39. HODKIN, D., The Dynamics of Vehicle Handling, Soc. of Automotive Eng., Preprint No. 957A, International Automotive Congress, January 1965.
40. LISTER, R. D., Retention of Directional Control When Braking, Soc. of Automotive Eng. Preprint No. 963A, International Automotive Congress, January 1965.
41. TOTI, G., Aerodynamic Effects on Vehicles Moving in Stationary Air and Their

Influence on Stability and Control, Soc. of Automotive Eng., Preprint No. 948D, International Automotive Congress, January 1965.

42. BARTH, R., Effect of Unsymmetrical Wind Incidence on Aerodynamic Forces Acting on Vehicle Models and Similar Bodies, Soc. of Automotive Eng., Preprint No. 948E. International Automotive Congress, January 1965.
43. BITZL, F., The Influence of Cross-Wind Forces on Road Traffic (Die Einwirkung von Seitenwindkräften auf den Strassenverkehr), *Zeitschrift für Vehrkehrssicherheit*, Vol. 7, No. 3, 1961, p. 171. and Vol. 8, No. 2, 1962, p. 105.
44. BOURKE, P. J., Rear-, Front- or Four-Wheel Drive?, *Automobile Engineer*, May 1965, p. 180.
45. KAMM, W., H. HONTSCHICK, and K. NIEMAN, Direction Holding of Vehicles—Model Tests and Calculations, Part I (Modellversuche und Berechmungen uber die Richtungshaltung von Kraftfahrzengen), *Automobiltechnische Zeitschrift*, Volume 67, No. 5, May 1965, p. 146.
46. KOESSLER, R., Enhanced Travel Safety Through Directional Stability (Fahrsicherheit durch Fahrtrichtungshaltung) *Automobiltechnische Zeitschrift*, Volume 66, No. 4, April 1964, p. 101.
47. BURMAN, H. J., Track Stability of Pneumatic-Tired Vehicles and Their Behavior in Sidewinds (Kursstabilität Luftheraifter Fahrzenge und deren Verhalten bei Seitenwindstorung), *Automobiltechnische Zeitschrift*, Volume 66, No. 6, June 1964, p. 175.
48. DEININGER, W., Effect of Driving Power on the Road Stability of Vehicles (Einfluss der Autriebaskraft auf die Fahrstabilität von Kraftfahrgengen), *Automobiltechnische Zeitschrift*, Volume 67, No. 7, July 1954, p. 205.
49. HASSELGRUBER, H., Vehicle Cornering Behavior (Verhaltness eines Kraftfahrgenges bei Kurvenfahrt), *Automobiltechnische Zeitschrift*, Volume 67, No. 7, July 1965, p. 213.
50. BRAESS, H. H., Comments on Direction Holding of a Vehicle when Traveling Straight with Special Reference to Kingpin offset (Beitrag zur Fahrichtungshaltung des Kraftwagens bei Geradeausfahrt Unter besonderer Berucksichtigung des Lenkrollhalfmessers), *Automobiltechnische Zeitschrift*, Volume 67, No. 7, July 1965, p. 218.
51. HOFFMAN, E. R., Note on Vehicle Stability, *Australian Road Research*, March 1964, p. 15.
52. CLARK, D. C. and L. SEGEL. The Steering and Drawbar-Pull Performance of Pneumatic-Tired Vehicles, *Proc.* 1st Int. Conf. Soil Vehicle Systems, Torino, Italy, June 1961.
53. JINDRA, F., Tractor and Semi-Trailer Handling, *Automobile Engineer*, October 1963, p. 438.
54. HALES, F. D., The Lateral Stability of a Simplied Articulated Vehicle, *Proc.* Symposium on Control of Vehicles During Braking and Cornering. The Inst. of Mech. Eng., London, 1963.
55. JINDRA, F., Lateral Oscillations of Trailer Trains, *Ingenieur-Archiv*, Vol. 33, 1964, p. 190.
56. JINDRA, F., Tractor and Trailer Handling, *Automobile Engineer*, Vol. 55, No. 2, 1965, p. 60.
57. MARQUARD, E., V. W. GIRARDET, Vibration Dynamics of the Fast Road-Vehicle (Schwingungsdynamik der Schnellen Strassenfahrzeuges), Essen, 1952.
58. ROCARD, Y., Dynamic Instability, London: Crosby, Lockwood & Sons, Ltd., 1957.
59. STEEDS, W., Mechanics of Road Vehicles, London: Iliffe & Sons, Ltd.
60. JANTE, A., The Fundamentals of Vehicle Stability (Die Grundlagen der Fahrstabilität), Abhandlungen der Deutschen Akademie der Wissenschaften Zu Berlin, Jahrgang 1963, No. 4; Berlin: Akademie-Verlag, 1963.
61. KURZ, K., The Lateral Control Force of the Wheels of Motor Vehicles for the Case of Alternating Wheel Loads (Seitenfurhrungskraft des Kraftwagenrades bei wechselnder Radlast), *Automobiltechnische Zeitschrift*, Vol. 60, No. 5 (1958).

SOIL AND SNOW MEASUREMENTS AND INSTRUMENTATION

A Technique for Measuring Deformations Within a Sand Under Controlled Wheel Loads

C. W. BOYD

S. J. WINDISCH*

ABSTRACT

In the past there has been much speculation surrounding the question of how soils behave under vehicular loads. This "speculation" has stemmed almost entirely from the mensurational problems involved. A new approach to this problem is described in this paper whereby a Ciné Flash X-ray technique is used to study the translations, rotations, and velocities of small objects buried within the sand, as a function of depth and relative wheel position. Some basic considerations associated with the development of this technique are included, together with a section on the accuracy of information derived from the radiographic results.

The specific case of a rigid-wheel loading pattern on a loose dry sand has been studied, and some results from one test series are reported. The influence of applied torque on the translation patterns is described.

Large horizontal deformations are found within the soil, where the ratio of horizontal to vertical movements can exceed positive or negative equality. The actual ratio is shown to be strongly dependent on applied torque.

Whereas many studies, in the field of soil-vehicle mechanics, have concentrated on the measurement of purely surficial information, few have been concerned with the study of the actual soil behaviour beneath the wheel-soil interface. The assumption that (*a*) rolling resistance can be equated to the vertical component of work expended in the formation of a rut, (*b*) that dynamic side effects may be ignored at low vehicle speeds, and (*c*) that the soil response may be examined on the basis of static limit equilibrium analyses, are a few of the topics that have provided continuous speculation. The physical problems of measuring stress and non-linear deformations in the soil have been largely responsible for this speculation.

In this paper a Ciné Flash X-ray technique is described, providing a means for studying individual particle translations, rotations, and velocities that occur under a variety of scaled wheel loading conditions, within a loose sand. The emphasis will be on the experimental method and its accuracy, with one section on the results, and further examples will be described in subsequent publications. The results described in this paper do, however, produce sufficient evidence to question some of the assumptions used in contemporary theories.

---

*Graduate Research Assistants, Soil Mechanics Laboratory, McGill University, Montreal, Canada.

THE FLASH X-RAY EQUIPMENT AND TECHNIQUE: SOME BASIC CONSIDERATIONS

In studying transient phenomena, in this case particle translocation in a deforming soil mass, it is essential that time-dependent reactions be "frozen" at any prescribed instant. To achieve this with X-rays, and at the same time have adequate penetration, one of two techniques may be adopted:

(i) employing a high intensity X-ray or γ-ray source that is at peak output for the duration of the test, and in conjunction with a high speed Ciné-Camera;

(ii) employing a high intensity photon source that can be pulsed at a prescribed rate. In this case, a series of stationary radiographic plates may be used.

Sequential events are recorded in the first technique by a suitable choice of frame-rate, and in the second technique by two or more successive pulses and the conversion of a "space," or geometric, base to a time base. The main limitations associated with (i) above, are first that extensive radiation screening is required for operator safety, with the associated inconvenience of operation, and secondly the high operating costs of X-ray Ciné Photography.

Recent advances in cold-cathode Flash X-ray techniques, with high-intensity pulses of short duration, not only make method (ii), above, possible but also provide a relatively safe working environment for the operator. These considerations led to an exploratory study and the development of the Ciné Flash X-ray technique that is described in this paper.

Description of X-Ray Equipment

The McGill unit was manufactured by Field Emission Corporation, model number 730/233. This unit has an output voltage of 300 kV and output peak power of 420 megawatts. The duration of the pulse is 100 nanoseconds, and with a Voltronics model BAC-32-25 power supply it is possible to recharge and fire two pulses/second, using the same tube to a maximum of 10 pulses. It is this short pulse duration that enables the effective instantaneous recording of particle displacements at any given time. The focal spot diameter of the tube is 4 mm. Film and screening requirements are discussed below.

Linear Attenuation in Soils

The absorption of X-rays is a function of the thickness, the density, and—most important of all—the atomic nature of the test medium. In this case, the medium was a fine grained crushed quartz sand with a specific gravity of 2.70. The choice of sand was made for a variety of reasons. From a similitude standpoint the discrete nature of sand appeared to offer a better material to work with than clay. The presence of clay-water forces makes it difficult to scale because resultant interactions are not gravity dependent. For practical reasons sand also appeared to be a more attractive alternative for this initial testing. Unfortunately, from an analytical view point, there is no available method for determining absolute values of absorption (attenuation), by soil. However, in order to obtain a first approximation, a quasi-theoretical approach has been developed based on Victoreen's results (4), (5), and on some simplifying assumptions. This method, (1) relates the mass attenuation coefficients of a material whose linear attenuation characteristics are known, to the attenuation coefficients of the soil. In this instance, the absorption of aluminum, as a function of source-to-film distance was known and the thickness of soil required to produce the same reduction in intensity as a known thickness of aluminum may be expressed as a non-dimensional ratio, often referred to as a Radiographic Equivalence Factor, R.E.F. A range of R.E.F.'s were calculated for quartz sand as a function of source to film distance, S.F.D., void ratio and moisture content and are plotted as in Figure 1. It is a logical extension of this approach to calculate the R.E.F. as a function of S.F.D. for a three-phase soil–air–water system. It may be seen from Figure 1 that the maximum penetration is obtained for dry sand at low S.F.D. values. The validity of these R.E.F. values has been proved by subsequent experimentation. Since maximum penetration consistent with acceptable limits of object definition was considered a prerequisite, a minimum S.F.D. of 16.5 inches has been used for all testing. The geometry of the radiographic cassette, the sand-filled region, and the X-ray source thus dictate a maximum thickness of sand absorber—perpendicular to the plane of the cassette—of 7.25 inches of sand. This gives acceptable radiographic contrast and marker definition, and takes into account the attenuation caused by the lateral confining plates (see Figure 2).

Choice of Embedded Object

The fundamental requirement that the embedded object possess properties identical to those of soil—size, compressibility, etc.—is difficult, if not impossible, to satisfy. The net result is invariably a compromise that attempts to satisfy as many of the fundamental requirements without detracting from the usefulness of the test technique.

Using a flash X-ray technique, it is possible to eliminate the intrusive effects of leads—associated with electrical mensurational devices. It is not possible, however, to satisfy the condition of soil-object density equality. The reason for this is based on the attenuation requirements.

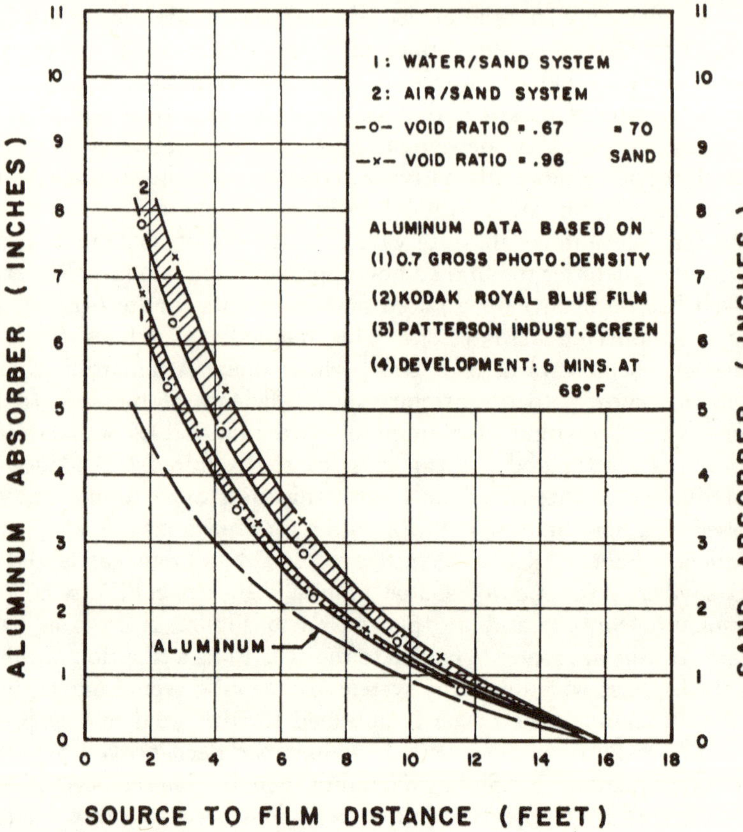

FIGURE 1. Calculated absorption ranges for sand as a function of s.f.d., void ratio, and moisture content (field emission model 730/233, 300kv.)

Within the effective wave-length range of the 300 kV unit, namely 0.0413 Å to ~0.25 Å, there is only a slight variation in the mass attenuation coefficients of all elements in the periodic table—and thus, implicitly, of all compounds. Thus it is only the influence of density, within this wave-length range, that creates the condition of differential linear absorption that is necessary for buried-object identification. It may be shown, again by using Victoreen's results, that under the above conditions a 1 mm. thickness of lead produces differential film densities greater than the minimum distinguishable visible limit of two per cent.

Since maximum penetration has been specified, the generally accepted maximum penumbral image width of 0.25 mm. has to be forgone, as this would dictate an S.F.D. greater than 8 feet.

However this does not present a major problem since the centre of a cross may be identified almost independently of the penumbral image. A

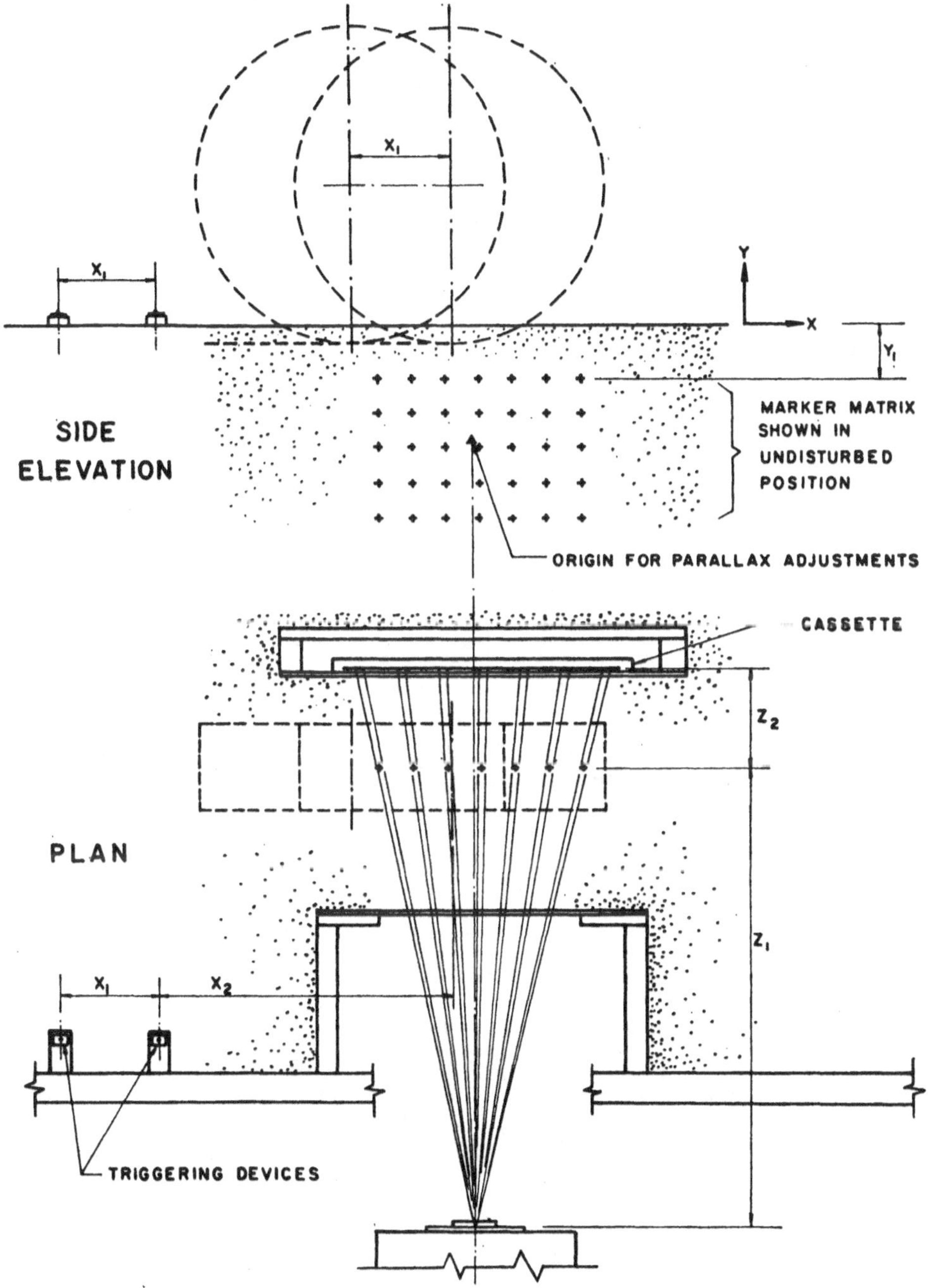

FIGURE 2. The source-marker-wheel-film geometry

second source of image unsharpness, caused by secondary radiation, or Bremsstrahlung, is less easy to control.

The outcome of these considerations was a three-dimensional cross with six orthogonal arms—each of length 3 mm.—giving a maximum dimension of 7 mm. Each arm is 1 mm. in diameter. This shape and size make it possible to measure both translations and rotations, with an accuracy that will be discussed below. These markers have been made from a composition of 94 per cent lead and, to reduce the malleability, 6 per cent antimony.

The possibility that differential inertia effects might occur, caused by an object to sand density ratio of 4.2, has been examined experimentally. Using a coloured sand tracer technique and gelatin fixation no measurable differential inertia effects were observed under controlled vertical loading at strain rates of 30 inches per second. This adds experimental support to the requirement that the buried objects move "with," rather than "through," the surrounding sand.

The Source–Object–Wheel–Film Geometry

The geometrical arrangement of the Flash X-ray pulser, the buried objects, and the film cassette is shown diagrammatically in Figure 2. Superimposed on this figure are the relative positions of the wheel for the two intermediate Ciné Flash X-ray pulses. In order to obtain the complete particle trajectory, and consequently the velocity and acceleration history, for each test, it is fundamental that the two consecutive pulses be synchronized with the appropriate wheel positions. This is achieved experimentally by adjusting distances x_1 and x_2. The time delay from triggering to firing is in the order of nanoseconds and thus may be ignored.

Film and Screening Requirements

The requirements of maximum penetration and good object definition are, in large measure, dependent on the film and screening techniques used. Experimentation with a variety of films and screens led to the choice of Kodak Royal Blue Medical Film and DuPont, calcium tungstate, fluorescent screens. General Electric Supermix Developer was used with development times corresponding to the maximum contrast listings.

For double exposures on the same radiograph a 0.002 inch Ilford Lead screen is placed in the cassette, on the source side of the X-ray film, to reduce the cumulative effects of Bremsstrahlung.

Sand Preparation and Marker Placing Technique

Reference to Figure 2 shows the position of the lead composition markers with reference to the geometry of the confining walls and the

X-ray equipment. The sand is poured in one-inch layers and the markers are superimposed on each layer with the aid of a template. Analyses of the accuracy of the marker placing technique show that the variability in the ordinate value for any row falls within ± 0.024 inches, at the 95 per cent confidence level. These limits include components of variation contributed by the data transfer and reduction methods described below.

THE SUPERIMPOSED WHEEL LOAD

Combinations of five basic rigid wheel parameters were used in order to study the response behaviour of the soil. These parameters are:

(i) Wheel size: a family of three rigid aluminum wheels were used in the first series of tests. A prototype 13.5 inches diameter by 3.75 inches wide, a 2/3 scale model, and a 9 in. diameter wheel with a width of 6.75 ins.

(ii) Linear Wheel Velocity: This can be controlled, with the present facility, up to linear speeds of 32 inches/second. All velocity scaling is based on the equivalence of the Froude number, with the implicit assumption that gravitational forces cannot realistically be scaled.

(iii) Slip rate: For driven wheels this is controlled by adjusting the input torque and may be used to vary the slip rate from low-negative to high-positive values.

(iv) Wheel load: This is controlled by adding additional weights to the wheel frame, directly above the axle.

(v) Wheel surface material: To date only two surfaces have been used, the first, a polished aluminum surface, the second, a natural hard rubber layer.

Since the behaviour of the soil in the prepared region, illustrated in Figure 2, must be considered to be influenced by the restriction in the third, z, dimension, a series of tests was conducted to establish the soil behaviour in the two dimensional case. In order to obtain radiographic continuity, the same effective thickness of sand was retained and a pair of additional glass plates was installed at a separation just greater than the width of the wheel under test. Analyses of the side wall effects, however, have not yet been made.

THE MARKER TRACING TECHNIQUE

The X-Ray Pulse Sequence

For each test a series of four X-ray pulses were fired. The sequence may be broken down as follows:

(i) The first pulse was fired before each test, in order to establish the initial position of each marker. After one exposure the first cassette was removed. At this stage the dynamometer carriage and wheel were stationary and at a distance of some 12 feet from the centre-line of the X-ray pulser.

(ii) A second loaded cassette was then placed in the holder and the wheel—under controlled loading conditions—was passed over the prepared region. The wheel carriage triggered two consecutive X-ray pulses that were synchronized to two relative wheel positions. The actual delay between consecutive pulses is dependent on the spacing, x_1—see Figure 2—and the velocity of the dynamometer carriage. This cassette was removed once the wheel was some six feet beyond the centre line of the X-ray pulser.

(iii) A fourth, and final, pulse was fired to establish the final position of all markers. For this purpose a third cassette was used.

Transfer of Radiographic Results to the Digital Computer

Because of the number of tests required for the over-all study, a semi-automatic data reduction system was developed. It was first necessary to combine all the information from the three radiographs on to a single frame of reference. Since a reference grid appeared on all radiographs, from the images formed by permanent identification markers on the inside of the Cassette holder, this involved the transfer of all data points to a single base. A "Matex" acetate sheet was used for this purpose, because of its translucent properties. Once these points were transferred, it was necessary to establish precisely the co-ordinates of each point. This was achieved by controlling the position of the carriage needle, on an XY plotter, by manually adjusting two heliopotentiometers. With the acetate sheet fixed in position on the plotter the voltages corresponding to the co-ordinates of each image location were then recorded. The Flow Chart (p. 191) shows the sequence followed in reducing these image locations to information that can be handled by the McGill 7040 computer.

The over-all accuracy of the data transfer and reduction is discussed below.

Preliminary Computations

The purpose of the X-ray technique is to determine the characteristic deformations occurring in the soil mass under a given loading condition as a function of position and time. It is thus necessary to relate the buried marker position to the locations of the wheels at any instant. By taking two consecutive pulses, the relative movements of each object in a row are recorded. However before displacement or velocity calculations are made it is necessary to reduce all of the image co-ordinates from the

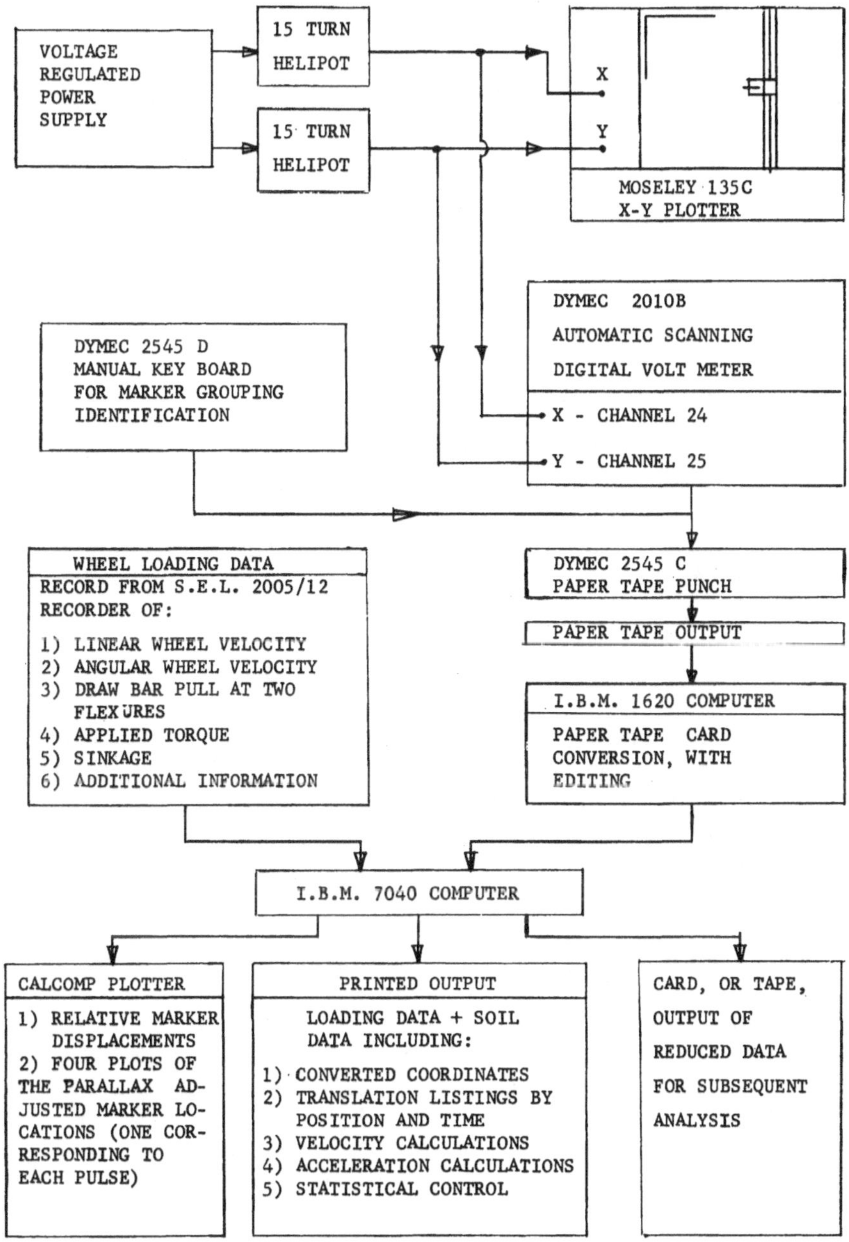
VOLTAGE REGULATED POWER SUPPLY
15 TURN HELIPOT
15 TURN HELIPOT
X
Y
MOSELEY 135C X-Y PLOTTER
DYMEC 2010B AUTOMATIC SCANNING DIGITAL VOLT METER
X - CHANNEL 24
Y - CHANNEL 25
DYMEC 2545 D MANUAL KEY BOARD FOR MARKER GROUPING IDENTIFICATION
DYMEC 2545 C PAPER TAPE PUNCH
PAPER TAPE OUTPUT
WHEEL LOADING DATA
RECORD FROM S.E.L. 2005/12 RECORDER OF:
1) LINEAR WHEEL VELOCITY
2) ANGULAR WHEEL VELOCITY
3) DRAW BAR PULL AT TWO FLEXURES
4) APPLIED TORQUE
5) SINKAGE
6) ADDITIONAL INFORMATION
I.B.M. 1620 COMPUTER
PAPER TAPE CARD CONVERSION, WITH EDITING
I.B.M. 7040 COMPUTER
CALCOMP PLOTTER
1) RELATIVE MARKER DISPLACEMENTS
2) FOUR PLOTS OF THE PARALLAX AD-JUSTED MARKER LO-CATIONS (ONE COR-RESPONDING TO EACH PULSE)
PRINTED OUTPUT
LOADING DATA + SOIL DATA INCLUDING:
1) CONVERTED COORDINATES
2) TRANSLATION LISTINGS BY POSITION AND TIME
3) VELOCITY CALCULATIONS
4) ACCELERATION CALCULATIONS
5) STATISTICAL CONTROL
CARD, OR TAPE, OUTPUT OF REDUCED DATA FOR SUBSEQUENT ANALYSIS

FLOW CHART

plane of the radiograph to the plane of the markers, to eliminate the effects of optical enlargement. Simple optics and a knowledge of the optical centre and the ratio $z_1(z_1 + z_2)$, (see Figure 2), are all that is required to make these adjustments. These adjusted pseudo-co-ordinates are then referred to the original sand surface level to obtain the final corrected co-ordinates. It is these final co-ordinates that are used as a basis for the subsequent finite difference calculations. Examples of these calculations will be described in the two sections that follow.

RESULTS: SELECTED FROM ONE TEST SERIES

1. The Test Series

The results described in the following paragraphs were all obtained using a rubber coated wheel with an effective diameter of 9.25 inches and width of 2.5 inches. The average linear velocity for the three tests was 7.66 inches/second and the wheel load was constant at 16 pounds.

The effect of altering the slip rate parameter on the behaviour of the soil, as recorded by the X-ray technique, was part of the overall experimental design. Two examples of these results, the first comparative, are described below.

2. Marker Translations as a Function of Applied Torque

The translations of markers, originally placed at a depth of 1.5 inches below the sand surface, are shown—for the three tests—in Figure 3. The slip rate corresponding to each curve is listed, together with other loading information, in the same figure.

For the particular conditions of the test it should be noted that in the plane under the centre line of the wheel:

(i) The paths followed by the markers and the soil were highly non-linear.

(ii) Employing the sign convention in Figure 3, the ratio of horizontal to vertical deformations along the paths either approached, or exceeded, negative equality in all of the tests. In the case of Test 16, with a slip rate of 35.5 per cent, this ratio also reached a positive net value of 1.2. The actual ratio was thus strongly dependent on applied torque.

(iii) The net vertical deformations of the compared layer may be assumed independent of the applied torque, yet it should be noted that the total wheel sinkage increased approximately linearly by 0.277 inches.

This additional net vertical deformation was shown quite clearly by the deformation patterns of markers placed below this layer, to a depth of 6.5 inches.

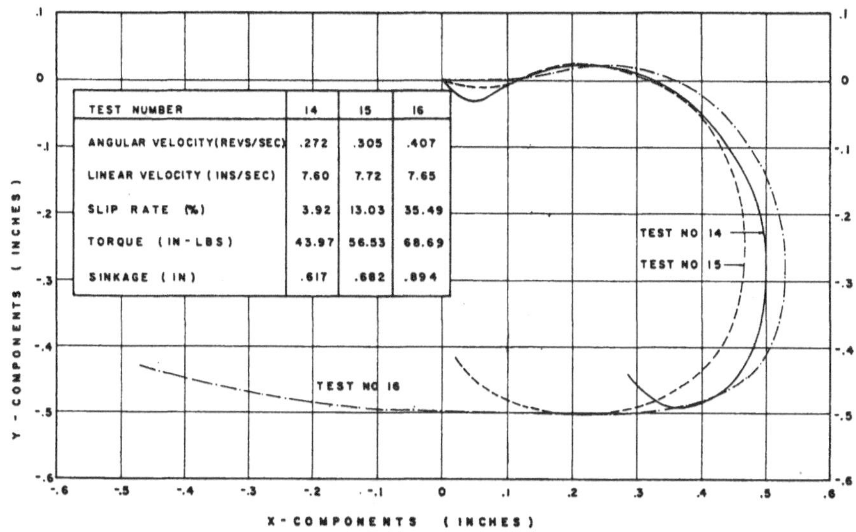

FIGURE 3. Translation patterns at 1.5 inch depth

3. Measurements of Velocity

The two planar components of velocity were calculated, using a finite difference technique, for a series of twelve positions along each displacement path. In order to compute velocity components at other locations along each path, the two commponents were plotted against a cumulative time base. One of the most satisfactory functional approximations for this purpose appears to be obtained by using a Chebyshev polynomial fit, since a minimum-maximum error technique is considered most desirable.

The velocity fields of both the x and y components have been plotted as Figure 4, from the results of test number 14. These component fields have been plotted separately, to avoid confusion, and are relative to the same instantaneous fixed wheel position. To obtain the $\Delta x/\Delta t$ velocity field, relative to the moving wheel reference, one simply subtracts 7.60 inches/second from each contour value. Thus the shape of this component field remains the same but all values become negative.

SOME ASPECTS OF THE RELIABILITY OF THE TECHNIQUE

Although the components of variability that can be ascribed to various sources have not yet been analyzed, one can obtain some measure of their algebraic sum. One measure of this is the reliability of the statistic

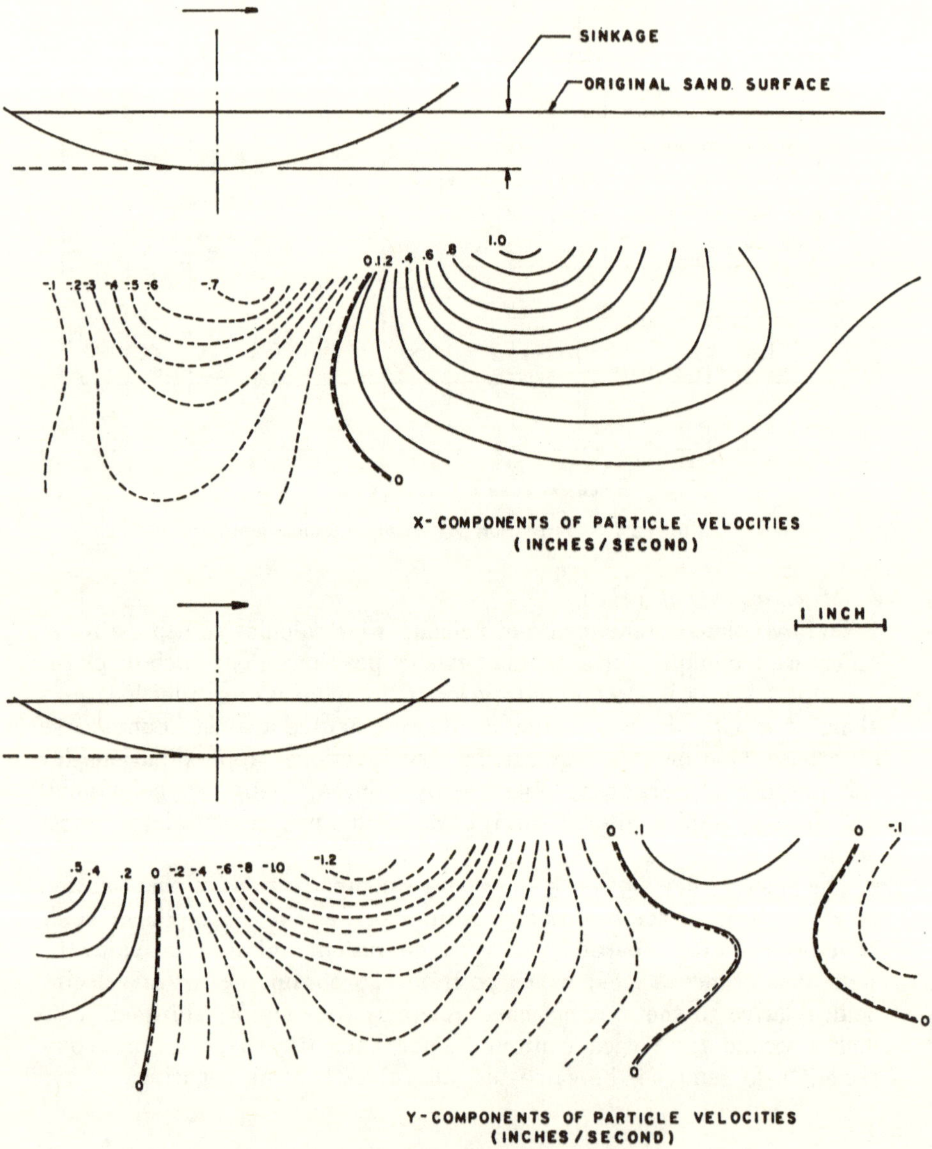

FIGURE 4. Particle velocity fields

of final marker position. Since markers placed in the same row have, theoretically, all been exposed to the same stressing cycle, they would be expected to "enjoy" the same net displacement, relative to their initial positions. The averaged relative co-ordinate values of the final marker positions have been entered by row, in Table I, for Test No. 14.

The variability of both relative co-ordinates has also been entered in this table by row, all values having been adjusted using the Bessel Correction term for small sample size ($n = 7$).

TABLE I

RELIABILITY OF FINAL MARKER POSITIONS, BY ROW AND BY CO-ORDINATE, FOR TEST NO. 14

1	2	3	4	5	6
		Averaged relative displacement (inches)		Reliability of $\bar{X}$ and $\bar{Y}$ measurements at 95% confidence level (inches)	
Row	Initial mean ordinate	$\bar{X}$	$\bar{Y}$	$\bar{X}\pm$	$\bar{Y}\pm$
1	−1.5	0.276	−.429	0.041	0.021
2	−2.5	0.175	−.249	0.025	0.016
3	−3.5	0.114	−.150	0.023	0.025
4	−4.5	0.057	−.084	0.008	0.020
5	−5.5	0.046	−.056	0.022	0.024
6	−6.5	0.027	−.025	0.016	0.011

Columns 5 and 6 indicate the reliability of the experimental facility. The inter-row variation is more a reflection of the variations in such items as the sand-pouring technique, the transfer of image centres from radiographs to acetate, and the conversion to D.V.M. values, than it is of the X-ray technique *per se*.

SUMMARY

Using the Ciné Flash X-ray technique, described in this paper, it has been possible to examine some aspects of the complex soil response to a moving rigid wheel. The technique makes it possible to study the movement of markers at any prescribed position under the centre line of a wheel, and thus to calculate translations, velocities, and accelerations in this plane.

A complete mathematical description of the soil response must not only account for such items as: (*a*) three-dimensional soil response, (*b*) large volume changes, (*c*) non-unique stress-strain relationships, (*d*) soil heterogeneity, to mention a few, but must also provide a solution.

It is thus evident that, for a practical solution to many problems in soil-vehicle studies, it will be necessary to rely on empirical relationships at the present time. The success that empirical solutions can attain is, however, contingent on the simplifying assumptions used, or—more directly—on an understanding of the physical situation.

The results, shown in Figure 3, may be used to illustrate one example of a case where a theory was developed without an adequate appreciation of the physical situation. The explicit assumption that rolling resistance, in sand, may be equated to the vertical component of work done in creating a rut is seen, from Figure 3, to neglect completely the horizontal component of both resistance and energy dissipation. This theory, and the spectacular discrepancy of the rolling resistance measurements, have been reported on by Wills *et al.* (6). The fact that horizontal soil deformations have been consistently ignored is even more surprising in the light of McKibben's experimental findings (2).

The results reported above demonstrate that, under the conditions of the test, one is concerned with deformations resulting from stress combinations beyond those corresponding to the static yield surface. For sands this results in a flow phenomenon. The choice or applicability of various yield functions, to both static limit equilibrium analyses and to flow phenomena, is beyond the scope of this paper. However, from an examination of the results of this limited test series, it would be difficult to delineate the active and passive Rankine zones associated with static limit equilibrium analyses. These analyses play an important role as a temporary expedient, however, more precise relationships can only be developed from mathematical models based on an understanding of the behaviour of the physical model. The development of the flash X-ray technique is considered as a step towards this objective.

ACKNOWLEDGMENTS

This study is being supported by the Defence Research Board of Canada, under a contractual arrangement with the Department of Defence Productions. The project officer is Mr. T. A. Harwood of D.R.B. The authors wish to thank Professor R. N. Young for his constructive suggestions and also to acknowledge the support of the De Havilland Aircraft Company of Canada. This is a contribution from the Soil Mechanics Laboratory of McGill University.

REFERENCES

1. Boyd, C. W., A Discussion of the Linear Attenuation Characteristics of Soils. Unpublished Report, McGill University, April 1965.

2. McKibben, E. G., *Some Kinematic and Dynamic Studies of Rigid Transport Wheels for Agricultural Equipment.* Research Bulletin 231, Iowa State College of Agriculture, Aimes, Iowa, March 1938.
3. Roscoe, K. H., J. R. F. Arthur, and R. G. James, The Determination of Strains in Soils by an X-ray Method. *Civil Engineering and Public Works Review,* July and August, 1963.
4. Victoreen, J. A., The Absorption of Incident Quanta by Atoms as Defined by the Mass Photoelectric Absorption Coefficient and the Mass Scattering Coefficient. *J.App. Phys.,* Vol. 19, 1948, pp. 855 *et seq.*
5. Victoreen, J. A., The Calculation of X-ray Mass Absorption Coefficients. *J. App. Phys.,* Vol. 20, 1949, pp. 1141 *et seq.*
6. Wills, B. M. O., F. M. Barret, and G. J. Shaw, An Investigation of Rolling Resistance Theories for Towed Rigid Wheels. *J. of Terra Mechanics,* Vol. 2, No. 1. 1965.

Plate Sinkage Tests in Sands and Clays

L. J. GOODMAN*

E. HEGEDUS†

R. A. LISTON†

ABSTRACT

Experimental and analytical investigations of the vehicle sinkage problem were conducted by means of plate sinkage studies on sand and clay soils. These studies were concerned with the influence of rates of plate penetration and size on the pressure-sinkage relations of loose sands and remolded clays, with particular attention given to scaling considerations.

Experimental results are analysed to predict pressure-sinkage relations for large-scale footings based on laboratory models. Sinkage rates as high as 1,750 inches per minute are included. Control measures involved constant soil density and water content for each soil type used and an adequate ratio of soil bin to footing size to eliminate interference from stresses induced by the sides or bottom of the soil bin.

Physical characteristics of the soil-footing system relative to pressure-sinkage relations are discussed. Dimensional analysis is applied to determine the relationship between the model and the prototype, with the results compared with model test results.

The resistance of the sandy soils and remolded clay soils to footing sinkage generally increased with increasing rates of sinkage. The resulting relationship from the dimensional analysis is only partially encouraging since it implies that it is only possible to scale plate sinkage with predominantly frictional soils.

---

*Department of Civil Engineering, University of Syracuse, Syracuse, N.Y.

†Land Locomotion Laboratory, Detroit Arsenal, Detroit, Michigan.

The Load Sinkage Equation in Theory and Practice

B. M. D. WILLS*

ABSTRACT

Plate tests carried out on sand, loam, and clay with a family of similar rectangular footings ranging in size from 24 sq. in. to 600 sq. in. show that current methods of analyzing and extrapolating load sinkage data are in many cases highly inaccurate and incorrect.

A pressure sinkage equation can be formulated in one of three ways: by a similarity or model technique, by a fundamental approach based on plasticity theory, or empirically. Load sinkage curves for large footings can be predicted accurately by similarity from small model footing tests but only for purely cohesive or purely frictional soils. For a composite soil only a general functional equation can be suggested.

A fundamental approach derived from civil engineering bearing capacity theories yields a similar general question. Unfortunately this is only strictly applicable to incompressible soils and requires much further development before it can be applied successfully to field soils.

From the two similar general equations derived by the above methods an empirical pressure sinkage equation is suggested. This is based on soil values obtained from model footing tests and proves to be extremely accurate for the range of footings and soil types used.

NOTATION

- F Vertical load (lb.)
- P_p'' Resultant passive earth force (lb.)
- l Characteristic length (in.)
- b Footing width (in.)
- z Sinkage (in.)
- s Footing perimeter (in.)
- h Soil depth (in.)
- p Pressure (lb. in.$^{-2}$)
- p_0 Equivalent free surface stress surface bearing capacity (lb. in.$^{-2}$)
- q Surcharge (lb. in.$^{-2}$)
- c Cohesion (lb. in.$^{-2}$)
- k' Pressure sinkage modulus (lb. in.$^{-2}$)
- γ Bulk density (lb. in.$^{-3}$)
- k Pressure sinkage modulus (lb. in.$^{-2-n}$)
- k_c Pressure sinkage modulus (lb. in.$^{-1-n}$)
- k_ϕ Pressure sinkage modulus (lb. in.$^{-2-n}$)
- v Velocity (in. sec.$^{-1}$)
- g Acceleration due to gravity (in. sec.$^{-2}$)
- ϕ Angle of internal friction (degrees)
- $\theta, \eta, \beta, \omega$ Angles (degrees)
- n Sinkage exponent; sinkage to width ratio exponent (dimensionless)
- k_c' Pressure sinkage modulus (dimensionless)

*This work was carried out at the Land Locomotion Laboratory during leave of absence from the Agricultural Engineering Department of the University of Newcastle upon Tyne.

k_ϕ' Pressure sinkage modulus (dimensionless)
N_c General bearing capacity factor due to cohesion (dimensionless)
N_q General bearing capacity factor due to surcharge (dimensionless)
N_γ General bearing capacity factor due to soil weight (dimensionless)
λ Linear scale ratio (dimensionless)
m Mobilization of shear stress in equivalent free surface (dimensionless)
N_{cq} Composite general bearing capacity factor (dimensionless)
$N_{\gamma q}$ Composite general bearing capacity factor (dimensionless)

TO THE OUTSIDE WORLD it would seem that all serious workers in the field of soil vehicle mechanics must pass through the initiatory rite of pressing footings into the ground before progressing to the more profound mysteries of the art. It would do well to study the genesis of this cult, attempt to understand its underlying philosophy and significance and evaluate its usefulness.

One of the most important and fundamental assumptions current in soil vehicle mechanics is that the rolling resistance of a vehicle is largely a result of the vertical work done on the soil which is manifest in the sinkage of the vehicle and the formation of ruts. Vertical work done on the soil, however, can be very easily measured by means of a plate test which on the basis of the above assumption now becomes a convenient procedure for evaluating rolling resistance.

The use of the plate test in this way provides a reasonably accurate model for rolling resistance studies and although the theoretical justification for this procedure, in many cases, is somewhat tenuous it remains the one and only usable method for evaluating rolling resistance and is likely to be so for some time to come. This work is only concerned, therefore, with the problem of obtaining accurate pressure sinkage data and presenting it in such a form as to be reasily usable in rolling resistance theories. The problem may be tackled in one of three ways: empirically, by means of similarity or model analysis, or fundamentally.

To be able to predict rolling resistance empirically for large wheels or tracks from small plate tests it is necessary to extract from the experimental load sinkage curves certain soil constants which are independent of footing shape and size to be used in a general pressure sinkage equation (Figure 1). If this could not be done accurately one would be faced with the highly impractical and ludicrous situation of using, for example, an "M-60" tank to predict the sinkage of the "M-60" tank. The measurement of the necessary soil constants and the extrapolation necessary constitutes a major problem in the use of the plate test in this way.

The general pressure sinkage equation which forms the basis of the present day empirical approach goes back to Bernstein, Letoshnev, and others and is of the form (1), (2):

$$p = kz^n. \qquad (1)$$

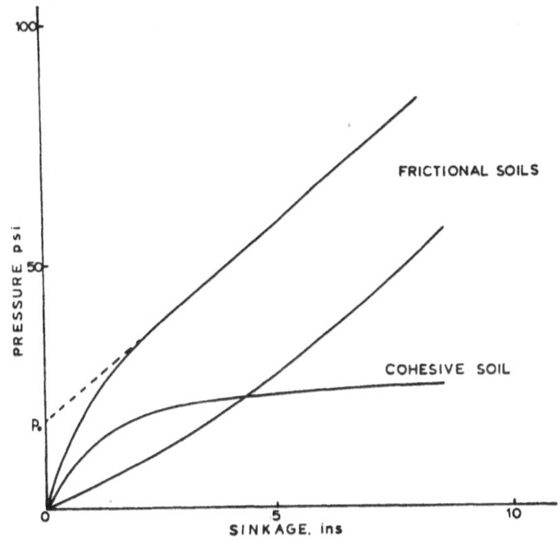

FIGURE 1. Typical pressure sinkage curves

Values for the constant and exponent were determined by plotting the experimental pressure sinkage results logarithmically. It was generally found that although the exponent appeared to be a soil value, the constant varied considerably with the size and shape of the footing. In an attempt to improve this situation Bekker suggested that the constant k was in fact a function of two soil value constants, $k_c z$ and k_ϕ, and footing width b, where

$$p = ([k_c/b] + k_\phi)z^n. \tag{2}$$

These constants together with the exponent can be evaluated as before by using two similar footings of different width instead of one footing (3). Despite some improvement in the accuracy of this new equation many workers discovered that k_c and k_ϕ were not in fact true soil values but that they varied appreciably even for a limited size range of footings. The value of the exponent n was also shown to vary to a lesser degree but significantly nevertheless.

According to the above equations pressure sinkage data should reduce to a linear function when plotted on logarithmic axes. This is rarely the case in practice as experimental values invariably exhibit a markedly non-linear relationship over the lower sinkage range (Figure 2). This often results in great difficulty in measuring the required constants as it becomes necessary to determine the average or best fitting values over the required pressure or sinkage range. It was suggested that this non-linearity was due to some form of elastic entry effect prior to soil failure and that in the case of those soils which showed some degree of bearing capacity the basic equation could be improved upon in the following way (4):

$$p - p_0 = ([k_c/b] + k_\phi)z^n. \tag{3}$$

The value of p_0, approximately the surface bearing capacity of the soil, is determined by extrapolating, the experimental curve back to the vertical axis (Figure 1). When plotted logarithmically in this equation experimental values usually become more linear over the normal sinkage range and it now becomes possible to measure the soil constants much more accurately and readily (Figure 2). This convenience, however, is at

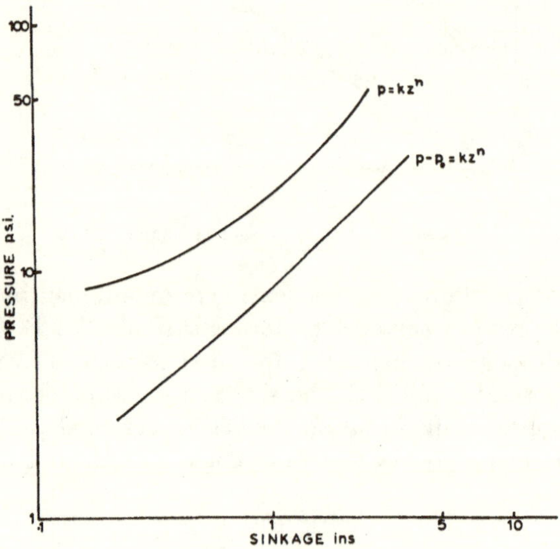

FIGURE 2. Logarithmic pressure sinkage values

the expense of the accuracy of the original equation. For small footings this discrepancy is negligible and is confined to fractional sinkages, but becomes increasingly important with the size of the footing. When predicting pressure sinkage data for large contact areas from small plate tests the value of p_0 can be obtained by means of civil engineering surface bearing capacity theory. If the experimental plate test values for p_0 are equated to $(cN_c + [\gamma^b/2]N_\gamma)$ the two constants can be evaluated and the expression used to predict p_0 values for larger footings.

Equation 3 is necessarily limited as it cannot be used for loose soils or clays. For clays however an exponential equation of the form

$$p = p_u(1 - e^{-k[z/b]}) \tag{4}$$

has been suggested where p_u is the ultimate bearing capacity equal to

8.28 c and k is an empirical constant (5). Although such an equation can be used to describe the pressure sinkage curve for a clay soil it is restrictive in that it lacks generality and it presents great difficulty in extracting the necessary constants. The use of a series to describe the pressure sinkage relationship has also been suggested (6). Such a method would involve so much complexity however as to be impractical for everyday use and is of purely academic interest.

The most fundamental and most recent change in the form of the empirical pressure sinkage equation has been suggested by Reece (6). He pointed out that the popular forms of the equation (1 and 2) were dimensionally unattractive as the dimensions of the sinkage modulus were a function of the value of the exponent. Influenced by work of a more fundamental nature in civil engineering he proposed a new equation:

$$p = \left(ck_c' + \frac{\gamma^b}{2} k_\phi'\right) z^n/b. \tag{5}$$

This equation is at once dimensionally acceptable and involves the width of the footing in a more meaningful way.

Apart from the problem of obtaining an accurate empirical equation there remains the difficulty of extracting the values for the constants from the experimental data. Statistical methods involving a breakdown of the experimental curves, numerous calculations and the almost certain use of calculating machines are quite impractical in this context, and the alternative method of describing the best fitting straight line has until recently been highly unsatisfactory. The minimum error method proposed by Reece, however, has established the curve fitting technique on a scientific basis and allows for the accurate and objective measurement of the necessary constants over the required pressure or sinkage range (6).

It is important to note at this point that although the empirical method implicitly assumes that the data from a small plate test can be used for large scale ground drive equipment no attempt has ever been made to verify this procedure experimentally by a series of plate tests covering a large size range.

When faced with a highly complicated problem involving many variables as is the case with most soil mechanics analyses it at first appears that the only methods of solution are either empirical or fundamental. An empirical solution will give reasonably accurate results quickly but without any insight into or understanding of the problem involved. A more fundamental approach promises a general solution but will most certainly be a formidable task of long duration and even then may require the use of some empiricism.

A third alternative exists, however, in the form of similarity or model analysis which combines many of the benefits of the other two methods but few of the drawbacks. Although such a method will not provide a general solution it does give some insight into the nature of a problem by reducing the number of variables involved and will provide a specific answer in certain cases. Similarity or model analysis techniques rest on a very substantial theoretical foundation and have been well tried and proven over many years. It is at first surprising that although the plate test is used as a model for vehicle sinkage the solution of the problem has not been carried out along the lines of classical model analysis. No doubt this was partly a case of expediency when a general empirical solution was required in the shortest possible time and when the limitations inherent in a similarity solution or the effort and time involved in a general solution could not be considered. This situation has long since passed and it should now be possible to achieve a more accurate understanding of the pressure sinkage problem by means of orthodox model analysis.

As far back as 1948, Taylor published a model law to represent loading tests on a sand surface in a dimensionally correct form (7). He showed that for dynamic similarity:

$$(z/l) = f(p/\gamma l) \tag{6}$$

where l is a characteristic length of the footing. It was not until the work of Nuttall and McGowan however that these techniques were used extensively in the field of soil vehicle mechanics and then solely on problems concerned with the performance of wheels and the use of scale models (8), (9).

Nuttall contributed a valuable chapter on scale model testing to Bekker's book, *Theory of Land Locomotion* (10). In it he traced the history of scale model testing with particular reference to vehicle and soil vehicle mechanics. He was able to show how dimensional theory could be applied in practice and in analysis. In the case of bearing capacity he suggested that the following dimensionless groups were applicable:

$$\phi,\ cl^2/F,\ \gamma l^3/F,\ z/l. \tag{7}$$

The characteristic dimension, l is taken as half the footing width.

In examining the feasibility of using model analysis for research into earthmoving equipment, Emori and Schuring considered the following functional equation as necessary to describe the soil working process in general (11):

$$F/\gamma l^3 = f(\phi,\ c/\gamma l,\ v^2/gl) \tag{8}$$

where $\gamma = \rho g$. When applied to the plate test where the velocity effect is negligible the last group can be neglected and the equation becomes:

$$F/\gamma b^3 = f(\phi, c/\gamma b) \tag{9}$$

where the characteristic dimensions is now b, the width of the footing. In order to compare various plate sizes it is essential that there is geometric similarity and that the ratio z/b is consistent. For a pure sand when the cohesive term is negligible:

$$F_1/F_2 = b_1^3/b_2^3 \tag{10}$$

for geometrically similar plates and for the same values of z/b on the same soil. For the same conditions on a clay soil when the friction term can be neglected the following relationship results:

$$F \propto b^2. \tag{11}$$

A limited number of experimental plate tests on both sand and clay showed these relationships to be correct (12). In the case of a composite soil where both friction and cohesion are present it was tentatively suggested that Equation 9 might be written in the form:

$$F/\gamma b^3 = (c/\gamma b)f_1(\phi) + f_2(\phi) \tag{12}$$

where f_1 and f_2 are functions of ϕ.

Liston and Hegedus carried out a general dimensional analysis of the load sinkage relationship for both soil and snow and supported their findings with a series of experimental plate tests (13). Together with the usual variables they considered both the effective soil depth and the plate perimeter as well as a characteristic plate length. The result of their analysis for a composite soil was of the form:

$$f(z/s, p/\gamma b, c/p, p/\gamma h, c/\gamma s, \phi) = 0. \tag{13}$$

The fourth term can be neglected for deep uniform soil. The data from experimental plate tests carried out on sand with families of rectangular and circular plates was shown to collapse on to a single curve in each case when plotted in the form (z/s) against $(\gamma b/p)$. Similar tests carried out on a Detroit clay, however, were unsatisfactory in this respect. This was thought to be due in part to the non-uniformity of the clay.

As a result of this analysis the following pressure sinkage equation was proposed for frictional soils:

$$p = (\gamma b/A^n B s^n) Z^n. \tag{14}$$

A and B are the co-ordinates of any point on the line obtained by plotting experimental values of (z/s) against $(\gamma b/p)$ logarithmically, and n is the inverse of the slope of that line. It can be seen that for a particular set of conditions this equation is equivalent to a constant multiplied by the sinkage factor and is identical to Equation 1. It was not thought to be possible to treat soils with appreciable cohesion in this manner. The prediction of the pressure sinkage curves for a 4 in. × 18 in. footing and a 6 in. diameter footing on sand from the experimental results of smaller rectangular and circular test footings gave very much the same results as those obtained by using Equation 2. On a basis of work done, however, the dimensional method gave values much closer to the experimental ones.

It can be seen then that at best the dimensional or similarity method can only be effective directly for the two extreme soil types. Indirectly, however, it can be most useful in influencing empirical solutions and in unravelling the complexities of fundamental solutions.

Mathematicians have so far been unable to provide a rigorous general solution for the soil failure problem. For the few isolated and specific cases for which a solution has been obtained it has usually been so complicated and unwieldy as to be impracticable. In order to achieve a useful and practical fundamental solution therefore it has been necessary to make certain simplifying non-rigorous assumptions. In the case of bearing capacity and retaining wall theory, for example, an acceptable plastic failure solution can be obtained by regarding the effect of cohesion and surcharge as distinct from the effect of soil weight. Terzaghi analysed surface bearing capacity for a rigid material in this way and expressed it in the form

$$p = cN_c + qN_q + (\gamma b/2)N_\gamma \tag{15}$$

where N_c, N_q, and N_γ are bearing capacity factors solely dependent on the value of ϕ (14). Terzaghi also attempted to predict the bearing capacity of shallow and deep foundations. He assumed that the effect of the over-burden need only be added to the surface bearing capacity value and that the failure planes for the surface case would remain unaltered with depth (Figure 3). In the case of a shallow foundation he considered only the weight of the over-burden as an equivalent surcharge. For a deep foundation he assumed the failure surface shown in Figure 3 and considered both the effect of shearing stresses acting on the over-burden and the weight of the over-burden itself. These theories proved to be unsatisfactory in practice and were shown to be quite inadequate in describing the shape and extent of the failure surface and in predicting the bearing load at any particular depth.

To avoid the limitations imposed by Terzaghi's theory and to treat the effect of the over-burden on shallow and deep foundations in a more realistic manner, Meyerhof proposed a theory in which the extent and

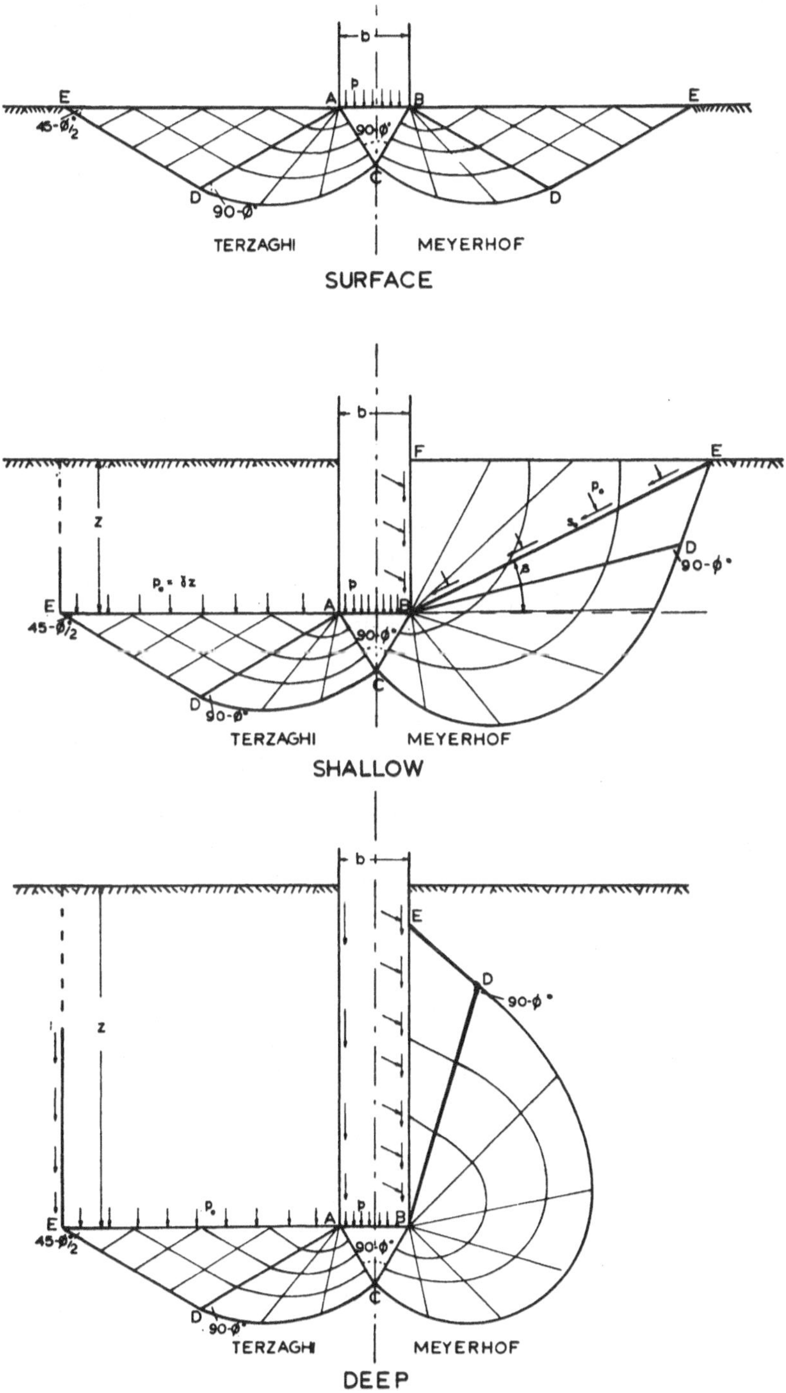

FIGURE 3. Plastic failure beneath foundations

shape of the failure zones are also a function of foundation depth (15) (Figure 3).

In the case of a surface foundation the failure zones remain the same as those suggested by Terzaghi. As the foundation sinks below the surface, however, the over-burden forms part of a composite failure zone BDEF resulting in a change in the shape and size of the over-all failure zone (Figure 3). This composite zone can be divided into two parts, a zone of radial shear BDE and a plane shear zone BEF. The boundary between these two zones, BE, can be considered as an equivalent free surface. The radial and tangential equivalent free surface stresses p_0 and s_0 can be obtained from a consideration of the equilibrium of the wedge BEF. In the case of a deep foundation the equivalent free surface becomes the vertical side wall of the foundation itself (Figure 3). If one now considers these stresses in place of a surcharge the bearing capacity can be written in the familiar form given by Equation 15. In this case, however, N_c, N_q, and N_γ are general bearing capacity factors dependent not only on ϕ but on depth and footing width as well.

Due to the theoretical difficulty of considering the weight of a material in a plastic equilibrium solution it is necessary to analyse the bearing capacity problem in two distinct parts, as did Terzaghi. The first part can be evaluated by considering a weightless material with cohesion and equivalent free surface loading in the failure zone shown in Fig. 4. The failure zone is bounded by a logarithmic spiral with its centre at A and a plane surface DE, determined by the inclination of the equivalent free surface and the degree of mobilisation of shearing stresses on this surface. A consideration of the equilibrium of this failure zone results in the following expression.

$$p' = c\left[\cot\phi\left\{\frac{(1+\sin\phi)e^{2\theta\tan\phi}-1}{1-\sin\phi\sin(2\eta+\phi)}\right\}\right] + p_0\left[\frac{(1+\sin\phi)e^{2\theta\tan\phi}}{1-\sin\phi\sin(2\eta+\phi)}\right] \quad (16)$$

or

$$p' = cN_c + p_0 N_q \quad (17)$$

The second part or component of the bearing capacity can be obtained by considering the failure zone shown in Figure 4 for a soil with weight but with no cohesion or surface stresses. The failure surface is determined by a semi-graphical method, first suggested by Ohde, by moving the centre of the logarithmic spiral, O, so as to give a minimum value, p_p'', for the passive earth pressure on BC. The equilibrium of the soil forces on the failure zone BCDE about the centre O results in the partial bearing capacity p'' given by:

$$p'' = \gamma b/2\left[[4P_p''\sin[45+(\phi/2)]/\gamma b^2] - \tfrac{1}{2}\tan[45+\phi/2]\right], \quad (18)$$

or

$$p'' = (\gamma b/2)N_\gamma. \quad (19)$$

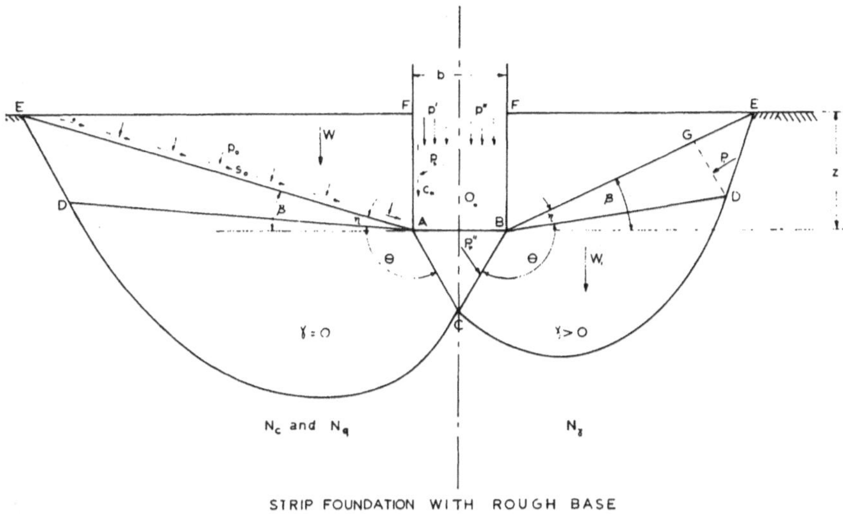

FIGURE 4. Determination of Meyerhof general bearing capacity factors

The total bearing capacity is now given by the algebraic sum of the two components:

$$p = p' + p''. \tag{20}$$

By evaluating the bearing capacity factors in this way Meyerhof was able to express them as functions of ϕ and the inclination of the equivalent free surface β. He did this in each case for the extremes of zero and full mobilization of the shear stress on the equivalent free surface (Figure 5).

Meyerhof carried out a number of field tests with piles and was able to show reasonable agreement between theoretical and practical results for both sand and clay. He finally suggested that bearing capacity could be more conveniently expressed in the form:

$$p = cN_{cq} + (\gamma b/2)N_{\gamma q} \tag{21}$$

rather than by the usual three-part equation. The composite factors N_{cq} and $N_{\gamma q}$ are derived by considering the free surface stresses as the sum of cohesive and weight components. N_{cq} is thus dependent on both N_c and N_q and represents the effect of cohesion. $N_{\gamma q}$ is likewise dependent on N_γ and N_q and is the weight term. Meyerhof's theory has been dealt with in some detail as it represents perhaps the most successful attempt in civil engineering to relate bearing capacity to depth. It should thus provide a basis and a starting point for the eventual fundamental solution of the plate sinkage problem.

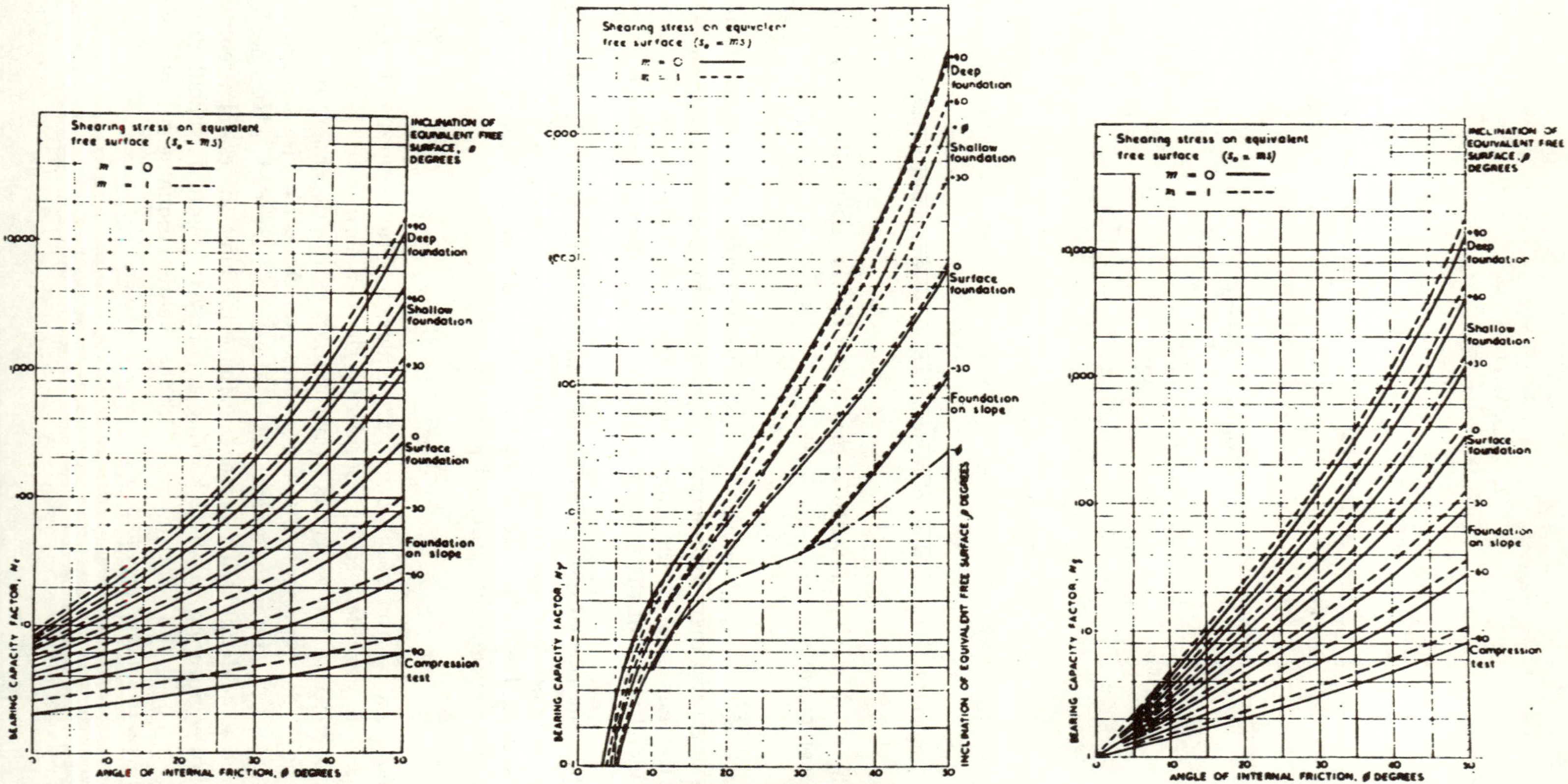

General bearing capacity factor N_c for strip foundation

A

General bearing capacity factor N_γ for strip foundation

B

General bearing capacity factor N_q for strip foundation

C

FIGURE 5. Meyerhof general bearing capacity factors (after Meyerhof)

In the past the many workers involved in the plate sinkage problem have each gone about their separate tasks by one or other of the three methods already discussed. This has led to three schools of thought completely independent of each other except for the occasional contact when the superiority of one method over another is demonstrated. Since the work of Bekker, advances have been made by individual workers but there has been no generally accepted advance in the state of the art as a whole. New work has been attacked or ignored by opposing schools of thought as a matter of principle, and in general, workers have been so busy repeating for themselves what has already been done many times before by others that it becomes quite evident why there has been so little progress. This situation of mutual suspicion and distrust has resulted in an enormous amount of repetitive and useless work. Unless there is a change in attitudes and acceptance of facts the practice of pressing footings into the ground will simply become an obscure but somehow important ritual from the past.

The purpose of this work is basically to provide an accurate and practical solution to the pressure sinkage problem. It can only be completely solved in the long term by means of a fundamental solution and it is hoped that an adaptation of Meyerhof's bearing capacity theory will provide the first step in this direction. The task is so formidable that one can only hope to make a beginning and there is need, therefore, for a short term solution. It will be shown how both fundamental and similarity theories can assist in the formulation of an accurate empirical equation which will provide an immediate solution to the problem.

No matter which method of solution is used the vertical work done theory for rolling resistance requires the prediction of pressure sinkage data for large contact areas either from soil values or model plate tests. A major criticism of all past pressure sinkage theories in soil vehicle mechanics is that not one of them has been verified by carrying out large plate tests comparable in shape and size to the vehicle ground drive equipment for which they are required. An important part of this work therefore is the series of experimental tests carried out with a family of similar rectangular footings ranging from 24–600 sq. in. in area.

APPARATUS AND TEST FACILITIES

1. *Penetrometer*

The penetrometer was designed to operate across a 12 ft. wide soil tank with a maximum loading of approximately 26,000 lb. It consists essentially of a very stiff frame to support the necessary ballast weights and the central penetrometer shaft which slides in bearings housed in an outer casing (Figure 6). The penetrometer is powered by an hydraulic

FIGURE 6. (U.S. Army photograph)

unit mounted on a soil tank carriage. This unit is connected to the two loading cylinders through flexible hose, self sealing couplings, and a manifold on the penetrometer frame (Figure 7). The two equally loaded hydraulic cylinders, which are anchored to the main frame, transmit the working load to the penetrometer by means of extension arms and a cross head on top of the shaft. The load applied to the soil is measured by means of a strain gauged load cell mounted in one of the loading arms.

A calibration of the penetrometer was carried out to ensure that the total load on a footing could be obtained accurately by doubling the load cell reading. This was done by comparing the load obtained by doubling up the load cell reading with corresponding loads obtained from a calibrated loadometer placed on the ground beneath the penetrometer shaft. The difference between the two calibrations was barely measurable over the full loading range.

The vertical movement of the penetrometer shaft was measured by means of a potentiometer and constant torque spring motor mounted on one of the loading arms (Figure 6).

Load and sinkage were recorded simultaneously on an X-Y plotter and all tests were carried out at a constant vertical speed of 0.075 ft/sec.

FIGURE 7. (U.S. Army photograph)

The reaction necessary to load the soil was provided by 18,000 lb. in ballast weights and by clamping the penetrometer frame to the soil tank (Figure 6). The penetrometer with the ballast weights in place could be moved quite easily along the soil tank by means of a 10 ton overhead travelling crane.

2. Pressure Plates

A family of six footings with a constant aspect ratio of 6:1 was used throughout the tests. The widths of the footings were 2 in., 3 in., 4 in., 6 in., 8 in., and 10 in. The smaller footings were used with a $1\frac{1}{2}$ in. diameter extension fitted to the bottom of the penetrometer shaft (Figure 8).

EXPERIMENTAL SOIL TYPES AND TEST PROCEDURE

1. Sand

A dry medium-fine Ohio sand at a density of 0.064 lb/in.^{3} was used in a soil tank 100 ft. long by 12 ft. wide by 5 ft. deep. The depth of the sand in the tank was 46 ins.

FIGURE 8. (U.S. Army photograph)

The angle of repose of the sand was found to be 31°. The shear strength of the sand as measured with the direct shear box at the above density was found to be purely frictional with $\phi = 33°$. The bevameter gave $\phi = 28\frac{1}{2}°$ after 180° rotation with a $6\frac{1}{4}$ ins. mean diameter annulus.

The sand was processed along the full length of the tank to a depth of 36 ins. by means of a large tine cut from 1 in. plate (Figure 9). The tine was mounted consecutively in a series of slots which were cut into one of the soil tank carriages and which extended across the full width of the tank on 8 in. centres. The chain driven carriage was then used to pull the tine through the sand.

Prior to a series of tests the sand was thoroughly worked to a uniform density by multiple passes of the tine and then levelled. It was then possible to carry out a complete series of tests with several replications for each of the six plates along the one length of the tank.

2. Loam

The loam soil was used at a density of 0.048 lb/cu. in. and a moisture content of 11.4 per cent on a dry basis on a soil tank 100 ft. long by 12 ft. wide by 5 ft. deep. The depth of the loam in the tank was 46 in.

The loam was found to have a cohesion of 0.25 p.s.i. and an angle of friction of 32.5°, as measured on the linear shear box for the conditions of moisture content and density given above.

The soil was processed in the first place by thoroughly breaking it up with a large tine to a depth of 36 in. in the same manner as for the sand. It was then thoroughly mixed to a depth of approximately 27 in. with the vertical mixer shown in Figure 9. The mixer could be mounted

FIGURE 9. (U.S. Army photograph)

in four positions on the soil tank carriage so as to cover the full width of the tank. Several slow passes were made in each position along the full length of the tank.

Prior to a series of tests the loam was prepared as described above but with a final pass of the mixer down the centreline of the tank. It was possible to carry out a complete series of tests with replications for each of the six plates along the one length of the tank.

3. Clay

A grey modelling clay was used to ensure homogeneity of the sample and complete freedom from foreign matter. Because of expense only a small sample filling a tank 6 ft. by 3 ft. by 28 in. deep could be used. Unfortunately this restricted the plate tests to the three smallest footings.

The clay was used at a moisture content of 31 per cent on a dry basis, and a density of 0.069 lb/cu. in. The shear strength of the clay as measured with the direct shear box was found to be almost purely cohesive with a cohesion of 1.1 p.s.i. and a friction angle of 0.2°. The triaxial test gave the same result.

The clay was processed by filling up the hole made by a footing with small lumps of clay individually kneaded by hand. The whole surface of the clay was then kneaded with bare feet after which it was levelled off with a pneumatic hammer. The surface was then accurately levelled by drawing a taut wire through the clay just below the surface after the fashion of a cheese cutter. It is interesting to note that the process of kneading a cohesive material with bare feet has now passed into the English language, as "todging."

Only one centrally placed footing test, was carried out between each preparation of the clay sample. The bevameter shown in Figure 10 was used to carry out the footing tests on the clay.

EXPERIMENTAL RESULTS

Experimental plate tests for the sand, loam and clay are shown in Figures 11, 12, and 13. Each curve represents the results of several tests and it should be noted that for all the plate tests on the three soils used exceptionally good correlation was achieved between replicated results.

THEORETICAL ANALYSIS AND DISCUSSION OF RESULTS

The experimental results for the three soils have been analysed by each of the three methods of solution already discussed. The order in which this is done is such that the solution or parial solution suggested by one method may be used to support or develop solutions by other methods.

1. Similarity Method

A careful consideration of the load sinkage problem and a review of

FIGURE 10. (U.S. Army photograph)

past work on this subject suggest the following six variables as being important and relevant:

$$F,\ z,\ b,\ \gamma,\ \phi,\ c. \tag{22}$$

These in turn can be expressed as independent dimensionless groups in the following way:

$$f(F/\gamma b^3,\ z/b,\ \phi,\ c/\gamma b) = 0 \tag{23}$$

This is essentially a slightly modfiied form of the groups suggested by Nuttall. In this case the characteristic length, l, is taken as the footing width, b. This analysis can also be shown to be identical to or in full agreement with the work of Taylor, Schuring, and Liston discussed previously.

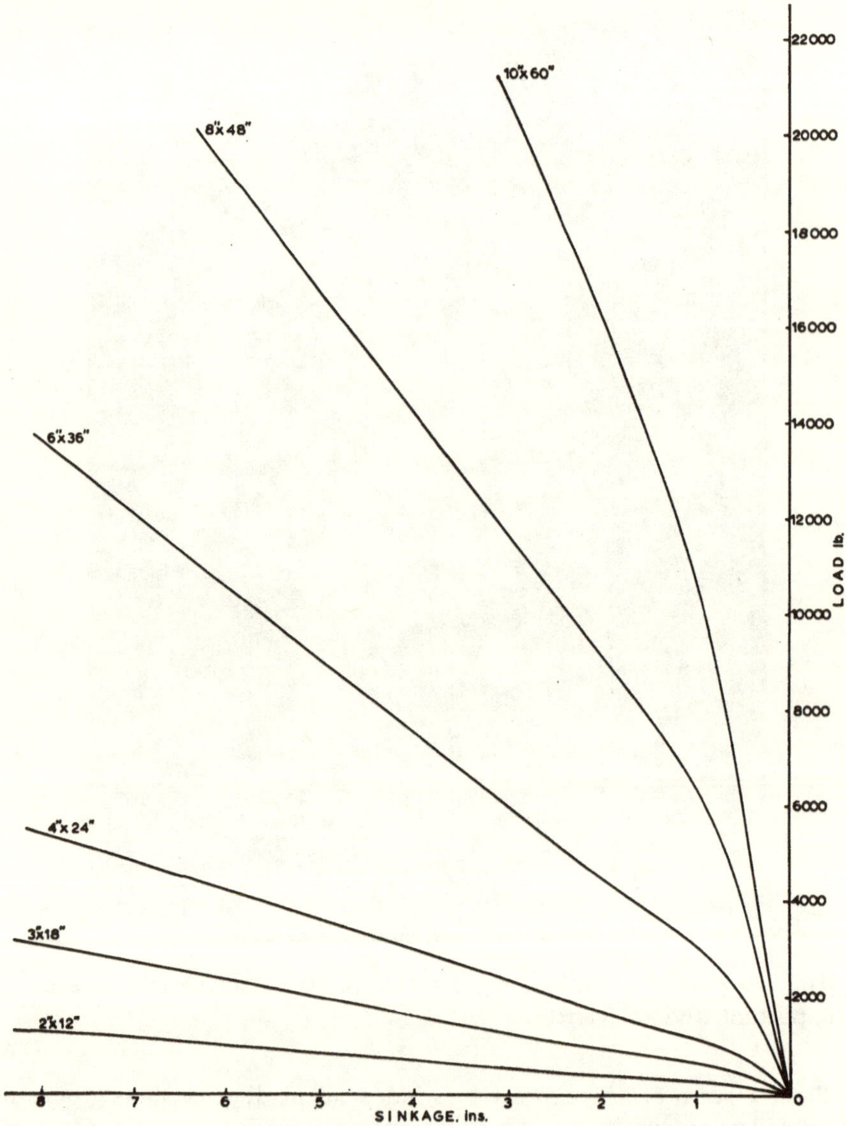

FIGURE 11. Load sinkage curves for sand

Taylor was solely concerned with surface bearing capacity on purely frictional soils. Equation 6 results, therefore, from the remaining relevant groups given in Equation 23 above. Schuring's analysis assumes geometric similarity in plate size and consistent depth to width ratios. The latter is in fact a group variable which when added to those in Equation 9 results in Equation 23 above. The result of the analysis carried out by Liston and Hegedus at first appears to be quite different from that suggested above. As the analysis is based on the use of similar plates the footing width b will be proportional to the perimeter s and can therefore

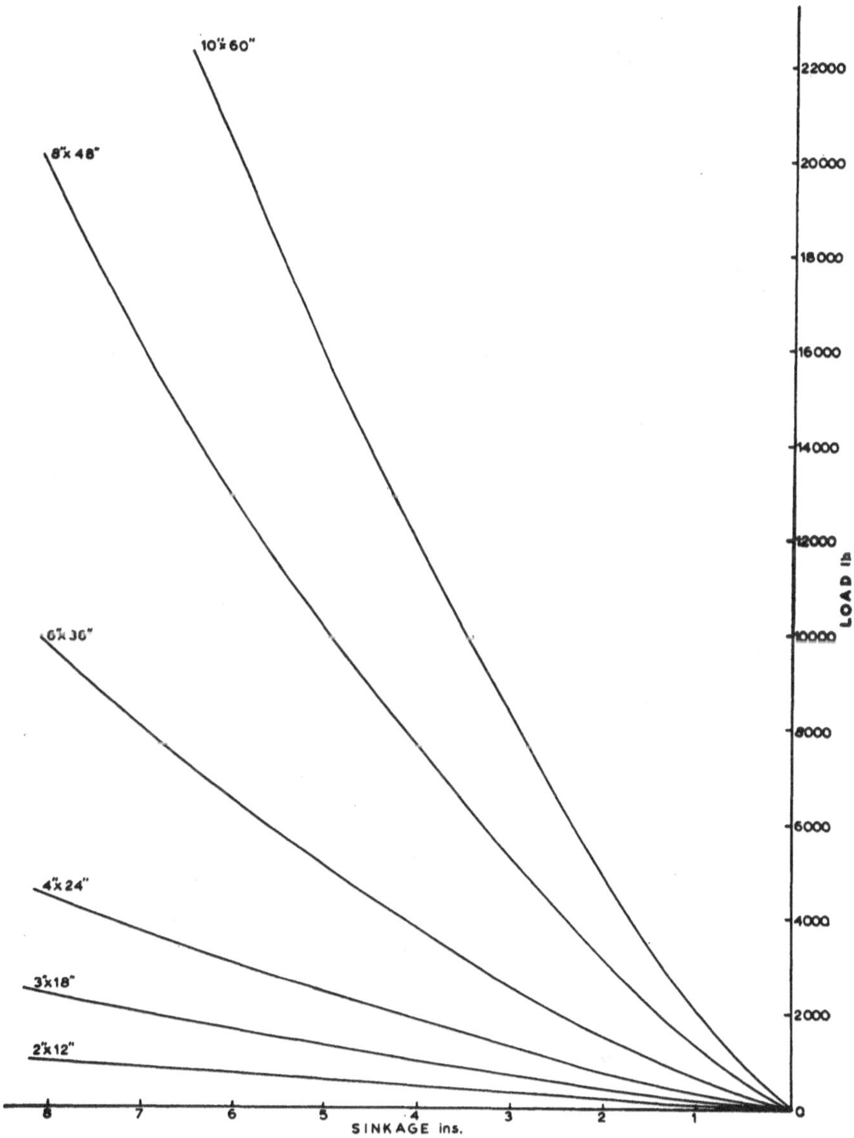

FIGURE 12. Load sinkage curves for loam

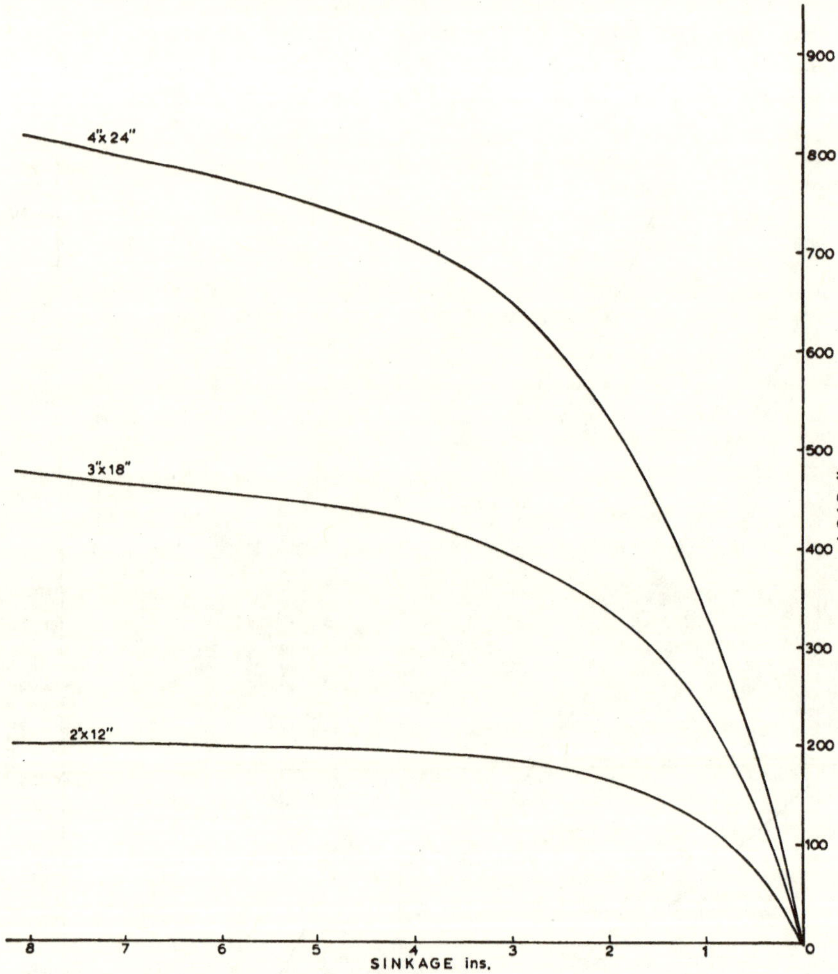

FIGURE 13. Load sinkage curves for clay

replace the latter in Equation 13. Discarding the dependent group which now appears and assuming the soil to have a large uniform depth this equation now reduces to a form identical to Equation 23.

With the established group variables it now becomes possible to predict load sinkage curves for a purely frictional or a purely cohesive soil by means of a model test. This can be done both simply and elegantly in the manner suggested by Emori and Schuring (11). The method of analysis suggested by Liston and Hegedus appears to be unnecessarily complicated and obscures the physical values involved. Thus for similar plates and for the same sinkage to width ratio the soil force F_1 on a footing can be related to the soil force F_2 on a model footing by the following:

$$\text{For } c = 0: \quad F_1 = F_2\lambda^3; \tag{24}$$

$$\text{For } \phi = 0: \quad F_1 = F_2\lambda^2, \tag{25}$$

where λ is the footing linear scale ratio.

The experimental sand results have been analysed in this way using the 3 in. × 18 in. footing as a model. The smaller 2 in. × 12 in. footing was not used as the small scale of the experimental curve made it difficult to measure off load values especially at low sinkages. The predicted load sinkage curves are compared with the experimental curves in Figure 14. It can be seen that the predictions are very accurate and especially so if an upper load limit of 30 p.s.i. is considered. The correlation between experimental and theoretical curves is extremely good over the lower sinkage ranges. As the loading increases however there is a progressive loss in accuracy. This is thought to be due to a change in the values of γ and ϕ from those values relevant to the model.

The 2 in. × 12 in. footing was used as a model on the clay. A comparison of experimental and predicted results in Figure 15 shows a fairly high degree of accuracy over the full sinkage range. The small discrepancy shown is obviously due to the difficulty in processing the clay after each test so that identical conditions are obtained throughout a series of tests.

Although the dimensional method will provide an accurate solution for the extreme soil types such as sand and clay it cannot be used to provide a direct solution for composite soils. No attempt has therefore been made to analyse the loam results in this way. Nor is there any theoretical or experimental justification for regarding such a soil as being either predominantly cohesive or predominantly frictional.

An examination of Equation 23 suggests the form of a general equation which could be applied to all soil types and which would provide an indirect solution to the pressure sinkage problem for composite soils. Emori and Schuring have expressed it in the form given by Equation 12. It can be expressed more accurately, however, in the form:

$$F/\gamma b^3 = (c/\gamma b)f_1(\phi, z/b) + f_2(\phi, z/b), \tag{26}$$

where f_1 and f_2 are functions of ϕ and z/b only. The existence of such an equation should prove to be extremely useful in the development of fundamental or empirical plate sinkage equations.

2. *Fundamental Method*

The eventual development of an accurate fundamental theory will obviate the need for the plate test. At present however the plate test

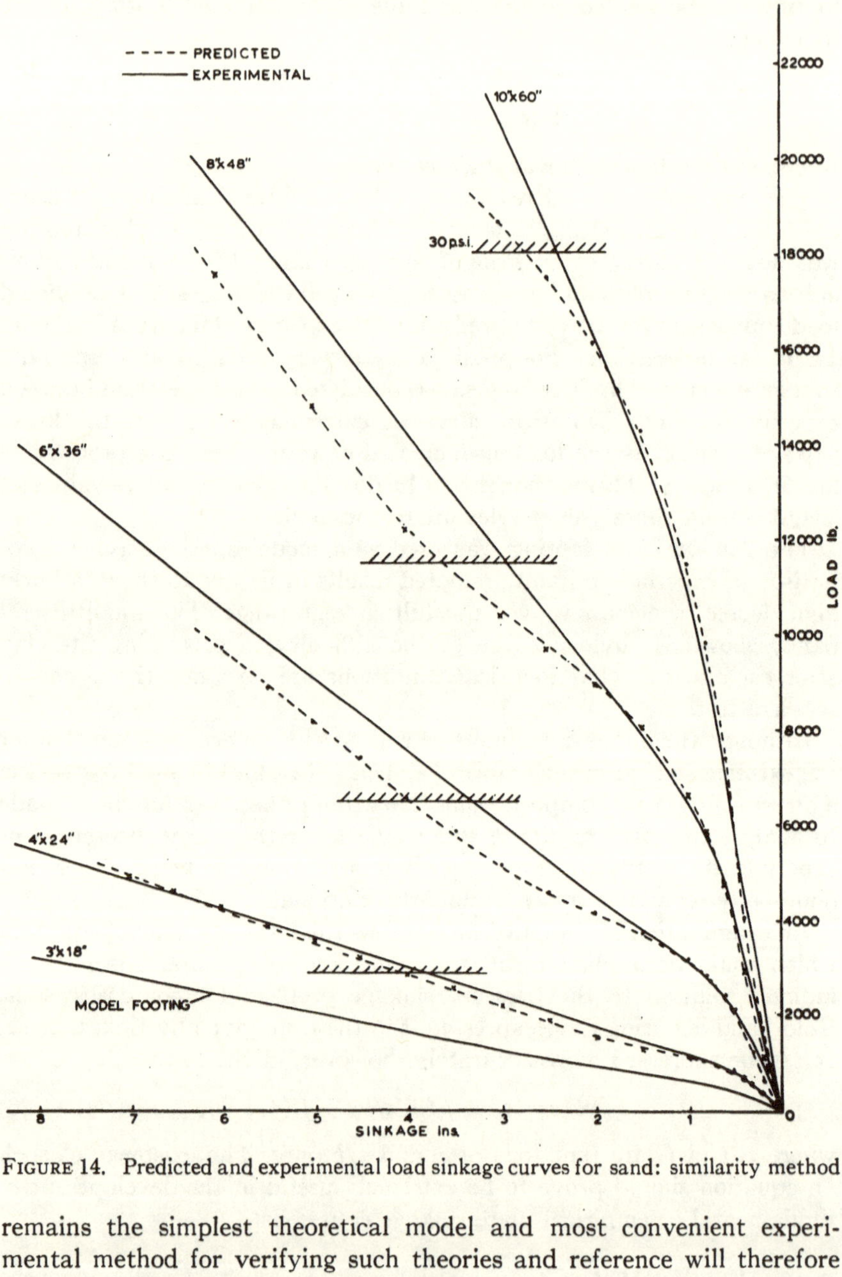

FIGURE 14. Predicted and experimental load sinkage curves for sand: similarity method

remains the simplest theoretical model and most convenient experimental method for verifying such theories and reference will therefore be made to it rather than to the vehicle.

A dimensional analysis of the plate sinkage problem suggests a general purpose pressure sinkage equation (26) which is similar in form to the

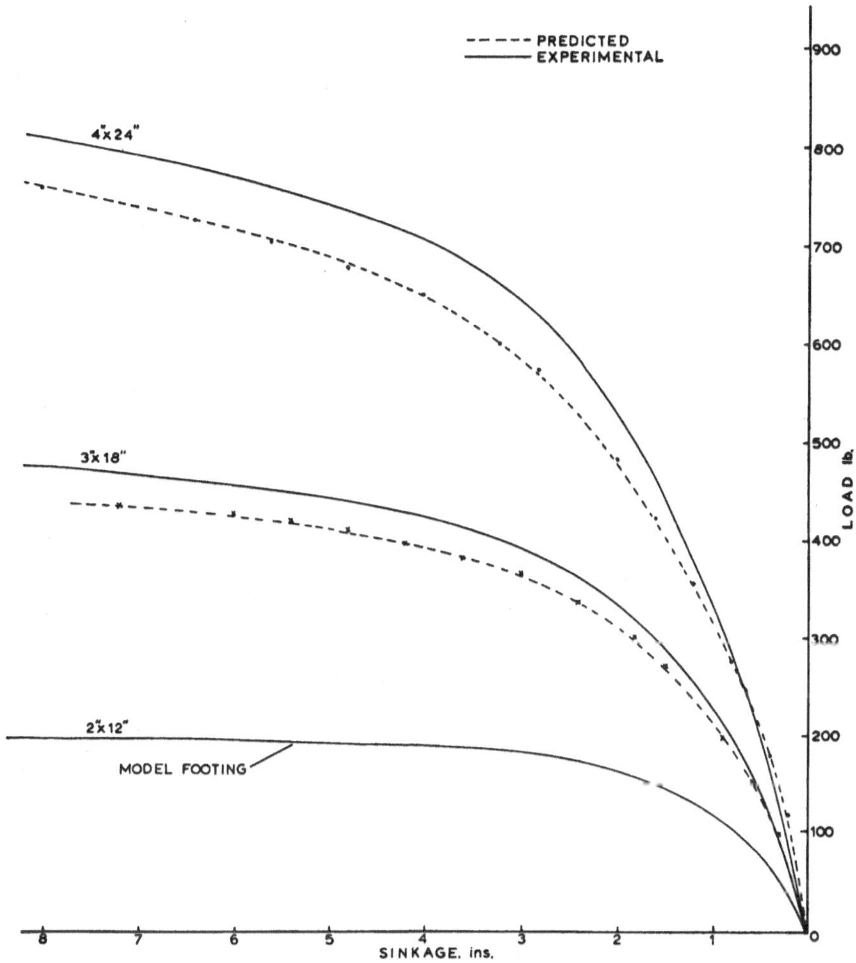

FIGURE 15. Predicted and experimental load sinkage curves for clay: similarity method

two part bearing capacity equation (21) suggested by Meyerhof. For this reason, and for the added convenience, bearing capacity will be expressed in this form rather than in the more popular three part form of Equation 15. The equivalent free surface stress p_0 and the two composite bearing capacity factors N_{pq} and $N_{\gamma q}$ will therefore need to be derived for the conditions and scale required for plate sinkage. Only the more important features of this analysis will be discussed below due to its complexity and the number of lengthy expressions involved. A full derivation and analysis of both Meyerhof's theory and its adaptation to plate sinkage is given elsewhere (16).

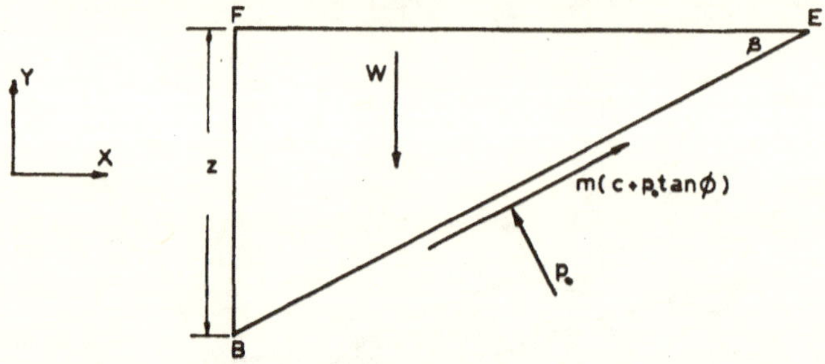

FIGURE 16. Determination of equivalent free surface stresses for plate sinkage

To derive the equivalent free surface stress p_o it is necessary to consider the equilibrium of the wedge BEF (Figure 3). In the case of plate sinkage where there will be no stresses on the vertical soil side walls it is necessary to assume, especially for frictional soils, that there will always be just sufficient cohesion present for the wedge to remain free standing.

Consider the equilibrium of the wedge BEF (Figure 16):

$$\sum Fy = 0$$

$$\therefore \quad (z^2\gamma/2 \tan \beta) - (p_0 z/\tan \beta) - mz(c + p_o \tan \phi) = 0$$

and

$$p_o = \gamma z/[2 - (1 + m \tan \phi \tan \beta)] - mc \tan \beta/(1 + m \tan \phi \tan \beta) \quad (27)$$

where $0 \leqslant m \geqslant 1$ and is the degree of mobilization of the shear stress on BE.

Substituting for p_o in Equation 15:

$$p = c\left[N_c - \frac{m \tan \beta \, N_q}{(1 + m \tan \phi \tan \beta)}\right] + \frac{\gamma b}{2}\left[N_\gamma + \frac{zN_q}{b(1 + m \tan \phi \tan \beta)}\right] \quad (28)$$

By substituting into this equation the values of N_c, N_q, and N_γ already given by Equations 16 and 18:

$$p = cN_{cq} + (\gamma b/2)N_{\gamma q}. \quad (21)$$

Where:

$$N_{cq} = \cot \phi\left[\frac{(1 + \sin \phi)e^{2\theta \tan \phi}}{(1 - \sin \phi \sin(2\eta + \phi))(1 + m \tan \phi \tan \beta)} - 1\right]. \quad (29)$$

$$N_{\gamma q} = \left[4P_p{}'' \frac{\sin}{\gamma b^2}\left(45 + \frac{\phi}{2}\right) - \frac{\tan}{2}\left(45 + \frac{\phi}{2}\right)\right.$$

$$\left. + \frac{z(1 + \sin \phi)e^{2\theta \tan \phi}}{b(1 + m \tan \phi \tan \beta)(1 - \sin \phi \sin(2\eta + \phi))}\right]. \quad (30)$$

The angle η is a function of m and has the values of $(45 - \phi/2)$ and 0 for $m = 0$ and $m = 1$ respectively.

The constant N_{cq} is determined for the same failure surface already suggested by Meyerhof for evalutinag N_c and N_q (Figure 4). The angle of inclination of the equivalent free surface, β, can be expressed as a function of z/b, ϕ, and m from the geometry of the failure surface by the equation:

$$z/b = [(\sin\beta\cos\phi\, e^{\theta\tan\phi})/(2\sin(45 - \phi/2)\cos(\eta + \phi))]. \qquad (31)$$

Values of β extracted for specific values of z/b, ϕ and m by means of a digital computer can then be used to solve for N_{cq} in Equation 29. The constant N_{cq} can now be expressed very conveniently for the required range of z/b and ϕ values. This is shown graphically for the case of $m = 1$ in Figure 17.

The evaluation of the first term of $N_{\gamma q}$ involves the calculation of the passive earth pressure p_p'' which acts on the wedge ABC (Figure 18). The correct solution requires the minimum value of p_p'' necessary to cause soil failure and can best be found by means of Ohde's logarithmic spiral method. Much discussion centres around the technique required to minimize the value of p_p''. Terzaghi maintained that the centre of the spiral could be restricted to a given line and that the minimum found in this way would not be significantly different from that obtained with an unrestricted pole (14). Meyerhof, on the other hand, states that any form of restriction is open to serious objects and in fact uses an unrestricted pole in the evaluation of N_γ (Figure 4) (15). As it is now thought that the use of a restricted pole is possibly more theoretically correct and as it represents an enormous saving in computation it was decided to calculate $N_{\gamma q}$ on this basis by restricting the centre of the spiral to the line DB or DB produced (17) (Figure 18).

For the case of $m = 1$ the angle $\eta = 0$ and the entire failure plane CDE becomes part of a logarithmic spiral (Figure 18). The centre of the spiral, O, is located by means of the angle ω, which can be related to z/b, ϕ, and β by the equation:

$$\frac{z}{b} = \frac{\sin\beta[\sin(45 + [\phi/2] - \beta)e^{(\pi/2+\beta+\omega)\tan\phi} - \cos(45 + \phi/2 + \omega)]}{2\cos(45 + (\phi/2))\cos(\beta + \omega)}. \qquad (32)$$

It is now possible to arrive at a value of P_p'' by considering the equilibrium of the soil in the failure zone about the centre of a spiral located in this way from values of z/b, ϕ, and β. By varying the value of β for specific values of z/b and ϕ the minimum value of P_p'' can be evaluated. N_{pq} can now be obtained for the required values of z/b and ϕ and their optimum values of P_p'' and β by Equation 30. The above work was necessarily carried out by means of a digital computer.

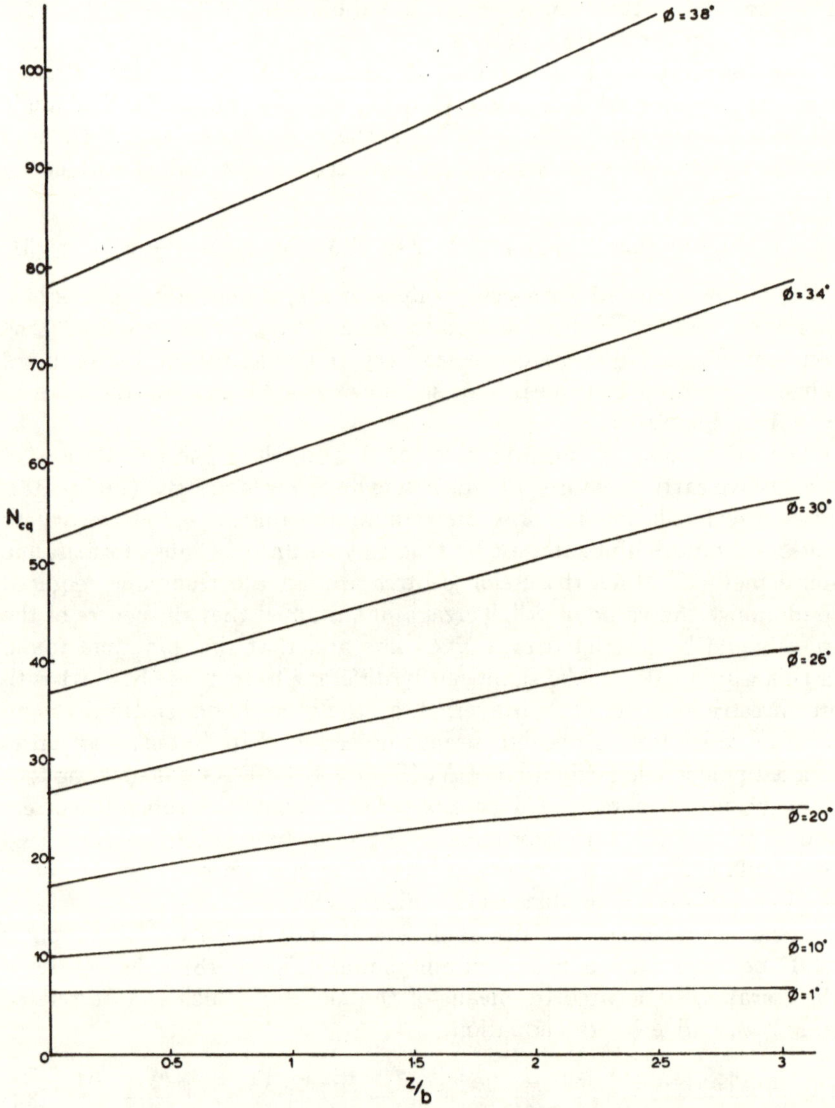

FIGURE 17. General bearing capacity factor N_{cq} (m = 1)

The constant can be expressed conveniently as a function of z/b and ϕ as shown in Figure 19. It should be noted that this has only been done for the case of $m = 1$.

Load sinkage curves for specific soil types and plate sizes can now be predicted by using the pertinent values of N_{cq} and $N_{\gamma q}$ in Equation 21.

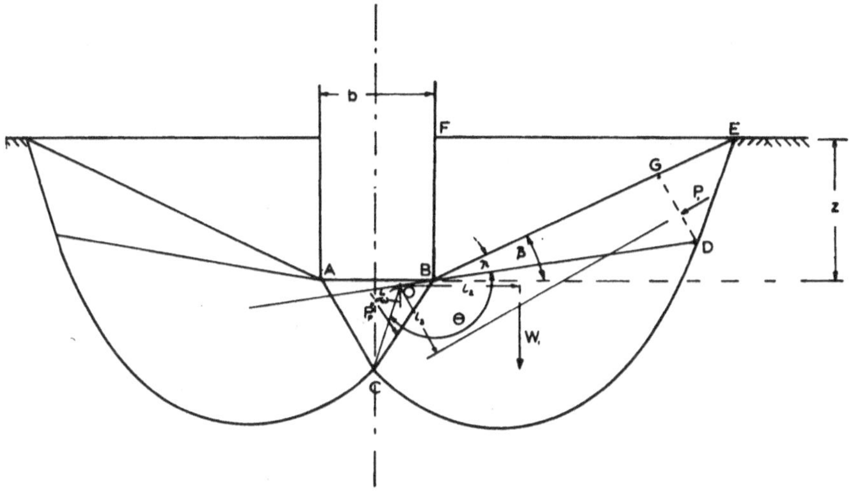

FIGURE 18. Determination of $N_{\gamma q}$ by restricted pole method

This has been done for both the sand and the clay and the results compared to experimental values (Figures 20 and 21). The theoretical sand results are obviously highly inaccurate for all the plate sizes concerned. It should be noted that the theory does agree closely in each case with what would be the assumed surface bearing capacity, and shows itself to be sensitive to the effect of footing width. It appears to fail miserably, however, in evaluating the effect of depth. The theoretical clay results are in much better agreement with experimental values (Figure 21). The loads for the larger sinkage values are predicted quite accurately in each case but there is some discrepancy at the lower values possibly due to the compressibility of the clay.

No attempt has been made to predict experimental load sinkage values for the loam as the theory at this stage in its development applies only to incompressible soils. This factor is the most probable cause of error with the sand results. To simulate the condition of incompressibility with frictional soils it is best to use either a closely packed sand with uniform grain size and shape or a common well-graded sand in a very dense state. It was not possible to use a single grain size sand, either natural or synthetic, due to the volume of the sample required. The Ohio sand which was in fact used was well-graded and showed an appreciable variation in density between being loose and compact. Due to the size of the sample, however, it was not possible to influence the density to any great extent. If the sand had been rolled or a vibratory rake had been used, the density could have been increased somewhat but at the expense of uniformity. It was thus thought best to process the sand in

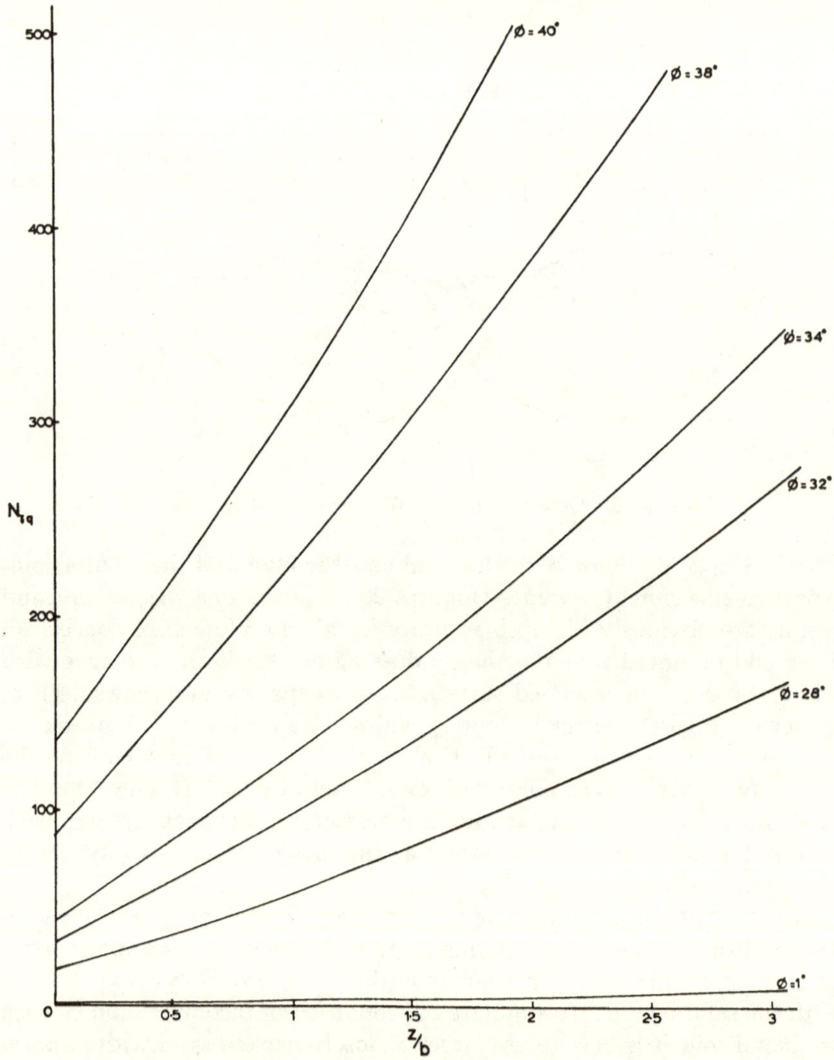

FIGURE 19. General bearing capacity factor $N_{\gamma q}$ ($m = 1$)

the way described so as to obtain a uniform condition and accept the resulting density.

The effect of density on the load sinkage curve and the value of ϕ can be seen in Figures 22 and 23. These results were obtained by using a steel box 30 in. × 30 in. × 26 in. deep standing on four legs with a vibrator attached to its underside. When filled with a known weight of sand the box was suspended several inches above the ground from an

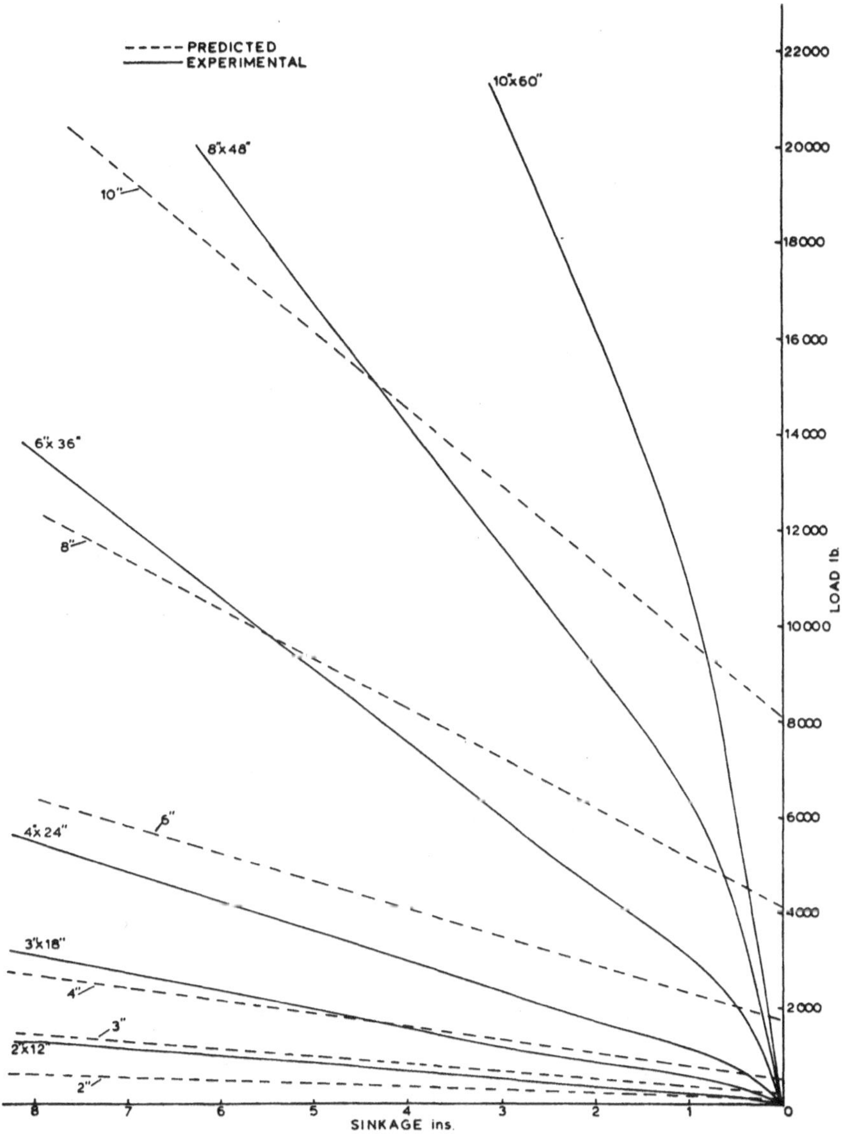

FIGURE 20. Predicted and experimental load sinkage curves for sand: fundamental method

overhead crane and vibrated until the sand reached its required density. Even for the limited range of densities shown there is a tremendous difference in the load sinkage values with the 2 in. × 12 in. footing (Figure 22). A change in density also has an appreciable effect on the

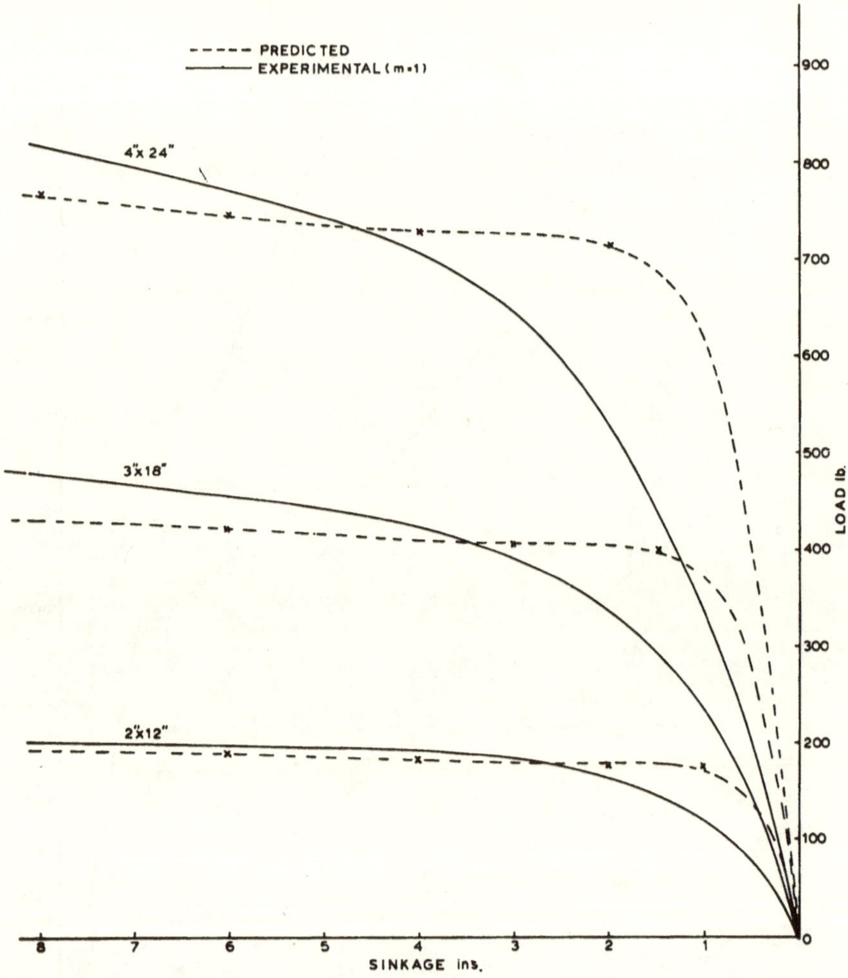

FIGURE 21. Predicted and experimental load sinkage curves for clay: fundamental method

value of ϕ (Figure 23). It can be appreciated that the density of a soil in general is of prime importance in determining bearing capacity not only in the sense of a weight factor but also in defining the basic frictional characteristics of the soil. If one considers the pressures developed beneath the test footings and the disturbance caused by failing the sand which was initially in a relatively loose condition it becomes obvious that some change in density must have occurred. Consequently it should be noted how sensitive the value of $N_{\gamma q}$ is to values of ϕ (Figure 19). For

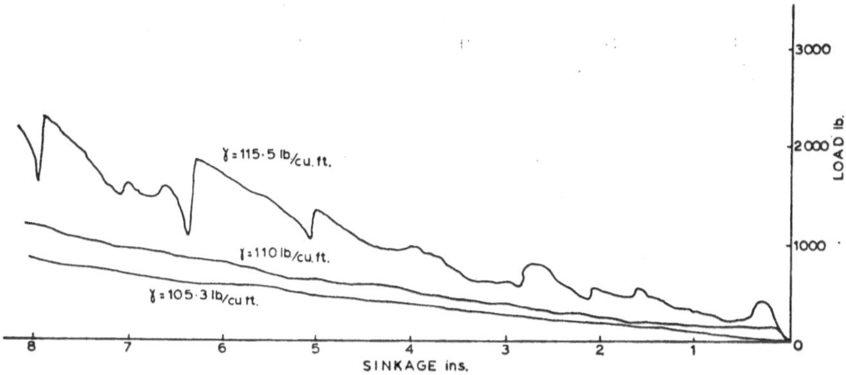

FIGURE 22. Effect of density on sand load sinkage values for 2 in. × 12 in. footing

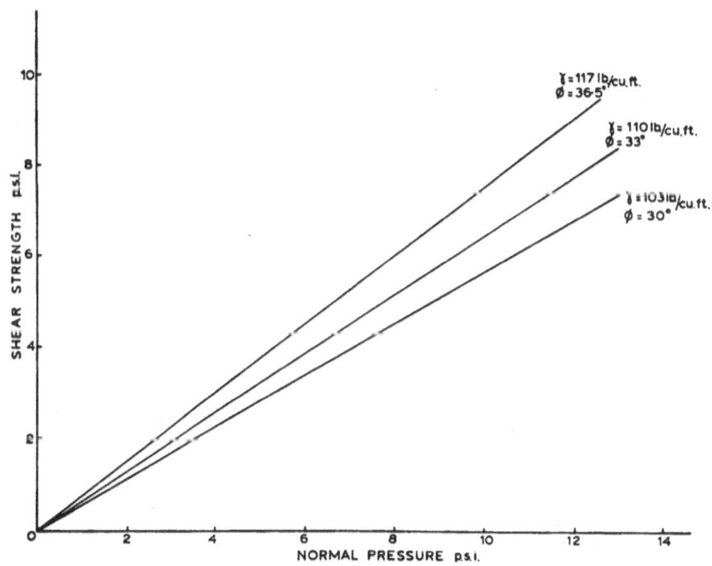

FIGURE 23. Variation in friction angle with density for sand: linear shear box values

example, values of $N_{\gamma q}$ for $\phi = 38°$ are almost exactly twice the corresponding values for $\phi = 34°$.

The angle of internal friction of the Ohio sand at a density of 0.064 lb./in.^{3} was obtained with a linear shear box from shear test results which covered a normal pressure range from 0–13 p.s.i. (Figure 23). This pressure range would seem to be entirely inadequate to predict load sinkage curves in the light of the pressures developed beneath the footings and the changes in density that must have necessarily taken place. Triaxial shear tests carried out on the sand at that time were unsuccessful due to practical difficulties. A fresh attempt is now being made, however, and it is hoped to obtain shear strength values for a normal pressure range comparable to that experienced by the footings.

An important part of Meyerhof's theory is based on the use of the m factor which is defined as the degree of mobilization of the shearing stresses on the equivalent free surface. Apart from the two extremes of zero and maximum shearing stress when $m = 0$ or 1 and for any soil type other than a purely frictional one it is not possible to determine the shear stress on a surface from the soil strength parameters, the angle of inclination of that surface and the normal pressure. In the case of a purely frictional material the shear stress can quite easily be expressed as a function of the angle of inclination and the normal pressure and there is no need for an m value. Meyerhof has shown that there is a measurable difference in the values of N_c and N_q for $m = 0$ and $m = 1$, but that any such difference in N_γ is negligible (Figure 5). N_{cq} and $N_{\gamma q}$ have only been evaluated above for the case of $m = 1$. In the case of $m = 0$, which is only strictly true for a surface condition, a workable solution for $N_{\gamma q}$ could not be obtained due to theoretical and practical difficulties. The use of the m factor gives two limiting sets of values which are obviously only correct at certain points yet it remains the only method of treating with most soils if the equivalent free surface concept is to be used.

The application of Meyerhof's theory to plate sinkage has raised many problems and doubts concerning normally accepted procedures in civil engineering bearing capacity theory. In the past, before the advent of computers, the complexity of a rigorous solution was often avoided by means of a simplifying assumption which resulted in a calculated error for the particular case in question. Over the years the limitations of such simplified solutions are often forgotten and when, as so often happens, they are applied in general their accuracy becomes highly questionable. What was, in fact, in the first place a matter of expediency and an intelligent but particular assumption now assumes the mantle of theoretical orthodoxy. Such is the case with the shape of failure surfaces, the size of the wedge beneath a footing and the two part solution bearing capacity with its two widely different failure planes. What constitutes an acceptable error in civil engineering soil mechanics is not necessarily so in soil vehicle mechanics. In general the opposite is usually more correct and one would do well therefore to study much of the original theoretical work in context if it is to be applied accurately to the problems of soil vehicle mechanics. In this field, where weak soils are of special interest, the problem of compressibility becomes extremely important. So far all theoretical work has been based on the premise of an incompressible soil with consequent limitation to certain soil types. It is felt that if shear strength itself could be expressed more adequately in terms of the fundamental properties of a soil relevant to compressibility the fundamental method could be applied to a much wider range of soils.

Unlike the other two methods, the fundamental method is based on the values of c and ϕ and consequently requires their accurate measurement. The measurement of field shear strength in soil vehicle mechanics, however, is in a very unhappy condition. On soft soils, which are really the only ones of importance as regards the problems of vehicle mobility, the shear annulus is quite inaccurate and the results obtained completely subjective to the operator. On a loose frictional soil it has a limited range of experimental normal pressures, and sinks rapidly with torsional deformation resulting in a drag effect. There is no maximum value on the experimental shear stress deformation curve and the point of soil failure cannot be defined. As a result almost any values of c and ϕ can be obtained from such a series of tests. In general the above objections also hold true for a clay. With a stiff clay there is often loss of surface contact between the lugs of the annulus and large side wall error, whereas with a weak clay there will be excessive sinkage and appreciable drag.

At present it is necessary to rely on the linear shear box and triaxial tests to measure soil shear strength. Both are laboratory tests on small restricted soil samples, however, and not really suited to the needs of soil vehicle mechanics. If soil vehicle mechanic theories are ultimately to be applied successfully in the field after being proven in the laboratory it is imperative that shear strength can be measured in the field under conditions pertinent to those of a vehicle.

3. Empirical Method

This method is based upon the use of an accurate empirical equation to describe pressure sinkage values for a wide range of similar footings. The problems arising from this approach are the selection of a suitable pressure sinkage equation and the extraction of the necessary empirical soil values from model footing results. The minimum error curve fitting technique, however, has now reduced the latter problem considerably when applied within the limits suggested by Reece (6).

Equation 1 is the simplest and one of the oldest empirical pressure sinkage relationships. The constants involved can be obtained by plotting logarithmically the pressure sinkage values of a model footing test and applying a curve fitting technique where necessary. For the sand this has been done with the 3 in. $\times$ 18 in. footing and the load sinkage values for the larger footings have then been predicted (Figure 24). The loam

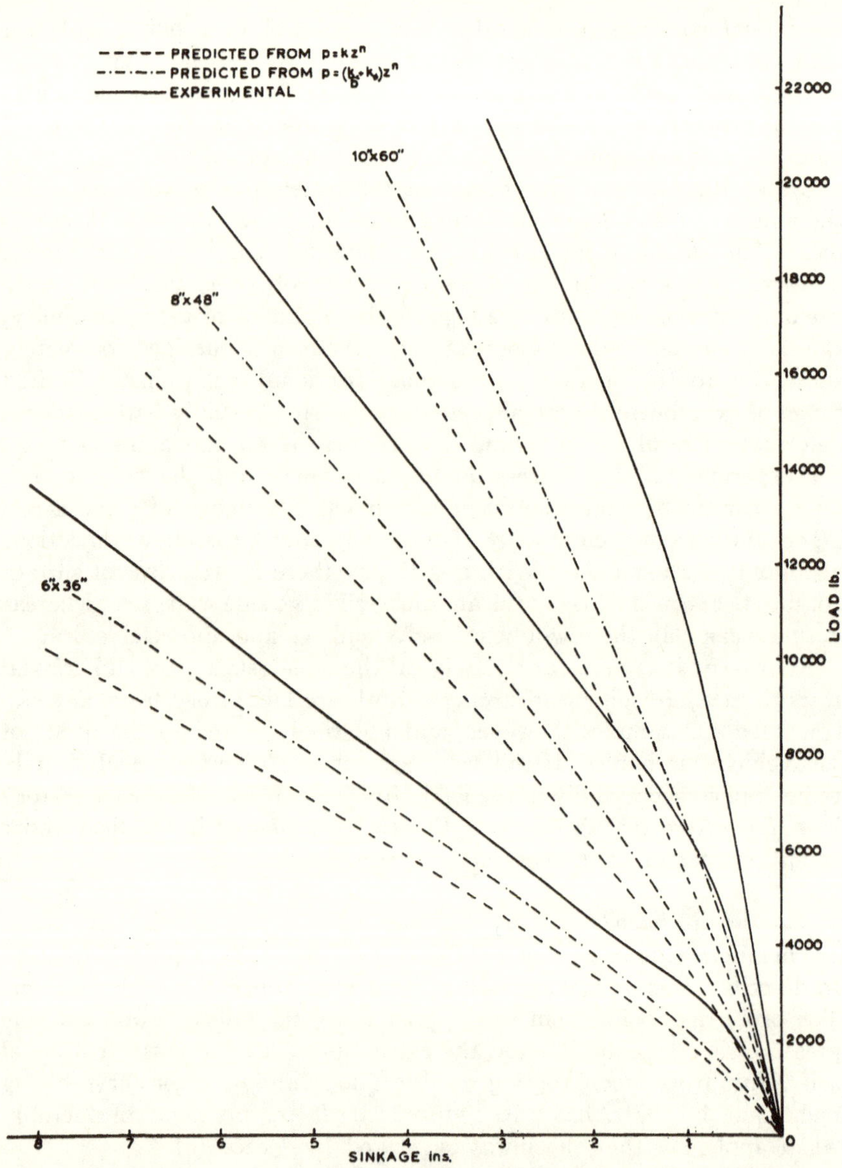

FIGURE 24. Predicted and experimental load sinkage curves for sand: empirical method

and clay results have been predicted in like manner using the 3 in. × 18 in. and the 2 in. × 12 in. footings respectively (Figures 25 and 26). The apparent accuracy of the predictions varies from being very poor for the sand to excellent for the loam with that for the clay falling somewhere in between. It should be remembered at this point that the use

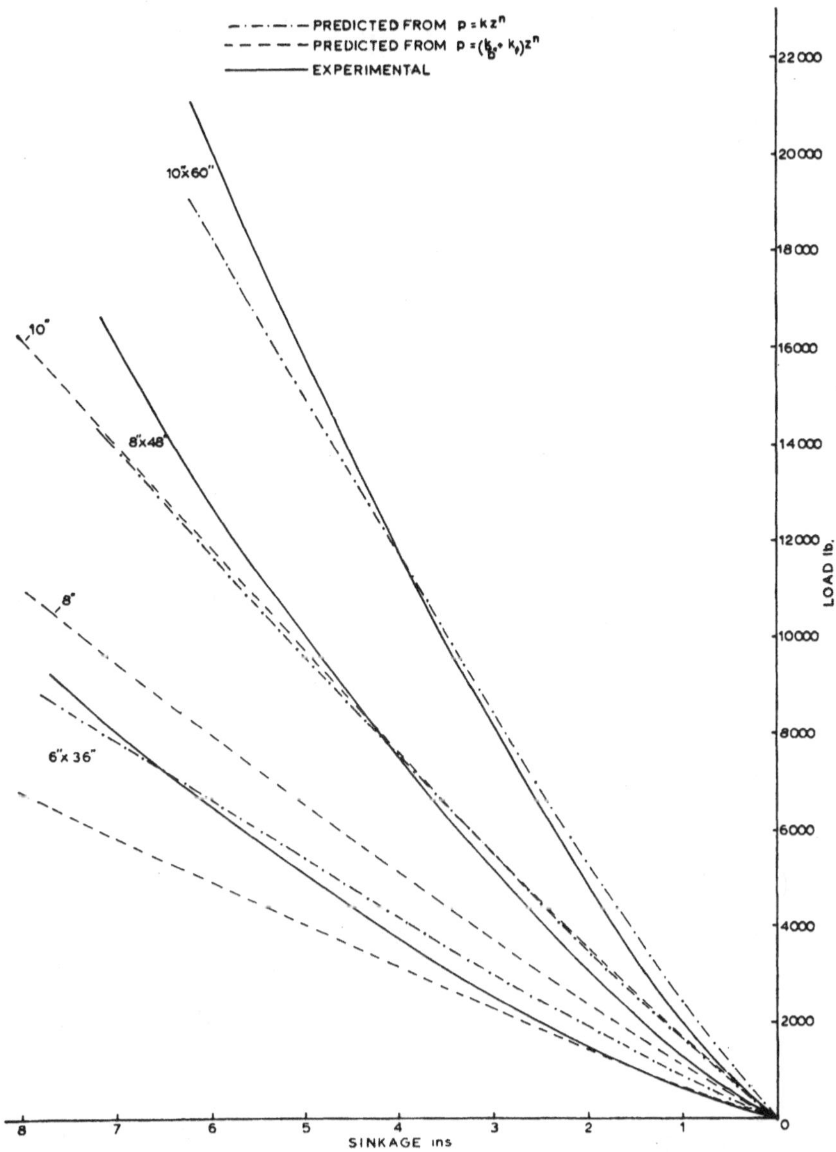

FIGURE 25. Predicted and experimental load sinkage curves for loam: empirical method

of Equation 1 is based on the premise that both n and k are true soil constants independent of both footing shape and size and that the values of these constants are determined from the model footing results. By plotting all the experimental pressure sinkage results on logarithmic axes experimental values of n and k can be obtained and it becomes obvious

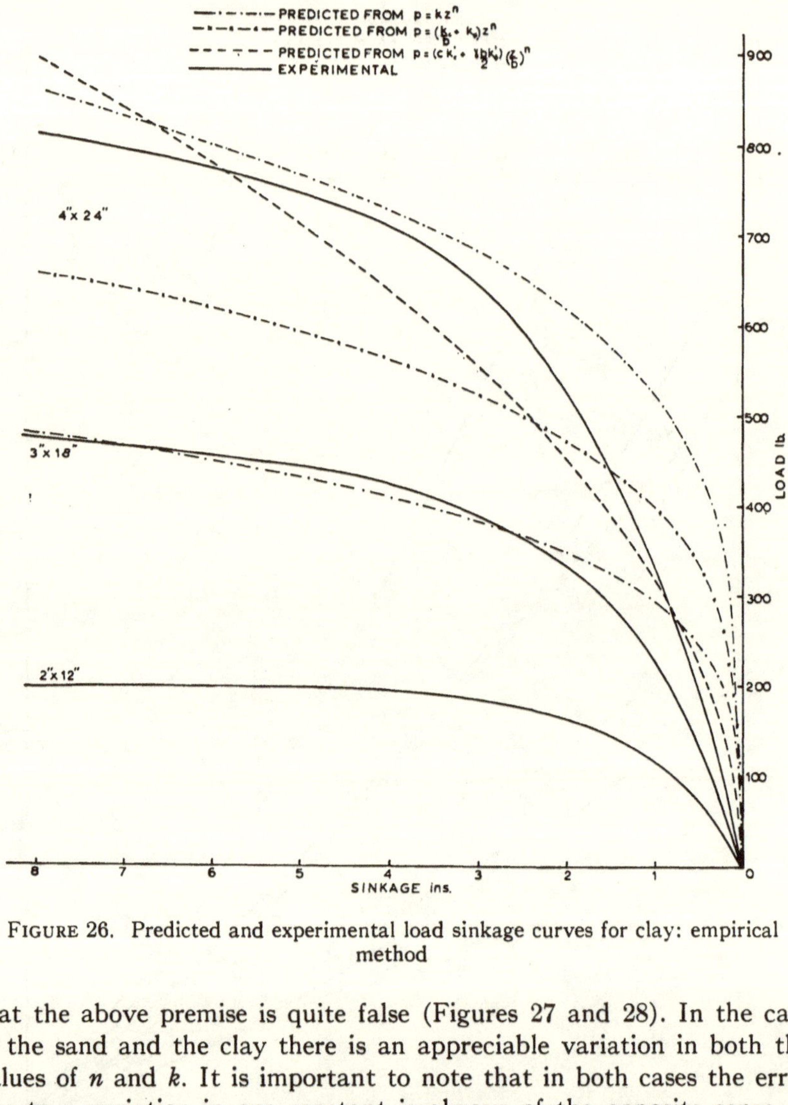

FIGURE 26. Predicted and experimental load sinkage curves for clay: empirical method

that the above premise is quite false (Figures 27 and 28). In the case of the sand and the clay there is an appreciable variation in both the values of n and k. It is important to note that in both cases the error due to a variation in one constant is always of the opposite sense to that induced by the variation of the other constant. This results in a partial compensation of the overall error of prediction dependent on the relative values of n and k. The loam is unique in that Equation 1 describes the load sinkage curves quite accurately, although there is still some variation in the values of the constants (Figures 27 and 28). The high degree of accuracy of the predicted values results from an almost complete compensation of the errors due to the variation in these two constants.

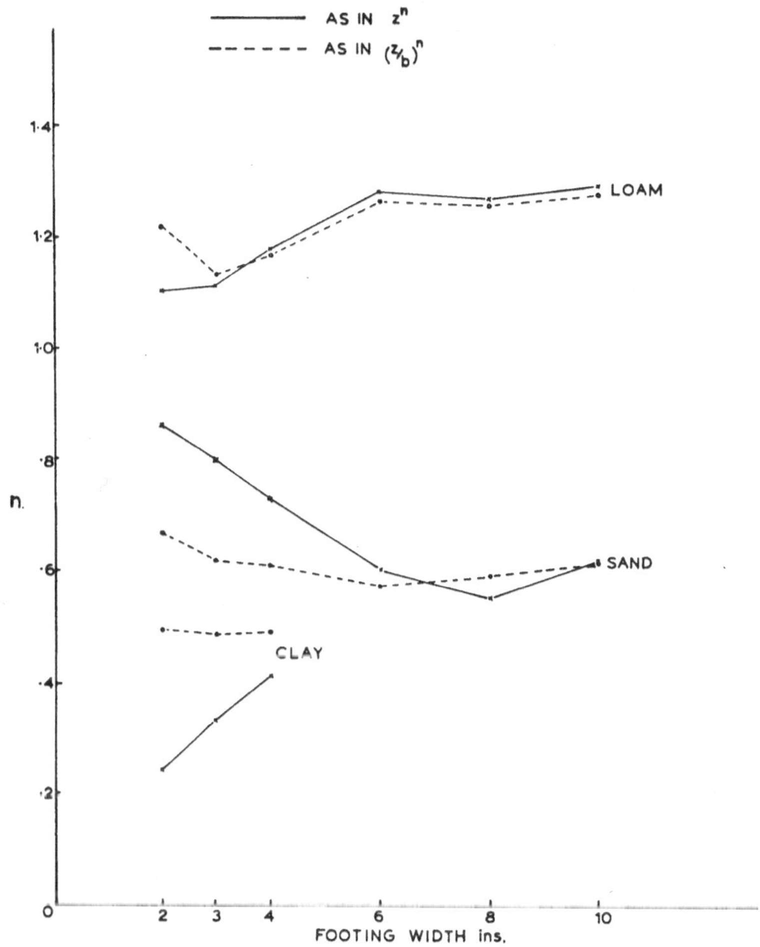

FIGURE 27. Experimental variation of empirical pressure sinkage exponents

In general it can be said that as Equation 1 is based on a false premise it cannot be taken seriously as a pressure sinkage equation. The apparent accuracy of any predictions based on this equation is the result of compensating errors and is therefore quite fortuitous.

Equation 2 attempts to improve on the accuracy of Equation 1 by modifying the value of the constant k. The value of n however remains unaltered. It had become obvious that k was in some way dependent on the size of the footing and was recognized as such by regarding it in the composite form of the constant in Equation 2. To evaluate k_c and k_ϕ two model footing tests are now required. These values together with footing width can then be used to predict the modified constant for a range of footings (Figure 29).

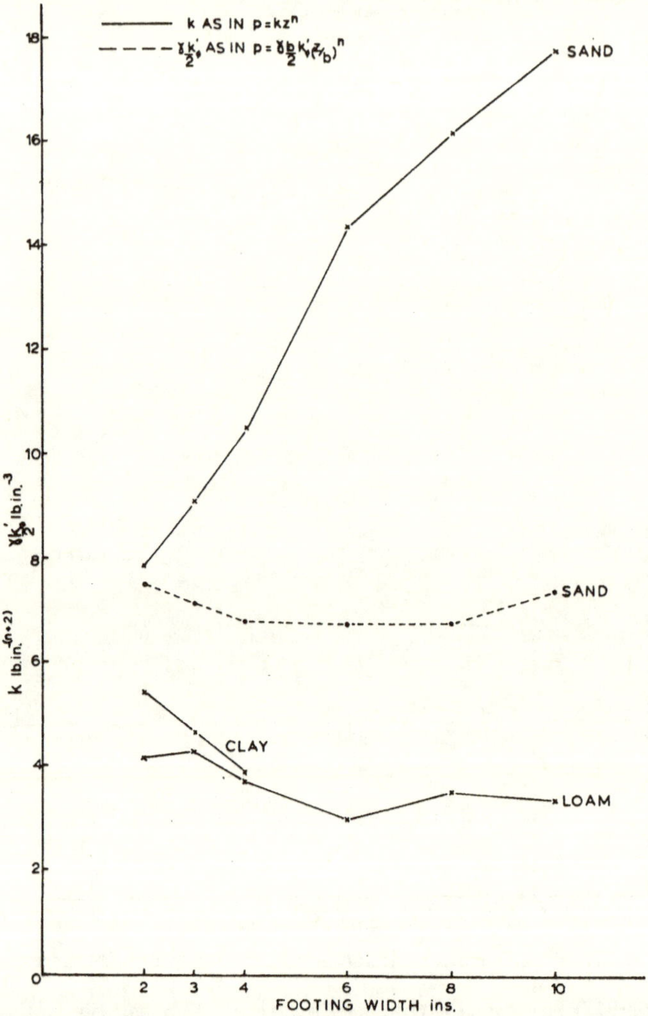

FIGURE 28. Experimental variation of empirical pressure sinkage moduli

In the case of the sand this has been done with the 2 in. × 12 in. and the 3 in. × 18 in. footings as models and the relevant constants used to predict the load sinkage curves for the larger footings (Figure 24). It can be seen that Equation 2 gives a more accurate prediction than Equation 1. This can be explained by the fact that although the *n* value remains the same the new *k* value now increases with footing width in the manner shown in Figure 29. Whereas in Equation 1 the *k* value remained constant

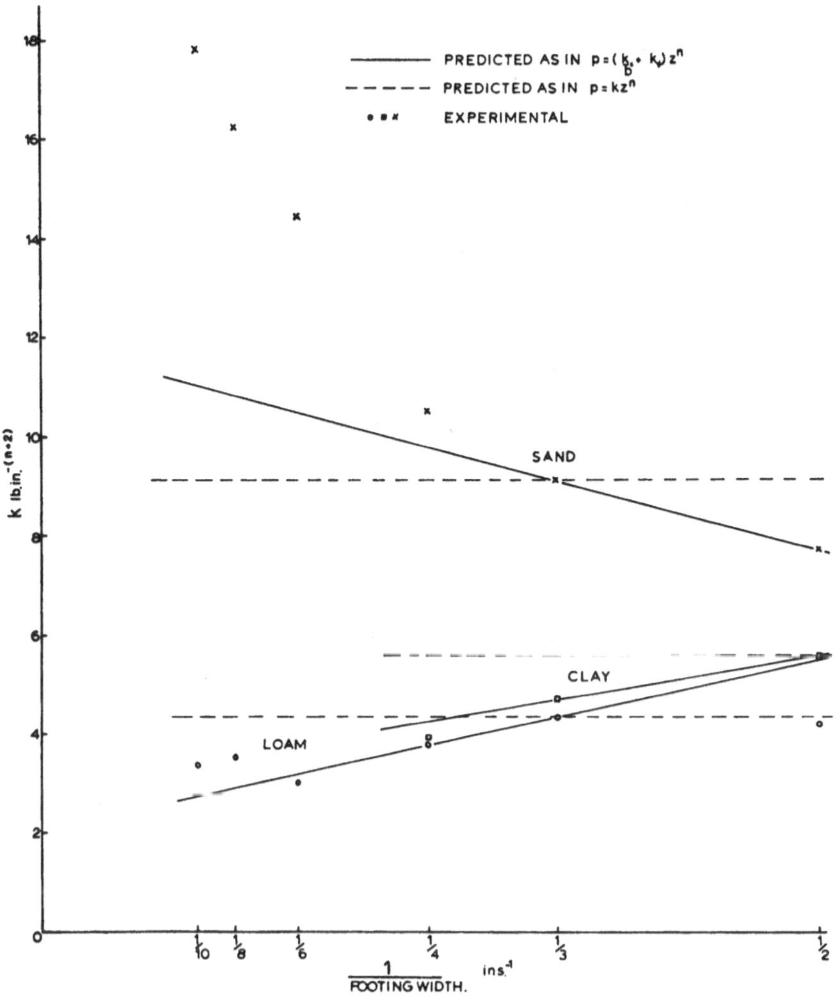

FIGURE 29. Experimental and theoretical pressure sinkage moduli

at the value determined by the one model footing, it now attempts to follow the experimental values more closely with the result that the predicted load sinkage curves fall closer to the experimental curves. In the same way for the loam the new k values have been calculated with the 3 in. × 18 in. and 4 in. × 24 in. footing as models and can be seen to fall in value with increasing footing width (Figure 29). As the model n value is already much lower than the experimental values the effect of the reduction in the k value is to lower the predicted load sinkage curves well below those of Equation 1 (Figure 30). Taking the 2 in. × 12

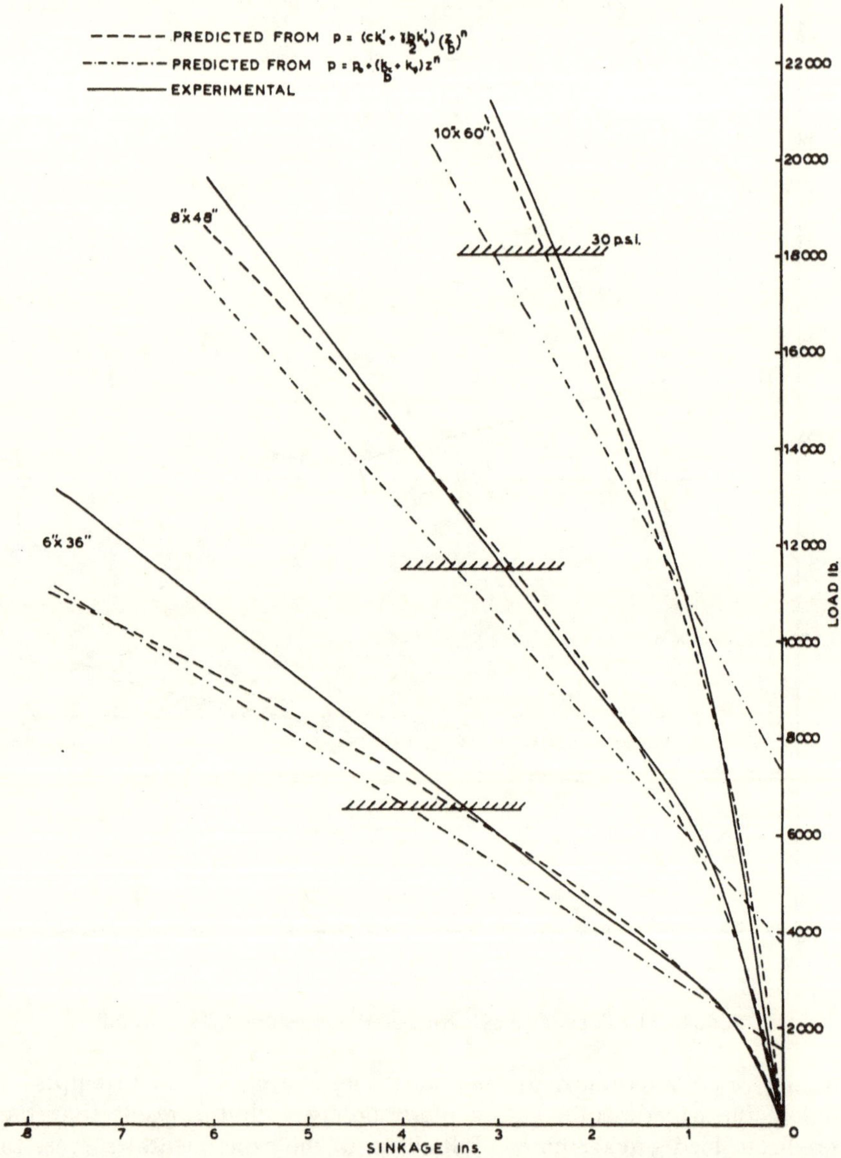

FIGURE 30. Predicted and experimental load sinkage curves for sand: empirical method

in. and the 3 in. × 18 in. footings as models for the clay, it can be seen that the new k values fall with increasing footing width. As in the case of the loam the model n values already fall well below the experimental values with the result that the predicted load sinkage curve for the

4 in. × 24 in. footing falls well below the values already predicted by Equation 1 (Figure 26). Although Equation 2 is partially successful in describing the k value more closely, all, in fact, that it effects is a change in the relative magnitude of the errors due to the variation of k and n. This is shown quite clearly in that it results in improved accuracy in some cases and has quite the opposite effect in other cases. The value of n is the same as for Equation 1 and varies in the same way with the result that there are two possible values of n for each pair of model footings. These two values vary appreciably for each pair of model footings considered and can affect the predicted load sinkage values accordingly. As there is a free choice of n value in each case the value for the smaller footing of each pair was used.

For soils which exhibit some form of bearing capacity such as the sand it has been suggested that Equation 3 is of a more suitable form. This equation assumes that the bearing capacity p_0 is instantly mobilized and that the load sinkage curve proceeds from this elevated point on the load axis (Figure 1). When plotted in the form shown in Figure 2, experimental values do, in fact, become much more linear and it is consequently easier to extract the soil constants more accurately. By using two model footings as before the values of k_c and k_ϕ can be evaluated. The value of p_0 can be obtained by making use of civil engineering bearing capacity theory. In the case of no surcharge p_0 can be given in terms of a cohesive component and a weight component dependent on footing width as in Equation 15. By using the two experimental p_0 values from the model footing tests the two components can be evaluated in a similar fashion to the k values and the p_0 values for larger footings predicted. This has been done for the sand with the 2 in. × 12 in. and the 3 in. × 18 in. footings as models and the load sinkage curves for the larger footings have been predicted (Figure 30). On the whole the predictions are fairly accurate although the use of an intercept introduces an error which becomes more important as the width of the footing increases. Equation 3, however, is essentially the same as the previous pressure sinkage equations other than that it has been adopted for particular use by the introduction of a constant and as such is extremely limited in its application.

It becomes obvious from the above that a pressure sinkage equation which involves sinkage raised to some power cannot adequately describe the physical phenomenon of plate sinkage and that some entirely new equation is required. A dimensional analysis of the problem has suggested an equation of the form:

$$p = cf_1(\phi, z/b) + \gamma b f_2(\phi, z/b). \tag{12}$$

A more fundamental approach has produced the equation:

$$p = cN_{cq} + (\gamma b/2)N_{\gamma q} \tag{21}$$

where N_{cq} and $N_{\gamma q}$ are functions of ϕ and (z/b).

Both equations are similar and it seems clear that any empirical equation should be of a similar form. By expressing each function in the above equation in terms of some exponent of (z/b) and a ϕ dependent constant the equations can be approximated in the form:

$$p = c[k_c'(z/b)^n] + \gamma b/2[k_\phi' z(/b)^n],$$

or

$$p = \{ck_c' + (\gamma b/2)k_\phi'\}(z/b)^n. \tag{5}$$

For a given soil and footing width this becomes:

$$p = k'(z/b)^n. \tag{33}$$

If the experimental pressure sinkage results are now expressed in the form of pressure against sinkage to width ratio on logarithmic axes it is possible to evaluate the experimental values of n and k' in each case (Figures 27 and 28). When (z/b) is used instead of sinkage the lower limit for curve fitting is taken at $(z/b) = 0.1$ whenever possible.

The value of n is seen to approach a constant value more closely when expressed as an exponent of (z/b) rather than z (Figure 31). This is noticeably so with the sand and clay and even with the loam there is some improvement. It is important to note that, even though n still varies slightly, the predicted load sinkage curves are much less sensitive to any change in n when expressed as a function of (z/b) as compared to z only.

To evaluate the constants ck_c' and $(\gamma/2)k_\phi'$ it is still necessary to use the experimental k' values from two model footings tests except in the case of a purely frictional soil when only the k value from one footing is required to evaluate $(\gamma/2)k_\phi'$. Thus, for the sand, only the 3 in. × 18 in. footing was used as a model whereas on the loam the 3 in. × 18 in. and 4 in. × 24 in. footings were used. The 2 in. × 12 in. and 3 in. × 18 in. footings were used as models on the clay. The experimental values of ck_c' and $(\gamma/2)k_\phi'$ together with footing width can then be used to predict the k' values for a range of footings. This has been done for each soil type in Figure 31 and the experimental and predicted k' values compared. It can be seen that the k' values increase linearly with footing width and that there is very good agreement between predicted and experimental values in each case. As might be expected, the value of k' is directly proportional to footing width for the sand as only k_ϕ' is involved. It is also possible in this one instance to isolate the experimental value of

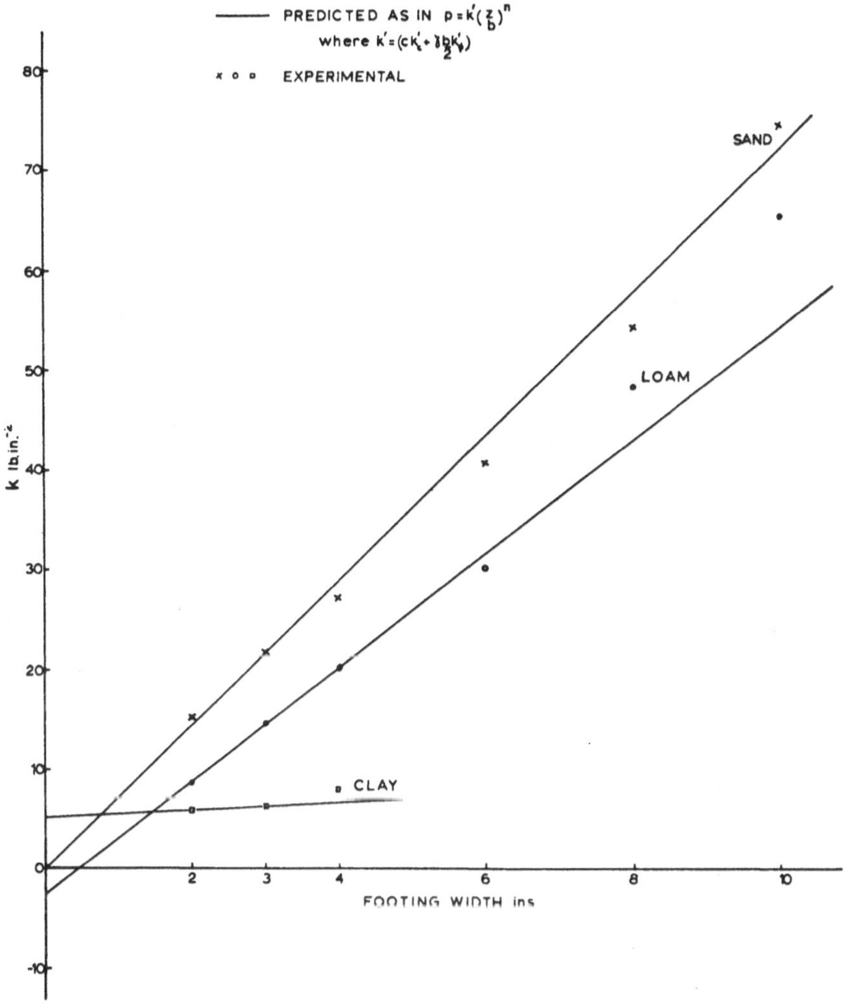

FIGURE 31. Experimental and theoretical pressure sinkage moduli

$(\gamma/2)k_\phi'$ for each of the footings and to demonstrate the constancy of k_ϕ' (Figure 28). With both the loam and the clay the value of ck_c' is given as the intercept on the vertical axis and for all three soils the slope of the line is equal to $(\gamma/2)k_\phi'$. When expressed in this way the value of k' becomes much more meaningful and illustrates the characteristics of soil type quite plainly. The small negative value of ck_ϕ' for the loam is thought to be due largely to curve fitting error.

The load sinkage curves for the larger footings have been predicted for each soil type from Equation 5 with soil values derived from the

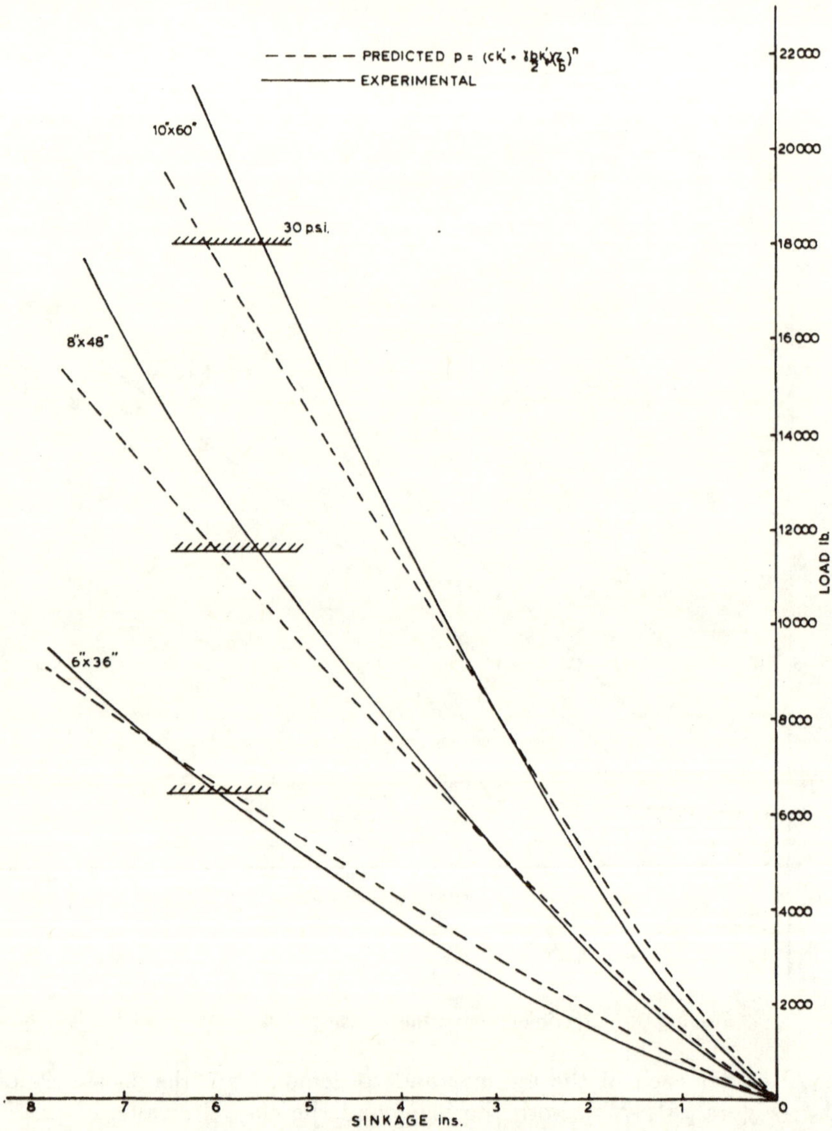

FIGURE 32. Predicted and experimental load sinkage curves for loam: empirical method

model tests (Figures 30, 32, and 26). The predicted values can be seen to be in excellent agreement with the experimental curves, especially for the sand and the loam. Although there is a close correlation over the lower sinkage values with the clay it tends to deteriorate somewhat as the sinkage increases. The particular distribution of the error is characteristic of cohesive soils. It is a result of fitting a minimum error straight line to experimental values which remain very much non-linear even on logarithmic axes. Curve fitting error is usually greatest for this type of soil and the predicted curves extremely sensitive to the curve fitting limits.

In general it might be argued that the use of Equation 5 results in compensating errors as before and that the accuracy of the equation is only apparent. This is not so, however, as the errors arising from any variation in n and k' are now very much reduced and for the soils in question are in the same sense. The equation can be seen to be accurate for a wide range of soil types and footing sizes. It could be improved, however, especially in the case of cohesive soils by a more accurate method of obtaining the empirical constants.

CONCLUSIONS

1. The evaluation of rolling resistance from vertical work theories has been almost exclusively accomplished to date by one or other of the two empirical general pressure sinkage equations: $p = kz^n$, and $p = [(k_c/b) + k_\phi]z^n$. Both equations are completely inadequate in describing plate sinkage phenomena. They are based on a premise of false soil values and in no way agree with any theoretical approach to the problem. The apparent accuracy of any predictions based on these equations is the result of compensating errors and is therefore quite fortuitous. There is no merit whatsoever in these equations and their use should be terminated forthwith.

2. The load sinkage problem can only be completely solved in the long term by means of a solution based on the fundamental properties of the soil and soil failure. Solutions based on the use of model analysis or general empirical equations are meanwhile required.

3. A dimensional analysis of the variables relevant to plate sinkage yields the functional equation of group variables: $f(F/\gamma b^3, z/b, \phi, c/\gamma b) = 0$ From this equation load sinkage curves can be predicted accurately for purely frictional or purely cohesive soils from model footing tests. For similar footings at corresponding z/b values the model law is given by:

$$\text{For } c = 0 \qquad F_1 = F_2\lambda^3;$$
$$\phi = 0 \qquad F_1 = F_2\lambda^2.$$

In the case of composite soils no direct solution is possible but the general functional load sinkage equation is: $F/\gamma b^3 = (c/\gamma b)f_1(\phi, z/b) + f_2(\phi, z/b)$.

4. The pressure sinkage equation derived from a fundamental plastic

solution of the load sinkage problem is of the form, $p = cN_{cq} + (\gamma b/2) \times N_{\gamma q}$; where N_{cq} and $N_{\gamma q}$ are functions fo ϕ and z/b.

Although this equation agrees qualitatively with the general similitude equation given above it is, so far, quite inaccurate in predicting load sinkage curves other than for cohesive soils. This is due to the effect of the compressibility of the soil and the inadequacy of the measurement and expression of soil shear strength.

5. The similitude and fundamental solutions of the load sinkage problem suggest a similar form for the pressure sinkage equation which can be approximated by the empirical equation: $p = (ck_c' + (\gamma b/2)k_\phi')(z/b)^n$.

The constants k_c' and k_ϕ' are valid soil values obtained from model footing tests. The equation will predict load sinkage values accurately for large areas, and for frictional, composite, and cohesive soil types. The accuracy of the equation could be further increased however by an improvement in the method of evaluating the empirical soil constants, especially for cohesive soils.

REFERENCES

1. BERNSTEIN, R., *Probleme zur experimentellen Motorpflugmechanik*, Vol. 16, 1913.
2. LETOSHNEV, M. N., Wheeled Carts and their Performance on Country Roads, *Teoria i proisvod-stvo Sielskohozaynih Mashin*, Moscow, 1936.
3. BEKKER, M. G., *Off the Road Locomotion*, Ann Arbor, Mich.: Univ. of Michigan Press, 1960.
4. KARAFIATH, L., An Analysis of New Techniques for the Estimation of Footing Sinkage in Soils, U.S.A.T.A.C., Land Locomotion Lab., Report No. 18, Warren, Mich., 1957.
5. EVANS, I., The Sinkage of Tracked Vehicles on Soft Ground, *J. of Terramechanics*, Vol. 1, No. 2, 1964.
6. REECE, A. R., Problems of Soil Vehicle Mechanics, U.S.A.T.A.C., Land Locomotion Lab., Report No. 97, Warren, Mich., 1964.
7. TAYLOR, D. W. *Fundamentals of Soil Mechanics*, New York: Wiley & Sons, 1948.
8. NUTTALL, C. J., Scale Model Vehicle Testing in Non-Plastic Soil, Exptl. Towing Tank, Report 394, Hoboken, N.J., 1949.
9. NUTTALL, C. J. and R. P. M'GOWAN, Scale Models of Vehicles in Soils and Snows, *Proc. 1st Int. Conf. Soil-Vehicle Systems*, Turin, 1961.
10. BEKKER, M. G., Theory of Land Locomotion, *The Mechanics of Vehicle Mobility*, Ann Arbor, Mich.: Univ. of Michigan Press, 1956.
11. EMORI, R. I. and D. SCHURING, Model Study in Earthworking Equipment, *A.S.A.E.* March, 1964.
12. SCHURING, D., The Mechanics of Rigid Wheels on Soft Ground, *J. of Terramechanics*, Vol. 1, No. 2, 1964.
13. LISTON, R. A. and E. HEGEDUS, Dimensional Analysis of Load Sinkage Relationships in Soils and Snow, U.S.A.T.A.C., Land Locomotion Lab., Report No. 100, Warren, Mich., 1964.
14. TERZAGHI, K., *Theoretical Soil Mechanics*, New York: John Wiley, 1956.
15. MEYERHOF, G. G., The Ultimate Bearing Capacity of Foundations, *Geotechnique*, Vol. 2, 1951.
16. WILLS, B. M. D., The Load Sinkage Equation in Theory and Practice, U.S.A.T.A.C., Land Locomotion Lab., Warren, Mich. (To be published.)
17. REECE, A. R., B. D. WITNEY, and D. R. P. HETTIARATCHI, The Basis of Soil Failure Theory, *Proc. 2nd Int. Conf. Soil-Vehicle Systems*, Quebec, 1966.

The Significance of Disturbance and Thixotropy in Mobility Problems

LOUIS J. GOODMAN*

ABSTRACT

The details of an investigation to evaluate the effects of disturbance on stress-strain relationships of cohesive soils in the problem areas of plate sinkage and shear strength are described in this paper. The problem presented is twofold:

1. The investigation of the effects of complete disturbance or remolding to clay soils and subsequent thixotropic regain characteristics as measured by empirical soil stress-strain parameters.

2. The establishment of a measuring system for identifying different degrees of disturbance imparted to clay soils.

Disturbance processes result in a decrease of the empirical vertical load sinkage soil values and the shear strength properties, with subsequent increase noted with increased time after disturbance. A general increase in resistance to sinkage of both a test footing and a rigid wheel is also noted with increased time after disturbance, demonstrating the significance of thixotropic regain characteristics of cohesive soils in mobility problems.

Partial disturbance was initially investigated by analyzing thixotropic regain characteristics of the soil in both wheel sinkage and shearing strength in an attempt to simulate measurable degrees of disturbance. This method did not provide a positive approach and was replaced by a constant volume shear box to uniformly introduce a predetermined amount of disturbance.

An analysis of force-angular displacement relationships obtained with a modified constant volume shear box resulted in establishment of a system for identifying different degrees of disturbance. This makes it possible to analyze the significance of partial disturbance on a mobility soil value system in cohesive soils.

THE FIELD OF OFF-ROAD transportation is one of the oldest known to man and yet the problems inherent in this field have attracted little attention from engineers and scientists until recently. The mechanics of off-road transportation or land locomotion are concerned with the complex problems of the relationship between wheeled or tracked vehicles and various types and conditions of natural soil surfaces. The need for basic research on the ground properties which affect vehicle performance has been emphasized by experience of land locomotion in the areas of military mobility, agriculture mechanization, timber transport, and earth moving equipment.

*Associate Professor of Civil Engineering, Syracuse University, Syracuse, New York.

The past ten years (1955–1965) have seen an increase in research efforts in the field of land locomotion mechanics to obtain a better understanding of the action of soil-vehicle systems. The purpose of these investigations is the prediction and evaluation of performance and designs for off-road or cross-country locomotion. These studies are generally directed toward the problems most frequently encountered in the field by land vehicles operating over natural ground surfaces that fall in the categories of loose or soft soil. The basic problems in these soil types are excessive wheel or track sinkage due to physical characteristics of both the soil and the vehicle, and excessive wheel or track slippage caused by a combination of insufficient shearing strength of the soil and vehicle characteristics.

It is clear that the study of soil behaviour under vehicle action embraces the two areas of wheel or track sinkage and slippage. These areas are related to one of the two principal problem groups of soil mechanics, the stability problem. This problem group deals with the conditions for equilibrium of soils immediately preceding ultimate failure by plastic flow. In order to solve the stability problem, it is necessary to know the stress conditions for the failure of the soil. Solutions to these problems must be consistent with the rate of loading which is not simply a problem of statics as in dealing with the foundation of a building, but rather a quasi-static problem. Another problem area in soil-vehicle relationships is encountered when the strength of the soil is exceeded and plastic deformation results, which is of practical significance in the case of a vehicle moving on soft ground.

Four approaches to the studies of land locomotion problems have been adopted by various groups: purely empirical, direct extension of soil mechanics, dimensional analysis and use of soil force-deformation characteristics (1).

The application of soil force-deformation curves to soil-vehicle relationships has proven to be the most general of the various approaches taken. This approach was developed primarily by Bekker (2), (3) and is based on his soil-value system which consists of a group of six parameters that are used in equations to approximate the horizontal and vertical force-deformation curves. The vertical force-deformation (or pressure-sinkage) curve is described by the equation:

$$p = (k_c/b + k_\phi)z^n \tag{1}$$

In this equation, k_c, k_ϕ, and n are soil values, p is the pressure, b is a dimension of the loaded area, and z is the depth of sinkage. No attempt is made to relate the soil values to cohesion, density, friction angle, or other soil properties. The values are determined by conducting a series of pressure-sinkage tests with sinkage plates of varying sizes on any particular soil of interest.

The horizontal force-deformation (or shear) curve is described by a modification of the Mohr-Coulomb equation:

$$s = (c + p \tan \phi)(1 - e^{-j/k}) \tag{2}$$

n which c, ϕ, and k are soil values, s is the unit shear strength, p the normal load and j is the deformation at any point. The values of c, ϕ, and k are established by conducting shear tests in which different normal loads are applied to an angular shear head or to a shear block.

Methods have been developed to quickly determine the value of these coefficients in the field. This approach has served a very useful purpose in current studies of land locomotion problems.

During the past six years, Syracuse University, through its Research Institute, has done research in the field of land locomotion mechanics, with particular emphasis on the effects of disturbance to clay soils on mobility problems. The principal objective of this research effort has been to evaluate the effects of various degrees of disturbance to cohesive soils on empirical soil stress-strain parameters proposed for use in the mechanics of vehicle mobility. The programme included a study of both the influence of the geometry of the soil-loading devices and the significance of thixotropic regain characteristics on the soil values. The concept and significance of partial disturbance in the analysis of soil-vehicle systems has been examined and a technique for identification of different degrees of disturbance is proposed.

The details of theoretical considerations, and both experimental and analytical phases of this programme are covered in Report No. 7782, LL76, Land Locomotion Laboratory of the U.S. Army Tank-Automotive Command (June 1962) and in the Final Reports, Third and Fourth Year's Study, Reports No. CE734-6312F3, F4, Syracuse University Research Institute (December 1962 and December 1963) (4), (5), (6).

This paper is primarily concerned with the determination of a measuring system for identification of different degrees of disturbance imparted to cohesive soils between the two limiting states of undisturbed and completely disturbed or remolded. Partial disturbance or degree of disturbance between the two extreme states was initially simulated by thixotropic regain time periods. The basic parameters of partial disturbance are strength-deformation characteristics which were investigated by means of measuring sinkage from successive passes of a rigid wheel, measuring moisture tension or negative pore water pressure, and determining in-place shearing strength with a laboratory shear vane. The above parameters were obtained in undisturbed, completely disturbed, and various thixotropic regain states of the cohesive soils selected for this phase. A constant volume shear box was then designed and constructed as a disturbance source to impart a predetermined degree of disturbance uniformly within a cohesive soil; this latter approach to the problem shows promise and has resulted in a proposed method of identifying different degrees of disturbance.

DISTURBANCE AND THIXOTROPY

Remoulding or completely disturbing a mass of saturated cohesive soil by a kneading-type action has been observed to result in a pronounced reduction in strength properties and an increase in deformation characteristics for the majority of natural clay deposits encountered on a world-wide basis. It has also been observed that these soils will acquire or regain cohesive strength if permitted to stand without further disturbance and without change in water content. If the soil mass is again disturbed at unaltered water content, its strength will decrease considerably, with an increase in strength properties and a decrease in deformation characteristics again noted with increased time after the disturbance. This phenomenon is known as thixotropy and the softening of the soil owing to disturbance and subsequent recovery are apparently due to a breakdown of the molecular structure of the adsorbed layers and subsequent gradual reorientation of the adsorbed molecules of water.

The practical significance of thixotropic regain characteristics of cohesive soils is related to the effect of multiple passes of a vehicle or vehicles over a surface cohesive soil on passability, the influence of the length of the time interval between the passes, and the change in soil conditions caused by the front wheels of a multiaxle vehicle which influences the action of the rear wheels. The consideration of a disturbed clay mass exhibiting increased strength properties and decreased deformation characteristics owing to thixotropic regain characteristics can be demonstrated by the determination of wheel sinkage resulting from successive passes of the wheel over a bed of clay in undisturbed, completely disturbed, or remoulded and various thixotropic regain or recovery states. This is illustrated in Figure 1, which shows the relation between number of passes of a rigid wheel and resulting sinkage for the two limiting states and thixotropic regain time intervals up to 14 days. Figure 2 is a replot of the data as per cent regain of resistance to sinkage for 1 to 5 passes and recovery periods of 1 to 14 days.

It is apparent from Figures 1 and 2 that it should be possible to simulate partial disturbance states or different degrees of disturbance between the two limiting states of undisturbed and completely disturbed by analysis of thixotropic regain characteristics.

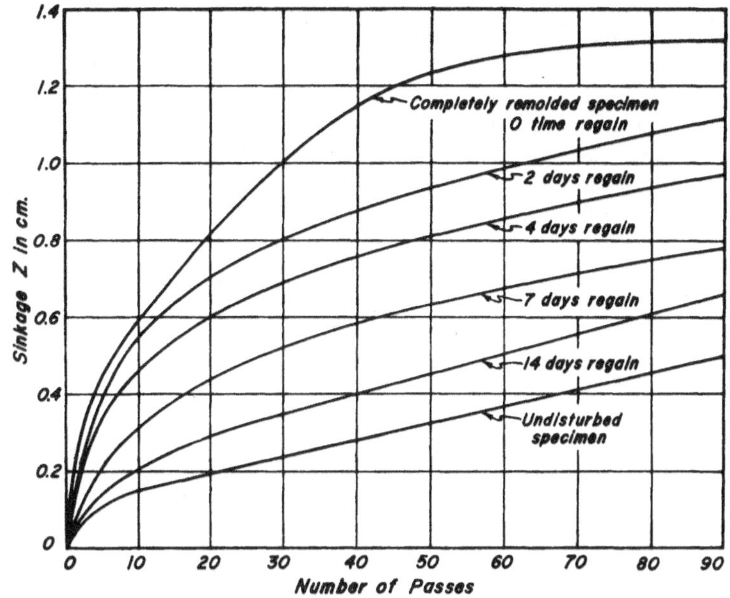

FIGURE 1. Resistance to rigid wheel sinkage for various thixotropic regain periods, Clay F

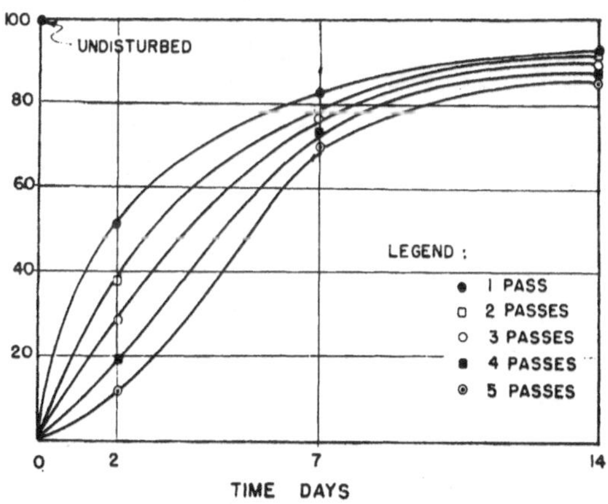

FIGURE 2. Per cent regain of resistance to sinkage produced by a rigid wheel, Clay F

The complex phenomenon of thixotropy is of quite general occurrence in fine-grained soils such as clays. This phenomenon by which a cohesive soil regains part or all of its original strength after complete disturbance is time dependent, and its practical applications warrant consideration in the evaluation of any soil value system for off-road vehicles.

Mitchell (7) attempted to explain thixotropy by analyzing the internal energy and the stress conditions in a thixotropic soil. When a thixotropic soil is remoulded or compacted, a part of the externally applied shearing energy is utilized in dispersing the platey clay particles into a uniform, parallel arrangement. That is, the externally applied energy assists the repulsive forces between particles, resulting from double layer interaction, in producing a dispersed system. The energy of interaction between particles is at a level commensurate with the externally applied forces, and the adsorbed water layers and ions are distributed in accordance with the high energy level. The net result is a structure similar to that shown schematically in Figure 3a. Such a structure is compatible with the conditions existing during the shearing process.

However, as soon as shearing ceases, the externally applied energy drops to zero. Thus, the attractive forces now exceed the repulsive forces for the particular arrangement of particles and distribution of water. As a consequence, the structure attempts to adjust itself to a new, lower energy condition. The energy dissipation may be accompanied by changes in particle arrangements, adsorbed water structure, and distribution of ions. Since structural changes of this type are dependent upon the actual

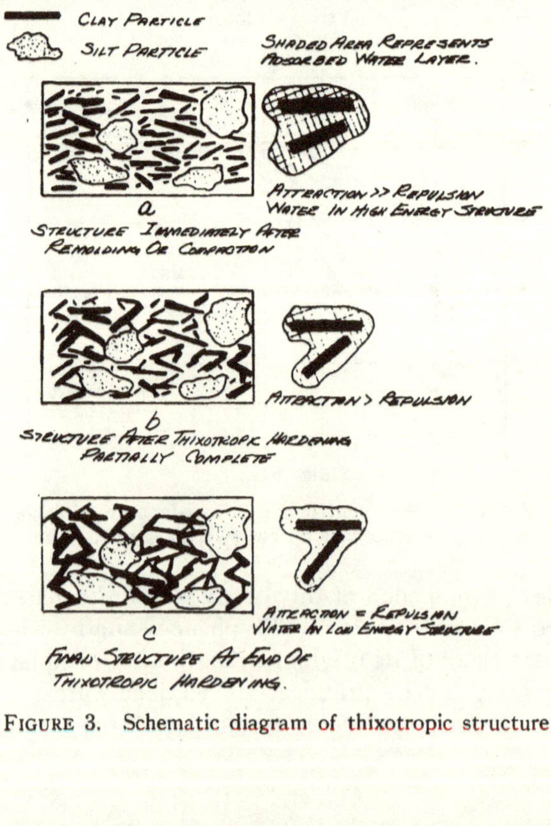

FIGURE 3. Schematic diagram of thixotropic structure

physical movement of the particles, water, and ions, they are time dependent. These internal structural changes are in turn reflected by soil behaviour, for example, an increase in the shearing strength. A schematic diagram of the structure at some time after remoulding is shown in Figure 3b; the final structure is shown in Figure 3c.

EXPERIMENTAL PROGRAMME

A. Soil Types

The soil types for the different phases of this programme were soft to medium lacustrine clays obtained from the general Syracuse area and a marine clay from Boston, Mass. (Boston Blue Clay). The index properties of the clays investigated are shown in Table I.

The Boston Blue Clay specimens are designated by F-1 to F-5. The F-1 samples were used in the experimental phase concerned with the influence of disturbance as measured by empirical soil stress-strain parameters. The F-2 and F-3 samples were used in the small scale footing studies. The phase concerned with the identification of different degrees of disturbance was generally based on the F-4 and F-5 series of the samples.

B. Special Test Equipment

1. Load Sinkage

The load sinkage machine developed and constructed to obtain the vertical load values (k_ϕ, k_c, and n) is shown in Figure 4. The maximum

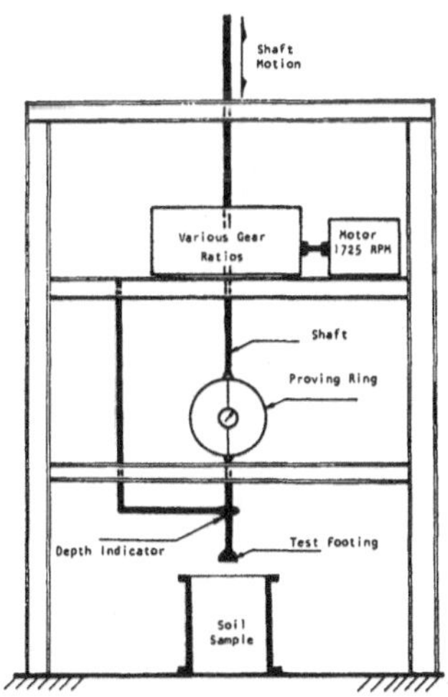

FIGURE 4. Test apparatus for sinkage rates up to 9.0″/min

TABLE I

CLASSIFICATION CHARACTERISTICS OF SOILS INVESTIGATED

Soil designation	Source	Mechanical analysis			Atterbergs		Natural water content	Cohesion		Sensitivity	Description
		% Sand	% Silt	% Clay	L.L.	P.I.		Undisturbed	Remoulded		
A	Syracuse	0	42	58	55	31	45+	0.88	0.18	4.9	Silty clay
B	Syracuse	10	66	24	26	5.0	26+	0.79	0.21	3.8	Clayey silt with trace sand
C	Syracuse	6	55	39	32	13.9	29+	—	—	—	Clayey silt with trace sand
F-1	Boston	12	40	48	35.3	13.9	26	0.44	0.015	29.4	Silty clay with trace to little sand
F-2	Boston	2	45	53	34.8	14.5	28.7	—	—	—	Silty clay with trace sand
F-3	Boston	14	40	46	38.2	15.6	28.7+	0.40*	0.04*	10	Silty clay with trace to little sand
F-4	Boston	6.0	40	54	34.9	13.8	30.5	0.22*	0.039*	5.6	Silty clay with trace sand
F-5	Boston	3	39	58	40.5	17.0	36.7+	0.44*	0.074*	5.9	Silty clay with trace sand
G	Syracuse	7.3	24.4	68.3	68.3	38.9	35.4	0.69*	0.87*	0.8	Silty clay with trace sand
H	Syracuse	3	31	66	33.5	13.3	25.4–35.4	1.19*	0.31*	3.8	Silty clay with trace sand

NOTE: L.L. = liquid limit; P.I. = plasticity index; unconfined compressive strength and laboratory shear vane measured in tons per sq. ft.; soil C is a mixture of soils A and B.

*Determined by laboratory shear vane.

sinkage is 12 inches with the load limited to a maximum of 350 pounds. The rate of vertical movement of the test footings can be varied by changing the size of the pulleys on the belt drive section of the machine.

Circular test footings with a diameter range of $\frac{3}{4}$ to $4\frac{1}{2}$ inches and rectangular footings with a breadth range of 1 to 4 inches have been investigated. The test footings have a constant side area to eliminate any variation that might result from side friction.

2. Different Degrees of Disturbance

Partial disturbance or degree of disturbance between the two extreme states of undisturbed and completely disturbed was initially studied by thixotropic regain time periods. Experimentation on partial disturbance based on this concept was investigated by means of the following techniques:

(*a*) Measuring sinkage from successive passes of a rigid wheel over the surface of a bed of clay;

(*b*) Measuring moisture tension or negative pore water pressure;

(*c*) Determining in-place shearing strength.

The use of a constant volume shear box to study relationships between the degrees of disturbance imparted to undisturbed clay and its shearing strength or strain was also investigated in an attempt to establish an identification system for partial disturbance. This shear device serves as a disturbance source to a clay specimen by means of simple shear deformation under constant volume. The degrees of weakening of the soil or the degree of disturbance can be related to the angular deformation by strength determinations or work input considerations. The details of the initial design and construction of the shear box are shown in Figure 5.

C. Influence of Disturbance on Soil Value Measurements

This phase of the experimental programme was primarily concerned with the effects of complete disturbance or remoulding on the vertical load sinkage soil parameters, k_c, k_ϕ, and n. Control measures involved constant soil density and water content and an adequate ratio of soil bin to footing size to eliminate interference from stresses induced by the sides or bottom of the soil bins.

D. Thixotropic Regain Characteristics

Thixotropic regain characteristics were determined for both the lacustrine and marine clay specimens in connection with shear strength parameters, resistance to sinkage of a rigid wheel and plastic resistance in sinkage. The basic approach was to measure these properties for the soil specimen which had been initially completely disturbed or remoulded and then tested at unaltered water content and density for thixotropic

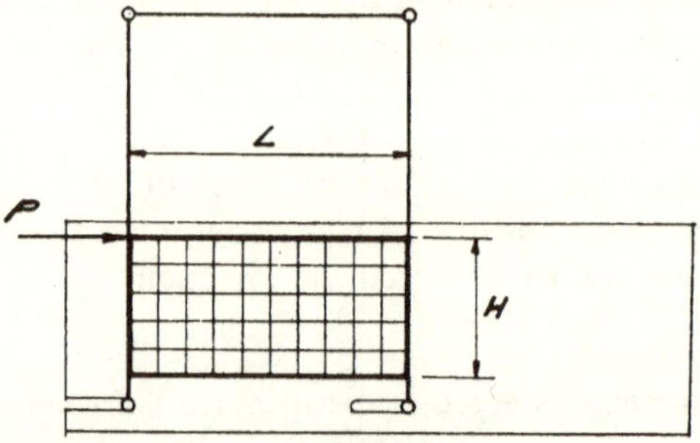

FIGURE 5A. Longitudinal section through constant volume shear box. Soil specimen in undisturbed state

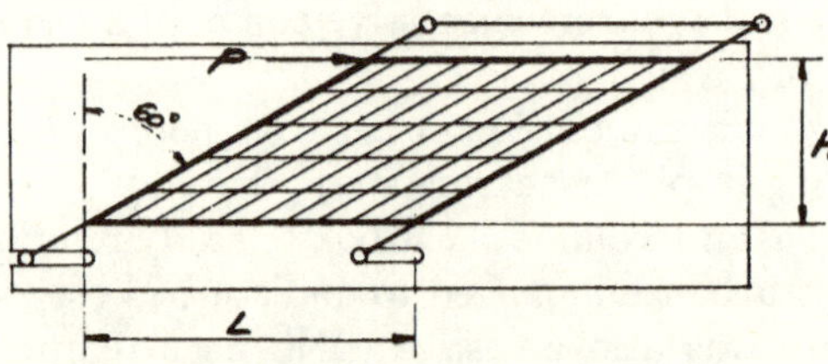

FIGURE 5B. Longitudinal section through shear box after reaching maximum deformation

regain time periods ranging from 1 to 14 days. The significance of this phenomenon was investigated for the cohesion parameter of soil strength by means of standard shear testing procedures, such as direct shear, triaxial compression, unconfined compression, and laboratory shear vane.

E. Different Degrees of Disturbance

1. Sinkage from successive passes of a rigid wheel

The significance of partial disturbance was initially investigated by means of passing a rigid wheel over the surface of a bed of clay. The procedure followed was to make passes on an initially undisturbed sample, using the same path for subsequent passes, and measuring the sinkage after each pass. Simultaneously, laboratory shear vane determinations were made in the wheel path after each pass. The cohesion of the soil as determined by the laboratory shear vane and the sinkage values were used as criteria for analyzing the different degrees of disturbance by comparing these values after each pass with those obtained on undisturbed and completely remoulded specimens. This preliminary work was done with a rigid wheel with a diameter of 3 inches and a width of $\frac{1}{2}$ inch.

2. Tensiometer (Moisture Tension)

Two preliminary series of experiments were conducted in an attempt to determine a relationship between soil moisture tension and the shearing strength of the soil as a basis for measuring the amount of disturbance in a soil.

The first experiment with the tensiometer showed an increase of pore water tension of about 30 increments over a period of 30 days. The sample was subjected to fairly large temperature fluctuations, resulting in a modest change in water contents which apparently have a significant effect on the tensiometer readings. In the second experiment two samples were used, with one placed in the constant temperature room (20°C) and the other in the moist room, also at constant temperature of 20°C. Each sample showed an increase in pore water tension of 2.5 increments in 20 days. The difference in readings between the first and second experiments, 30 and 2.5 increments respectively, is apparently due to the temperature fluctuations and the resulting sample moisture content fluctuations that occurred during the first experiment.

3. Laboratory shear vane

A preliminary study was also made of intermediate shear strengths between the limits of the undisturbed and the completely disturbed states by use of the laboratory shear vane in an attempt to identify partial disturbance states. The results were not encouraging, so a new approach to the problem was investigated.

4. Constant volume shear box

A constant volume shear box was then designed and constructed in order to find a relationship between a predetermined amount of shearing strain imparted to a cohesive soil in terms of angular deformation and the resulting degree of weakening. Preliminary experimental data were obtained for soil specimens 3 × 3 × 7 inches in the following manner:

An undisturbed sample was deformed repeatedly from 0 to 45° angular displacement. During each straining cycle the resistance to deformation of the specimen was measured, for equal intervals, with a proving ring.

It was found, as expected, that the resistance to deformation or the total work input for a cycle decreased as the number of cycles increased and that the total work input for one cycle approached the value that was obtained when a specimen of a completely remoulded soil sample was deformed in the constant volume shear box.

An attempt was also made to measure the uniformity of disturbance within a soil specimen that was subjected to several cycles of straining in the constant volume shear box. The laboratory shear vane was used for this purpose and it was found that the degree of disturbance was a maximum along the top and the moving sides, decreasing towards the centre of the sample. In order to increase the uniformity of disturbance within a sample the constant volume shear box was modified to accommodate a sample size 3 × 3 × 3 inches.

Preliminary tests with the modified constant volume shear box showed much better uniformity of disturbance within a test specimen. In these preliminary tests, specimens of undisturbed Boston Blue Clay were used and subjected to repeated cycles of shear deformation extending from 0 to 40° angular deformation. During the tests the force required to deform the specimen was measured with a proving ring, for angular deformations of 10, 20, 30, and 40°.

When the area under the curve of minimum force is integrated, the work required for moving a completely remoulded sample from 0 to 40° angular deformation is obtained. The effective work for each cycle is obtained by deducting the aforementioned work from the total work done during each cycle of deformation. This is the work which is effective in imparting disturbance to the undisturbed or partially disturbed soil.

The total effective work is obtained by summing up the effective work for the number of cycles for which the effective work becomes zero and is equivalent to the energy needed to completely remould the soil. The effective work done on the soil when plotted versus the amount of angular deformation lends itself as a possible scale for measuring the degree of disturbance of a cohesive soil.

RESULTS AND DISCUSSION

A. Soil Types

The soil types selected for the initial phase of this study were secured from soft to medium lacustrine clay deposits. A medium marine clay and a stiff deposit of lacustrine clay were selected for the intermediate and final phases. The lacustrine clay specimens were obtained from the general Syracuse area and the marine clay from Boston, Massachusetts (Boston Blue Clay).

Lacustrine clays are lake deposits and generally consist of alternate relatively thin strata of silt and clay. Deposition was such that these soils might be quite erratic with respect to homogeneity. It was observed that undisturbed specimens of this type of clay from the Syracuse area contained variable layers of silt and some pockets of fine sand which would account for some of the erratic experimental results. Remoulding has a homogenizing effect, which would tend to produce more uniform experimental data. Clay deposits of marine origin are as a rule much more uniform in composition than lake deposits. This had been our experience in past investigations with Boston Blue Clay and it was decided to work with this type of clay at an early date. As a point of interest, deposits of lacustrian origin occupy an intermediate position between marine deposits and deposits of fluvial (river) origin, with fluvial soils tending to be most erratic.

B. Sinkage Problem

The sinkage problem was studied from two basic standpoints, (1) the influence of complete disturbance or remoulding on the sinkage soil parameters, and (2) sinkage footing geometry and rate efforts. This section will discuss the results of the effects of complete disturbance on the soil value measurements.

Attention was given to the spacing of the test footings in the early stages of this phase of the experimental programme to ensure that each reading would not be affected by stress overlap from adjacent tests or from confinement due to the specimen container. Observations showed that a ratio of specimen diameter to test footing diameter of 5 or 6 is necessary to eliminate both overlapping and lateral confinement effects. It was also observed that bottom effects from soil samples with the above diameter ratios began to appear when the test footing was 8 to $8\frac{1}{2}$ in. from an unyielding base and the entire sample was in a homogeneous state.

1. Effects of complete disturbance on soil values in sinkage

To determine the effects of complete disturbance on the values of k_ϕ, k_c, and n, two general series of tests were conducted, one in the undisturbed state and the other in the remoulded state. Typical results from this phase for Clay F are shown as cumulative distribution functions in Figures 6, 7, and 8. It can be noted from these figures that the data follow fairly close to a normal distribution. Figures 6 and 7 show the most significant effect of remoulding to be a reduction in the dispersion of the values of k_ϕ, and k_c. There is also a reduction in the value of the mean for k_ϕ and k_c and an increase of the value n by remoulding. These effects are probably due to the homogenizing effect and the removal of concentrated resistance to penetration from the remoulding process.

An extensive series of load sinkage tests was then conducted with Clay F in a completely disturbed state to insure homogeneity of the specimens. This phase of the investigation included both circular and rectangular test footings with contact areas ranging from approximately $\frac{3}{4}$ to 16 sq. in. Frequency distribution curves for values of the cohesive and frictional moduli of soil deformation, k_c, k_ϕ, and the sinkage exponent, n for Clay F also show that the data approach the form of a normal distribution (5).

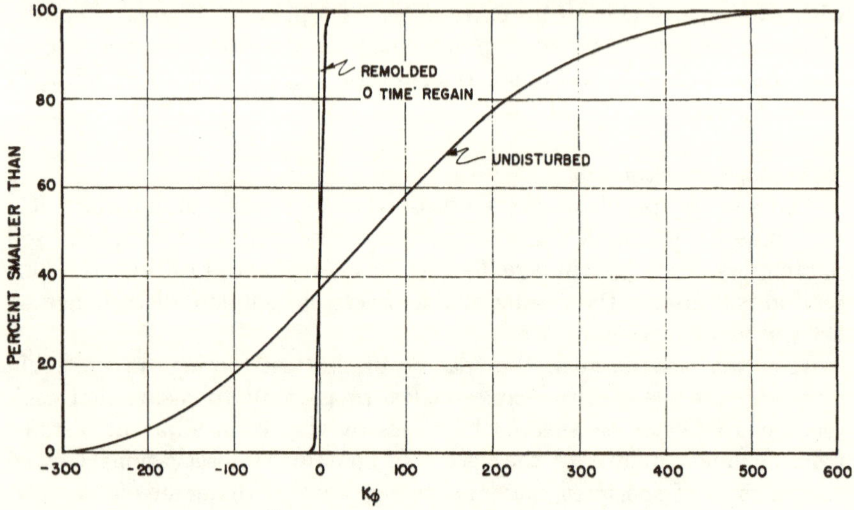

FIGURE 6. Cumulative distribution function of K_ϕ, Clay F

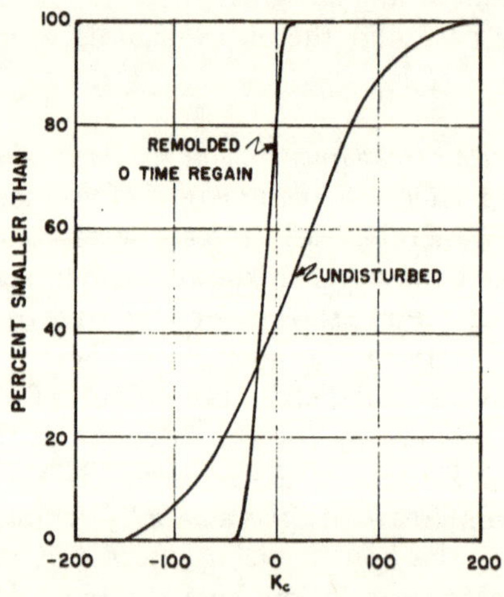

FIGURE 7. Cumulative distribution function of K_c, Clay F

C. Disturbance and Thixotropic Regain Characteristics

The significance of disturbance processes to strength and deformation characteristics of cohesive soils with subsequent thixotropic regain or age hardening characteristics was investigated for shear strength parameters, resistance to sinkage of a rigid wheel, and resistance to plate or footing sinkage.

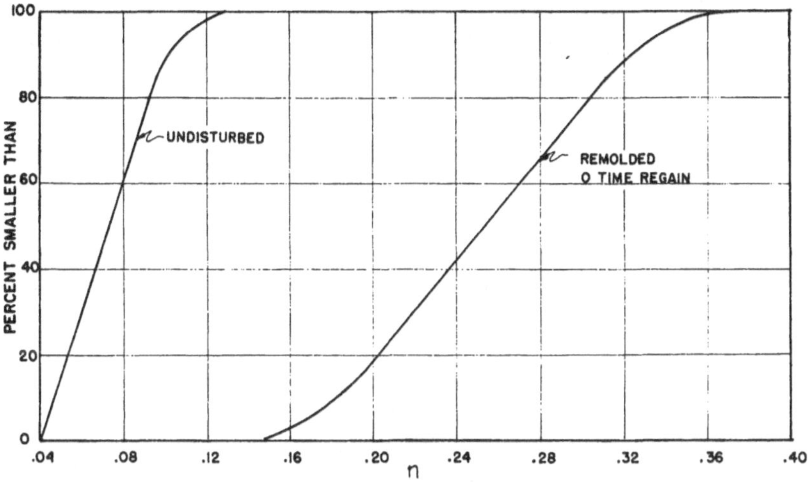

FIGURE 8. Cumulative distribution function of n, Clay F

1. Laboratory shear tests

The curves of Figure 9 show the effects of remoulding with subsequent thixotropic regain characteristics for the mean value of cohesion for Clay F obtained by various shear testing methods. This included the use of a special horizontal shear device developed during the early stages of this research study (4). It can be noted from this figure that an increase in the mean value of cohesion occurs with increased time after complete disturbance, regardless of the shear testing method used.

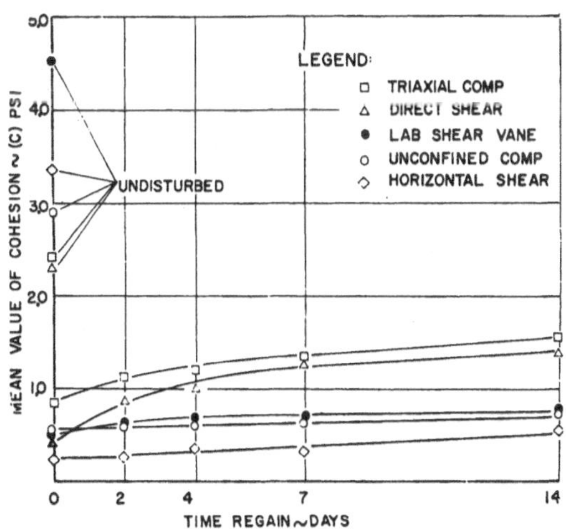

FIGURE 9. Thixotropic regain characteristics of cohesion from various shear testing methods, Clay F

2. Resistance to wheel sinkage

Figures 1 and 2 show the effects of complete disturbance or remoulding on the resistance to sinkage produced by a rigid wheel passing over a bed of marine clay. It can be noted from these curves that the subsequent thixotropic regain in resistance to sinkage is significant, which is of much practical value. The significance of this phenomenon will be further emphasized in part D of this section.

3. Resistance to plate or footing sinkage

The influence of a thixotropic regain period of one day in connection with resistance to sinkage of a $3\frac{1}{2}$ in. diameter footing in Clay F is shown in Figure 10.

It can be observed from Figure 11 that an increase in both k_c and k_ϕ values occurs with increased time after complete disturbance (remoulding) to Clay F. It can also be observed that a decrease in n values occurs with increased time, as expected. The observed thixotropic regain values of p for plastic flow are shown in Figure 12. The dashed lines in this figure have

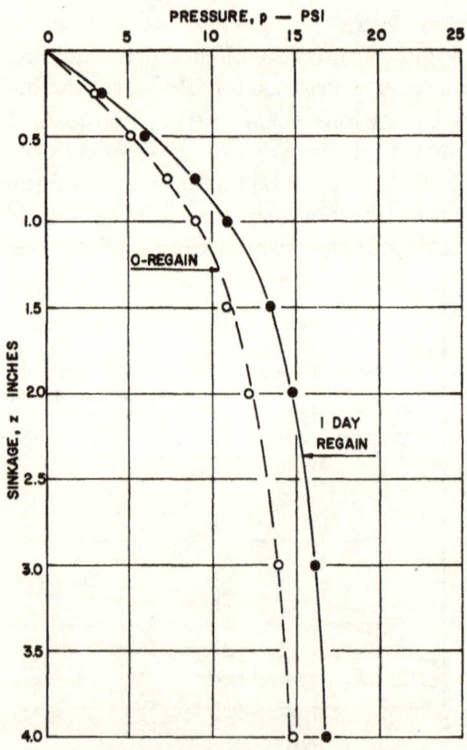

FIGURE 10. Thixotropic regain characteristics of resistance to footing sinkage, Clay F

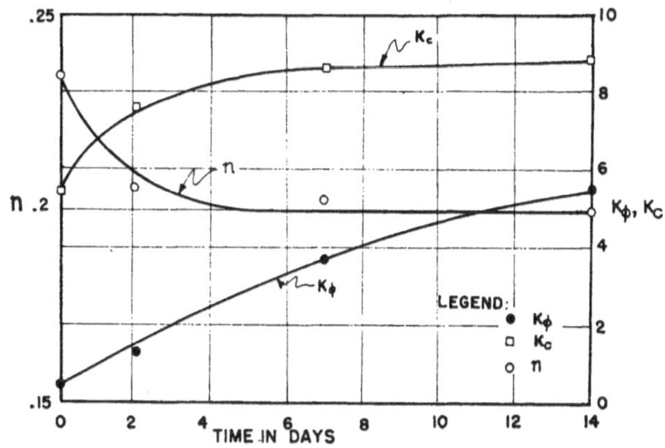

FIGURE 11. Thixotropic regain characteristics for K_c, K_ϕ, and n, Clay F

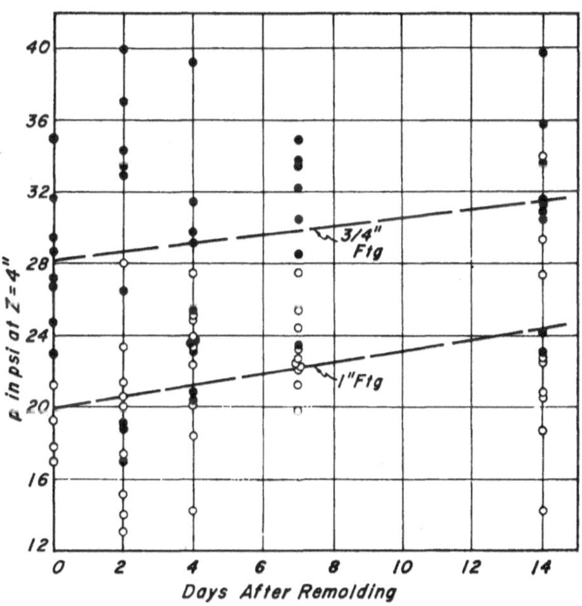

FIGURE 12. Thixotropic regain characteristics of plastic resistance in sinkage, Clay F

been determined by the method of least squares. There seems to be no significant regain of the value of p for plastic flow. It is interesting to note that although there are increases of k_ϕ and k_c with time and therefore changes in the shape of the p vs. z curves, the values of p at Z remain statistically unchanged.

D. Measuring Partial Disturbance

1. Sinkage from successive passes of a rigid wheel

Tests were made with undisturbed, partially disturbed, and completely disturbed or remoulded specimens of clay soil to determine if a correlation could be developed between wheel sinkage and disturbance. The partial disturbance was simulated by completely disturbing a sample and then aging it in the moist room with constant water content and density control of the sample. Samples of 1, 2, 4, 8, 15, 30, and 60 days were used to obtain pertinent experimental data.

The data obtained by rolling the test wheel over the clay specimens in various states of disturbance were plotted on both arithmetic and semi-logarithmic paper. The arithmetic plot of sinkage versus number of passes shows a family of curves each having a decreasing rate of sinkage with an increasing number of passes of the wheel. This is illustrated in Figure 13 for Boston Blue Clay.

The semi-logarithmic plot of sinkage versus the log of the number of passes (Figure 14) shows a more or less distinct straight line portion in the middle region of each curve. When the slope of each straight line portion is plotted versus the days after remoulding of the soil sample, an indication of the regain characteristics is obtained, as shown in Figure 16.

A similar plot has been prepared by working with a focal point which was arrived at by obtaining the intersection of the straight line portions of the two extreme curves (undisturbed and completely disturbed) which is shown in Figure 15. Through this point of intersection a line is drawn at a tangent to the upper part of each curve. The numerical value of each tangent line, which is the rate of sinkage, is then plotted against the days of regain after remoulding. This plot is also shown in Figure 16.

It can be noted from Figure 16 that the rate of wheel sinkage after complete disturbance tends to approach the undisturbed case owing to thixotropic regain. This phenomenon indicates that it should be possible to produce and measure intermediate sinkage values with this technique and therefore affords promise as one of the bases for identifying partial disturbance states.

2. Constant volume shear box

The results of preliminary tests with the modified constant volume shear box are shown in Figure 17 for Boston Blue Clay. The resulting force for as many as 10 cycles of straining is plotted against the angular displacements. The minimum force required for deforming a completely disturbed specimen is also shown in this figure. The total effective work is obtained by summing up the effective work for the number of cycles for which the effective work becomes zero and is equivalent to the energy needed to completely remould the soil. The effective work done on the

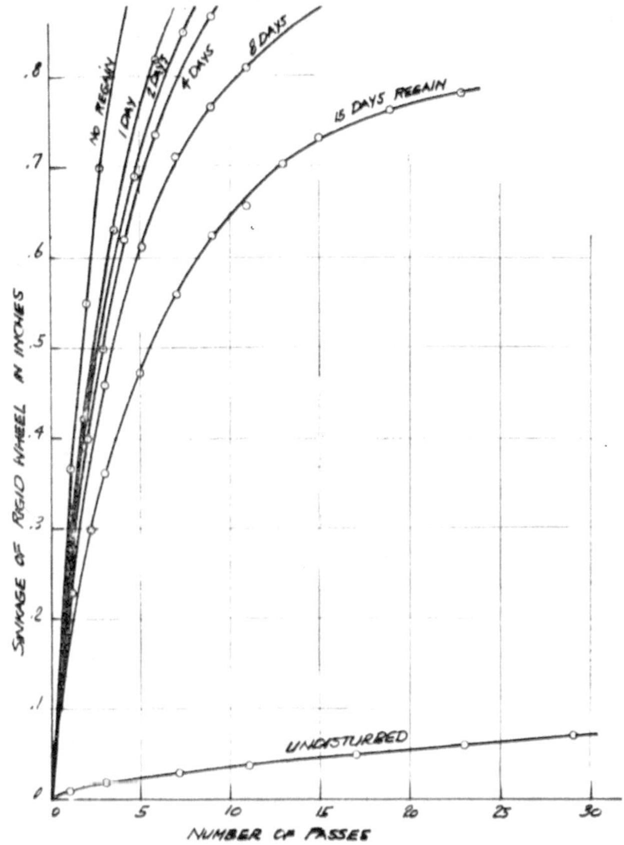

FIGURE 13. Effects of thixotropic regain on wheel sinkage, Clay F

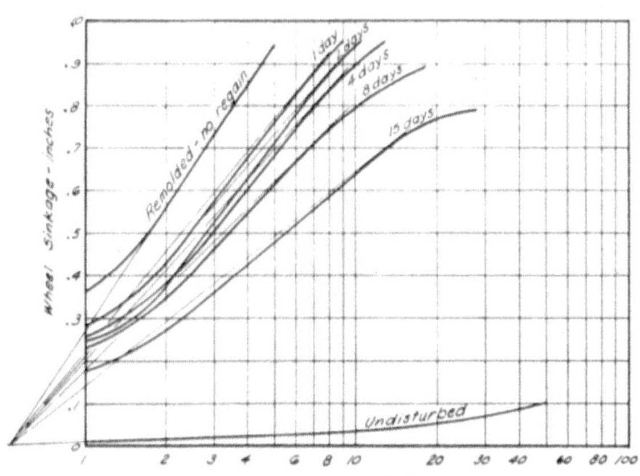

FIGURE 14. Effects of thixotropic regain on wheel sinkage, Clay F

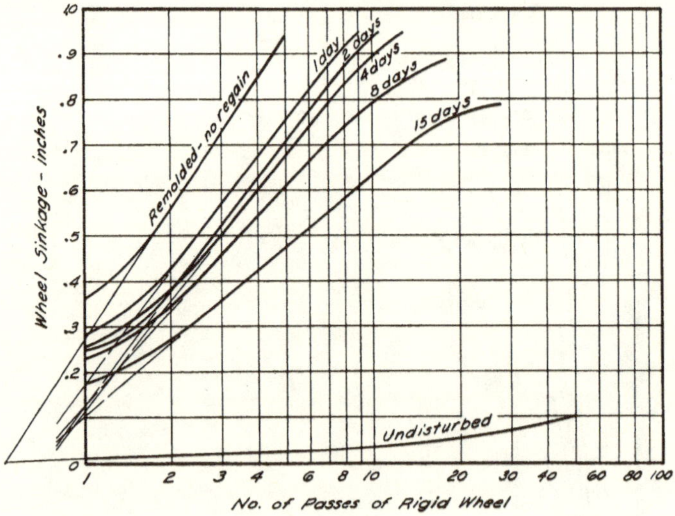

FIGURE 15. Effects of thixotropic regain on wheel sinkage, Clay F

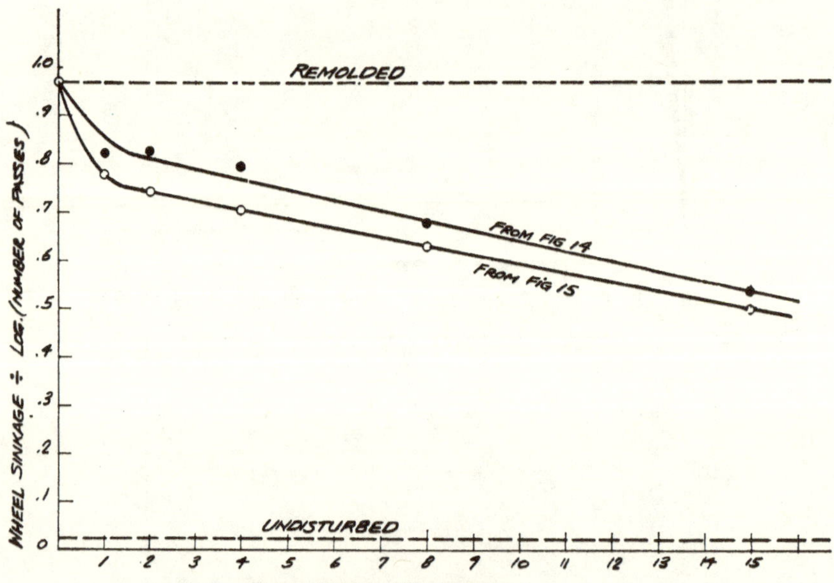

FIGURE 16. Rate of sinkage vs. days after remoulding, Clay F

soil when plotted versus the amount of angular deformation lends itself as a possible scale for measuring the degree of disturbance of a cohesive soil as can be seen from Figure 18.

The force-angular displacement relationships for Boston Blue Clay in a

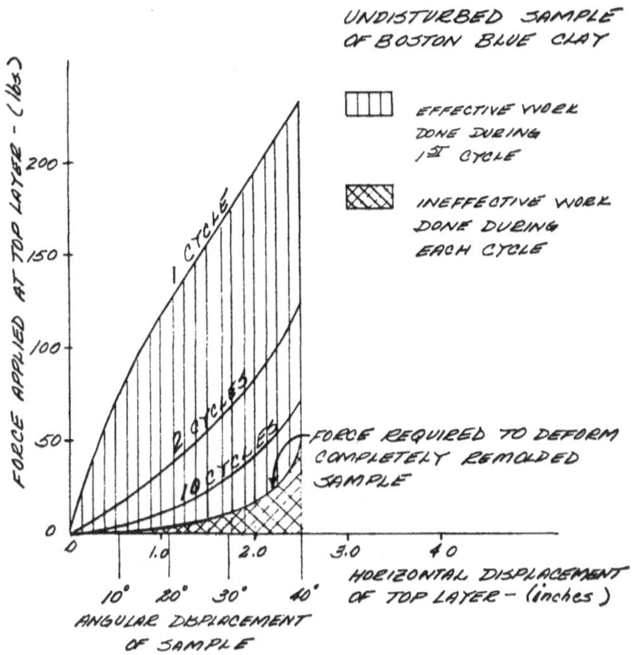

FIGURE 17. Force-angular displacement relationships using constant volume shear box, Clay F

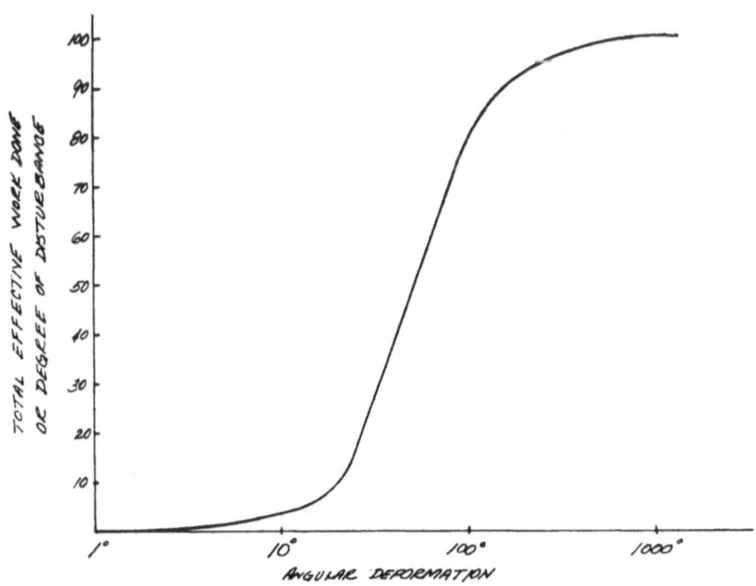

FIGURE 18. Disturbance scale vs. angular displacement, constant volume shear box method, Clay F

subsequent series for 20 cycles of loading and a maximum angular deformation of 50 degrees per cycle are shown in Figures 19 and 20. Figure 21 shows the relationship between the accumulated effective work imparted to the soil and the number of deformation cycles, with a suggested disturbance scale. Figure 22 shows a similar plot for the earlier series of cyclic loading (Figure 17) for comparison purposes.

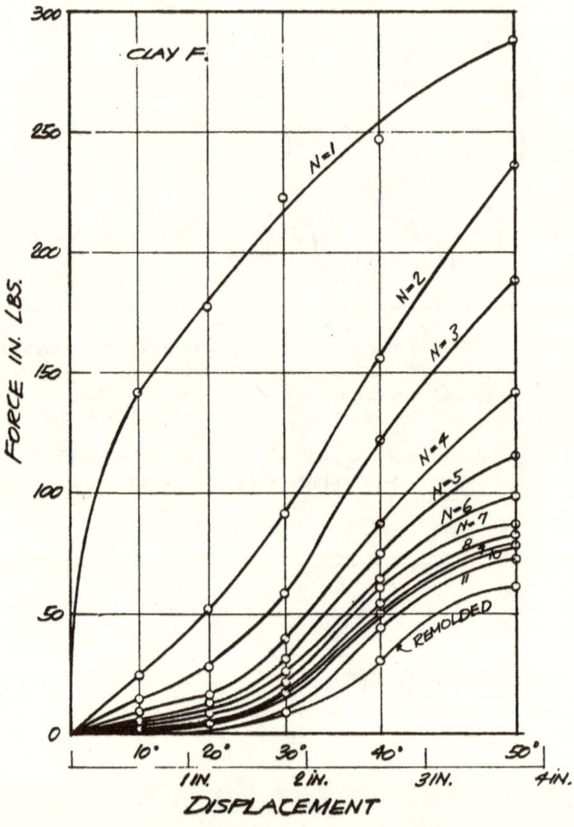

FIGURE 19. Force-displacement rleationships for angular deformations from 0–50°, Clay F

The constant volume shear box was used with a surface deposit of a cohesive soil (lacustrine clay) in an attempt to obtain a relationship between the degree of disturbance and angular deformation. Experimentation with this technique and analysis of data obtained with partially saturated specimens of a surface cohesive soil indicate that a heavier constant volume shear box should be used, particularly at angular deformations between 40 and 50 degrees. This is explained by

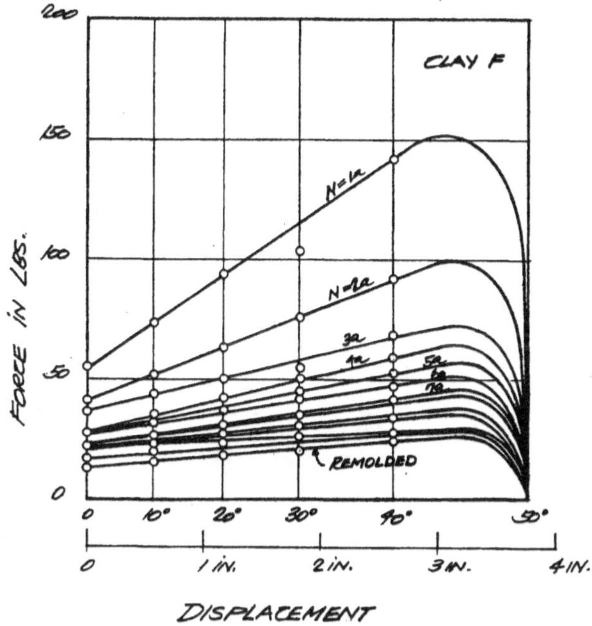

FIGURE 20. Force-displacement relationships for angular deformations from 50°–0°, Clay F

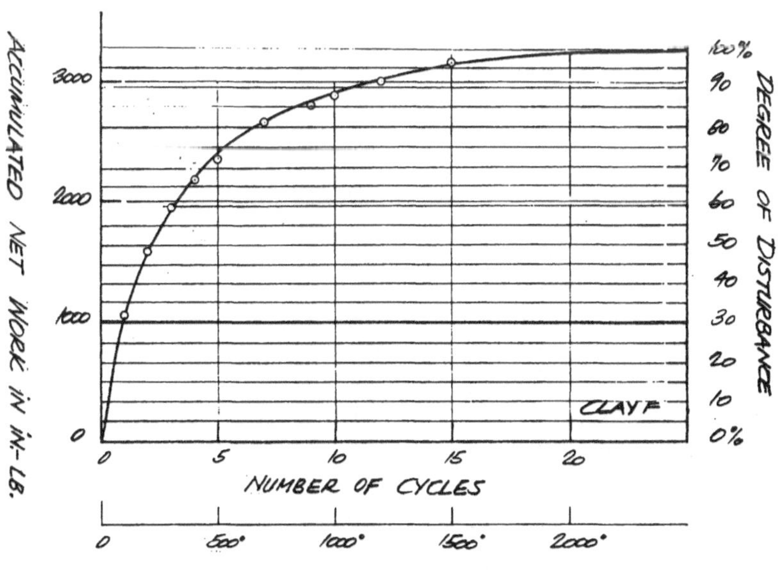

FIGURE 21. Accumulated angular deformation in degrees, Clay F

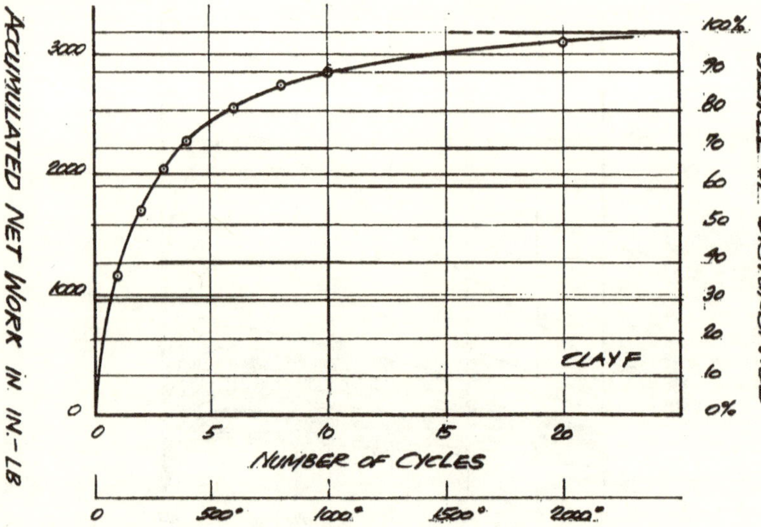

FIGURE 22. Accumulated angular deformation in degrees, Clay F

the fact that stiff clays afford more resistance to deformation than soft to medium clays, resulting in significant squeezing out of the soil at the higher angular deformations with a sensitive shear box.

It was therefore decided to investigate the possibility of identifying the degree of disturbance for stiff surface cohesive soils by means of the initial slope of the stress-strain curve obtained from the unconfined compression tests for undisturbed and completely disturbed or remoulded states. Intermediate states of disturbance were accomplished by the use of the constant volume shear box, using cyclic loading as necessary to induce various degrees of shear deformation or disturbance. Specimens were secured from each sample subjected to the predetermined amount of disturbance for testing in the unconfined compression machine. Figure 23a shows the range of initial slopes for the two limiting states of disturbance. It was expected that the results of the unconfined compression tests on specimens obtained from the shear box would plot a family of curves as shown in Figure 23b. Work input considerations based on the horizontal force-displacement data from the constant volume shear box could then be used to supplement shear deformation data in the establishment of a partial disturbance identification system for this type of soil.

The final phases of the experimental programme investigated the dual approach of the use of a constant volume shear box and unconfined compression testing to establish a relationship between the degree of weakening or disturbance and induced deformation for a surface cohesive

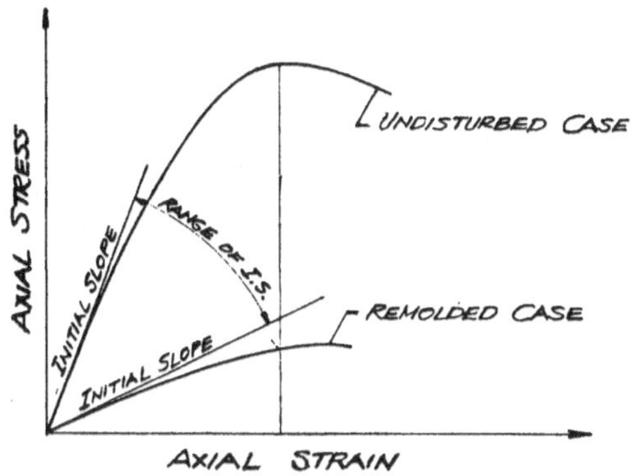

FIGURE 23A. Range of initial slopes as determined from unconfined compression tests

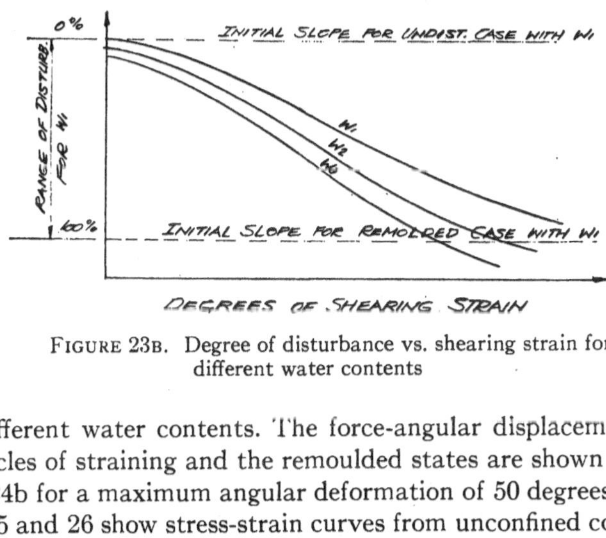

FIGURE 23B. Degree of disturbance vs. shearing strain for different water contents

soil at different water contents. The force-angular displacement curves for six cycles of straining and the remoulded states are shown in Figures 24a and 24b for a maximum angular deformation of 50 degrees per cycle. Figures 25 and 26 show stress-strain curves from unconfined compression tests on the surface clay in both undisturbed and completely disturbed states and two states of partial disturbance imparted to the soil in the shear box. Figure 26 is a "blow-up" of the initial portions of the curves for initial slope purposes. The range of disturbance plotted against degrees of shearing strain for two significantly different water contents is shown in Figure 27.

Investigations of the influence of partial disturbances on the shear and sinkage properties of both a soft to medium clay and a stiff surface clay were then commenced. The results of laboratory shear vane testing of the two clays in undisturbed, partially disturbed, and completely

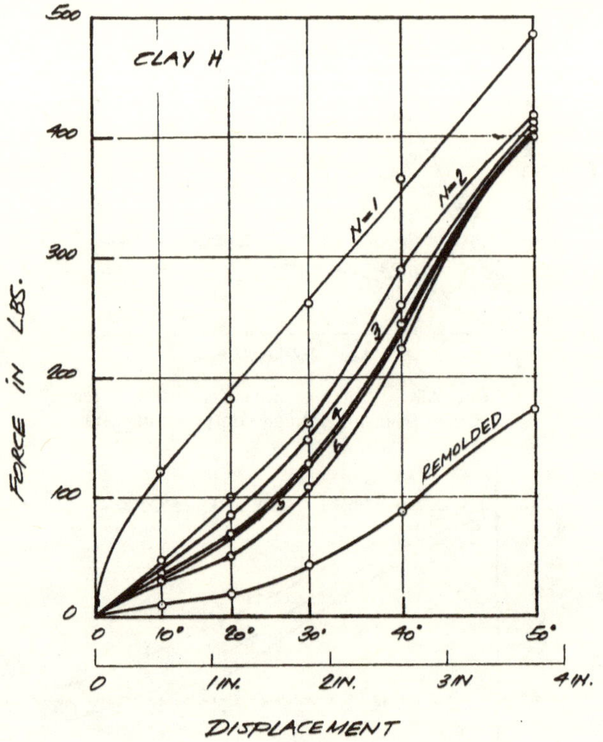

FIGURE 24A. Force-displacement relationships for angular deformations from 0–50°, Clay H

disturbed states are shown in Table II. Figure 28 shows the sinkage of a solid wheel plotted against the number of wheel passes for a medium clay in four states of disturbance.

Experimental and analytical data with the constant volume shear box indicate that this method is feasible for the identification of partial disturbance states when working with a soft to medium clay. However,

TABLE II

SUMMARY OF LABORATORY SHEAR VANE DATA

Different degrees of disturbance

	Laboratory Shear Vane Data, Tons per Sq. Ft.				
Soil type	Undisturbed	1 Cycle	3 Cycles	11 Cycles	Completely Disturbed
F	0.44	0.21	0.15		0.074
H	1.19	—	—	0.5	0.31

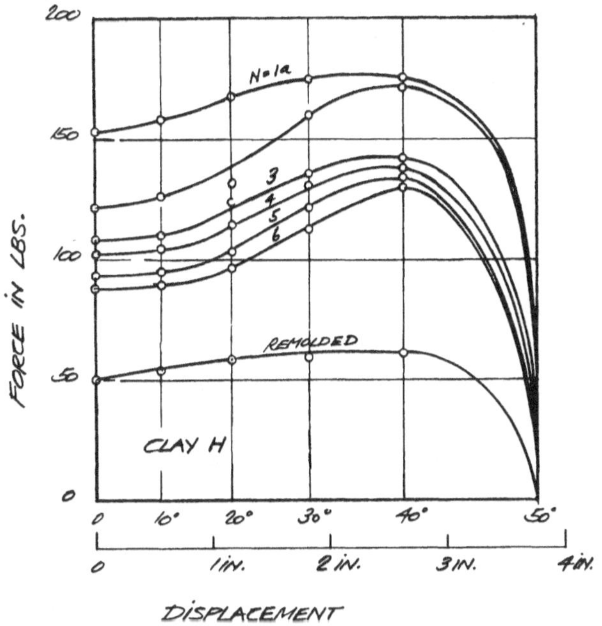

FIGURE 24B. Force-displacement relationships for angular deformations from 50°–0°, Clay H

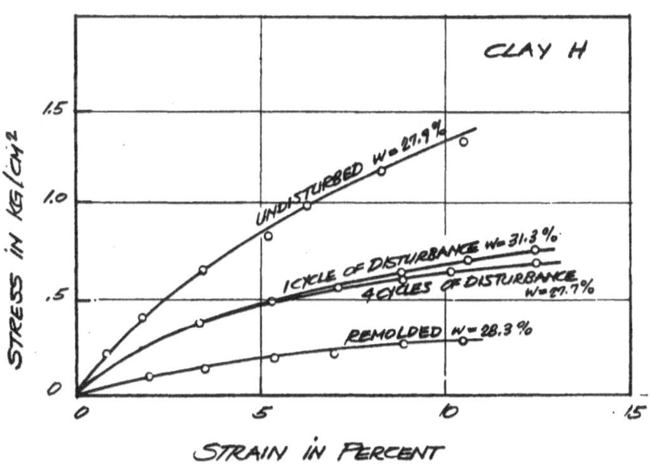

FIGURE 25. Unconfined compression tests for different degrees of disturbance, Clay H

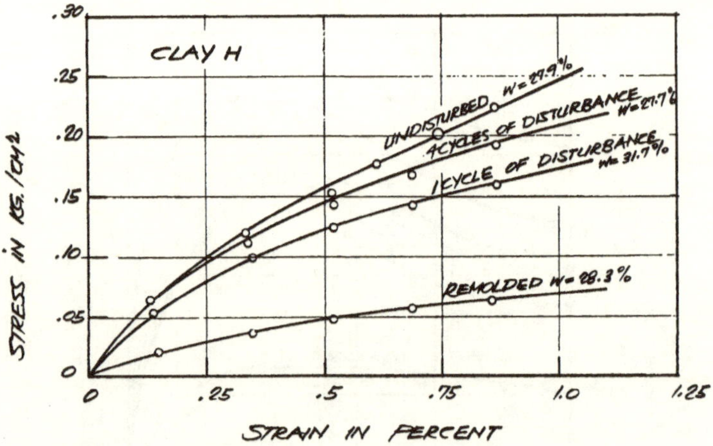

FIGURE 26. Unconfined compression tests for different degrees of disturbance, Clay H

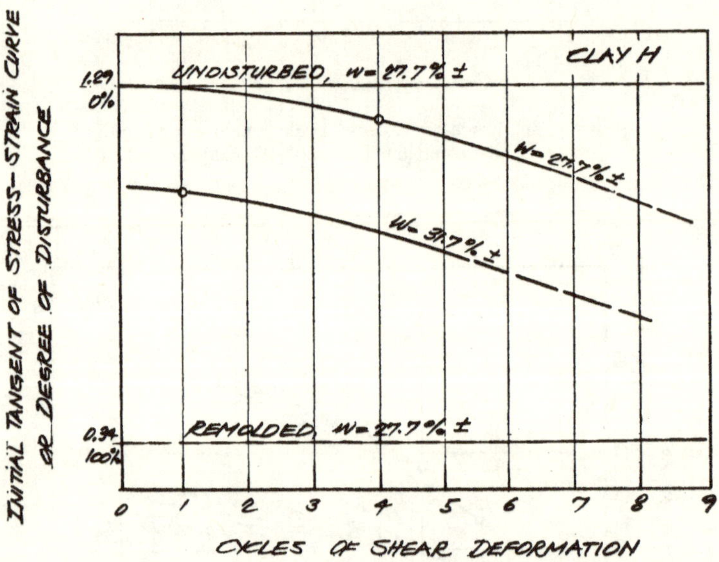

FIGURE 27. Degree of disturbance vs. the number of cycles of shear deformation, Clay H

it appears that the combined approach of the initial slope of shear stress-strain curves and the constant volume shear box must be used with surface cohesive soils owing to complications of partial saturation and a stiffer consistency.

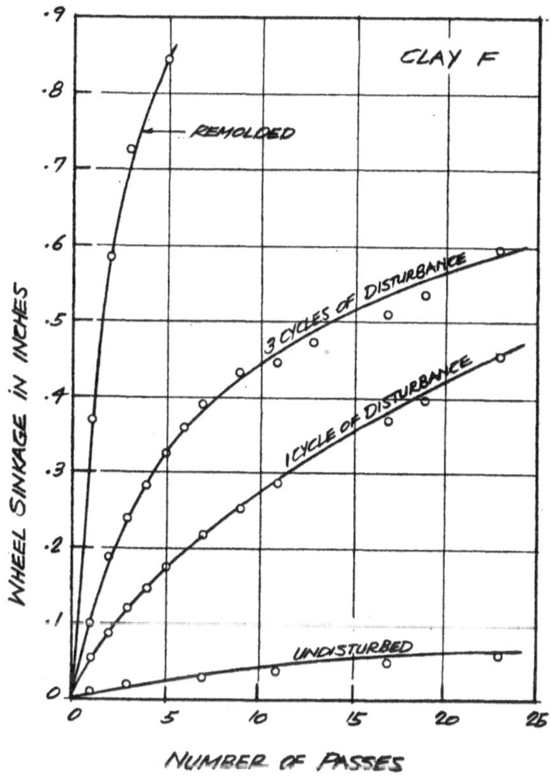

FIGURE 28. Effects of disturbance on wheel sinkage, Clay F

The ability to impart a predetermined amount of disturbance to a cohesive soil by means of the shear box once the disturbance scale has been established for a given soil type and condition has much practical significance. It can be noted from Table II and Figure 28 that it should be feasible to obtain shear and sinkage data for given soils at different degrees of disturbance. This should afford a basis for a more comprehensive analysis of the applicability of a mobility soil value system in soil-vehicle relationships. However, a detailed experimental and analytical study of general mobility parameters must be undertaken to determine statistical variations between different degrees of disturbance and disturbance source parameters. This study would include such parameters as disturbance source, soil type, soil moisture and density, strength-deformation characteristics and thixotropic regain characteristics.

CONCLUSIONS

The following conclusions are based on the results of this investigation:

1. Disturbance processes in cohesive soils result in a decrease of mobility soil values, with subsequent increase noted with increased time after disturbance.

2. A general increase in resistance to sinkage of both a test footing and a rigid wheel is noted with increased time after disturbance, demonstrating the significance of thixotropic regain characteristics of cohesive soils in mobility problems.

3. Measurements and analyses of soil moisture tension data indicate that currently available tensiometers are unsuitable for use in identifying partial disturbance states owing to the fact that a relatively small change in water content has a much larger effect on the reading than a slight change in the degree of disturbance.

4. A correlation between intermediate shear strengths of the soil as measured by a laboratory shear vane and partial disturbance states has not been satisfactory, apparently due to the fact that an unknown amount of disturbance is introduced when the shear vane is forced into the soil sample.

5. The relationship between rate of wheel sinkage and thixotropic regain time periods indicates that the phenomenon of thixotropy lends itself to the simulation and identification of different degrees of disturbance between the two limiting states of undisturbed and completely disturbed or remoulded.

6. An analysis of force-angular displacement relationships obtained with a modified constant volume shear box resulted in establishment of a system for identifying different degrees of disturbance.

7. The feasibility of imparting a predetermined degree of disturbance quite uniformly through a mass of cohesive soil by means of cyclic loading with a constant volume shear box has been established. This makes it possible to analyse the significance of different degrees of disturbance on a mobility soil value system.

RECOMMENDATIONS

It is recommended that a range of surface cohesive soils be investigated to establish specific disturbance scales for given soil types under different conditions of initial moisture and density. A detailed experimental and analytical study of general mobility soil parameters should then be undertaken to determine statistical variations between different degrees of disturbance and mobility soil parameters. This study would include such parameters as soil type, soil moisture and density, strength-deformation characteristics, and thixotropic regain characteristics.

ACKNOWLEDGEMENTS

This paper describes a research study conducted at Syracuse University under the sponsorship of the Land Locomotion Laboratory, U.S. Army Tank-Automotive Center, Warren, Michigan. The author wishes to thank Mr. Ronald A. Liston, Chief of Land Locomotion Laboratory and Mr. E. Hegedus, formerly Research Engineer at Land Locomotion Laboratory and now with Herron Testing Laboratory for their interest and co-operation in this study. Messrs. Hegedus and Haley arranged for the experimental results at the higher footing sinkage rates to permit a more comprehensive evaluation of this parameter. Appreciation is also expressed to Dr. Paul J. Brennan, Chairman, Dr. Wen H. Li, Professor in the Department of Civil Engineering, and Prof. Charles N. Lee of the College of Forestry at Syracuse University for helpful discussion given during the course of the investigation.

REFERENCES

1. LISTON, R. A., Methods and Goals of the Mechanics of Off-the-Road Locomotion, SAE Paper 782A, 1964 SAE Congress, Detroit, Michigan, January, 1964.
2. BEKKER, M. G., A Proposed System of Physical and Geometrical Terrain Values for the Determination of Vehicle Performance and Soil Trafficability, Vehicle Mobility Symposium sponsored by the Office of Ordnance Research and held at Stevens Institute of Technology, April 18–20, 1955; Reprinted in Land Locomotion Research Report No. 4, 1956.
3. BEKKER, M. G., *Theory of Land Locomotion*, Ann. Arbor: University of Michigan Press, 1956.
4. GOODMAN, L. J., Effects of Disturbance to Clay Soils as Measured by Empirical Soil Stress–Strain Parameters, U.S. Army Tank-Automotive Center, Research Division, Land Locomotion Laboratory 7782 (LL-76), June, 1962.
5. GOODMAN, L. J., The Influence of Disturbance to Clay Soils on Mobility Problems, Syracuse University Research Institute, Final Report, Third Year's Study, December, 1962.
6. GOODMAN, L. J., The Influence of Disturbance to Clay Soils on Mobility Problems, Syracuse University Research Institute, Final Report, Fourth Year's Study, December, 1963.
7. MITCHELL, JAMES K., Fundamental Aspects of Thixotropy in Soils, J. of the Soil Mech. and Found. Div., *Proc. of the Am. Soc. of Civil Engineers*, Vol. 86, No. SM3, June, 1960.
8. MICKLETHWAIT, E. W. E., Soil Mechanics in Relation to Fighting Vehicles, School of Tank Technology, Military College of Science, Chertsey, England, March, 1944.
9. JANOSI, Z. and B. HANAMOTO, An Analysis of the Drawbar Pull vs. Slip Relationship for Track Laying Vehicles, U.S. Army Ordnance Tank-Automotive Command, Research Division, Land Locomotion Laboratory, Report No. RR-47, November, 1961.

10. BERNSTEIN, R., Problems of the Experimental Mechanics of Motor Ploughs, *Der Motorwagen*, Vol. 16, 1913.
11. GORIATCHKIN, B. P., *Theory and Development of Agricultural Machinery*, Moscow, 1936.
12. TAYLOR, D. W., *Fundamentals of Soil Mechanics*, New York: John Wiley & Sons, Inc., 1948.
13. MURAYAMA, S. and S. HATA, On the Remolding Effect of Clay, Proc. 4th Int. Conf. on Soil Mech. and Found. Eng., London, 1957.
14. DAY, PAUL R., Soil Moisture Tension Measurements: Theoretical Interpretation and Practical Application, Proc. of the 3rd Nat. Conf. on Clays and Clay Minerals, National Academy of Sciences, Publication 395, Washington, D.C., 1955.
15. ORESTE, MORETTO, Effect of Natural Hardening on the Unconfined Compression Strength of Remolded Clays, 2nd Int. Conf. on Soil Mech. and Found. Eng., Vol. 1, Rotterdam, 1948.
16. GRAY, HAMILTON, Field Vane Shear Tests of Sensitive Cohesive Soils, Trans. of the Am. Soc. of Civil Eng., Vol. 122, 1957, p. 844.
17. TERZAGHI, K., Theoretical Soil Mechanics, New York: John Wiley & Sons, 1943.
18. CZAKO, T. and M. G. BEKKER, *Determination of Vehicle Sinkage Parameters by Means of Rigid Wheels*, Part I, May, 1958. Department of Army Ordnance Tank-Automotive Command, Research Division, Land Locomotion Laboratory.
19. HANAMOTO, B., Artificial Soils for Laboratory Studies in Land Locomotion, Ordnance Corps, Land Locomotion Research Branch, Research and Development Division, Ordnance Tank-Automotive Command, November 15, 1957.
20. SEED, H. B. and C. K. CHAN, Thixotropic Characteristics of Compacted Clays, Proc. Am. Soc. of Civil Eng., Vol. 83, No. SM4, November, 1957.
21. KARAFIATH, L., An Analysis of New Techniques for the Estimation of Footing Sinkage in Soils, Land Locomotion Research Branch, Research and Development Division, OTAC, Report No. 18, October, 1957.
22. HEGEDUS, E., A Simplified Method for the Determination of Bulldozing Resistance, Department of the Army, Ordnance Tank-Automotive Command, Research Division, Land Locomotion Laboratory, Detroit Arsenal, Report No. 61, May, 1960.
23. HANAMOTO, B. and Z. JANOSI, Determination of K_c, K_ϕ, n-Values by Means of Circular Footings, Modified Procedure, Department of the Army, Ordnance Tank-Automotive Command, Research Division, Land Locomotion Laboratory, Report No. 57, November, 1959.
24. HARRISON, W. L., Z. JANOSI, CAPT. R. A. LISTON, CAPT. L. S. LODEWICK, Mobility Studies, Ordnance Tank-Automotive Command, Detroit Arsenal, Centerline, Michigan, Report No. 59, December, 1959.

Determination of Shear Stress of A Fine Cloddy Soil with A "Guarded" Shear Head*

MOHAMMED A. GHANI

ABSTRACT

A new shear annulus with guard ring plates on both sides has been constructed and developed to the extent that satisfactory values of shearing strength characteristics after failure could be measured in a closed sample. This device has eliminated almost all important drawbacks of most torsion shear instruments. Both maximum and ultimate values of shear stress obtained with the new shear head are compared with those from some well known shear instruments. The reproducibility in shear stress values obtained with each shear instrument is also shown over a long shear displacement. The comparison shows that the new shear device has provided a larger testing range and less scattered results.

In shear test, a cloddy soil apparently behaves like two types of soil at shearing displacements corresponding to maximum and ultimate shear stresses.

An empirical relation is found for the specific shearing energy dissipated to an unsaturated cloddy loam as a function of moisture content and normal stress for a loose condition of the soil at the start of shear test.

IT HAS BEEN INCREASINGLY recognized that soil mechanics as a branch of mechanics of solids should be based on the relation between stress quantities and compatible or geometric quantities. Nevertheless, such basic knowledge of stress-strain relations or their equivalent, specially after failure, requires improved methods of measuring technique and improved measuring instruments. Further, knowledge of energy dissipated to a soil and more exact knowledge of the type of soil under investigation which provides experimental data, are useful and may be essentially required for a basic study or for a comparison.

This paper will first deal with a shear device which was developed to provide more accurate results than existing devices over large ranges

*This is an abstract from the doctoral thesis (not yet published) of the author. The work was initiated by Professor Dr.-Ing. Hans von Sybel, former director of the Institute of Agricultural Machines and Laboratory of Mechanical Processing in the Technical University Munich. The chief financial aid came from various sources interested generally in strength of loose granular materials.

of normal stress and shear displacement, specially after failure. The results of some comparative tests are presented to clarify the relative merits and demerits of the new shear device.

It is further shown that a cloddy soil under shear test represents two types of soil according to texture. One of them is apparently related to the peak values of shear stress of failure, while the other is primarily related to the ultimate shearing strength. The change in texture of a cloddy soil during a shear test is briefly discussed.

The study of energy dissipation is confined to the shearing energy which is by far the largest component of energy dissipated to a soil during a long shearing displacement. An empirical relation of specific shearing energy as a function of normal stress and moisture content is described in the last section of this paper.

THE TORSION SHEAR DEVICE*

General outline

The novel features of the new shear device lie chiefly in the shear head (Figure 2). The success of the shear head, however, depended primarily on the design of the split between measuring annulus and each guard ring plate and its packing, only which will be described in detail. The test specimen is of disc shape. The stresses are measured only on a dentated ring plate which will be called a measuring annulus and is separated from the outer and the inner guard ring plates by a rubber packing arrangement developed for this purpose. The measuring annulus and the guard plates rotate at the same angular velocity during shear about the axis of the shear head. The differential movements among them are negligibly small due to the steel-elastic suspension of the measuring annulus on the heavy frames of the guard plates.

Figures 1A and 1B show the general outline of the shear device. The vertical load is applied hydraulically and the torque from a motorized drive. A mechanical device records both the twist of shear head and the change in sample height during shear. An oscilloport (an electrical instrument for continuous and automatic recording of several electrical signals) records signals from vertical load and torque continuously by means of two "carrier frequency amplifiers" connected with the strain gages which are attached with the measuring annulus. The sample receiver is only free to move in a vertical axis by means of hydraulic pressure, and the shear head is only free to rotate in the horizontal plane.

*All test results shown in this paper were obtained with this shear device except where the contrary is stated. The sizes of clods employed were smaller than 2 mm.

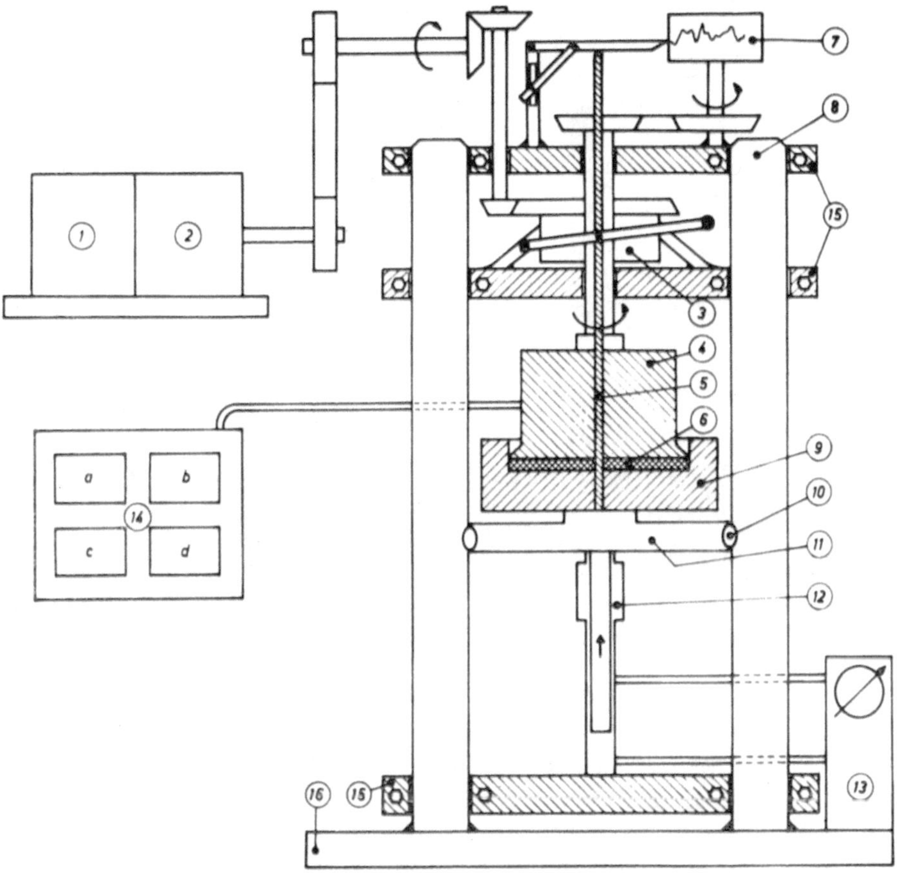

FIGURE 1A. Torsion shear device. 1, D.C. motor; 2, gear box; 3, coupling; 4, new shear head; 5, height measuring rod; 6, test specimen; 7, sample height recorder; 8, steel pipe frame; 9, sample receiver; 10, guide wheels; 11, sliding cross head; 12, hydraulic cylinder; 13, hydraulic aggregate; 14, wagon (*a*, *b*, *c*, *d*, electrical instruments); 15, cross bar frame; 16, base

SYMBOLS

ϕ_u, c_u = Coulomb's ϕ and c obtained from ultimate shearing strength.

σ_m = Precompaction stress applied before the application of normal stress [kp/cm²].

σ_a = Normal stress at the beginning of shear displacement [kp/cm²].

e_1 = Void ratio before the application of normal stress. [−].

M = Torque [m.kp].

E_v = Specific shearing energy = shearing energy per unit volume of sample [cm.kp/cm³].

Kilogram-force = unit of force [*kp*].

FIGURE 1B. Torsion shear device

Shear head

Figure 2 shows a composite section of shear head and the sample receiver. The test specimen is confined between the sample receiver and the shear head. Both torsion pipe (15) and the double ring plates (14) were selected to be strong enough not to cause a deflection larger than 0.1 mm under a maximum normal stress of 5 kp/cm^2 during a shear test in the average agricultural soils and weak enough to cause sufficient readings in the scales of carrier frequency amplifiers under the influence of a minimum shear stress and a normal stress equal to 0.1 kp/cm^2 each. This was accomplished chiefly through trial tests. The double ring plates (14) and the steel sheet (24) bend under the influence of normal stress, and this deflection changes the electrical resistance of strain gauges (17) which measure the vertical load on the measuring annulus. The inductive strain gage (18) which is fixed in a holder (26), measures the torque produced by the test specimen on the measuring annulus during shearing displacement. The bottom surfaces of the guard ring plates were chosen approximately equal in area and the ratio inside diameter to outside diameter of measuring annulus 0.9. The new shear head in this paper will be called "Plastorupt."*

*The name "Plastorupt" is proposed by Prof. Hans von Sybel.

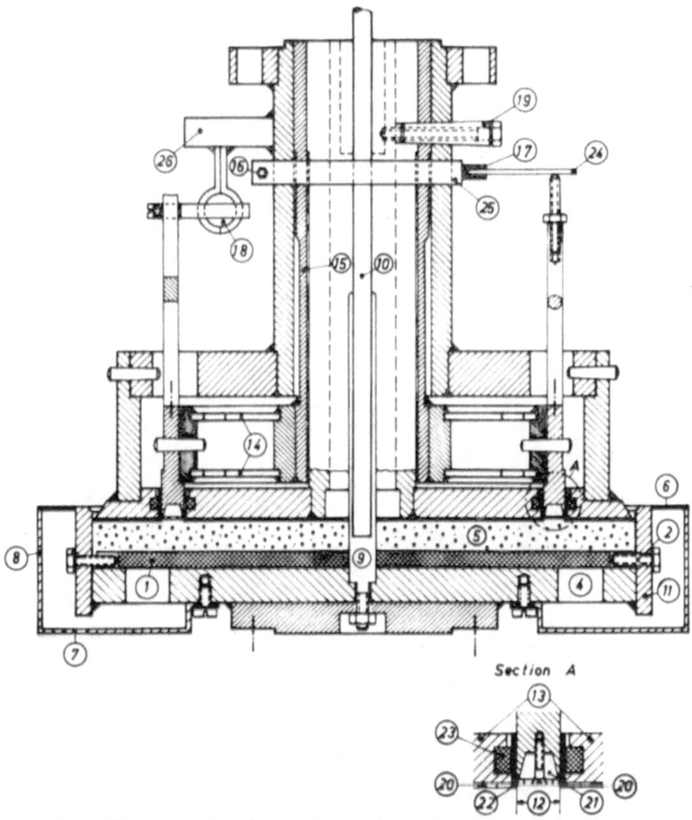

FIGURE 2A. Shear head and sample receiver. 1, porous plate; 2, stop; 4, holes; 5, specimen; 6, lid; 7, water outlet; 8, water jacket; 9, central rod, lower part; 10, central rod, upper part; 11, sample receiver; 12, measuring annulus; 13, guard ring plates; 14, double ring plates; 15, torsion pipe; 16, clamp fixing; 17, resistance strain gauge; 18, inductive strain gauge; 19, tapered pin; 20, dentated surfaces; 21, dentated conical ring; 22, raw rubber; 23, moss rubber; 24, steel sheet; 25, clamp; 26, holder, inductive strain gauge

Split-packing problem and its solutions

The design of the split and its packing should be such that the state of each stress in the measuring zone is almost perfectly homogeneous during the whole period of a test.

The chief factors governing these problems in shear tests are:

(1) The split and its packing should be so designed that the guard elements (surrounding) carry over only a negligible part of force exerted on the measuring element. In the new shear head the maximum limit of this transmission was 5 per cent of the aforementioned force.

FIGURE 2B. Shear head and sample receiver

(2) The width of the split at the narrowest zone should be as small as possible. One millimeter split was selected in the new design.

(3) The width of the packing at the plane of contact with test specimen should be slightly larger than the largest size of particles in the sample.

(4) The packing in the split should withstand considerably high normal stress during shear in materials such as agricultural soils without renewal of packing before a few thousand shear tests. The maximum normal stress was selected to be 5 kp/cm^2 in "Plastorupt."

All these conditions were satisfied approximately by the shape of split and its packing provided in the new shear head. To meet the aforementioned requirements, a large number of trial tests were carried out.

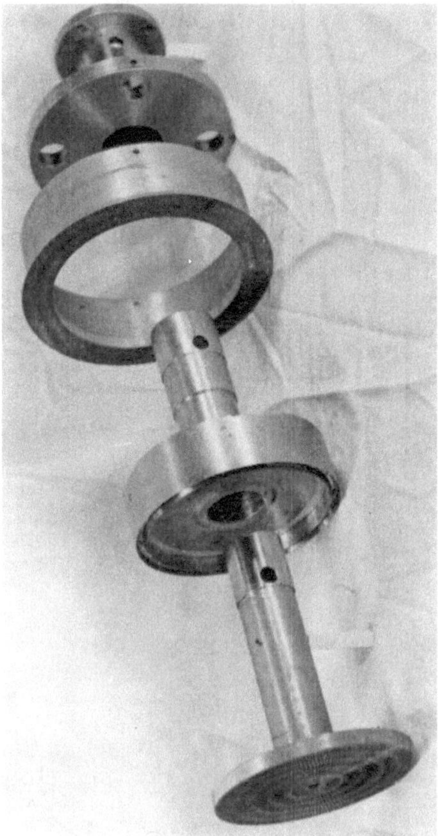

FIGURE 2c. Some details of shear head

Figure 3 shows some arrangements of packing and the split between measuring annulus and guard ring plates (Figure 2). Most arrangements of packing in the split were tested in an apparatus built for trial tests. The tests with various thicknesses of sponge rubber were found unsuitable for the purpose.

Figure 3a shows a packing arrangement in which rubber was vulcanized in the split. Figure 3b shows a very elastic raw rubber sheet glued in the split. Figure 3d shows a folded thin rubber sheet glued on both sides of the split. All of these arrangements were found unsuitable for the purpose. Figure 3e shows a proposal in which an angular "moss rubber" section was to be glued on both sides of the split. The inclination of the section might vary from 30° to 60°. This might be one of the best solutions, but could not be tested due to high cost. Figure 3g shows a novel design which was

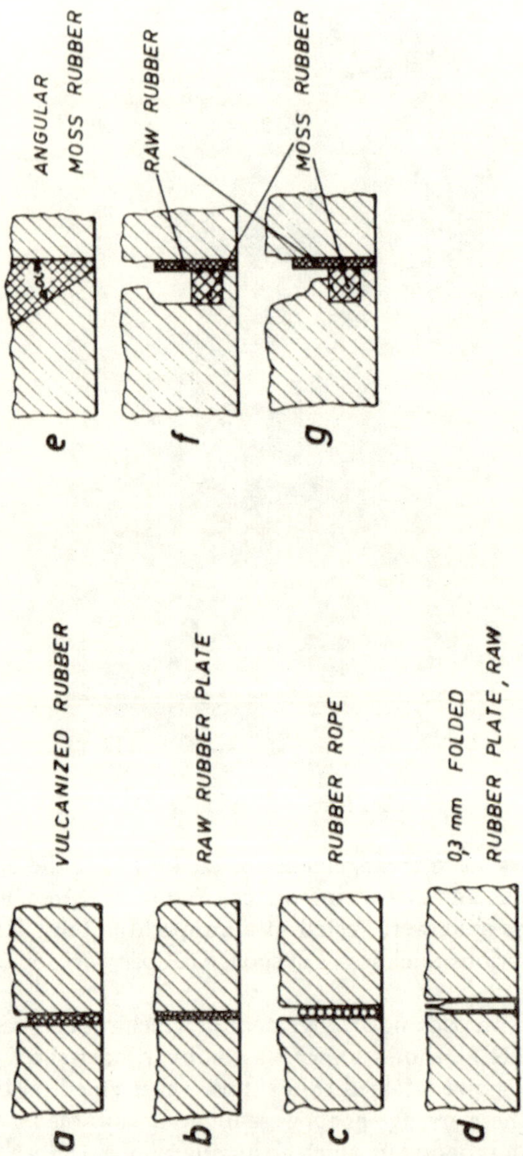

FIGURE 3. Some arrangements for packing in split

found very satisfactory for the prescribed requirements. In this case, a very elastic raw rubber sheet of 1 mm thickness was glued with a rectangular "moss rubber" section. Araldite was used as the glue material both for rubber to rubber bond and for rubber to metal bond. The forces carried by the guard plates through the packing were determined by loading the measuring annulus with weights in two series of tests with and without packing in the split. The stresses carried through the packing were included in the calibration of the shear instrument satisfactorily. The same method for the elimination of boundary effect as mentioned above can also be used in compression or consolidation tests in granular materials.

Dentated measuring annulus

The bottom of the measuring annulus (Figure 2) was machined to a circular groove where a dentated conical ring could be fixed. This arrangement made it possible to find out the appropriate height and spacing of grouser teeth so that only a minimum progressive failure in shear tests took place, and a maximum uniformity in the stress distribution on the plane of failure could be achieved. By changing the dentated conical ring which is screwed on the measuring annulus, the design of grouser teeth can be changed if required. Similarly the distance between the bottom plane of the dentated conical ring and that of guard ring plate

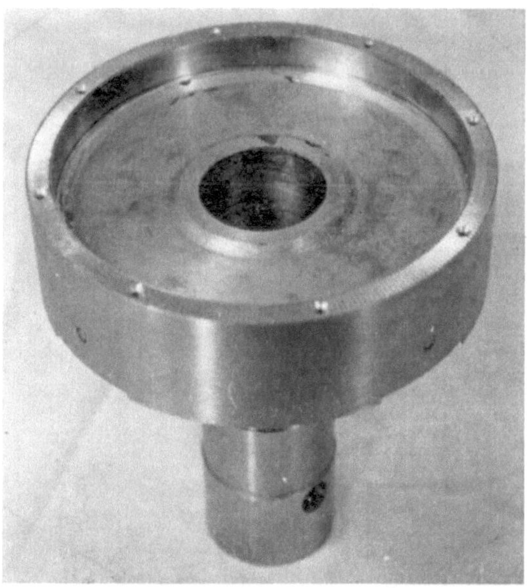

FIGURE 4. Measuring annulus with dentated ring

can also be varied. Figure 4 shows a dentated conical ring fixed in the groove of the measuring annulus. This dentated surface was found to be very satisfactory with respect to minimum progressive failure and maximum uniformity in the normal stress distribution.

COMPARATIVE SHEAR TESTS

Testing procedure

In all shear heads shown in Figure 5 except in the case of the direct shear apparatus, the same method of loading and recording was employed

	SHAPE OF SLIPPING PLANE	BOUNDARY CONDITION OF SAMPLE	SKETCHES OF SHEAR HEADS	TYPE OF SHEAR LOADING	DESCENT
a	CIRCLE	OPEN	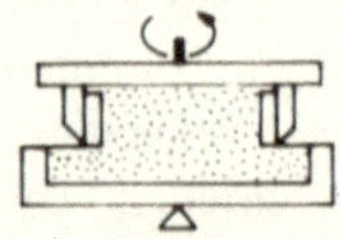	TORSION	PAYNE FOUNTAINE
b	RING PLATE	OPEN	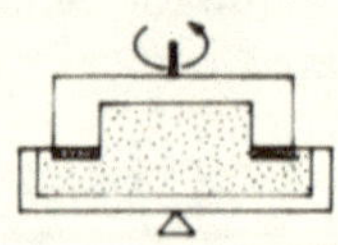	TORSION	BEKKER SÖHNE
c	RING PLATE	CLOSED	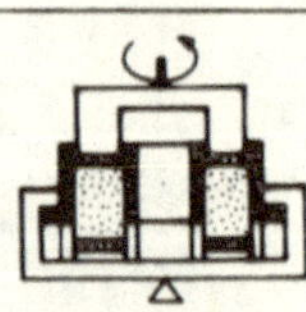	TORSION	HVORSLEV
d	RING PLATE	CLOSED	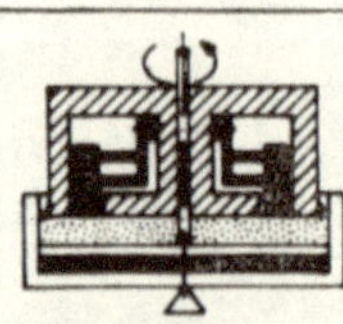	TORSION	NEW DEVELOPMENT
e	RECTANGLE	CLOSED		DIRECT SHEAR	DIRECT SHEAR APPARATUS

FIGURE 5. Sketches of shear heads

as described in the section "Torsion shear device." Only the shear heads were different (Figure 5). The application of the same loading device and recording method with various shear heads eliminated the errors arising from the uses of various loading arrangements and recording systems. Therefore, there was only the influence of the shear heads on the test results in the comparative tests. A rectangular shear box was employed in the case of direct shear test, and its usual testing procedure was followed.

In the new shear head, the following expression provides the average shearing stress, τ, corresponding to the torque M applied, when it is assumed that the shearing stresses in the soil under the measuring annulus are uniformly distributed over the horizontal plane of failure, $\tau = F/A$, where F is the horizontal reaction of soil on the annulus and A is the area of cross section of the latter covering half of inner and outer splits. The horizontal reaction F is computed from the torque M which acts on the mean circumference of the measuring annulus. The differences between the following mean diameters in this case are negligible:

$$D_{\mathrm{m}} = (D + d/2) \approx (2/3)(D^3 - d^3/D^2 - d^2) \approx \sqrt[3]{(D^3 + d^3)/2},$$

where D_{m}, D, and d are mean, outer, and inner diameters of annulus, respectively.

Test results

A comparison of test results of some well known shear instruments is made with those of the new shear device. Various sample heights had to be used in comparative tests. The sample heights during shear tests in closed samples were almost constant, while the same in shear tests on open samples changed significantly with increasing deformation. But this change in each of the repeated tests in one single shear instrument remained almost constant.

A cloddy soil (loam) was always used except in Figure 8a. A fresh sample was employed in each test. It was observed in earlier tests (not shown here) that the influence of precompaction on Coulomb's ϕ and c could not be detected in this cloddy soil, and a small deviation of moisture content had no significant influence on ϕ and c, and a lot of work was required to maintain exactly the same moisture content and precompaction stress in all tests to gain only a small improvement in accuracy. On these grounds, a considerable variation of precompaction stress and a small deviation of moisture content in these comparative tests with some shear heads were maintained (Figures 6 to 8). There were also small variations of normal stress during long shearing displacements in some tests probably due to unavoidable inhomogeneity in the specimen or due

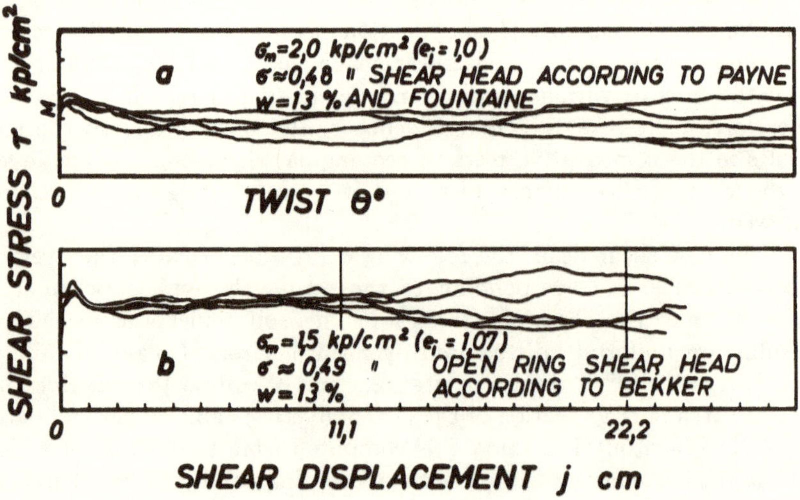

FIGURE 6. Reproducibility in stress-deformation curve with various shear heads

to an unknown disturbance in the recording or measuring systems (Figures 7 and 8). Only the void ratio at the beginning of shear test has been shown, because the actual void ratio in the immediate neighbourhood of slipping plane seemed to deviate very much from the rest of the sample due to dilatation and mutual readjustments of soil particles and clods during shear displacement.

Reproducibility in test results

Figures 6 to 8 show shear stress-deformation diagrams in repeated tests with various shear instruments. The normal stress-shearing deformation curve is not shown when normal stress was almost constant with increasing deformation. All stress-deformation curves shown in this section were traced from automatic continuous records. However, this was not the case with the test results from direct shear apparatus (Figure 7c). Figure 6a shows the torque-twist curves from repeated tests on open samples with the shear head shown in Figure 5a. The scattering in maximum shear stress was not large, but the values of shear stress after failure were very much scattered. This was confirmed by several series of repeated tests. The diameter of the shear head and the maximum twist during shear were 127 mm and approximately 180 degrees respectively. Figure 6b shows shear stress-deformation diagrams from repeated tests on open samples with an annular shear ring shown in Figure 5b.

The scattering in the maximum shear stress was negligible, and the same was not large in the values of shear stress up to a shearing displacement of 12 cm., beyond which the values of shear stress were unreliable. Figure 7 (a, b) shows stress-deformation curves in repeated tests in a

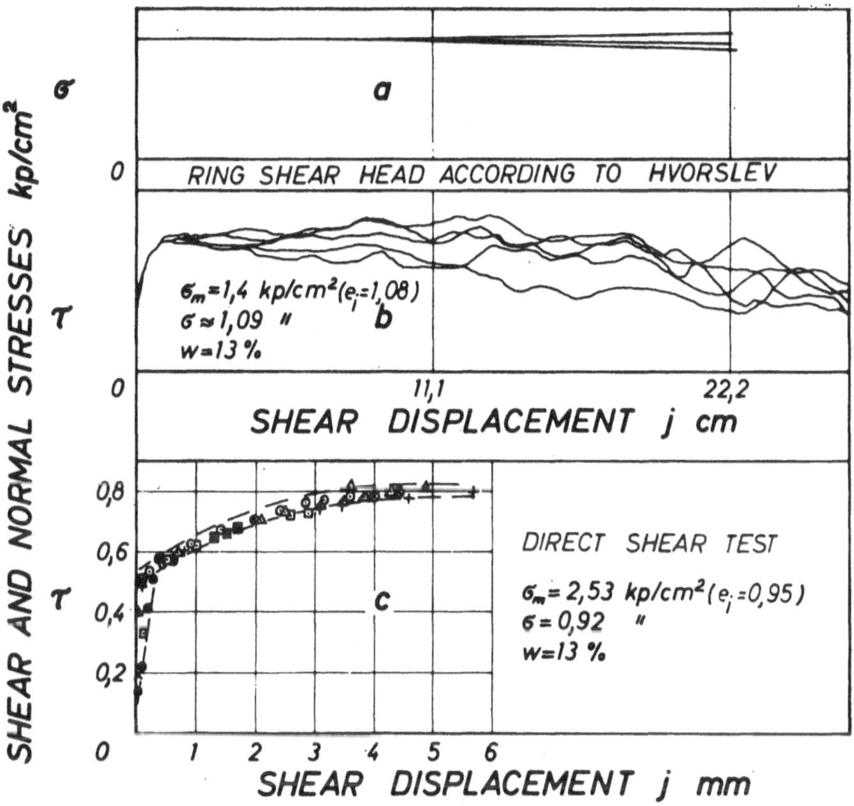

FIGURE 7. Reproducibility in stress-deformation curve with various shear heads

closed sample with an annulus shown in Figure 5c. The normal stress was constant up to a shear displacement of 7 cm, beyond which it changed very slowly with increasing displacement. The scattering in the maximum values of shear stress was small, but the corresponding shear displacements were much scattered. Beyond a shear displacement of 2 cm, the values of shear stress were unreliable. The shear stress-deformation diagrams of five repeated tests in a direct shear apparatus are shown in Figure 7c. The scattering in the values of maximum shear stress was small, but no comparison could be made after failure as the loading was stress-controlled, and after failure no measurement of shearing stress or displacement was possible. The stress-deformation curves of repeated tests in the new shear device are shown in Figure 8. The peak in the normal stress-displacement curve appeared in all tests and was due most probably to hindered dilatation. The maximum values of shear stress could be accurately determined because the peak of shear stress-strain curve was sharp. The ultimate values of shear stress could be determined more accurately and satisfactorily than in the case of any other shear head used. The comparison of Figure 8a with Figure 8c shows that the cloddy soil required considerably larger shearing deformation to reach the ultimate values of shear stress due apparently to size reduction in the slipping plane. This size reduction seemed to cause many complexities which did not exist in the case of sand. Repeated tests in sand with the new shear head also show the dependence of reproducibility in results on the bulk density, texture, and structure of testing materials (Figure 8).

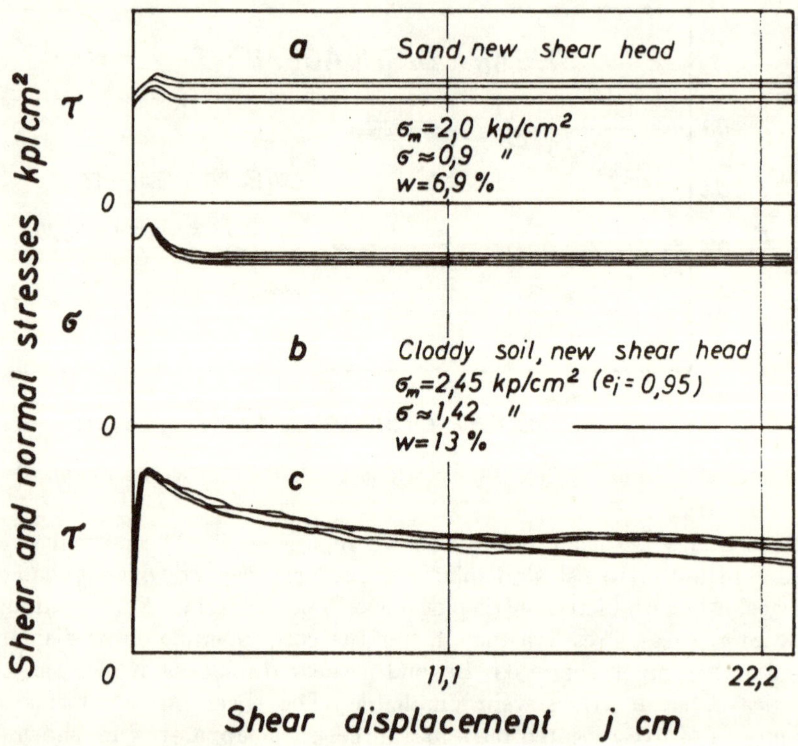

FIGURE 8. Reproducibility in stress-deformation curve with various shear heads

Typical shapes of curves

Some typical shapes of stress-deformation diagrams are shown in Figures 9 to 14. Most curves except those in Figure 13 did not run even approximately parallel to each other with increasing displacement at various normal stresses when all other variables remained constant, and frequently they crossed one another at a shearing deformation after failure. In most curves except those in Figure 13, the ultimate values of shear stress could not be determined satisfactorily. Figure 13 shows the shapes of some stress-deformation diagrams from soil tests in the new shear device. The curves at various normal stresses run approximately parallel. They are smooth and without a break. Not only the maximum values of shear stress but also the shear stress after failure, especially the ultimate shearing strength, could be satisfactorily determined. This was also shown by the results of tests in Figures 8 and 11.

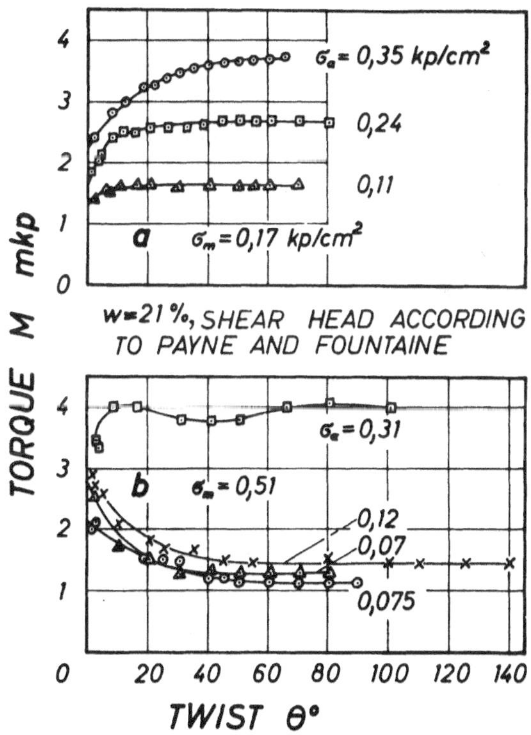

FIGURE 9. Typical shapes of stress-deformation curves with various shear heads and normal stresses

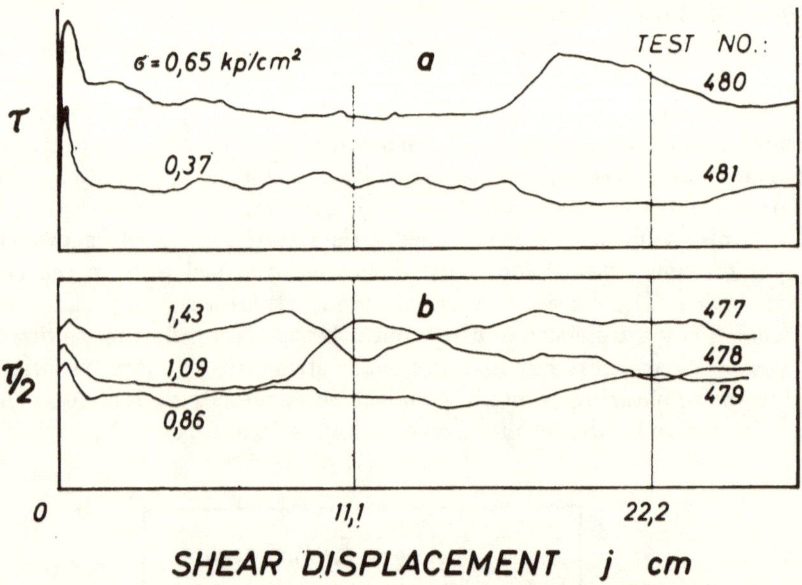

FIGURE 10. Typical shapes of stress-deformation curves with various shear heads and normal stresses. Open ring shear head according to Bekker $\sigma_m = 3.0$ kp/cm² ($e_j = 0.9$); $w = 13\%$

KEY TO FIGURE 11

$\sigma_m = 0$, without precompaction; A: Average normal stress; B: Average shear stress

Serial No. Oscillogram	Normal* stress σ kp/cm²	Moisture content %	Soil type
1	0.46	2.23	Cloddy loam
2	2.86		
3	2.50	5.89	
4	3.29		
5	0.38	9.58	
6	1.95		
7	1.95	14.10	
8	3.50		
9	0.43	19.10	
10	3.16		
11	0.66	25.50	
12	1.82		
13	0.60	6.90	Fine sand
14	2.10		
15	0.76	0.60	Coarse sand
16	1.85		

*Normal stress shown here corresponds to the peak or yield values of shear stress. Figure 19 shows size analysis of the cloddy loam used.

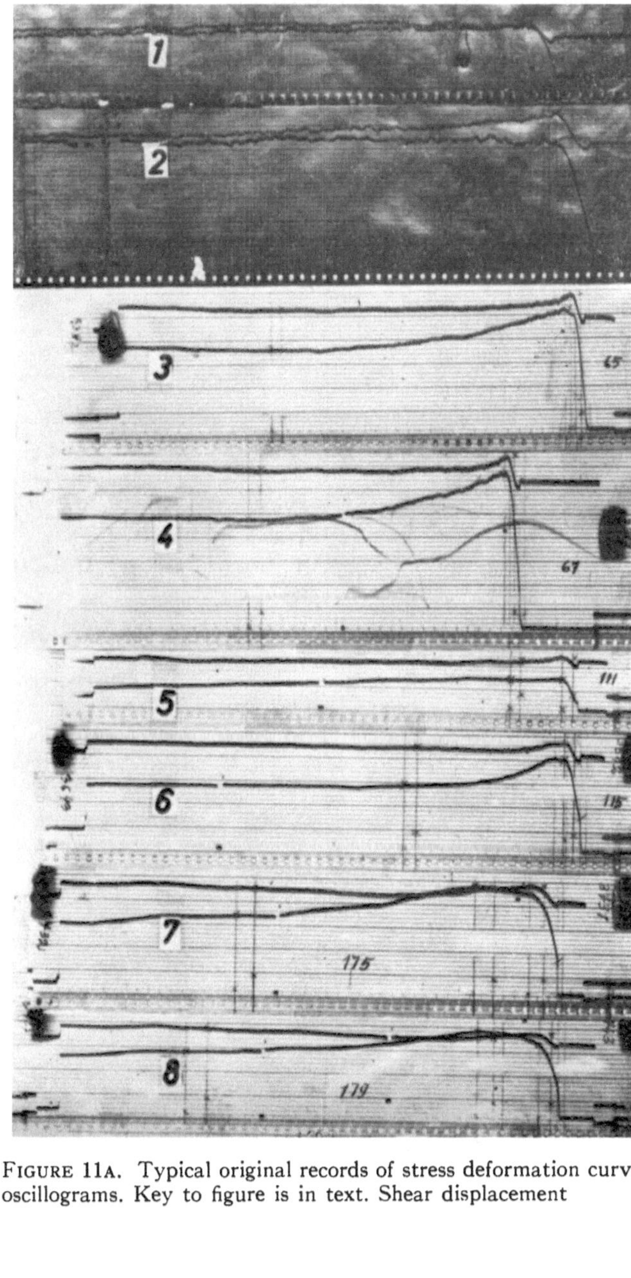

FIGURE 11A. Typical original records of stress deformation curves, oscillograms. Key to figure is in text. Shear displacement

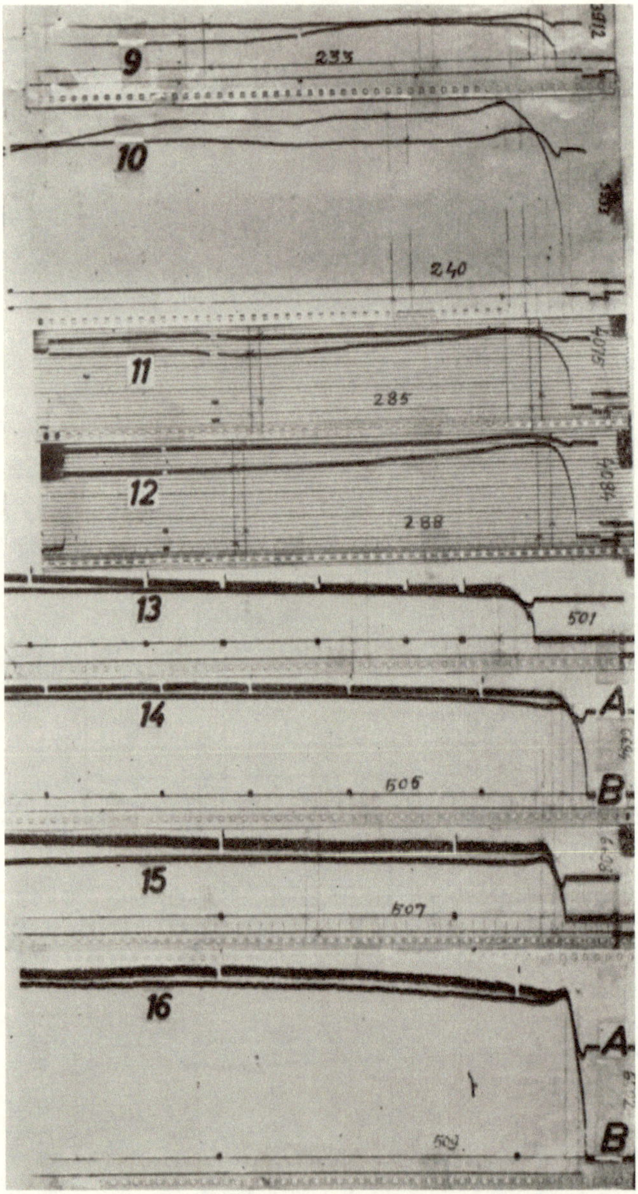

FIGURE 11B. Typical original records of stress deformation curves, oscillograms. Key to figure is in text. Shear and normal stresses

Maximum and ultimate values of shear stress

The maximum and ultimate values of shear stress plotted against normal stress have been shown in Figures 15 to 17 for various shear heads. The scattering in the maximum shear stresses was small irrespective of the type of shear instrument. Either ultimate shearing strength could not be determined, or its values were much scattered, or the working range of normal stress was very much limited in tests with various shear heads except the new one which provided the less scattered values (Figure 17) over a large range of normal stress. In all these shear instruments, the ultimate shear stresses were more scattered than the maximum shear stresses, apparently due to complex behaviour of the cloddy soil

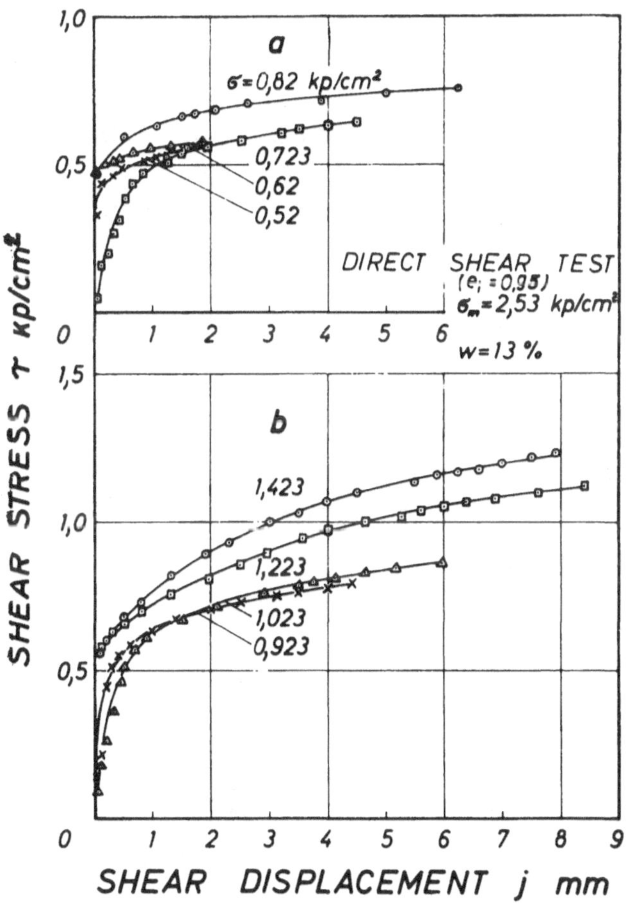

FIGURE 12. Typical shapes of stress-deformation curves, various shear heads, and normal stresses

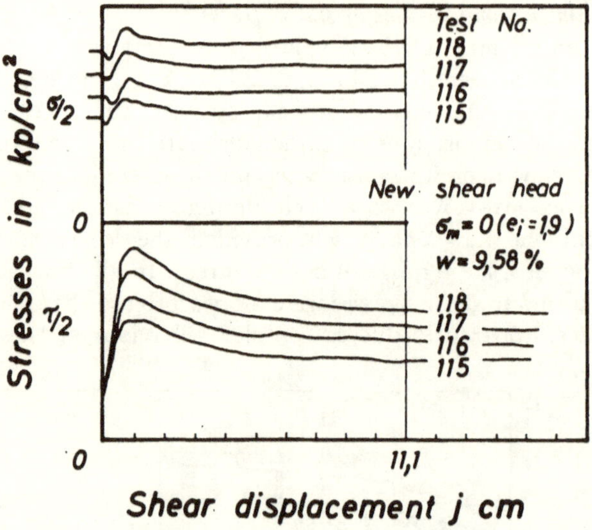

FIGURE 13. Typical shapes of stress-deformation curves, various shear heads, and normal stresses

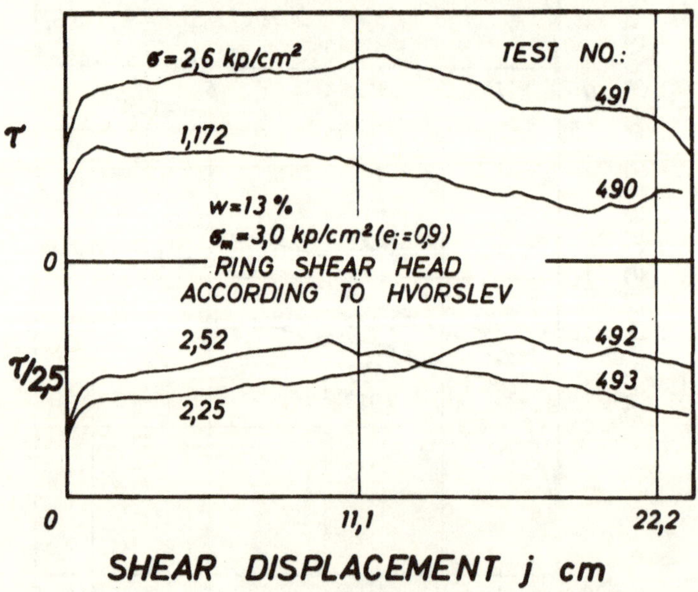

FIGURE 14. Typical shapes of stress-deformation curves, various shear heads, and normal stresses

---

FIGURE 15. *Opposite.* Shear head according to Payne and Fountaine (*a*, *b*). Open ring hear head (*c*). Maximum shear stress ⊙ — ⊙. Ultimate shear stress × — ×

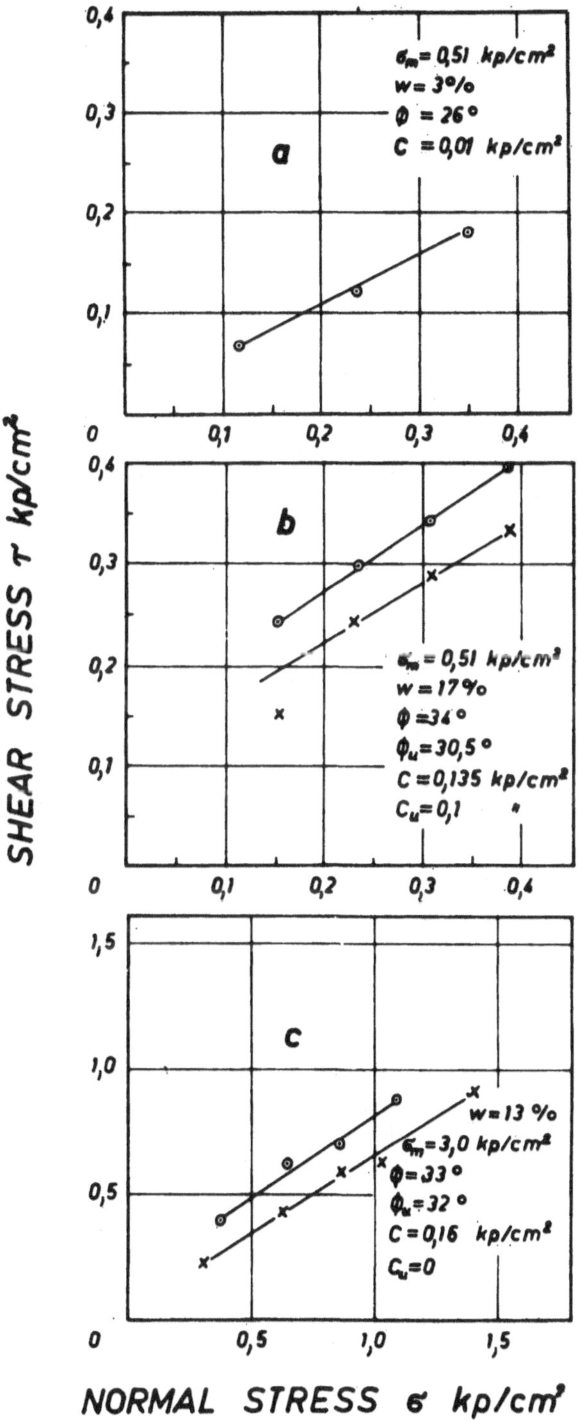

a
σ_m = 0,51 kp/cm²
w = 3%
φ = 26°
C = 0,01 kp/cm²
b
σ_m = 0,51 kp/cm²
w = 17%
φ = 34°
ϕ_u = 30,5°
C = 0,135 kp/cm²
C_u = 0,1 "
c
w = 13 %
σ_m = 3,0 kp/cm²
φ = 33°
ϕ_u = 32°
C = 0,16 kp/cm²
C_u = 0
SHEAR STRESS τ kp/cm²
NORMAL STRESS σ kp/cm²

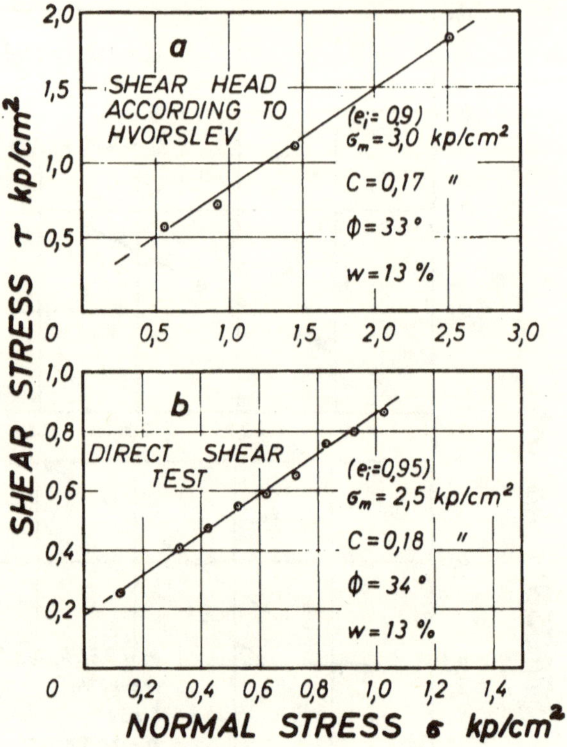

FIGURE 16. Maximum and ultimate shear stress vs normal stress with various shear heads

during shearing displacement. Figure 18 shows c- and ϕ-values obtained with some shear instruments in a cloddy soil (loam). The c-values were very small and might not be so accurate as ϕ-values. The ϕ-values from maximum shear stresses were slightly larger than those from ultimate shear stresses. This was primarily because of size reduction of clods during the required long shearing deformation, and also due apparently to higher bulk density produced by precompaction and retained at the initial failure. The influence of precompaction on bulk density in the shearing zone seemed to disappear, first rapidly and then slowly, due to dilatation during long shear displacement after failure.

The results of comparative tests show that the new shear device was significantly better than the others for the investigation of shearing strength characteristics after failure, and the values of shear stress particularly after failure were more satisfactory.

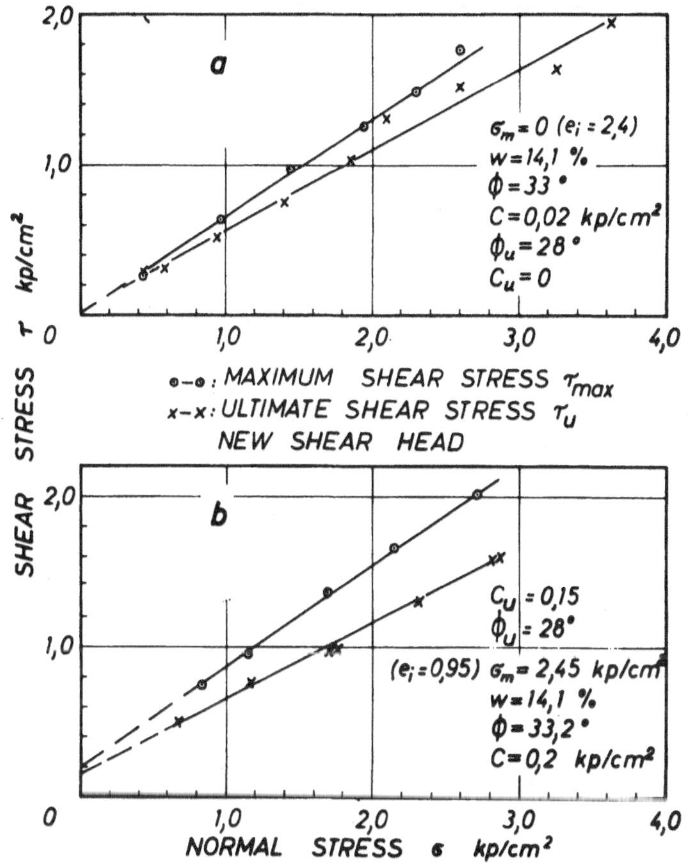

FIGURE 17. Maximum and ultimate shear stress vs normal stress with various shear heads

CLASSIFICATION OF THE CLODDY SOIL

Since the shearing strength of soil varies significantly with texture, shear test results lose significance, to a large extent, if they are not accompanied by the results of size analysis. Not only the size analysis of soil particles (sedimentation analysis) but also the dry sieve analysis is required for some fundamental investigations in soil mechanics if the soil is a cloddy one. Frequently, the moisture content, normal stress, and soil type in the surface soils are such that the clods largely retain their size and shape during failure in a shear test, while the clods in the shearing zone lose their size and shape nearly completely at a shearing deformation

	DESCENT, SHEAR HEADS	MOISTURE CONTENT %	VOID RATIO e_i	PRECOMPACTION STRESS σ_m kp/cm²	FROM ULTIMATE SHEAR STRESS ϕ_u°	c_u kp/cm²	COULOMB ϕ°	c kp/cm²
a	PAYNE FOUNTAINE	17	1,2	0,51	30,5	0,1	34	0,13
b	BEKKER SÖHNE	13	0,90	3,0	32	0,02	33	0,16
c	HVORSLEV	13	0,90	3,0	—	—	33	0,17
d	NEW DEVELOPMENT	14	2,4	0,0	28	0,02	33	0,0
			0,95	2,45	27	0,15	33,2	0,2
e	DIRECT SHEAR APPARATUS	13	0,95	2,53	—	—	34	0,18

FIGURE 18. Table for c and ϕ values obtained with several shear instruments

corresponding to the ultimate shearing strength. This was observed qualitatively in the new shear device. If the shearing deformation must be carried out up to the ultimate shearing strength, the same behaviours of a cloddy soil may be expected, at least qualitatively, in all well known shear instruments. Thus the texture of a cloddy soil in the shearing zone may vary during a long shearing deformation. The bulk density of a cloddy soil in the shearing zone may also vary significantly during shearing displacement because of size reduction as well as dilatation. Usually the interior of a clod possesses a density considerably higher than the bulk density of a cloddy soil. A similar difference may be expected between the structure of soil particles inside each clod and the structure of a cloddy soil, considering each clod as a particle of soil mass. Thus, the density, the texture, and the structure of soil mass inside each clod seem to be chiefly related to the ultimate shearing strength of a cloddy soil in a shear test, while the same of whole soil mass may be primarily related to the maximum shear stress of the same soil.

Figure 19 shows that the size analysis of soil particles (sedimentation analysis) of a cloddy soil, and its dry sieve analysis in the cloddish condition, indicated two different types of soil, which differed from each other to a large extent. According to sedimentation analysis, this soil can be called loam (curve B), and according to dry sieve analysis the same soil can be called coarse sand (curves A). The peculiar difference, however, lies in the shapes of the curves in two size analyses (Figure 19). The same curves further show that the influence of moisture content on the size distribution was significant only at a high range of moisture content, apparently because volume change of soil is negligible at a moisture content below the shrinkage limit. Since the difference in strength between a clod and a mineral particle of soil was very large, a clod of soil could not always behave like a single mineral particle. This fact explains why two different curves of size distribution might be expected in the cloddy soil at shear displacements corresponding to maximum and ultimate shear stresses. Further, about 70 to 90 per cent of clods at various moisture contents were larger than 0.2 mm and smaller than 2 mm, and about 63 per cent of soil particles were larger than 0.01 mm and smaller than 0.1 mm (Figure 19). The aforementioned facts explain why a cloddy soil behaves differently from a soil consisting of no clod.

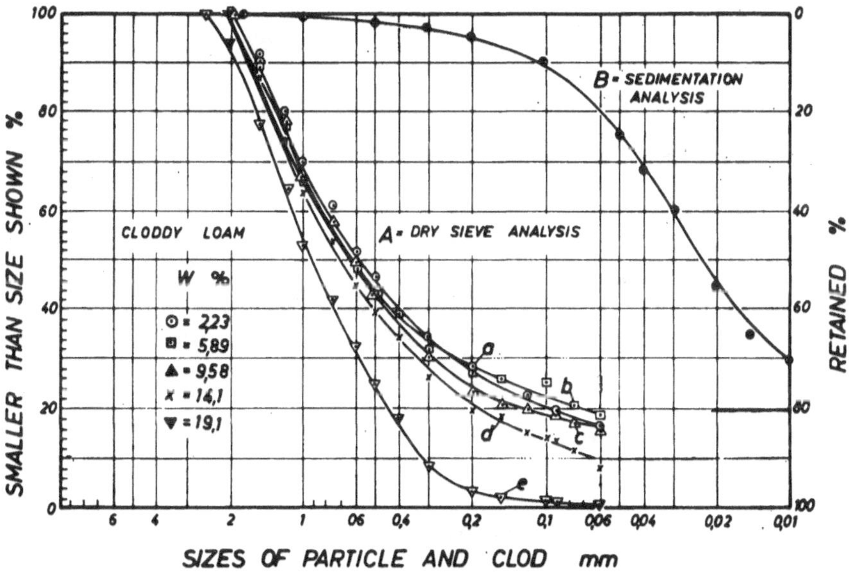

FIGURE 19. Soil classification data, clods and particles of a loam soil before a shear

SPECIFIC SHEARING ENERGY

Up to this point mechanical properties of granular materials as derived from our flow curve $\tau(j)$ were described in conventional terms: for instance, shearing strength, peak and ultimate values, and corresponding shear displacements. In order to obtain a general "flow rule" from the function $\tau(\gamma)$, we have to transform $\tau(j)$ by converting j into γ. Since the function $\gamma(j)$ is yet unknown, the only solution has been to restrict the sample height to a value of 1–2 cm and the shear deformation to a small value up to collapse. In this case a linear j–h profile could be expected, and $\tan \gamma \approx \gamma$ could be approximated following Figure 20. Since this has been confirmed by some photographic records (Figure 20),

FIGURE 20. Symmetrical deformation of a soil in shear

we felt justified to plot τ vs. (j/h) curve, τ, j, and h having been the average values as usual in annular ring shear test. Now the area under the stress-strain curve, which we regard as a flow rule, is represented by:

$$E_v = \int \tau \mathrm{d}\gamma, \tag{1}$$

the shearing energy per unit volume, which a soil element is able to sustain up to any interesting value of shearing deformation. A more complete description of flow behaviour results from this specific shearing energy because this energy includes all flow characteristics which occurred during a shearing deformation between two distinct points on stress-strain curve, while strength and corresponding strains represent mere point values. By this integrating method, energy reveals such mechanical properties as stiffness and softness, offering a measure which integrates stress with strain over a certain interesting range of deformation. Moreover, this integrating method permits a comparison between the mechanical properties of materials in the same range of deformation with different nature of flow: elastic or plastic, reversible or irreversible. After collapse, when shear strain γ theoretically ceases to proceed, shearing energy is dissipated as mere friction in the sliding plane or in the shearing zone. Here shearing energy per unit area of sliding plane is simply determined by:

$$E = \tau \times j = \sigma \tan \phi \times \text{displacement } j \tag{2}$$

Equation 2 is, however, a rough approximation in a soil where considerable progressive failure takes place after collapse.

The shapes of stress-strain diagrams up to failure have been shown in Figures 21 to 23. The elastic range of shearing deformation is clearly evident in each curve in all tests. The shearing strain at elastic limit did not change significantly with the variation of normal stress or moisture content except at 19 per cent and at a normal stress higher than 2 kp/cm², when positive neutral stress was suspected in the sample. At a moisture content of 5.9 per cent, the shearing strain at failure did not increase considerably with increasing normal stress (Figure 21a). The energy area

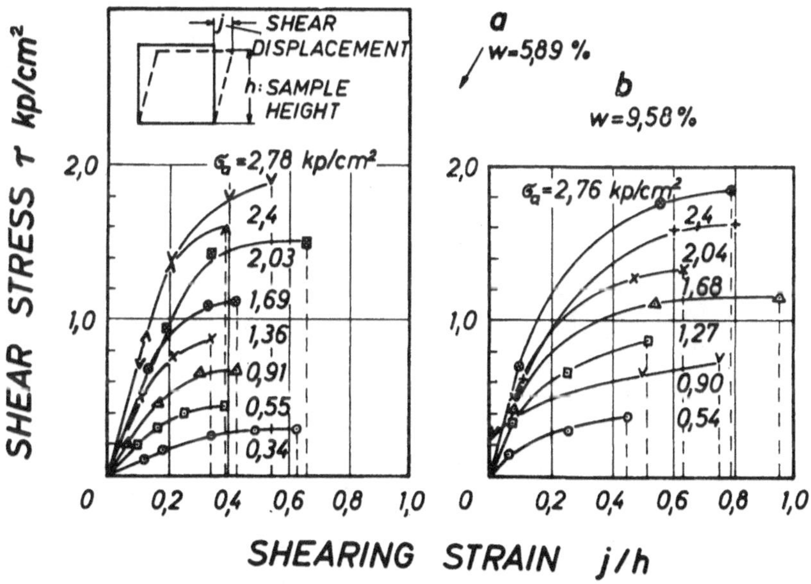

FIGURE 21. Shear stress-strain relation for various normal stresses and moisture contents

and the shearing deformation prior to failure were small. At higher moisture contents, the soil showed greater ability to take up energy and to be more deformed before failure (Figures 21 to 23) because shearing strain at failure increased with increasing moisture content for a constant normal stress. The increase of shearing strain occurred apparently

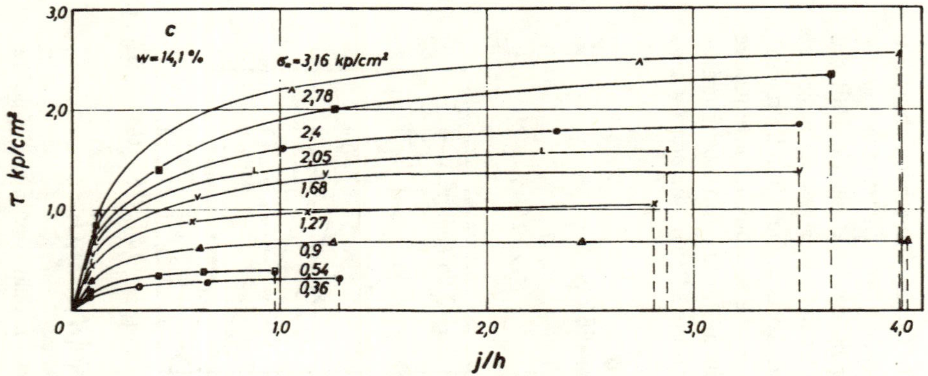

FIGURE 22. Shear stress-strain relation for various normal stresses and moisture contents

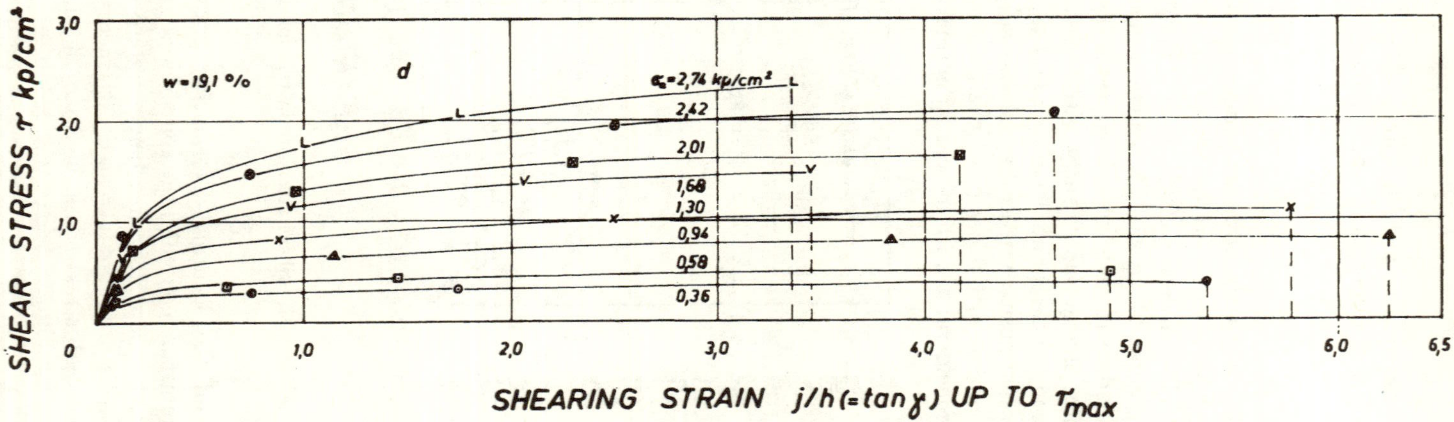

FIGURE 23. Shear stress-strain relation for various normal stresses and moisture contents

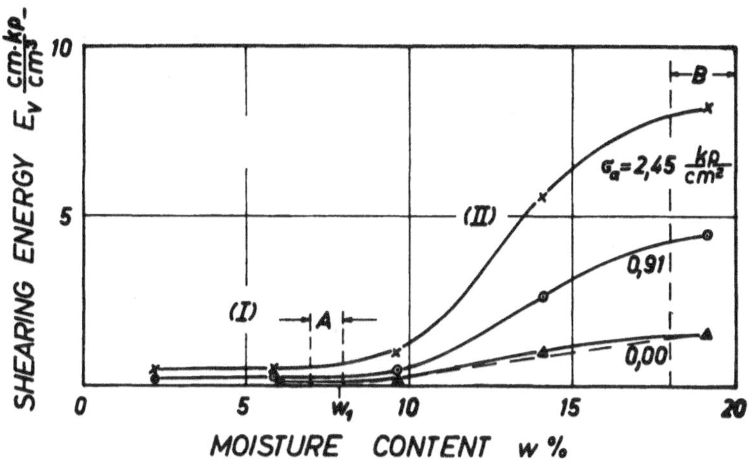

FIGURE 24. Variation of shearing energy with moisture content at three normal stresses

because this cloddy soil changed from a weak brittle condition to a plastic one and obtained capacity for greater deformation prior to failure with increasing moisture content. Two adjacent soil particles inside a clod seemed to possess capacity for increasing relative movement with increasing moisture content before they experienced the maximum shearing resistance against this movement. The peak values of shear stress also increased, first slowly and then rapidly (Figure 27), with increasing moisture content at a constant normal stress, since shearing at failure seemed to take place increasingly inside individual clods whose interiors possessed undisturbed soil particles. Experimental results (not yet published) showed that Coulomb's ϕ also increased slightly with increasing moisture content from 9 to 19 per cent due probably to increasing packing during deformation before failure. At a constant moisture content the shearing strain at failure increased significantly with increasing normal stress presumably because the soil particles were more strongly attached to the metal part which conveyed the tangential stress (τ) to the soil. Figure 24 shows specific shearing energy plotted against moisture content for various normal stresses. At a moisture content below 8 per cent (probably shrinkage limit), the specific shearing energy was almost independent of moisture content, since the shearing strain and stress did not increase significantly with increasing moisture content. At a moisture content between 8 per cent and 19 per cent, the specific shearing energy increased with increasing moisture content, first rapidly and then slowly, due chiefly to increased shearing strain at failure. Figures 25 and 26 show shearing energy relation with initial normal stress for various

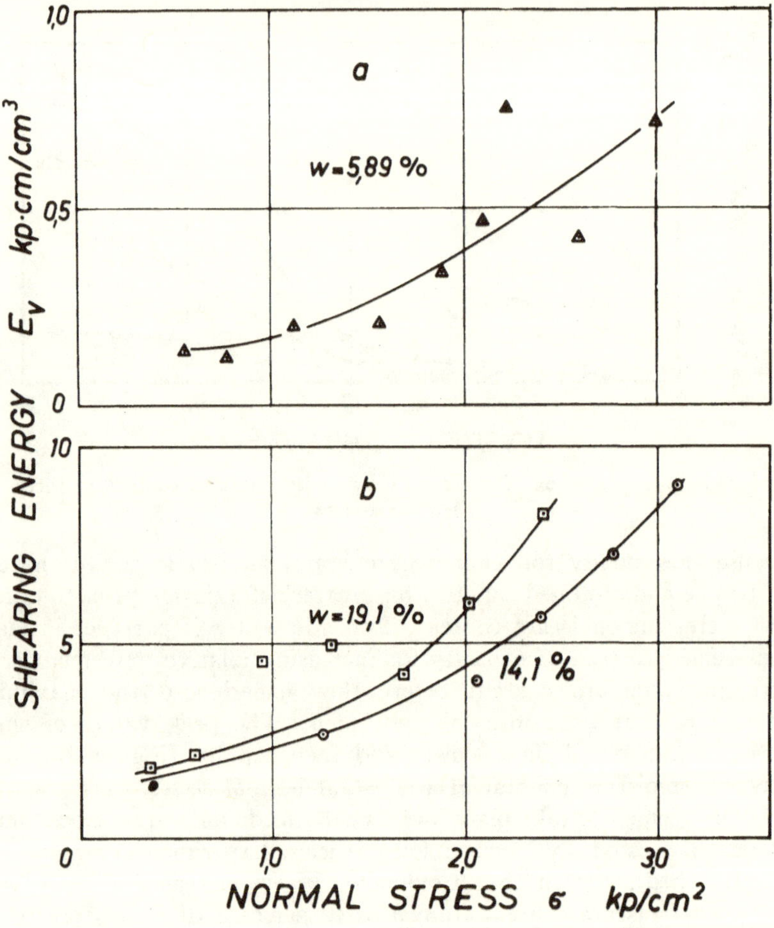

FIGURE 25. Specific shearing energy-normal stress relation at three moisture contents

moisture contents. The increase of shearing energy with increasing normal stress was more than linear because of increasing frictional resistance and increasing shearing deformation prior to failure as explained above. More precisely, the results showed that the shearing energy maintained an exponential relation with the normal stress (Figure 26).

Shearing energy is of course not the only component of energy involved in a shear test. If a complete energy balance at the boundary surface of the sample is desired, change in volume as well has to be considered. In our shear tests, total volume change recorded mechanically was very small and could be neglected. In cases where considerable compression

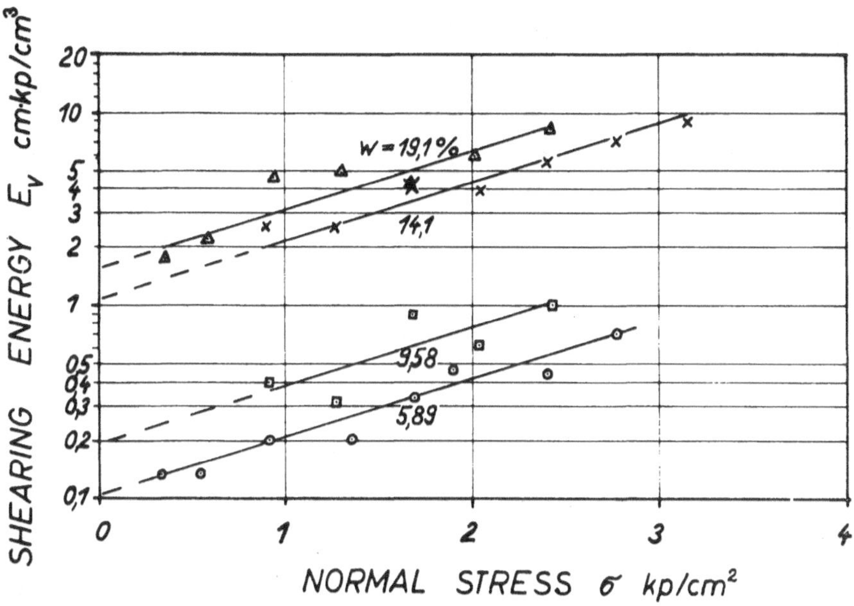

FIGURE 26. Specific shearing energy-normal stress relation at three moisture contents

in total volume of sample and volume increase (dilatation) only in the shearing zone occur during a shear test, both effects should be determined separately. This is, however, not easy to perform. In any case, an empirical or a theoretical relation among E_v, σ, w and e may be useful if an analysis of stresses, strains or size reduction in a cloddy soil involves energy balance or energy concept. The results indicate that such an

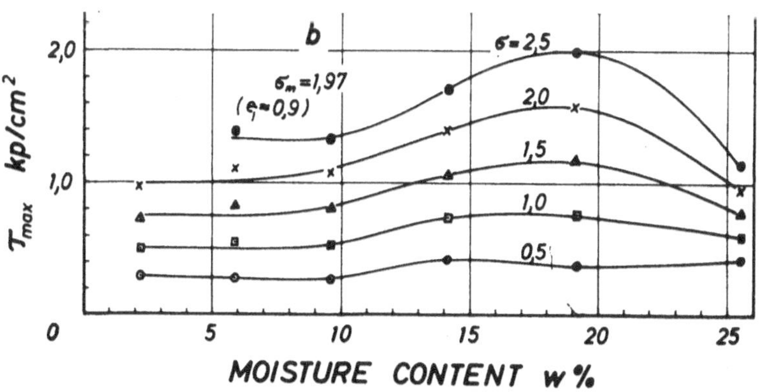

FIGURE 27. Maximum shear stress vs moisture content for various normal stresses

empirical relation may be found without involving unsurmountable difficulties (Figures 24 and 26).

It may now be concluded that the application of a guard zone in a shear head may be useful and desirable in some investigations and does not involve large complicacy or difficulty. On the contrary, the test results show that such a guard zone provides less scattered results in large ranges of normal stress, shear displacement, void ratio, and moisture content. However, it does not exclude the possibility of further development or modification of this shear head, specially for such uses as shear test on saturated soil, compression or consolidation test. The soil classification data indicates that the maximum and ultimate values of shear stress are not exactly comparable in a cloddy soil. A cloddy soil behaves differently from a soil consisting of no clod during shear.

REFERENCES

1. Hvorslev, M. J., *Torsion shear apparatus*, Bull. 38, Waterways Exp. Station, Vicksburg, Miss.
2. Söhne, W., *Proc. 1st Int. Conf. Soil-Vehicle Systems*, Turin, 1961.
3. Bekker, M. G., *Off-the-Road Locomotion*, Ann Arbor: University of Michigan Press, 1960.
4. Payne, P. C. J. and E. R. Fountaine, A field method of measuring the shear strength of soils, *J. Soil. Sci.*, 1952.

Measuring Trafficability of Snow

BENGT AGER*

ABSTRACT

In connection with studies carried out in Sweden on snow stabilization and vehicle mobility, the problems of measuring trafficability of snow have been touched upon.

A simple method of measuring the trafficability of the virgin snow cover, referred to as the "foot plate test," was studied in a pilot study. In this test the observer pressed a plate attached to his foot vertically into the snow cover. Plate sinkage and shear strength of the compressed snow was then observed. The results of the pilot study are briefly reported.

The trafficability of stabilized snow may generally be satisfactorily described by a simple "hardness" figure. Among the instruments tried the "Rammsonde" seems to be preferable for this purpose.

IN SWEDEN, several studies have been carried out on vehicle mobility both in or on virgin snow and on stabilized snow. The studies were sponsored by the forest research organization Forskningsstiftelsen SDA (now Forskningsstiftelsen Skogsarbeten, in English "Logging Research Foundation") partly in co-operation with the Swedish Army and the Royal College of Forestry. During the field tests density, grain type, grain size etc. were described on a routine basis. Other, more direct, measures of trafficability, were sought. The results of the investigation are briefly reported here.

TRAFFICABILITY OF VIRGIN SNOW COVER

General

Logging operations in winter in northern Sweden are often carried out when the snow is 60–100 cm deep. The tractors which are used for the extraction of timber generally open the strip road themselves. The tractors should possess two qualities as far as mobility is concerned:

—sufficient mobility in virgin snow, empty or with empty sleds or trailers;

—ability to pull loaded sleds or trailers on the hard-packed snow of the strip roads and main roads.

*The Royal College of Forestry, Stockholm, Sweden.

To unite these qualities in one vehicle means making a compromise between many different requirements. The machines used hitherto for extraction have mainly been tracked vehicles with a total weight of 3–5 tons, pulling sleds or trailers. When these machines were introduced in the middle of the fifties (gradually replacing the horse), vehicle weight was lower, the tracks were quite "dense" and the height of grousers moderate. The ground pressure was of the order of 0.10–0.20 kg/cm^2 of track contact area. Many of these vehicles had a relatively good flotation in snow, but were far from being "over-snow" vehicles. During the investigations and the practical tests it was gradually found that "open" tracks, high grousers (cf. Figure 1) and high ground clearance

FIGURE 1. Example of "open" track commonly used in Swedish logging operations today

resulted in a better performance by the types of vehicle used. Mobility in virgin snow was generally the same or slightly better, and the drawbar pull on the hard-packed snow increased.

During the first phase of development, when flotation was considered a more important quality of the vehicles than it is at present, and when research concentrated on finding efficient tracks with this quality, the author tried to develop a field method of measuring the trafficability of the snow cover. The objective was twofold:

(1) to find a simple and inexpensive way of studying the dependence of trafficability characteristics on the commonly used snow characteristics (density, grain type, etc.);

(2) to find a simple field method of predicting trafficability.

Owing to the change in vehicle development, this method was later abandoned. Only a pilot study was carried out, the results of which were never published. A brief description of the method—which will be referred to as the "foot-plate test"—and the few test results are presented here.

The "foot-plate test"

The following aids were used in observing trafficability:

(*a*) *a rectangular plate* (Figure 2*a*) which was attached to the foot of the observer (Figure 2*b*). In the pilot study a plate of 14 × 33 cm = 462 cm² was used as a standard. For an observer with ordinary body weight this gave a ground pressure of approximately 0.15 kg/cm², i.e. within the range of ground pressures of the vehicles tested. The following sizes were also tried occasionally: 12 × 29 cm = 348 cm²; 17 × 41 cm = 697 cm²; 24 × 58 cm = 1,392 cm².

(*b*) *a shear box* and *a dynamometer* (Figure 2*a*). This type of shear box was developed by A. Roch (de Quervain 1950). The size of the shear box was 10 × 10 = 100 cm². The top of the shear box was covered in order to allow loading. It was also painted with a plastic paint in order to prevent the snow from sticking to the box. The scale of the dynamometer allowed reading to a tenth of a kilogramme.

The following procedure was used in observation (Figure 2):

The observer pressed the plate vertically into the snow by gradually increasing the pressure on the plate from zero to full body weight. The observer was instructed to do this in 2–3 seconds. A vertical wall was then dug into the snow touching the compressed area and the plate sinkage was measured. The shear box was inserted into the compressed area, the snow was removed from the front of the shear box, and the dynamometer was attached to it. The observer then pulled horizontally at a rate of approximately 0.5–1.0 kg/sec. The maximum shear stress was then observed.

This was the standard procedure. It was carried out continuously and took about one and a half minutes for a trained observer. In the standard procedure, shear stress was observed at normal pressure 0 kg. Occasionally loads of 0.5, 1.0, and 2.0 kg were used.

At a later stage of the pilot study another observation was added. This

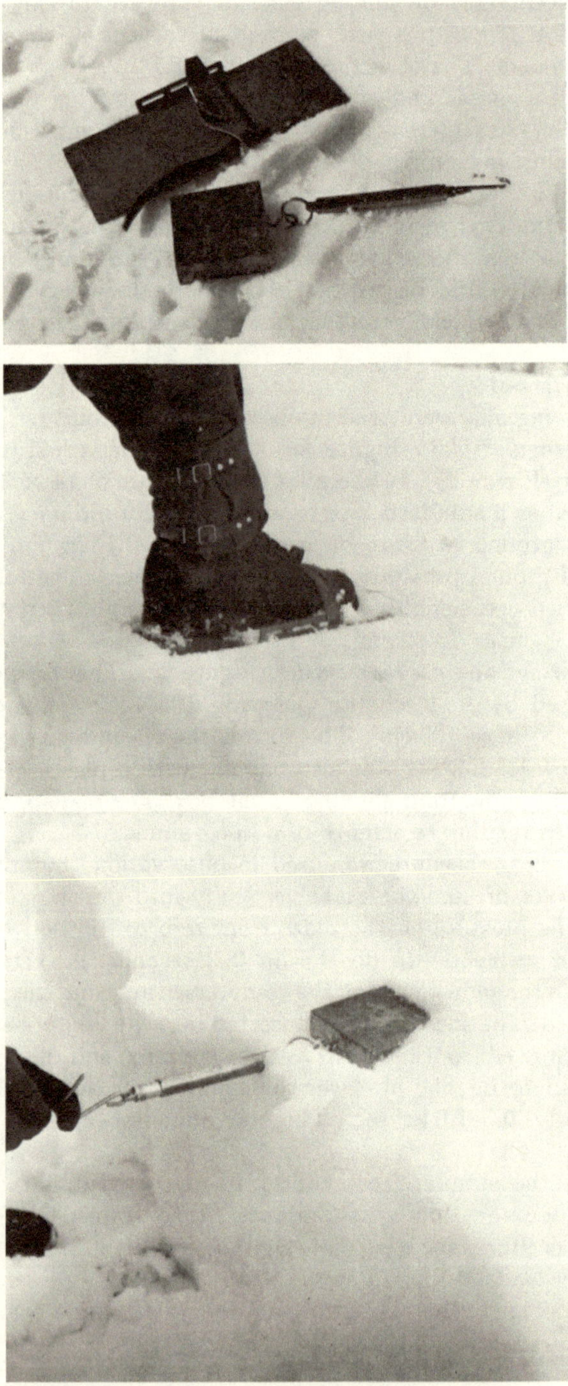

A

B

C

was measurement of the resistance of the snow wall above the compressed area to a long, narrow rectangular plate pressed horizontally against the wall. A proctor needle was used as an instrument for this, the ordinary plates being replaced by narrow rectangular plates of different lengths.

Pilot study

After some preliminary tests in the winter of 1956/57, a pilot study was carried out in 1957/58 and 1958/59. In 1957/58, 84 observations were made between January 4, and April 11, and in 1958/59, 58 observations between January 26 and April 7. One observation generally consisted of the following measurements:

(*a*) Plate sinkage and maximum shear stress according to the foot-plate test. Ten observations were taken in an area of about 2 × 2 m;

(*b*) hardness measured with the Rammsonde;

(*c*) hardness according to the "hand test" (de Quervain 1950);

(*d*) snow (crystal) type according to the International Association of Hydrology (1954);

(*e*) grain size;

(*f*) density at 5, 15, 25, etc., cm. depth in the snow pack;

(*g*) temperature at 0, 5, 15, 25, etc., cm depth.

Measurements (*c*)–(*g*) were taken in one profile dug vertically in the snow in the middle of the area where the foot-plate test was carried out.

The 1957/58 observations were made by one observer (body weight 70 kg) and the 1958/59 ones by another (body weight 75 kg). Both were men aged 20, doing their military service.

The most important observations were transferred to punched cards, (investigation No. 60.10.00 at the Royal College of Forestry, Department of Forest Biometry.)

Snow cover during the studies

Snow conditions in the test areas during the winters of 1957/58 and 1958/59 are roughly portrayed in Figure 3, which shows depth, snow type, and approximate density during the test periods.

In the winter of 1957/58 snow conditions were quite typical for these parts of the country (cf. Ager, 1965 and 1966). There were only two thin crust layers in the snow until the beginning of April. Ram hardness generally was between one and two kilogrammes. The winter of 1958/59

FIGURE 2. *Opposite.* Equipment and procedure for the "foot-plate test." A, plate, shear box and dynamometer; B, the plate is pressed into the snow cover; C, plate sinkage and maximum shear stress in the compressed area is observed

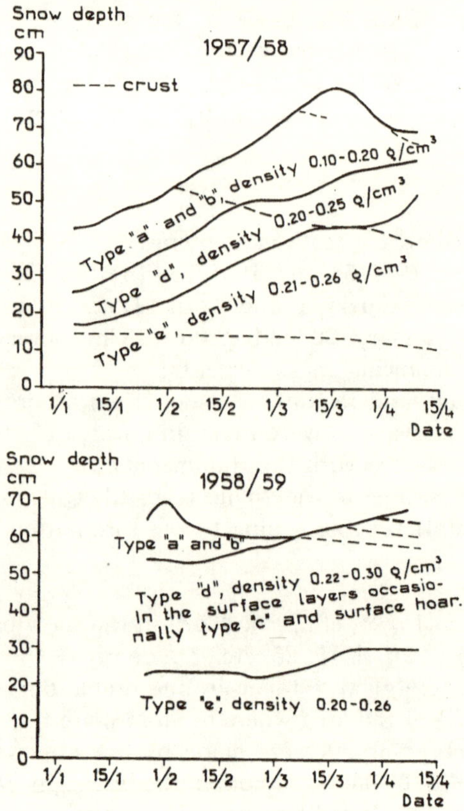

FIGURE 3. Snow conditions during the tests

was extraordinary in several ways. First, a period of heavy snowfall at the end of December and beginning of January resulted in a thick layer of snow with a density and hardness above the normal. Secondly, from the end of January until the end of March the temperatures were unusually high, often about freezing point. The snow was, therefore, denser and harder than normal, and the depth hoar layer was quite thin. Ram hardness often fell within the range of 2–5 kg.

Test results

The observations of 1957/58 were carried out simultaneously with vehicle tests. In these tests, track shape and load were varied on a test vehicle. Owing to these continuous variations, and to the fact that the test procedure was slightly changed during the course of the winter, this material was not suitable for studies of the relationship between vehicle performance and the foot-plate test observations. The foot-plate test was, therefore, only investigated in respect of variability of observations, and correlation with different snow properties.

For the 1957/58 observations, the coefficient of variation was around 3–5 per cent for plate sinkage and 15–20 per cent for maximum shear stress. For the 1958/59 observations the corresponding values were 20–40 per cent and 30–40 per cent. The spread of observations was thus considerably greater in the 1958/59 snow, which was harder and denser and much more stratified than the 1957/58 snow.

For the plate sinkage a correlation was found with density (Figure 4) and ram hardness, which might have been expected. The other snow

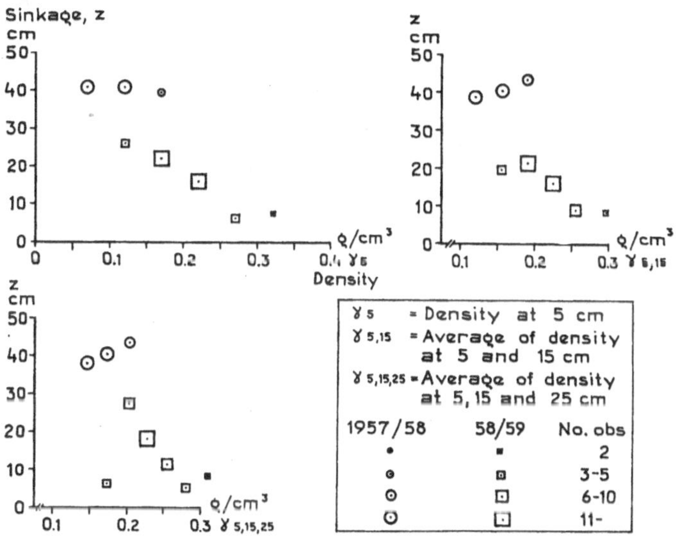

FIGURE 4. Sinkage of 462 cm² plate against density of snow. Weight of observers: 1957/58.70 kg; 1958/59.75 kg

characteristics observed did not seem to influence plate sinkage in the material collected.

Temperature was responsible for most of the variation of maximum shear stress (Figure 5). There were high shear strength values when the snow was at freezing point or slightly below (high cohesion), and when the snow was very cold (high friction). Initial density did not influence shear strength significantly, which might indicate that the density in the compressed area, within certain limits, does not vary with the initial density.

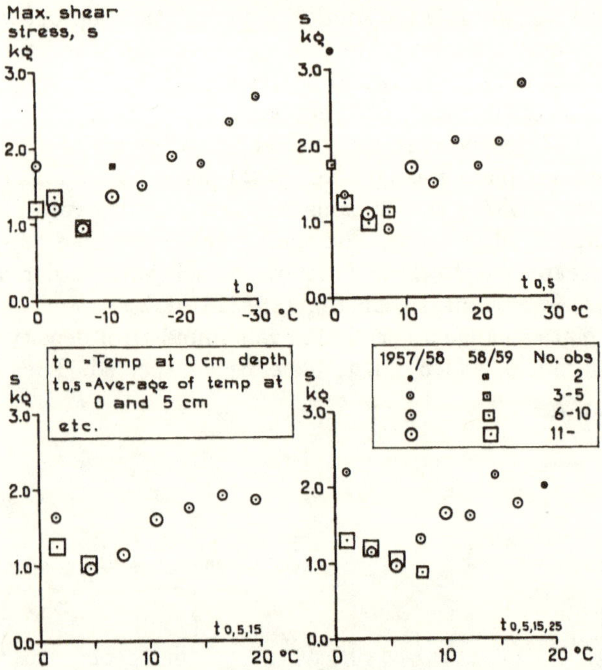

FIGURE 5. Maximum shear stress for a 10 × 10 cm shear box against snow temperature

Figure 6 shows the average (of 15 observations 1957/58) plate sinkage and maximum shear stress for different plate sizes. For the 462 cm² plate the pressure at full body weight was 0.15 kg/cm², for the 697 cm² plate 0.10 kg/cm² and for the 1,392 cm² plate, 0.05 kg/cm².

Some Comments

The methods of measuring trafficability described above are out-dated today. Both theory and experimental techniques have developed considerably since these tests were carried out. However, the general approach was in good agreement with the approach suggested by Reece (11): "However, it has led to a clear method of approach to the problem. This is to examine both theoretically and experimentally a soil loading system that imitates in a very simplified way, the manner in which a vehicle stresses the soil." Therefore, the author has found it justifiable to publish the results, even if there is only a small chance that anybody will find any of the ideas, results, and experiences useful.

The vehicles employed in extraction of logs in Sweden to-day are 3–7 ton tracked vehicles with "open" tracks and high grousers, and

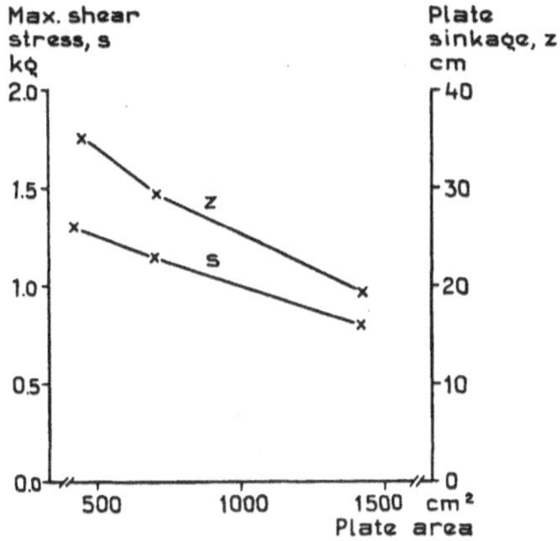

FIGURE 6. Plate sinkage and maximum shear stress vs plate size. Weight of observer 70 kg

4–10 ton wheeled vehicles with four-wheel-drive and articulated steering. None of these types of vehicles has a significant flotation in snow under the snow conditions encountered in Sweden. The most important trafficability characteristic of the snow cover is, in this case, the resistance of the snow pack to horizontally applied forces. We are not using sophisticated methods to measure this resistance (the method described on page 315 may be one way to measure it, however). When vehicle tests are carried out, the Swiss "hand test" (10) generally gives sufficient information on the trafficability of the snow cover for our purposes. If this method is to have a more widespread international use, it should be possible to standardize it more by recommending certain pressures against the snow for each hardness class.

TRAFFICABILITY OF STABILIZED SNOW

Snow stabilization generally means disaggregation, mixing, and compressing. This usually results in a substratum which, compared to the virgin snow cover or most natural soils, is comparatively homogeneous. Furthermore, the surface is generally quite smooth. For the present types of traffic in Swedish logging operations trafficability can, therefore, generally be sufficiently defined by a simple "hardness" figure, derived from the resistance to penetration of a plate, cone, or similar object.

The first instrument used to determine hardness of stabilized snow in the Swedish studies was the Proctor needle (8). However, the results of the Proctor measurements were found to vary considerably with the observer. It was therefore replaced by a drop cone penetrometer, which was then used during most of the studies (2), (3). The standard procedure was to drop a 30-degree 3 kg cone from a height of 10 cm and to observe the penetration. "Cone hardness" was then calculated according to the following formula (5):

$$\text{Cone hardness} \quad H_c = [Q(h + d)]/[0.07516d^3]\ \text{kg/cm}^2;$$

where Q = weight of the cone (kg),
h = fall (cm),
d = penetration (cm).

Cone hardness was generally observed at the surface and at 20 cm depth in the stabilized snow. In the latter case the 20 cm level had to be exposed by digging.

Bender (6), Abele (1), Wuori (12), and others have used the Rammsonde to measure trafficability of stabilized snow. The greatest advantage of the Rammsonde compared to a drop cone penetrometer is that hardness can be determined at any depth without digging. The Rammsonde is easier to transport and handle, but is not as quick for surface observations as the drop cone penetrometer. One disadvantage is its limitation to ram hardnesses less than 800 kg in its present shape. Hardness in trafficated snow roads sometimes exceeds this figure. The drop cone penetrometer has in practice no limitations with respect to hardness. Some standardization of the use of the Rammsonde, for example, recommendations with respect to hammer weight and height of fall for different hardness ranges, is also desirable, in order to increase the comparability of different observers. However, the Rammsonde in its present or modified shape seems to be preferable as an international standard instrument for measuring trafficability of stabilized snow. Its use is also already quite widespread.

In a recently published report (3), the author summarized those results of the Swedish studies on snow stabilization which were considered to be of more general interest. In that paper, trafficability requirements—expressed in cone hardness—of different types of traffic encountered in Swedish logging operations was reported. In order to enable a rough comparison of our results with the ones obtained by the CRRRL (Wuori [12] and other publications) a minor (27 observations) comparative study of ram and cone hardness was carried out. The correlation found between these hardness measures is shown in Figure 7.

The ram hardness was calculated according to the formula:

$$H_r[pnh/\Delta d + Q]k;$$

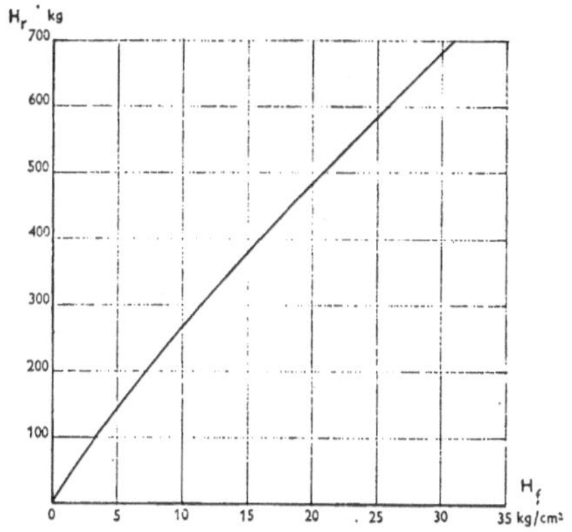

FIGURE 7. Ram hardness H_r vs cone hardness H_f (from Ager[3])

where p = weight of hammer (kg)
n = number of blows,
h = fall of hammer (cm),
Δd = penetration (cm),
Q = weight of penetrometer plus hammer (kg),
k = correction factor (cf. Brunke [7]) depending on the depth of measurement.

The ram hardness was measured from 0 to 5 cm depth in the stabilized snow. According to Brunke (7) the factor k is then 4.7. The one kilogramme hammer was used for all Rammsonde observations.

REFERENCES

1. ABELE, G., A Correlation of Unconfined Compressive Strength and Ram Hardness of Processed Snow. U.S. Army Cold Regions Research and Engineering Laboratory, Technical Report 85, January, 1963.
2. AGER, B. H., Packad snö som transportunderlag. Norrlands Skogsvårdsförbunds Tidskrift, 1956: III. Forskningsstiftelsen SDA medd. nr 56, Stockholm. (Compact snow as transport substratum. In Swedish, summary in English).
3. AGER, B. H., Om snöns egenskaper och snöstabilisering. En sammanfattning av 1950-talets svenska försök. Forskningsstiftelsen Skogsarbeten, medd. nr. 3, Stockholm. (On snow properties and snow stabilization. A summary of investigations carried out in Sweden during the fifties. In Swedish, summary in English).
4. AGER, B. H., Snow cover properties and climate in North Sweden. To be published in the proceedings of a conference on physics on snow and ice in Sapporo, Japan, August, 1966.
5. ANGERVO, J. M., Snöundersökningar i Finland. Teknisk Tidskrift 1951, s. 747, 1951, Stockholm. (Snow studies in Finland. In Swedish).
6. BENDER, J. A., Testing of a Compacted Snow Runway. American Society of Civil Engineers. J. of the Air Transport Division, Paper 1324.
7. BRUNKE, H., A Correlation of Crushing Strength and Hardness Values of Processed Snow. Michigan College of Mining and Technology. M.S. Thesis.

8. ERIKSSON, R., Den snöpackade vägen, I. Snöns egenskaper. Svenska Skogsvårdsföreningens Tidskrift 1954:2. Stockholm. Forskningsstiftelsen SDA, medd. nr 52. (The compact snow road, I. Properties of snow. In Swedish, summary in English).
9. INTERNATIONAL ASSOCIATION OF HYDROLOGY, The International Classification for Snow (with Special Reference to Snow on the Ground). Memorandum No. 31, Associate Committee on Soil and Snow Mechanics, National Research Council, Ottawa.
10. DE QUERVAIN, M., Die Festigkeitseigenschaften der Schneedecke und ihre Messung. Geofisica Pura E Applicata Milano. Bd XVIII, 1950.
11. REECE, A. R., Principles of Soil–Vehicle Mechanics. The Institution of Mechanical Engineers, Automobile Division. Proc. 1965–66, Vol. 180, Part 2A, London.
12. WUORI, A., Snow Stabilization Studies. Ice and Snow Properties, Processes, and Applications. Proceedings of a Conference held at the Massachusetts Institute of Technology, February 12–16, 1962. Cambridge, Massachusetts, 1963.

Measurements of Shearing Stress in Earth Under a Moving Vehicle

KENKICHI SAWADA*

ABSTRACT

The shearing stress in earth was measured by means of a test gauge embedded in earth. This gauge was so made as to catch the deformation of a proving ring both ends of which were respectively fixed to one each of the two layers of steel pieces which were lapped up and down to cause a relative dislocation because of a shearing stress in the earth. Although the earth and the gauge had a different deformation factor about this measurement, the finding of this test was interesting:

1. The reason for the good movability of a tracked vehicle on soft earth was due to a favourable distribution of shearing stress in addition to low contact pressure with earth.

2. The linear weight value of a flat smooth steel wheeled vehicle had an interesting meaning in dynamics.

THE TEST DESCRIBED IN THIS PAPER was undertaken to find out the running mechanism of a moving vehicle on soft earth, measuring the stress produced in the earth due to a force transmitted to earth surface as the reactive force against a wheel thrust. As the test finding, some interesting characteristics were uncovered: (1) the good movability of a tracked vehicle was found to be due to the fact that there was a favourable shearing stress distribution produced in the earth as well as low contact pressure with the earth in a test conducted with two different diameter wheels with the same linear weight corresponding to a wheel weight/the wheel width and a bulldozer; (2) the meaning was found of the linear pressure used for the index to represent the compacting performance of a tandem and macadam rollers in comparison with an earth stress measured with a shearing stress gauge under a flat smooth steel wheeled vehicle with different tractive forces applied on and the one calculated in the Froehlich formula.

TESTING ARRANGEMENT AND PROCESS

A testing runway was made a little wider than the width of the testing vehicle in the Chiba Branch of the Institute using a concrete pit 20 m long

*Public Works Research Institute, Ministry of Construction, Japan.

by 0.4 m deep, and sectionalizing frames 12 cm × 21 cm as seen in Figure 1. The earth applied in this test was of the properties as shown in Figure 2, adjusted to a predetermined water content, then placed

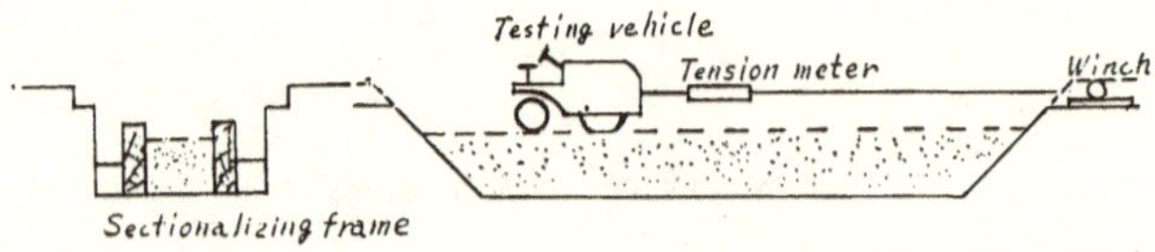

FIGURE 1. Testing runway

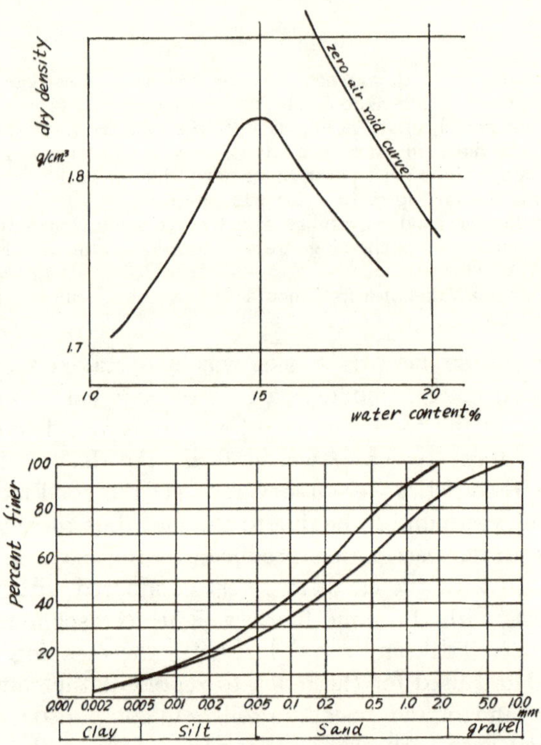

FIGURE 2. Property of soil

soft at 0.4 m thickness on the runway. On this earth, testing vehicles A, B, C, and D were driven. Vehicle A was a tandem roller for practical service, vehicle B a testing roller with different radii and an equal linear weight, vehicle C equal to B but for a less linear weight with a decreased additional weight, and vehicle D a small tracked type bulldozer. The

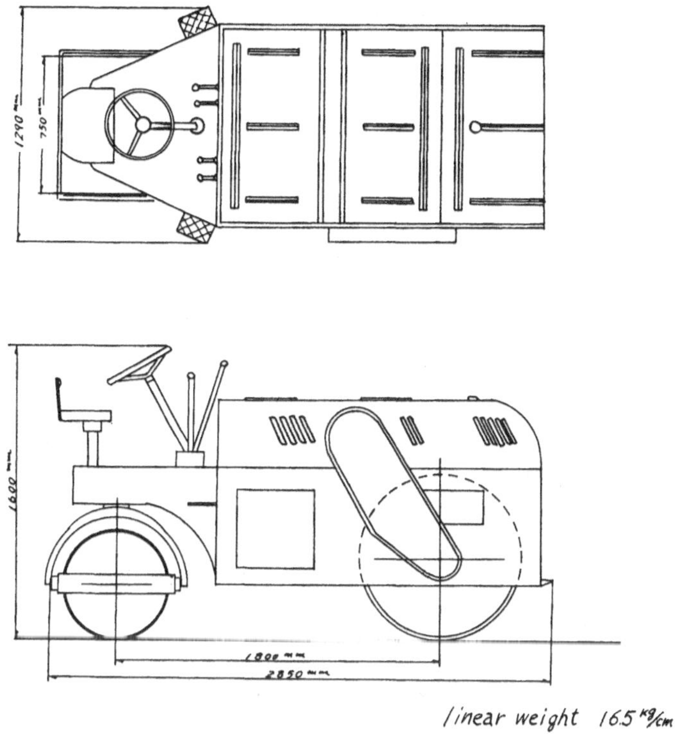

FIGURE 3A. Testing vehicle A

major parameters and whole pictures of these were as shown in Figure 3.

Measurements were made for the dry density and shearing stress of the earth as well as the earth pressure every time after 2, 4, 8, and 16 times passages of vehicle in principle. For this test, a specially constructed gauge was applied for shearing stress measurements for the earth. It was made of a proving ring and two layers of steel pieces lapped up and down as shown in Figure 4. One end of the ring was fixed to the upper steel piece and the other end fixed to the lower one so that any relative dislocation of the two lapped steel pieces occurred due to a shearing stress of the earth could be sensed by the proving ring. Such a gauge was embedded in the earth at depths of 5, 10, and 15 cm each under the ground surface with a preliminary compaction applied for fear that its position should be disturbed by a vehicle passage if it were embedded without such compaction. The depths 5, 10, and 15 cm mentioned above were of course approximate due to some inevitable variations.

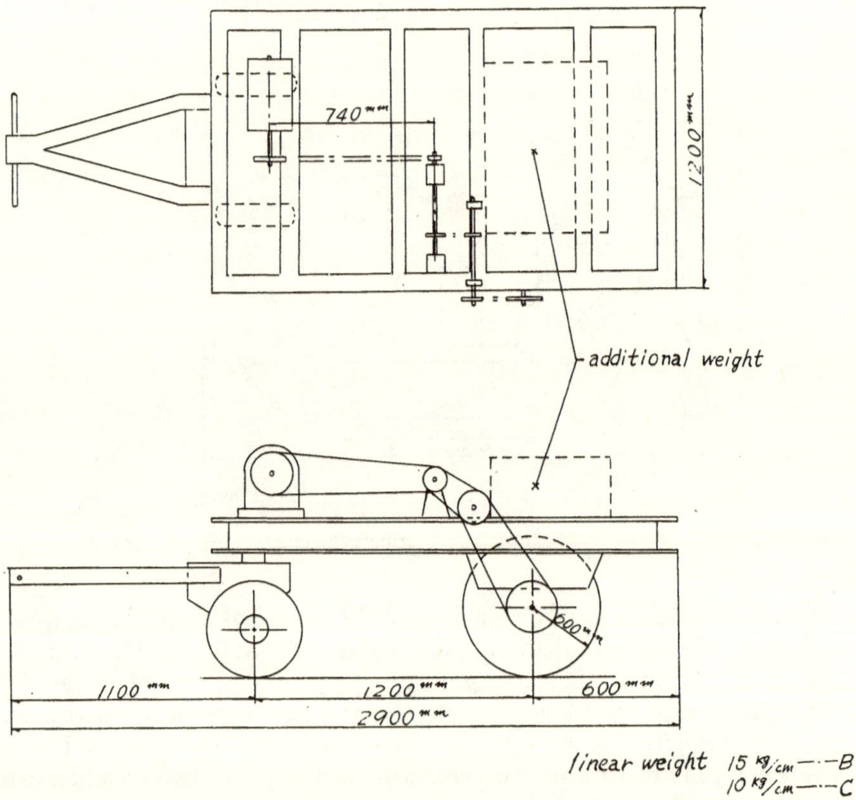

FIGURE 3B. Testing vehicles B and C

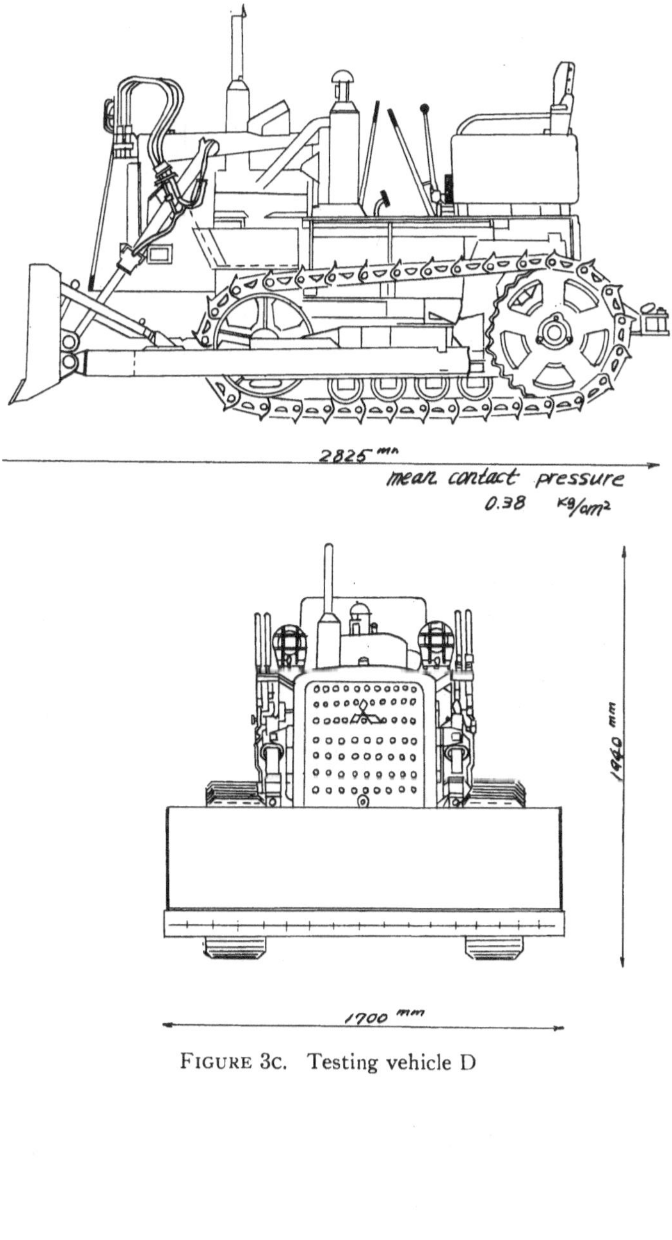

FIGURE 3C. Testing vehicle D

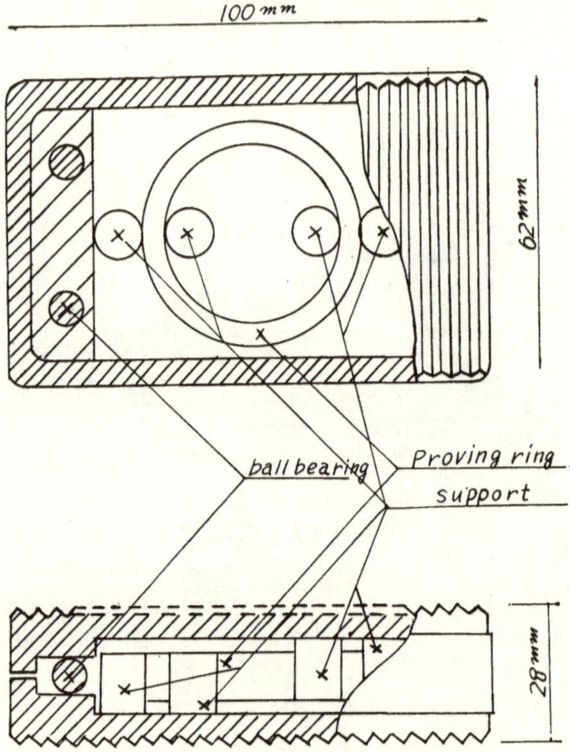

FIGURE 4. Shear stress gauge

TEST FINDINGS AND OBSERVATION

Figure 5 shows one type of the wave forms drawn on a pen oscillogram in the case of vehicle A. The shearing stress gauge especially made for

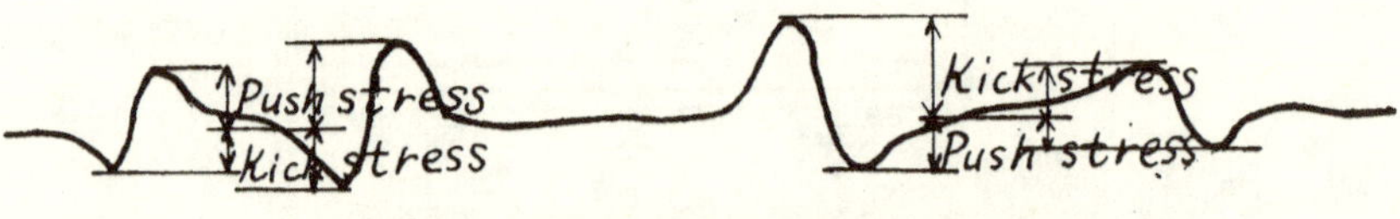

FIGURE 5. Model wave form of shear stress

this test also had a similar objection as to the disagreement in deformation coefficients between the earth and the gauge to the one experienced in the case of an earth pressure cell. In addition, there were a number of matters, such a residual stress, to be taken into account when a value was read on the oscillogram. The values with an arrow were taken up for the measured ones with a special attention paid to the relative magnitude of measured values in this paper.

1. A Different Shearing Stress by a Running Vehicle

In the test finding shearing stress was found almost constant with a little variation after five vehicle passes unless additional load was externally applied, so that an average value after five passes was taken up for the shearing stress at a certain depth and under a certain condition. The test results of the vehicles *A*, *B*, *C*, and *D* are compared in Figure 6 in which the shearing stress is shown in the ordinate and the test condition and loading direction in the abscissa. When the wheel was just to one side of the gauge, the soil was pushed forward with the result that the upper steel piece moved ahead of the lower piece. When the wheel passed over the gauge, the earth was moved backwards by the reactive force of the thrust with the result that the upper piece moved backward relatively. Here, by the "push" is meant the shearing stress shown in the gauge before the wheel reached directly above the gauge while by the "kick" is meant the value in the reverse direction.

The largest shearing stress was in the case of Vehicle *A* followed by *B*, *D*, and *C*, in the order shown in Figure 6, presenting no marked difference among these latter three. In a close checking of this comparison only a little difference was noticeable between self-propelling or towing or between the gauge embedded shallowly or deeply, in the case of vehicle *D*. A value at a time of kicking divided by a value at a time of pushing is shown in Figure 7 in order to make the characteristic clear. Such a quotient, less than 1.0, means that concentrated forces were acting on the ground surface in the same direction as the progress of the vehicle while one larger than 1.0 means that concentrated forces were acting in the reverse direction.

From these facts, it may be concluded that when a vehicle, inclusive of a tracked-type one, is running on earth, a horizontal force, in either positive or negative direction to the running direction, depending on whether self-propelling or towing, acts on the earth in addition to the vertical stress due to the weight of the wheel. Although the magnitude of horizontal force depends upon the type and weight of the wheel and the like, it has attracted little attention so far. A tracked-type vehicle is well known for its excellent movability on soft ground because it exerts low contact pressure on earth. However, based on the findings of this present test, the characteristics of this type of vehicle eliminates the phenomenon of the wheel leaning forward with a low shearing stress in pushing in comparison with the shearing stress in kicking and a low shearing stress in contrast to the tractive resistance in addition there is

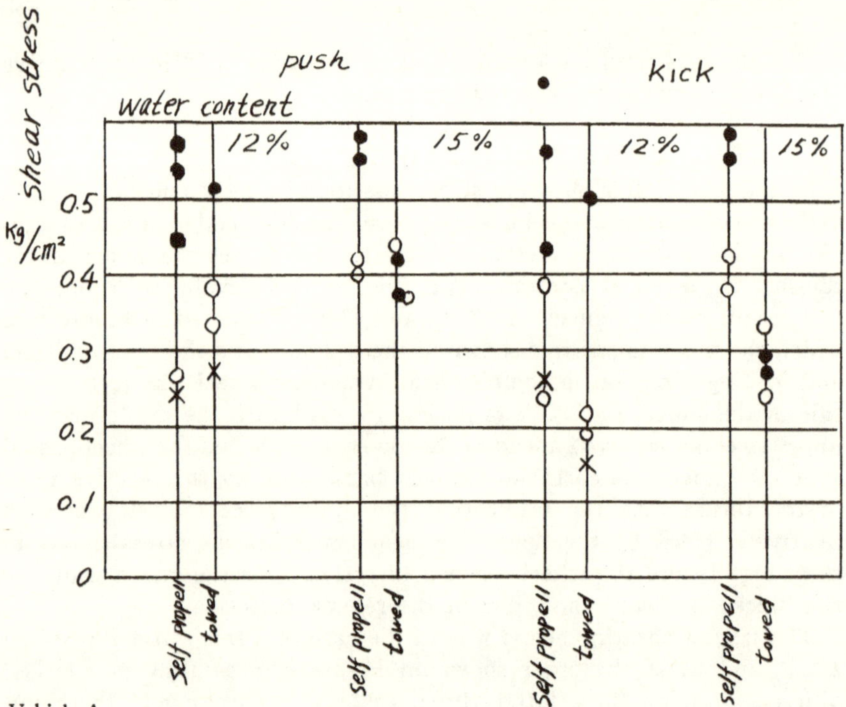

Vehicle A

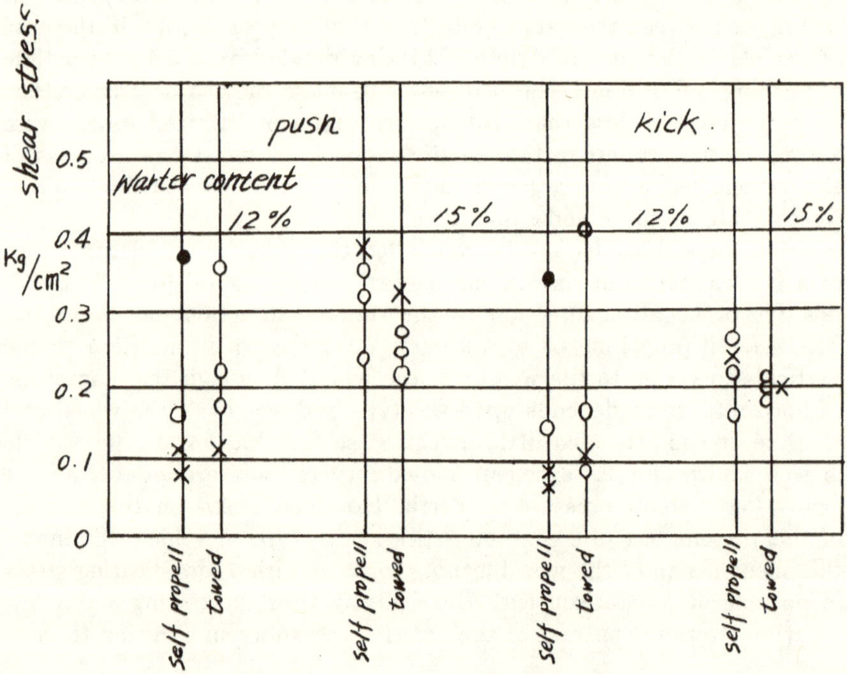

Vehicle B

FIGURE 6. Comparison of shear stress on testing vehicles
● = depth 0–5 cm; ○ = depth 5–10 cm; × = 10 cm or more.

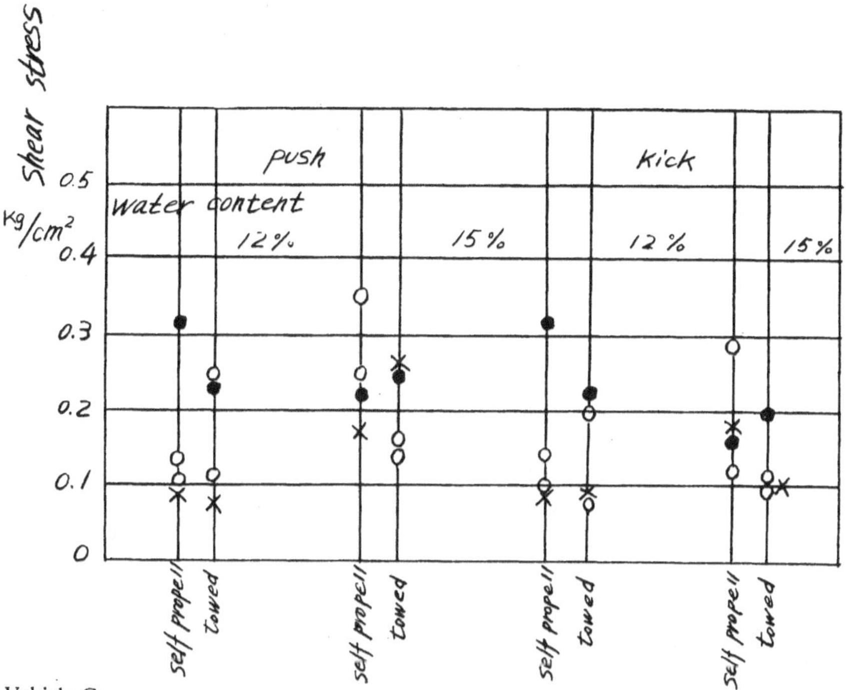

Vehicle C

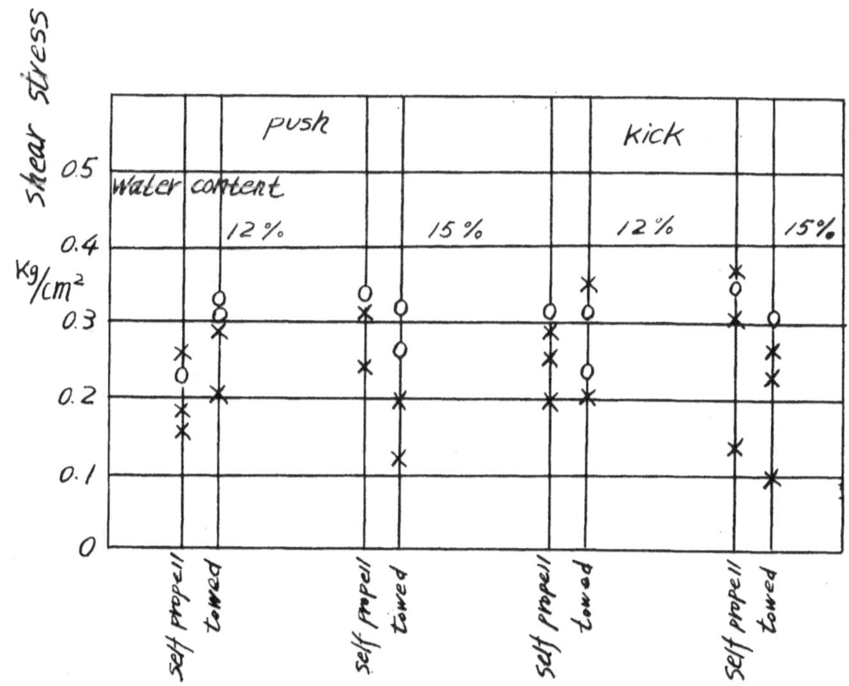

Vehicle D

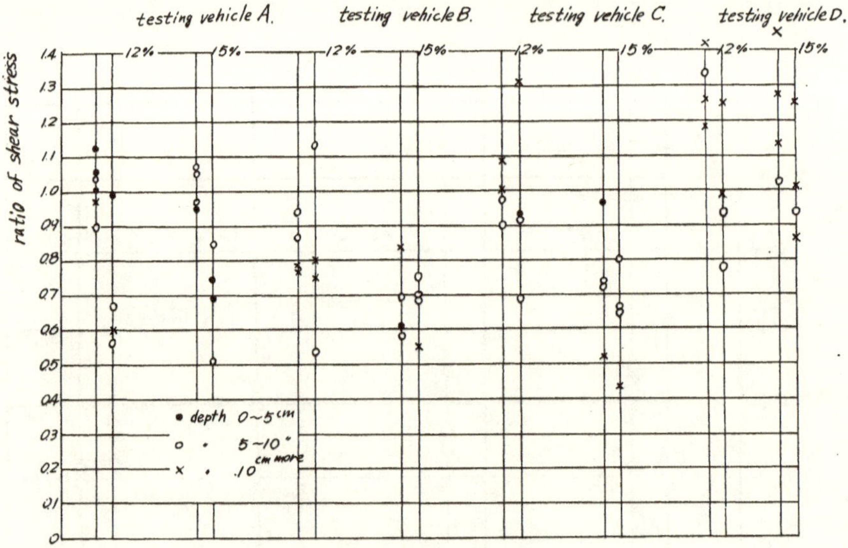

FIGURE 7. Ratio of shear stress

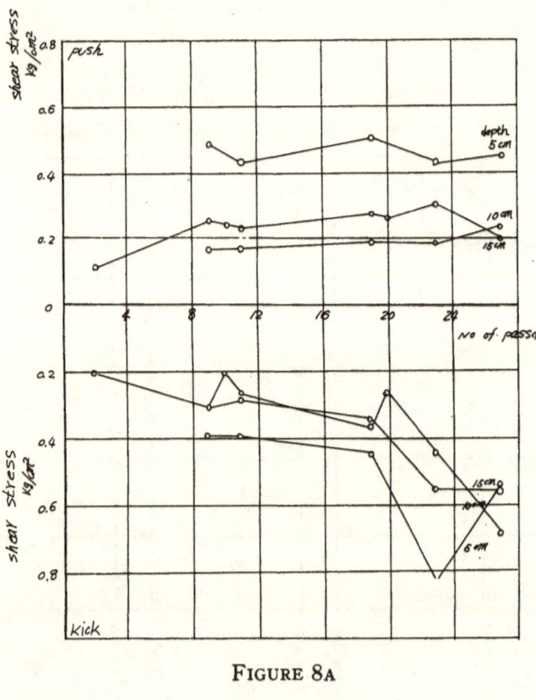

FIGURE 8A

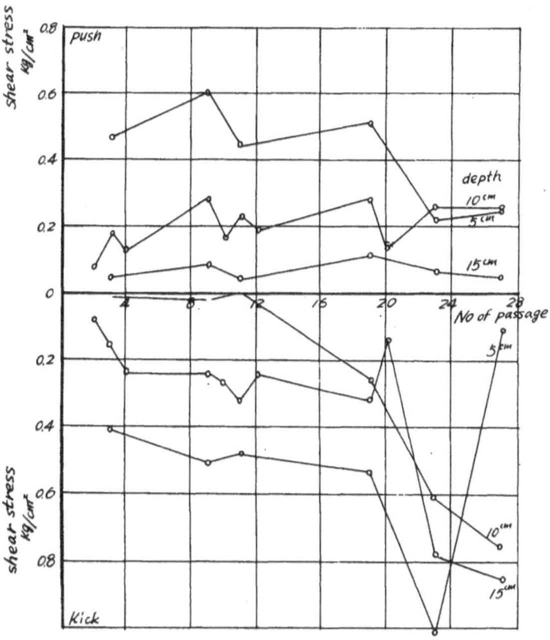

FIGURE 8B

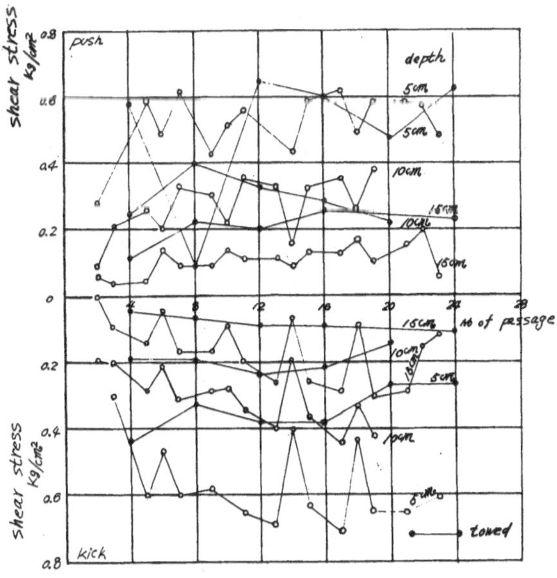

FIGURE 8C

FIGURE 8. A, shear stress: water content 13 per cent, towing load 1,000 kg at 23.27 passage; B, shear stress; water content 15 per cent, towing load 1,000 kg at 23.27 passage; C, shear stress; water content 15 per cent the simple merit that it causes little sinking on soft ground as it exerts a low contact pressure on earth. This characteristic would be excellent for a running vehicle.

2. *A Different Shearing Stress by a Running Condition*

One end of a wire rope was tied to vehicle A, the other end to a winch drum; the rope was pulled by the vehicle; a brake on the winch was applied to increase the tractive load to give a strong horizontal force exerted on the earth surface. The result of this attempt was examined by the gauge embedded in the earth as shown in Figure 8. It was found that the deeper into the earth the smaller was the value of shearing stress. Up to three vehicle passes, the stress seemed somewhat low and thereafter constant practically. Such a characteristic with only a few factors to affect the stress was found advantageous in research in Terrain Mechanics. With such constant stress, the reason why earth was compacted by vehicle passage resulting in higher density and strength was because there was a different mechanism of deformation, a difference in the lattice work of earth grains to bear stress while the stress remained the same.

The author's hypothesis for an explanation of the various phenomena of this matter, is that earth grains would move to be compacted in the first passage of a road roller, but the bearing force of the earth grains would not have been distributed uniformly yet. The density of the lattice-work of earth grains to bear the load would be higher than the average density. The second passage of the roller would crush down the lattice-work to form a new one to bear the load. In the case where the void ratio of compacted earth by the first passage of roller is lower than the critical void ratio without a volume increase to follow due to a shearing dislocation, the density would be increased by the second passage. Although the relation between the load and the density in the first passage is similar to the relation in the case where earth is statically compacted in a mould with a compressing machine, the dislocation would be different in the second passage and thereafter.

When tractive force of a wheel increased, the shearing stress increased. As a wheel was towed, the shearing stress came down. In application of the Frohelich formula, the ratio of shearing stresses before and after vehicle passages was found to be 1:3 when the vertical and horizontal stress components were equal at concentration coefficient 3 in the case where a uniform linear load acted on the ground surface. This ratio was found 1:1.5 when the coefficient was taken to be 5. Such coefficient of the earth used in this test had been found to be nearly 3. The sum of the moving resistance and tractive effort was approximately equal to or more than the weight of the wheel. It was found as the result of this test that the shearing stresses before and after the vehicle passage were about 1:3 in ratio with some variations and that the shearing stress before applying a tractive load could be approximated by the value calculated in the Froehlich formula, so that the foreign force given by a flat steel-tired wheel rolling on earth might be represented by vertical and horizontal linear loads, that is, a linear weight as mentioned before. Such a result was found, in the estimation made for a load from an earth density compacted in the test having been conducted before, also agreeable with the conclusion that load had to be distributed over a very narrow band of a roller circular surface.

CONCLUSION

The above treatment was the explanation of some interesting contents and conclusions of the test conducted in this Branch Centre. The test result of a vehicle with a pneumatic tire was omitted in this paper because it was assumed that the test result of a pneumatic-tired vehicle would fall between those of a tracked type and steel-tired vehicles and that manoeuvring such a vehicle to pass right on the rut defined on a soft ground deemed hard in the test.

THEORY AND DEFORMATION OF SOILS

On the Analysis of Soil Deformation Under a Moving Rigid Wheel

R. N. YONG*

J. C. OSLER†

ABSTRACT

In the problem of soil deformation under a moving rigid wheel, experimentation has shown that elemental soil units beneath the surface suffer both horizontal and vertical movements in addition to volume distortion. By examining the stress–strain behaviour of soils under various conditions of strain rate and sample confinement, it becomes evident that the relationship between shear stresses and strain is sensitive to both strain rate and confinement effects.

This paper examines the over-all problem of soil deformation under a moving rigid wheel relative to movement, rate effects, soil properties, and environmental constraints. The concepts of yielding and flow in soil mechanics are presented to provide the basis for an appraisal of the problem of the mechanics of soils in a dynamic regime.

The standard analytical strength parameters such as c and ϕ derived from soil statics are shown to be most limited and do not find valid application in situations where dynamical soil behaviour is concerned.

IN THE STUDY of soil deformation under a moving rigid wheel, experimental evidence shows that the soil response under the forcing function includes both yielding and flow extending to some significant depth below the wheel-soil interface. In the integrated two-part study being conducted in the Soil Mechanics Laboratory of McGill University, an attempt is being made to understand both the mechanics of soil under dynamic loadings (as would be anticipated from transient surficial loading), and the behaviour of the soil itself under the forcing function. In the former (soil dynamics study) triaxial tests on soils are being conducted at rates of sample straining ranging from creep rates to 1000 per cent per second. The usefulness of the study on soil dynamics for the description of operative strength parameters and the formulation of a yield function is dependent on one's ability to describe the stress

*Professor of Civil Engineering, Director of the Soil Mechanics Laboratory, McGill University, Montreal, Canada.

†Assistant Professor of Civil Engineering, Department of Civil Engineering and Applied Mechanics, McGill University, Montreal, Canada.

and strain-rate field in the soil under the loading function. Thus the second part of the two-fold study is designed to provide information on the behaviour of the soil in terms of elemental particle motion which may be used to establish a velocity vector field. In the visioplasticity method employed, a grid pattern is established using the X-ray technique (17), (18), (2).

Because of the obvious magnitude and complexity of the problem, and also because of the amount of work remaining to be performed in this study, no attempt will be made to provide an answer to the problem. Instead, it is the intent of this paper to provide some of the reasons for the examination of the mechanics of soil-vehicle systems in terms of a study of the soil response to the forcing function. Thus this paper examines, firstly, the present-day concepts of failure and yield in relation to soils, derived in general on the basis of quasi-static tests. It then proceeds to discuss the shear strength parameters now generally used in design and analysis of soil response for soil-vehicle studies. The significant advantages to be gained by the evaluation of soil response under a forcing function as a means of studying the soil-vehicle system are then presented. The omission of a study of the mechanics of the vehicle system is intentional since it is not within the scope of the over-all study. It is, however, recognized that the vehicle system is an important constituent and that this is being studied by other research groups. Taking heed of the remarks by Bekker (1), an attempt is made to reconcile the fact that less rigorous methods must be used to provide simple solutions in the absence of definitive theories.

FAILURE AND YIELDING OF SOILS

The concept of failure in the definition of the over-all state of soils under a forcing function is most generally confined to some limiting strain. "Failure" so defined, may include both fracture and flow. While fracture may be identified by the appearance of distinct surfaces of separation in a material undergoing straining, yielding, as a phenomenon, describes the onset of plastic deformation and the resulting successive deformations are generally defined as flow.

Soil deformation under a moving rigid wheel ranges from a minimum at the lower regions of the affected layer to a maximum at the wheel-soil interface (see Figure 1). The types of deformation are dependent on loading, sinkage, and slip. Thus in the affected zone shown in Figure 1, it is evident that the types of deformation will be a combination of flow beyond yielding and deformation below shear failure (i.e. strains that are small because the stress state is below critical or yield).

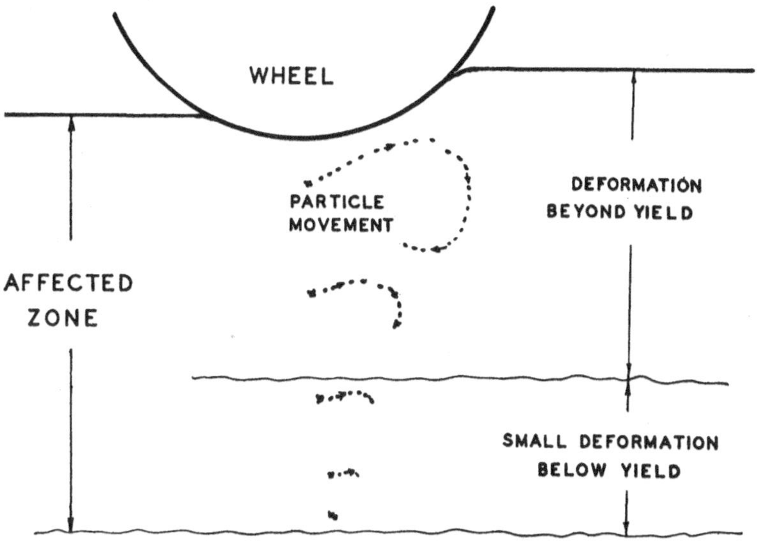

FIGURE 1. Deformation zones beneath a moving wheel

Defining the deformation-beyond-yield zone and the small-deformation-below-yield zone in the affected region as primary and secondary zones respectively, it is evident that if one utilizes only the characteristics of the primary zone for thrust and sinkage considerations, the analysis if formulated on the basis of limiting equilibrium of the soil within this zone. An example of this type of analysis is the one provided for plate bearing and footing studies. One therefore ignores the deformation below the primary zone, that is, in the secondary zone. The consideration of energy input at the wheel-soil interface, and the work and energy relationships within the soil, show that it is necessary to combine the work performed in the process of shear and flow within the primary zone with the work necessary to create deformation in the secondary zone. The problem becomes quite complex since what is needed is the establishment of appropriate stress-strain relations for the soil throughout the depth of the affected region. For simplicity, one may assume that the integrity of the wheel is determined by the primary zone if an appropriate correction for the response of the secondary zone is made in the analysis at some suitable time. Thus, one may now examine the most often used failure theory both in terms of what the theory requires to assure its validity and the type of information which it provides.

In the strength theory developed by Mohr, failure by both yielding and fracture is considered; the basic mode of failure assumed is that of slippage. The theory asumes that at failure the normal and shear stresses on the failure plane are connected by some functional relation,

$$\tau = f(\sigma) \tag{1}$$

where τ = shearing stress along the failure plane and σ = normal stress across the failure plane. The two-parameter nature of the theory is at once apparent. The curve defined by this functional relationship may be plotted on the τ, σ plane (see Figure 2). Since changing the sign of τ

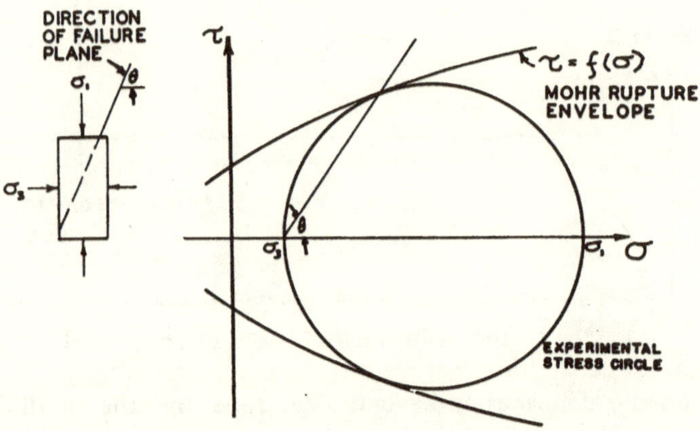

FIGURE 2. Mohr rupture envelope

merely changes the direction of failure but not the condition for it, the curve must be symmetrical about the σ axis, and is often termed the Mohr Rupture Envelope. It is the locus of all points defining the limiting values of both components of stress (τ and σ) in the slip planes under the different states of stress σ_1, σ_2, σ_3, to which the material may be subjected. The Mohr envelope should therefore be thought of as a property of the material and as independent of the stresses imposed on the material. It is not required that the material obey Hooke's law or that Poisson's ratio be constant. It is further noted that the strength and stiffness of the material in tension and compression do not need to be equal. For many of the above reasons, the theory is attractive for use in studying the shear strength of soils.

Since any state of stress may be represented in the τ, σ plane by its three principal stress circles (Mohr's circles) it follows that any limiting state of stress σ_1, σ_2, σ_3 may be represented by three circles. With the usual assumption that σ_2 is the intermediate principal stress, the largest of three circles representing the limiting state of stress will be of diameter $(\sigma_1 - \sigma_3)$ and centred at $(\sigma_1 + \sigma_3)/2$ along the σ axis, taking due account of the algebraic sign of the stresses. Since the two parallel sets of slip planes which occur when an isotropic specimen has been stressed slightly beyond the plastic limit by a state of homogeneous stress are symmetrically inclined with respect to the directions in which the major and minor principal stresses act, and the two plane systems intersect each other along the direction in which the intermediate principal stress acts, Mohr assumed that the intermediate principal stress is without influence on the failure of a material. Accordingly, some point on the perimeter of the circle of diameter $(\sigma_1 - \sigma_3)$ must represent the limiting stress condition.

Actual Mohr Rupture Envelopes are often curved. For soils, the curvature is usually not great and it has proved useful to approximate the envelope by a straight line, at least over a limited range of normal pressure. The equation of a straight line in the τ, σ plane can be of the form attributed to Coulomb, namely $\tau = S = c + \sigma \tan \phi$. Considerable progress has been made in the field of soil mechanics by the use of the Coulomb line to express the equation of the Mohr envelope and the resulting failure theory is referred to as the Mohr-Coulomb theory. It was established by Terzaghi and some of his early co-workers, particularly Rendulic and Hvorslev, that the above relationship only produced good correlations for soils when the normal stress was taken as the effective stress and not the total stress. For the purpose of this paper the relationship between the total stress and the effective stress will be taken as:

$$\sigma' = \sigma - u \tag{2}$$

where σ' = effective stress,
σ = total stress,
u = pore pressure.

The validity of this equation for granular materials has been established but the difficulty of defining pore pressure in clay soils, where the pore water is subjected to various interparticle forces and other surface and osmotic pressures, makes this simple equation difficult to apply. This topic is beyond the scope of this paper. Because of the obvious dependency of the shear strength of soils on effective stresses, their use throughout this paper is implied unless stated to the contrary.

The fitting of a straight line envelope to a series of Mohr stress circles permits an algebraic formulation of the Mohr-Coulomb theory, such as is commonly used in the field of soil mechanics. From Figure 3, the following formula may be derived:

$$R = (\sigma_1 - \sigma_3)/2 = c \cos \phi + ([\sigma_1 + \sigma_3]/2) \sin \phi \tag{3}$$

The terms c and ϕ, as used in Equation 3, refer to the "cohesion" and

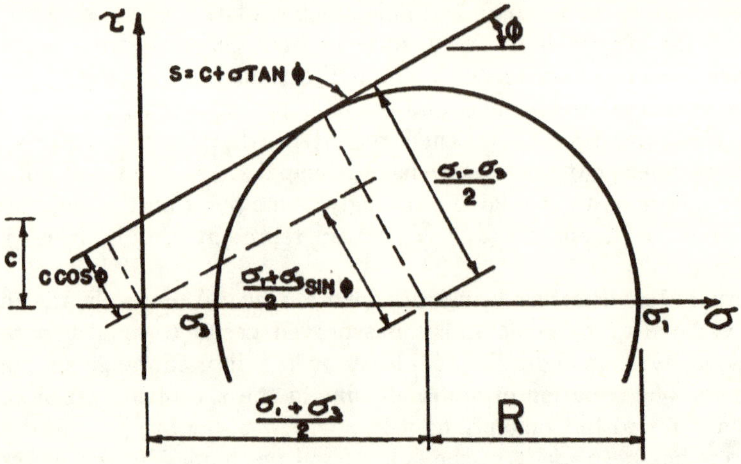

FIGURE 3. Geometric relationships involving Mohr–Coulomb rupture envelope

"angle of internal friction" of the soil. These shear strength parameters have been the subject of much research and it is now well established that they are not material constants but vary with the void ratio or water content as well as for the type of test considered. While these variations affect the general study of failure criteria, they will not be considered in this paper. The fundamental reasons for their variation, related to such things as soil structure, fabric, stress history, etc. have been stated in many research studies (see, for example, (19)).

Equation 3 may be manipulated in many ways to state the criterion in various forms. Two such variations, as given by Hvorslev (7) are,

$$(1/2)(\sigma_1 - \sigma_3)(1 - \sin\phi) = c\cos\phi + \sigma_3 \sin\phi, \tag{4}$$

which was used by Skempton and Bishop (15) and gives straight lines when $(1/2)(\sigma_1 - \sigma_3)$ is plotted against σ_3 and,

$$\sigma_1(1 - \sin\phi) = 2\,c\cos\phi + \sigma_3(1 + \sin\phi), \tag{5}$$

which gives straight lines when σ_1 is plotted against σ_3. This last equation has been used as a plotting method by Rendulic and more recently by Henkel (6).

Expressed in its most general form, the failure surface corresponding to the Mohr-Coulomb condition of failure is:

$$[(\sigma_1 - \sigma_2)^2 - (2c\cos\phi + (\sigma_1 + \sigma_2)\sin\phi)^2][(\sigma_2 - \sigma_3)^2 - (2c\cos\phi + (\sigma_2 + \sigma_3)\sin\phi)^2][(\sigma_3 - \sigma_1)^2 - (2c\cos\phi + (\sigma_3 + \sigma_1)\sin\phi)^2] = 0. \tag{6}$$

This equation is given in terms of effective stresses by Hvorslev (7) and by Roscoe, Schofield, and Thurairajah (11) in a form where c equals zero. The failure surface so defined is a pyramid with the space diagonal $\sigma_1 = \sigma_2 = \sigma_3$ as axis and a cross-section which is an irregular hexagon with non-parallel sides of equal length. The projection of this irregular hexagon on the plane $\sigma_1 + \sigma_2 + \sigma_3 =$ const. (i.e. a plane at right angles to the space diagonal or an octahedral plane) is shown on Figure 4. For

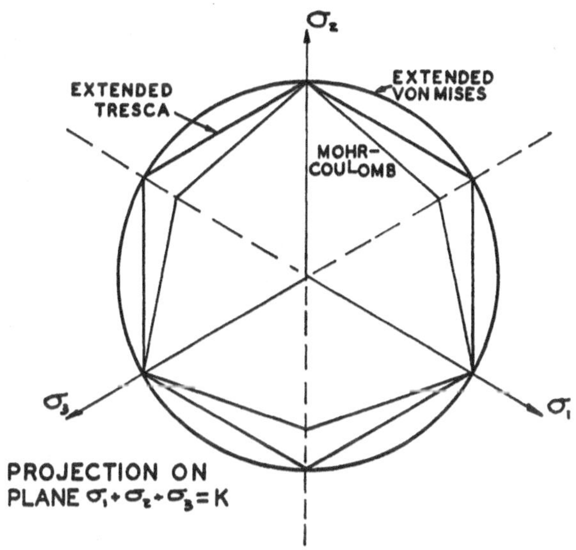

FIGURE 4. Octahedral representation of Mohr–Coulomb yield surface

reference, the circle and regular hexagon described by the extended von Mises and extended Tresca criteria are also plotted on Figure 4. The three criteria are seen to coincide for compressive tests but the strength in a tensile test is seen to be less for the Mohr-Coulomb failure theory.

From recent research studies, several modifications for Mohr-Coulomb and other existing yield criteria have been suggested. Both Newmark (9) and Kirkpatrick (8) suggest a slightly rounded shape for the yield surface following quite closely the outline of the Mohr-Coulomb criterion. This would be a three parameter criterion, as will most likely be the case for the problem at hand, soil stressing under a moving rigid wheel. Haythornthwaite (4) suggests a triangular section as the limiting shape which still satisfies the condition of convexity of the yield surface (see Figure 5).

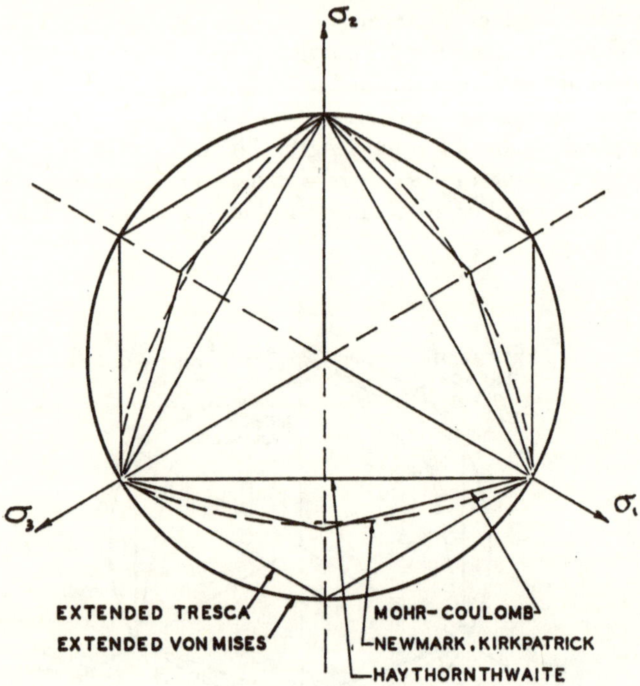

FIGURE 5. Octahedral representation of several yield surfaces

SHEAR STRENGTH PARAMETERS AND SOIL STRENGTH

The common usage of shear strength parameters in soil mechanics, identified as cohesion, c, and angle of internal friction, ϕ, stems from the analysis of test results using the Mohr-Coulomb theory. The correspondence between the values derived for c and ϕ based on this theory and whatever values one might obtain using a mechanistic formulation is generally assumed to be close. Where there is reason to doubt such agreement, the usual practice in an applied problem is to simulate field conditions in laboratory testing, assuming that the test results can be interpreted by the Mohr-Coulomb theory. In this manner, obvious success has been achieved in the solution of essentially static problems involving bearing capacity and stability, without requiring a close examination of the deficiencies and limitations of the theory. It is likely that the proven mathematical similarity between a general von Mises-type yield surface and the Mohr-Coulomb theory for plane strain cases (3) is a major factor in the successful application of the latter theory in many major soil mechanics problems. It is significant to note that although analyses are made at limiting equilibrium, the use of safety factors in modifying the quantitative relationships ensures that the actual stresses experienced are below yield or failure values.

For the case of soil support under a moving wheel, some obvious disparities with conventional techniques used in soil mechanics analyses exist. Firstly, failure or yielding of the soil actually occurs in addition to the existence of a zone of soil which is stressed below yield. Therefore specification of a yield criterion becomes a necessity rather than a design expedient. Secondly, the three dimensionality of the problem involving a variable intermediate principal stress and significant rotation of the principal stress axes throughout the stressed mass represents a departure from conventional soil bearing analyses. Thirdly, the problem is dynamic and this requires a knowledge of rate effects on soil strength. It also requires a knowledge of the influence of inertia effects on the geometry of the affected zone. Finally, the dynamic nature of the problem requires that the pore pressures generated in the soil by the wheel, their distribution and degree of dissipation within the stressing period be known so that the proper stress relationship may be used.

Because of the conditions used in the formulation of the Mohr-Coulomb analysis, the theory is not meant to yield information on the actual stress-strain relationship below or after yielding. There is ample evidence, both theoretical and experimental (12), (13), to show that the mobilized "angle of internal friction" varies with strain. The assumption that stresses at a point can be accurately represented by average stresses on a failure plane restricts the analysis to considerations without inertia terms.

Under microscopic considerations, the term "cohesion" and "friction" in clay soils lose most of their meaning. In many cases actual particle-to-particle contact is established through layers of water. The existence of clay-water forces dictating the response behaviour of the clay soil does not therefore allow for an easy definition of the soil strength in terms of friction and cohesion as mechanistic quantities (19). The role of pore pressures and the generation of pore pressures under varying strain rates have been reported previously (17), (18). The significant point is that the soil response will not be given solely in terms of total or effective stresses, as extreme cases of zero or full pore pressure development, but will be governed by some partial value of pore pressure.

The dependency on rate of straining is sufficiently strong to require a better understanding of the inter-relationships before attempting a rigorous analytical solution. However, design expediency, requiring simple solutions, will in all probability tolerate analyses made on the basis of total stresses. It must be realized that the variability of the stress and strain-rate field below the moving wheel will never allow for complete rigorous solutions.

SOIL RESPONSE

In this study, the visioplasticity technique is used to provide information for the determination of the instantaneous velocity vectors. The experiment permits the variation of wheel size, wheel velocity, slippage, and sinkage as well as soil type, density, and degree of saturation. Traces of particle motion in the deforming soil mass under the moving wheel load can therefore be obtained under appropriate constraints as a function of all the important prototype variables. Using available electronic data processing devices, correlations between the experimental variables will be obtained for ultimate theoretical analyses.

A typical trace of the motion of a particle embedded in loose, dry sand is given in Figure 6, where the trace is represented on the vertical plane of symmetry against elapsed time.

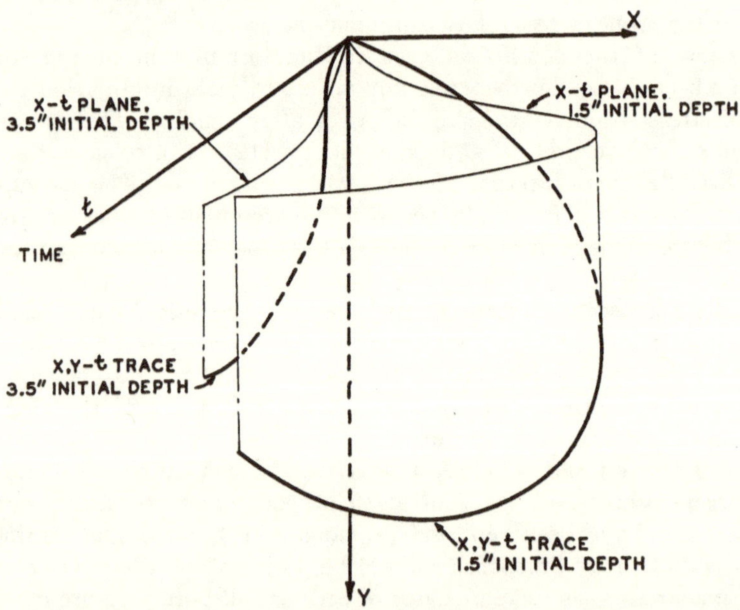

FIGURE 6. Trace of particle motion in a vertical plane versus elapsed time

The instantaneous velocity vectors may be used to draw an instantaneous velocity vector field, such as is shown in Figure 7 using data obtained in loose, dry sand. A simplified analysis can then be applied by assuming that there is a two-dimensional problem and that the

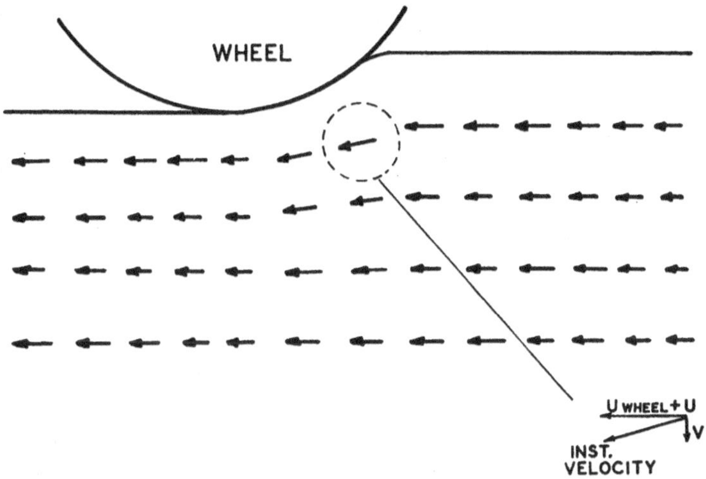

FIGURE 7. Instantaneous velocity vectors relative to moving wheel

deformation due to sinkage is irreversible. Momentum equations may then be written as follows:

$$(\partial\sigma_x/\partial x) + (\partial\tau_{xy}/\partial y) + u(\partial\rho_u/\partial x) + v(\partial\rho_u/\partial y) = 0; \qquad (7)$$

$$(\partial\sigma_y/\partial y) + (\partial\tau_{xy}/\partial x) + u(\partial\rho_v/\partial x) + v(\partial\rho_v/\partial y) = -\rho g \qquad (8)$$

where u and v are the velocities in the x and y directions.

The continuity relationship is shown in Equation (9)

$$(\partial\rho_u/\partial x) + (\partial\rho_v/\partial y) = 0. \qquad (9)$$

By using phase plane relationships (i.e. plotting u versus x, u versus y, v versus x, and v versus y) and measuring ρ, all but the first two terms in Equations 7 and 8 can be determined from the experimental results. Several possibilities exist for the determination of the stress terms. The conventional approach, following the practice in applied soil mechanics, would be to obtain appropriate analytical shear strength parameters and stress-strain-time relationships from dynamic soil tests conducted in triaxial cells or shear boxes. An important requirement of such tests would be that they duplicate the stress and strain-rate fields as well as the dynamic constraints experienced under the prototype loading. This is not easily achieved. It is likely that this method would produce successful results for loose, dry sands since it is known that the variation of ϕ with strain rate is small (16), although the influence of rate-of-strain on dilation (5) would require careful attention. For moist or saturated sands and clays, the ability to obtain applicable stress-strain-time relationships is questionable, because of the extreme difficulty in scaling pore pressure response from the test specimens to *in situ* conditions beneath a moving wheel.

The fact that volumetric strains can be measured in the deforming soil mass by the X-ray technique might also be exploited to solve the momentum equations. This would require the establishment of a stress-volume change relationship obtainable in separate laboratory shear tests under corresponding dynamic constraints.

Another method is available which, although more complex in the first instance, would resolve the latter complications of the first method. By an analysis of particle motion through the stressed soil mass, a flow relationship under actual prototype restraints could be formulated. This would require the use of principles of micro-mechanics and a large number of experimental observations would be necessary to allow statistically significant quantities to be obtained. One would envisage a function in the following form,

$$f(\sigma, \tau, t) > 0 = g(u, v, t). \tag{10}$$

It is evident that much work remains to be done before the equations may be solved. Apparent correlations for sinkage and thrust with plate-load tests using stability analyses based on static limiting equilibrium (see, for example, Reece (10) may in all probability be due to the fact that most of the tests conducted have been performed with dry sand as the supporting medium. As shown previously, a mathematical similarity between a general von Mises condition and the Mohr-Coulomb criterion exists for plane strain in dry sands. Thus, one could expect a certain degree of correlation despite the fact that the supporting dry sand exhibits flow characteristics under the forcing function in the primary zone. Departure from predicted values will be due in part to the three-dimensional aspect, the influence of the intermediate principal stress, the rotation of the principle stress axes, inertia effects, and rate-dependent properties of the material.

In a related problem involving gravity flow of loose cohesionless materials in a conical channel, Savage (14) has shown the Mohr-Coulomb criterion to be applicable for the flow of loose, dry sands. Thus, for example, in an analogous case for plane strain the momentum equations will assume the following form, for constant ρ:

$$(\partial\sigma/\partial y)(\sin\phi \sin 2\theta) + (\partial\sigma/\partial x)(1 - \sin\phi \cos 2\theta) + 2\sigma \sin\phi \cos 2\theta(\partial\theta/\partial y) + \rho u(\partial u/\partial x) + \rho v(\partial u/\partial y) = 0; \tag{11}$$

$$(\partial\sigma/\partial y)(1 + \sin\phi \cos 2\theta) + (\partial\sigma/\partial x)(\sin\phi \sin 2\theta) - 2\sigma \sin\phi \sin 2\theta(\partial\theta/\partial y) + \rho u(\partial v/\partial x) + \rho v(\partial v/\partial y) = \rho g \tag{12}$$

where $\sigma = (\sigma_x + \sigma_y)/2$,

θ = angle between major principal stress and y axis.

SUMMARY

It is clear that the mathematical complexities associated with a three-dimensional analysis of dynamic response behaviour of soil under a moving wheel will be such that no exact solution may be found. However, by assuming a two-dimensional problem and with an integrated study, as has been described, a less empirical approach may be obtained. The actual behaviour of physical models may be examined to provide the basis for informed simplified mathematical models. Until such time as sufficient information can be provided on stress-strain relationships under a dynamic regime one is forced to rely on approximate empirical relationships and correlative testing.

ACKNOWLEDGEMENTS

The over-all two-part study, under the general direction of R. N. Yong, is being performed under contract arrangement with the Defence Research Board of Canada through the Department of Defence Production, with Mr. T. A. Harwood of DRB as Project Officer. This paper is a contribution from the Soil Mechanics Laboratory of McGill University.

REFERENCES

1. Bekker, M. G., Summation and Closing Remarks, Proc. 1st Int. Conf. Soil-Vehicle Systems, Torino, Italy, 1961.
2. Boyd, C. W. and S. J. Windisch, A Technique for Measuring Deformations Within a Sand Under Controlled Wheel Loads, paper to be presented to the 2nd Int. Conf. Soil-Vehicle Systems, 1966.
3. Drucker, D. C. and W. Prager, Soil Mechanics and Plastic Analysis or Limit Designs, *Q. of Applied Mathematics*, Vol. 10, 1952, p. 157.
4. Haythornthwaite, R. M., Mechanics of the Triaxial Test for Soils, *J. of the Soil Mechanics and Foundations Division*, ASCE, Vol. 86, No. SM5, 1960, p. 35.
5. Healy K. A. The Dependence of Dilation in Sand on Rate of Shear Strain, Report No. 13, Soil Dynamics Series, MIT, 1963.
6. Henkel, D. J., The Relationships Between the Strength, Pore-Water Pressure, and Volume-Change Characteristics of Saturated Clays, *Geotechnique*, Vol. IX, No. 3, 1959, p. 119.
7. Hvorslev, M. J., Physical Components of the Shear Strength of Saturated Clays, *Proc. ASCE Research Conference on Shear Strength of Cohesive Soils*, 1960, p. 169.
8. Kirkpatrick, W. M., The Condition of Failure for Sands, *Proc.* 4th Int. Conf. on Soil Mechanics and Foundation Engineering, Vol. 1, 1957, p. 172.
9. Newmark, N. M., Failure Hypotheses for Soils, *Proc. ASCE Research Conference on Shear Strength of Cohesive Soils*, 1960, p. 17.
10. Reece, A. R., Principles of Soil-Vehicle Mechanics, Proc., Inst. of Mechanical Engs., Vol. 180, 1965, Part 2A.

11. ROSCOE, K. H., A. N. SCHOFIELD, and A. THURAIRAJAH, An Evaluation of Test Data for Selecting a Yield Criterion for Soils, *Laboratory Shear Testing of Soils*, ASTM, STP No. 361, 1963, p. 111.
12. ROWE, P. W., The Stress-Dilatency Relation for Static Equilibrium of an Assembly of Particles in Contact, *Proc., Roy. Soc.*, Series A, Vol. 269, 1962, p. 500.
13. ROWE, P. W., Stress-Dilatency, Earth Pressures, and Slopes, *J. of the Soil Mechanics and Foundation Division*, ASCE, Vol. 89, No. SM3, 1963, p. 37.
14. SAVAGE, S. B., Gravity Flow of a Cohesionless Bulk Solid in a Converging Conical Channel, *Int. J. of Mechanical Sciences* (in press).
15. SKEMPTON, A. W. and A. W. BISHOP, Chapter 10, "Soils", in *Building Materials, their Elasticity and Plasticity*, edited by M. Reiner, Amsterdam: North Holland Publishing Co., 1954.
16. WHITMAN, R. V. and K. A. HEALY, Shear Strength of Sands During Rapid Loading, *J. of the Soil Mechanics and Foundations Division*, ASCE, Vol. 88, No. SM2, 1962, p. 99.
17. YONG, R. N., R. D. JAPP, and H. C. LEITCH, On the Strength of Soil Under Dynamic Loading, paper presented to Fifth Meeting, Quadripartite Working Group on Ground Mobility, 1965, Report No. D Phys (G) Misc. 20, Directorate of Physical Research, Defence Research Board, Ottawa.
18. YONG, R. N., C. W. BOYD, and S. J. WINDISCH, Soil-Vehicle Interaction Study, paper presented to Fifth Meeting, Quadripartite Working Group on Ground Mobility, 1965, Report No. D Phys R (G) Misc. 21, Directorate of Physical Research, Defence Research Board, Ottawa.
19. YONG, R. N. and B. P. WARKENTIN, *Introduction to Soil Behaviour*, New York: The Macmillan Company, 1966, 448 pp.

The Basis of Soil Failure Theory

B. D. WITNEY

D. R. P. HETTIARATCHI

A. R. REECE*

ABSTRACT

Soil machine mechanics must be based upon the failure analysis of classical soil mechanics. Unfortunately, the theoretical basis for this type of analysis is obscure. The paper attempts to present a reasonable procedure for the solution of problems in soil mechanics. The determination of the correct failure surface, the methods and assumptions necessary to calculate the force to produce failure, and the means whereby the results can be generalized are described.

THE ORIGINAL ATTEMPTS to develop an understanding of the relation between machine performance and soil properties were empirical in nature. At first performance was related to qualitative descriptions of soil and then to soil value measurements by such devices as cone penetrometers and finally by semi-empirical equations. On the other hand civil engineers appear to base the design of their structures on a firm body of theoretical analysis generally called soil mechanics. This led the writers to attempt to apply soil mechanics to machine problems, at first with encouraging results. For example, bearing capacity theory immediately pointed out that existing pressure sinkage equations should have a slightly different form which greatly improved their dimensional status and their ability to describe a wide range of soil behaviour (1). Similarly it was found that retaining-wall theory could accurately predict the forces acting on simple cutting blades (2).

This initial success led to a vision of the future in which quite rigorous analysis would be applied to increasingly complicated machine problems to yield accurate solutions based on fairly complete understanding of the reaction of soil to loads and movements. Unfortunately it has not turned out quite like this. The second problem that was attempted after

*Department of Agricultural Engineering, University of Newcastle upon Tyne.

the retaining wall was the study of the bearing capacity of deep strip foundations with a view to the development of pressure sinkage equations for dense soils. In this work it was hoped to modify existing soil mechanics theories to yield results in a more readily usable form. However, this modest endeavour has run into major difficulties which have revealed that soil mechanics is at present not a very satisfactory tool. It has in fact led to the understanding that soil mechanics cannot really be said to have adequately solved one single problem.

At this point it is perhaps worthwhile to define what one means by soil mechanics. Soil mechanics is concerned with the estimation of the loads which structures exert on soil at the point at which complete soil failure begins. This in turn depends upon the maximum stress that the soil can withstand, which depends upon the state of drainage in the soil. The theory of water flow through porous media has therefore come to be accepted as a part of soil mechanics. It is also necessary that a building should not settle into the soil too much and this depends upon the way in which the soil consolidates under the weight of the building which in turn depends upon the stresses applied and the nature of the drainage. Since the stresses are much less than those required to produce failure they can be adequately described using elastic analysis. It is therefore seen that, while the fundamental basis of soil mechanics is failure theory, the subject will also include elastic stress analysis, drainage theory, and the study of soil consolidation. It is with the basic failure theory that soil machine mechanics is concerned.

The main problem in soil mechanics arises because a rigorous theoretical analysis is not possible; this is generally agreed and is noted in most contributions to the subject. However, having admitted that the rules applying to rigorous solutions are not fulfilled, it is unusual to find that the rules of the actual game of earth pressure calculation are described. Neither the basic properties of the idealized soil nor the assumptions and approximations used are properly defined. The result is that particular problems have numerous possible solutions in the literature and these solutions have very little point of contact with each other. Papers in soil mechanics often make reference to various types of rigorous analyses, for example plasticity theory, limit analysis, and static analysis, but does not make clear in what way these are relevant.

It is normally assumed that the theory of plasticity has some relevance to soils. The truth is that weightless, frictionless clay would be quite close to an ideal plastic material, but that taking into account the weight of the clay leads to considerable mathematical complexity. However, when friction is involved, the theory of plasticity loses its relevance. A perfectly plastic material failing, according to the von Mises-Coulomb criterion, does not represent a real soil, if for no other reason than that this material dilates greatly with shearing, which soil does not. The truth is that it has not so far proved possible to describe the stress-strain relations that exist in soils and until this is done a plastic analysis is not possible (3).

Soil mechanics often makes use of minimizing processes and it is often implied that this is something to do with the limit analysis theorems used in the theory of plasticity. There are two of these theorems and they state:

1. That a statically admissible state of stress satisfying the equation of equilibrium and the boundary stress conditions will yield a lower bound solution.

2. That a kinematically admissible velocity field will yield an upper bound solution.

The actual solution lies at the point at which these two approaches have the same value for the applied force. The use of these theorems would be tedious and time consuming and there is no guarantee that the upper and lower bound solutions can be made to draw sufficiently close together. Furthermore it does not seem to be generally realized that these theorems require extensive modification before they can be applied to frictional materials (4).

The minimizing methods used in soil mechanics would seem to be better based upon the following simple common sense proposition: if a family of failure surfaces can be found which everywhere satisfies the conditions of equilibrium, the boundary conditions, and the idealized soil properties, then the soil will fail on that surface which requires the minimum force. It will be shown later that the minimizing used by Ohde (5), Terzaghi (6), and Meyerhof (7) does not satisfy this proposition in that the failure surfaces violate one or more of the above requirements. The minimum obtained will then be considerably less than the failure force obtained in practice.

Perhaps the most relevant rigorous approach to soil mechanics is through the method of static analysis particularly associated with the work of Sokolovski (8). This technique can be used to determine the slip line fields in a soil which is on the point of failing under a known load. It is based on the properties of a rigid, heavy, Coulomb material and combines the known relations between the stresses at failure with the normal partial differential equations of equilibrium. These can be used in a step by step numerical method to follow the slip line fields from known boundary conditions. This is a powerful tool that must be treated seriously; at the moment, however, it has only been applied to certain rather simple problems.

It is clear therefore that soil mechanics is not plasticity nor limit analysis nor static analysis and the writers have not come across a clear statement of what soil mechanics actually is. It is the purpose of this paper to attempt a tentative statement of the way in which an adequate solution to a problem in soil mechanics can be obtained. The ideas in the paper have been developed in relation to the problem of the bearing capacity of deep foundations and this problem will be used to illustrate the argument.

THE NATURE OF A SOIL MECHANICS PROBLEM

There are three groups of variables involved in a soil mechanics problem. The soil and soil to structure properties are described by the cohesion c, the angle of internal resistance ϕ, the soil density γ, the soil to structure adhesion c_a, and the angle of soil to structure friction δ. The structure is described by its shape which in turn is described by angles and other ratios of dimensions. Finally, the size of the structure and associated soil masses is important. Normally in soil machine mechanics a general solution extending over the full range of all these variables is required, and this is also the case for the more common civil engineering problems. A computer programme cannot be considered a complete solution to the problem. To be generally useful to designers in industry, the solution has to be such that an actual problem can be solved by means of a set of graphs or tables and a slide rule. In general a single calculation for a particular set of variables would not be worth the effort involved.

SCIENTIFIC SOIL MECHANICS

Because the mechanical properties of soil are so complex, soil mechanics cannot be a rigorous discipline and cannot be treated as an exercise in mathematics. An adequate solution to a problem can therefore only be achieved by a scientific combination of theory and experiment. It is proposed that nothing less than the following procedure constitutes an adequate basis for a solution.

First, a statement must be made describing the properties of the idealized soil that is to be considered. This must be followed by a careful description of the simplifying assumptions, concerning the behaviour of such a soil, that it is necessary to make in order to overcome mathematical difficulties. These soil properties and simplifying assumptions must be combined with the normal conditions of equilibrium in order to determine the shape of the failure surfaces within the soil. Ideally every part of the soil involved in the failure should be in equilibrium. This may prove to be impossible but the departure from equilibrium must be small and must be described quantitatively. Furthermore at every point on the boundaries the correct stress conditions must exist. Once again this may not be possible but divergencies from the ideal must be small, must be shown to cause but small changes to the final answer, and, if possible, should be shown to represent an approximation to a more accurate but more complex failure pattern. The failure pattern so developed must be verified by carefully controlled experiments on a limited number of soils which cover a large part of the possible soil range. Once the failure pattern has been established the forces involved can be calculated and a suitable computer programme developed. The results of these force calculations must be experimentally verified over the same wide range of soil types. Finally suitable general describing equations must be developed to enable the results of the computer programme to be communicated in a simple way. The errors involved in this generalization must be stated.

The whole of this process of interlocking theoretical and experimental methods is necessary because any single part is subject to such large possibilities of error. Experiments with soil are peculiarly difficult to carry out. It is difficult even to measure the basic soil parameters as they exist in the problem being investigated, and it is difficult to repeat experiments with the same soil conditions. The process described above would require such a total corroboration of theory and experiment that it would leave little room to doubt that the right answers were achieved and that they were achieved for the right reasons.

THE PROPERTIES OF AN IDEAL SOIL

The soil is assumed to be rigid, heavy, and incompressible and it does not dilate when sheared. It fails according to Coulomb's equation, from which it results that, in a plane problem, failure surfaces occur in pairs making an angle of $90 + \phi$ degrees to one another. This is an idealization since real soil is not quite rigid and does dilate slightly as it shears.

The stresses at the soil-structure interface, when the soil slides along the surface of the loading structure, are assumed to be described by a modification of Coulomb's equation. In this the cohesion (c) is replaced by the adhesion (c_a) and the angle of internal shearing resistance (ϕ) the angle of soil-structure friction (δ). Under these conditions the soil-structure interface is not a soil failure plane and if these exist they must intersect the interface at an angle θ which can be determined from a Mohr diagram, such as is shown at Figure 1 for a cohesionless soil.

In most soil machine problems the soil flows away from the machine

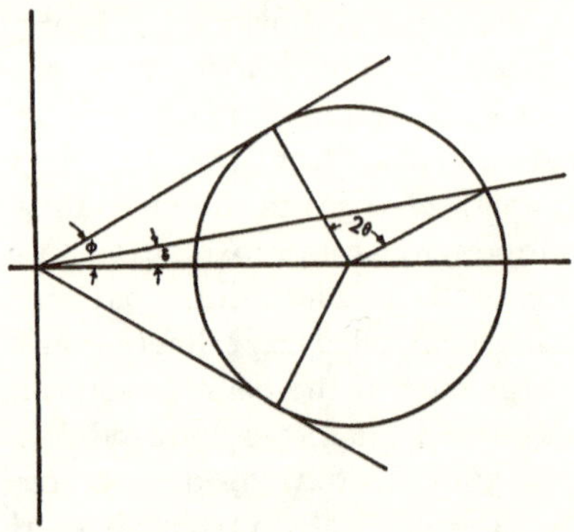

FIGURE 1

element and even in a foundation problem the point of failure represents the loading at which flow begins. There are certain characteristics of soil flow which are common to the flow of any other incompressible material. The motion will arrange itself so that the energy lost in friction in the flow is minimized. This requires that the paths of individual particles are as straight as possible and only deviate from straight lines by making the smallest angle of deflection on the largest radii. Soil will not flow along paths with sharp corners except on a plane of symmetry.

The kinematics of soil flow in response to a small movement of the structure can be simplified by defining only two possible types of flow. In the first, large masses of soil move as rigid bodies. It is not possible for the soil flow to result in the opening up of cavities within the soil; this is a common observation which depends upon the low strength to weight properties of soil. It is only possible for rigid bodies to slide one over the other without a hole appearing if the motion is bounded by two straight lines or a single circle. If the soil does not move as a rigid body then it is convenient to describe the other type of movement as a shearing flow. Zones of shearing flow will generally be bounded by non-circular curves. A given displacement can be achieved with a minimum energy input by a rigid body motion. This is therefore a preferred mode of movement but may not be possible due to the boundary conditions of the problem.

The static analysis of rigid Coulomb materials leads to the idea of continuous slip line fields. In practice these are only continuous in zones of shearing flow. The slip line fields are non-existent over the main area in which rigid body motion is taking place. The absence of complete slip line fields and the occurrence of rigid body motion is made possible because the soil is not quite rigid. Definite deformations along the slip surfaces are necessary in order to develop slip lines and soil failure.

SIMPLIFYING ASSUMPTIONS

Zones of shearing flow have continuous slip line fields in which the slip lines make angles of $90 + \phi$ to each other and in general these fields are made up of curved lines. It is a great mathematical simplification to replace such zones by radial shear zones in which one set of slip lines is a family of radii and the other a family of logarithmic spirals. The pole of such a radial shear zone represents a major discontinuity in the soil stress system, and it therefore can only exist at some real discontinuity such as a corner of the loading structure. However since the radial shear zone is being used as a simplification of what is really a doubly curved zone this may justify the placing of the centre elsewhere, for example, it may even be located outside the soil mass altogether. However, this can only be done if it results in a better approximation to a doubly curved zone the nature of which can be established by theoretical and experimental methods.

The forces involved in the shearing of the soil have two components, one due to the cohesion and the internal friction of the soil and the other due to the weight and internal friction. In practice these occur together simultaneously. The relative magnitude of these two components can vary and the solution to the problem should be correct when either is zero. It therefore seems logically justifiable to divide the force on the structure into two parts, one part due to the cohesive frictional soil forces and the other due to the weight frictional soil forces.

The equilibrium of the soil masses at failure depends upon the distribution of normal and frictional forces on the straight or circular failure surfaces of the rigid body zones. This distribution is difficult to obtain by rigorous methods. It is therefore proposed to assume that the stress due to the cohesion and friction forces is distributed uniformly along the straight or circular failure surfaces and, further, to assume that the normal and frictional stress due to weight and friction are distributed in a hydrostatic manner, that is, proportional to the depth from the nearest part of the free surface of the soil vertically above the failure surface. These two assumptions are described by the following equations:

$$P_{\phi c} = K_{\phi c} \tag{1}$$

$$P_{\phi \gamma} = K_{\phi \gamma} z \tag{2}$$

The two constants $K_{\phi c}$ and $K_{\phi \gamma}$ are assumed to have a constant value over the failure surface boundaries of any single body zone. The actual values of $K_{\phi c}$ and $K_{\phi \gamma}$ are obtained from the static equilibrium of the rigid body zone. If they are determined from the vertical and horizontal equilibrium there will remain an unbalanced couple. This couple could be used to modify one or both of the K factors to attain complete equilibrium, or it can be used as a measure of the error involved in using this assumption. For a passive or active Rankine zone Equations 1 and 2 are exact and the out-of-balance couple is zero.

STREAMLINING SOIL BODIES

The soil displaced by a moving structure will go to considerable lengths to minimize the energy lost in flow. It is quite prepared to modify the shape of the structure to make it more streamlined, by arranging a fixed soil body between the structure and flowing soil. Examples of such bodies are the wedges formed beneath foundations and in front of tines and the filling in of concave surfaces, particularly when they are rough. The existence of such a body completely covering the structure surface will greatly simplify the problem if for no other reason than that it eliminates two of the variables (i.e. because $\delta = \phi$ and $c_a = c$). The determination of the existence and shape of such a body is therefore the starting point for the solution of a problem.

A fixed body will form if it can reduce the energy losses. This it will do if it can reduce the angles through which soil particle paths turn or increase the radii of turn. It can also do this by allowing the soil to fail as a rigid body rather than by shearing flow. These considerations will suffice to determine the general form of the body. Its actual shape and size will be determined by the fact that it is made of soil and must satisfy the requirements of the ideal soil properties, the soil-structure properties, and the conditions of equilibrium.

GENERALIZING EQUATION

Terzaghi (6) proposed the following generalizing equation and used it particularly in relation to the bearing capacity of a strip foundation.

$$Q(lb/ft) = zb\, c\, N_c + 2b^2\gamma N_\gamma + 2bqN_q \qquad (3)$$

In this equation the three N factors take into account the effects of soil weight, soil cohesion, and surcharge pressure. The N factors are functions only of ϕ.

One of the authors (9) has pointed out that the most complete form of this equation is as follows.

$$F\,(lb.) = b^2cN_c + b^3\gamma N_\gamma + b^2qN_q + b^2c_aN_a \qquad (4)$$

This equation can be used to describe the force required to fail soil by any kind of structure whatsoever. The additional term takes into account the adhesion between the soil and the structure. The N factors are now functions of not only ϕ but also the soil to structure friction and the shape of the structure. It should be noted that the surcharge term must not be confused with the weight of the soil involved in the soil flow. It will generally refer to some loading on the soil which does not contribute any shearing, for example the loading applied by the tracks of a vehicle or the smooth foundation of some adjacent buildings.

The additive nature of this equation follows from the addition of the various components of soil force and the dimensional parts are necessary in order to provide dimensional homogeneity. The weakness of this equation is that the N factors are not independent of the values of the dimensional terms in the equation. The use of the equation will therefore involve an error, which can be computed and must be presented in a complete analysis of the problem. Another possibility is to compute the N values for different proportions of the four sources of force on the structure. In the case of the pressure on the base of a shallow foundation this is quite feasible because surcharge and adhesion are not involved and a completely accurate generalized solution can be presented in the form of graphs of N_c and N_γ as functions of z/b at different values of $c/\gamma z$.

In order to use equation 4 some arbitrary decisions will have to be made about the proportion of the various components of the force. In Reference 10 the authors have used the following scheme.

N_γ calculated with $c/\gamma z, q/\gamma z, c_a/\gamma z = 0$.

N_c calculated with $c/\gamma z = 1, q/\gamma z = c_a/\gamma z = 0$.

N_a calculated with $c_a/\gamma z = c/\gamma z = 1, q/\gamma z = 0$.

N_q calculated with $q/\gamma z = 1, c_a/\gamma z = c/\gamma z = 0$.

The N_γ and N_q are therefore correct for a dry sand, while the N_c is computed for a soil with weight but no adhesion or surcharge. The N_a is calculated for a soil with weight and cohesion, which is reasonable since it is not possible to have adhesion without cohesion. The actual values of $c/\gamma z$, $c_a/\gamma z$, and $q/\gamma z = 1$ was chosen for computational simplicity and it was shown later that over the complete range of likely combinations of these properties the errors would not exceed 5 per cent.

PREVIOUS THEORIES

There are many possible reactions to the preceding ideas. Better assumptions and idealizations could be used or they could be avoided by more rigorous methods. Another would be to enquire in what way the proposed approach differs from that of the currently accepted theories. In order to make this clear it is proposed to discuss briefly the bearing capacity theories of Terzaghi (6), Meyerhof (7), and Balla (11).

Terzaghi (6)

The failure geometry assumed by Terzaghi for a surface foundation is shown in Figure 2. Beneath the foundation is what we would call a

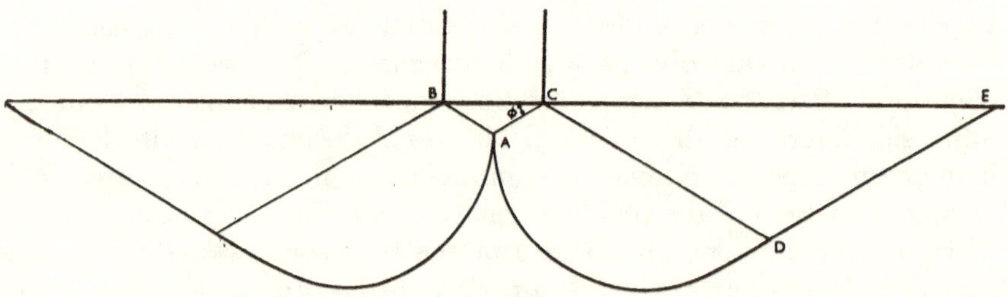

FIGURE 2

streamlining body. Terzaghi arrived at the existence of this triangle not from the idea of streamlining a flow process but from considerations of stress equilibrium. The four failure surfaces which intersect at the point A do not make angles of $90 + \phi$ and $90 - \phi$ with each other. While the angle CAD and its equivalent on the other side of the wedge are $90 + \phi$ the angle BAC is twice that which is necessary in a Coulomb material. While the angle between the spirals is zero instead of being $90 - \phi$.

The generalizing procedure that was carried out by Terzaghi was along the lines previously described. However, he chose to calculate N_γ with $c = 0$ and N_c with $\gamma = 0$. The first of these is reasonable since it is the condition for a dry sand. The second is a rather odd choice in that it is only relevant to a weightless soil, a material only to be found on very small planets. The reason for Terzaghi's choice is in fact perfectly clear; it greatly simplified the computation of the N factors. With digital computers readily available it is no longer necessary to do this.

In the calculation of N_c Terzaghi assumes that the pole of the logarithmic spiral is at the point C. This is entirely satisfactory since it gives the correct orientation of stress all the way along the face AC and makes DE a tangent to the spiral AD. For calculating N_γ Terzaghi allows the pole of the spiral to move anywhere along DC extended. The actual point is chosen so that the resulting bearing capacity factor is a minimum. The authors (10) have repeated Terzaghi's calculations and they have found that the spiral centre approaches quite close to the point C when the bearing capacity factor is a minimum. However, as long as it is not actually at C the spiral failure surfaces make the wrong angle as they intersect the radial failure surface AC.

Since the minimizing procedure is not carried out for a family of failure surfaces which all satisfy the conditions of equilibrium, the boundary conditions, and the idealized soil properties there seems to be little logical justification for using it. According to the authors' ideas the pole of the spiral should be located at C for both calculations.

Meyerhof (7)

Meyerhof extended Terzaghi's bearing capacity theory to deal with deep foundations and he assumed the failure pattern shown in Figure 3. He has minimized with an even greater abandonment of principle than

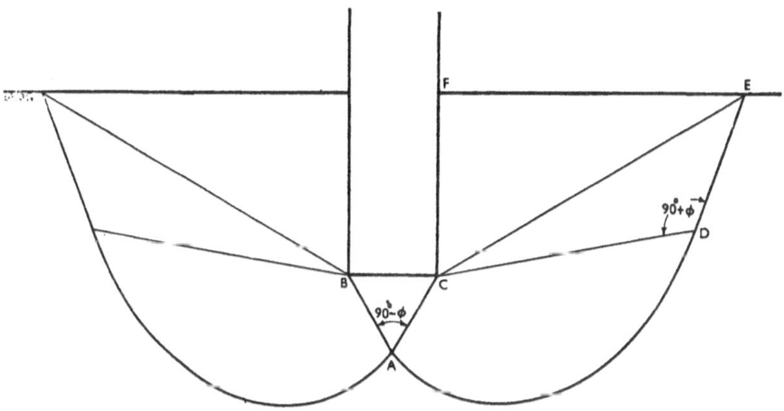

FIGURE 3

Terzaghi and has written a paper attempting to justify this (12). This paper proposes that in the case of surface foundations Terzaghi was wrong to restrain the pole to the line CD extended as in Figure 2. He shows that, by allowing the pole to wander about on either side of this line, the bearing capacity can be minimized even further. This of course also leads to an incorrect stress situation along the line AC and produces another such situation along the line CD. It is also unacceptable on the grounds that the failure surface ADE has a sharp corner in it at the point D. This is most unlikely if the soil movement is looked upon as a flow process, since materials do not flow around sharp corners within themselves. Meyerhof is minimizing with considerable disregard for the properties of the idealized soil he is considering.

In order to obtain the bearing capacity factors Meyerhof was obliged to assume a certain distribution of pressure on the faces CA, CE, ED, and

CD. However, he did not use the assumptions that we have proposed in equations 1 and 2. For the face *CD* he assumed that the frictional pressure would always have its resultant one third of the way from *A* to *C*. According to equation 2 this would be correct for a surface foundation but as the depth increased the resultant would move towards the centre of *AC*.

On the sides of the triangle *CED* Meyerhof has assumed that the pressures due to both cohesion and weight are equally distributed on all the sides. This startling assumption is quite hard to discover, it comes from using a Mohr circle to determine the equilibrium of the triangle *CED*. The Mohr circle is of course a construction that is correct for the stresses at a point at which uniform distribution of stress is a reasonable assumption.

Meyerhof's generalization procedure has confused the concept of a surcharge with that of a force due to soil weight. He considers the triangle *EFC* as a surcharge. There seems to be no justification for this; it is perfectly possible to include the effect of this triangle in the weight term. This leaves open the possibility of considering any surcharge applied to the surface *FE*.

Balla (*11*)

Balla considered the same problem of a deep foundation with the avowed object of obtaining higher values for the bearing capacity factors than the previous theories of Terzaghi and Meyerhof. The failure pattern used by Balla is shown in Figure 4. He has observed that it is not possible for the two spiral failure surfaces that meet at *A* to be parallel and

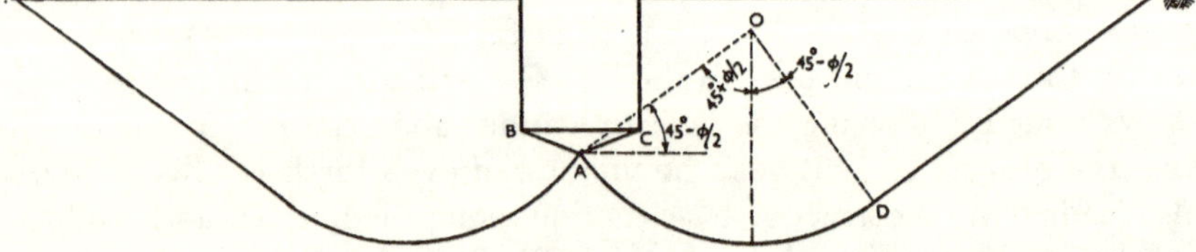

FIGURE 4

coincident with each other at this point as proposed by Terzaghi. He therefore opens them out to the correct angle; unfortunately he does not notice that the angle *BAC* is also not correct. There still remains then an intersection of four failure planes at the point A which is not possible for a rigid Coulomb material.

According to Balla the failure surface from A to E is made up of a circle AD and a straight line DE. This he justifies by the following sentence: "Kinematically only plane or circular, cylinder-formed sliding surfaces are possible. Therefore, in the present case, the sliding surface shall also be composed of plane and circular cylinder-formed surfaces."

Since this is the whole foundation for the paper it would surely have merited a little elaboration. This is certainly not a failure surface that is compatible with the assumptions that have been proposed in the first part of this paper. It was pointed out there that rigid body motion without the opening up of cavities was only possible by means of the squeezing out of wedges between two straight surfaces or the rotation of parts of circular bodies. This does not support Balla's failure surface. It is not possible for a body bounded by a circle and a straight line to move without holes opening up. Neither is it possible for the circular segment OAD to rotate, pushing the triangle OED upwards, without the opening up of cavities along the line OD. It seems that Balla's main contribution is to show that in the realm of soil mechanics a reasonable answer can be achieved by a large number of different methods, particularly if the answer is known beforehand.

General observations

The preceding discussion of three bearing capacity theories is not meant to imply that any part of them is necessarily wrong. It is intended to show that the theories do not have the same basis as is proposed in the earlier part of this paper. It may well be that soil does not behave as an ideal Coulomb material, and, for example, it may be justifiable to have failure planes intersecting at other than angles of $90 \mp \phi$. However, if this sort of thing is done the theoretical or experimental justification must surely be provided. The three theories referred to do not make their basis clear and do not justify their inconsistencies.

CONCLUSIONS

1. Much of contemporary earth pressure theory appears to be based on unstated or unsatisfactory idealizations, approximations, and assumptions.

2. A satisfactory solution to many problems of general interest is probably possible by means of the combination of theory and experiment described in the paper.

3. The basic idealizations and assumptions described in this paper are tentative. They have been developed in relation to the analysis of the problem of the force on retaining walls and cutting blades and the bearing

capacity of surface foundations. In the future it is hoped to show that they work well for the problem of the bearing capacity of deep foundations which is the same problem as that of pressure sinkage relations in soil vehicle mechanics.

ACKNOWLEDGEMENT

Published by permission of H.M.S.O. Crown Copyright Reserved.

REFERENCES

1. Reece, A. R., Principles of Soil Vehicle Mechanics. *Proc. Inst. Mech. Eng.*, 1965–66, Vol. 180, Part 2A.
2. Osman, M. S., The Mechanics of Soil Cutting Blades. *J. Agric. Eng. Res.*, vol. 9, 1965.
3. Drucker, D. C., On Stress–Strain Relations for Soils and Load Carrying Capacity. *Proc. 1st Int. Conf. Mech. Soil–Vehicle Systems*, 1961.
4. Drucker, D. C. and Providence, R. M., Coulomb Friction Plasticity, and Limit Loads. *J. Appl. Mech.*, Vol. 21, 1954.
5. Ohde, J., Zur Theorie der Erddruckes unter besonderes. Beruckrichtigung der Erddruck Verteilung. *Die Bautechnik*, Vol. 16, 1938.
6. Terzaghi, K., *Theoretical Soil Mechanics*. New York: Wiley, 1943.
7. Meyerhof, G. G., The Ultimate Bearing Capacity of Foundations. *Geotechnique*, Vol. 2, 4, 1951.
8. Sokolovski, V. V., *Statics of Soil Media*. London: Butterworth, London, 1960.
9. Reece, A. R., The Fundamental Equation of Earthmoving Mechanics. *Symposium on Earthmoving Machinery. Inst. of Mech. Engrs.* 1965.
10. Hettiaratchi, D. R. P., B. D. Witney, and A. R. Reece., The Calculation of Passive Pressure in Two-Dimensional Soil Failure. *J. Agric. Engng. Res.*, Vol. 11, 2, 1966.
11. Balla, A., Bearing Capacity of Foundations. *J. Soil Mech. & Found. Div., A.S.C.E.*, Vol. 88, 1962.
12. Meyerhof, G. G., An Investigation of the Bearing Capacity of Shallow Foundations on Dry Sand. *Proc. 2nd Int. Conf. on Soil Mech. and Found. Eng.*, Vol. 1, 1948.

The Effect of Tire Tread on The Distribution of Ground Contact Pressure

SHOZO MASUDA

TAKASHI TANAKA*

ABSTRACT

Ground contact pressure of a tractor tire is one of the most important factors concerning tire sinkage problems and the traction produced by shearing action of the tire on the soil. Up to this time, a mean value calculated from the vertical load on a wheel and its contact area has been usually accepted; but it has recently been disclosed by many researchers that the distribution of the ground contact pressure could not be indicated by such a mean value. In fact, this varies over a wide range even under the same foot, and the values of the shearing resistance acting upon the tire (which are affected by the ground contact pressure) also cannot be indicated by uniform values.

It is a well known fact that even under a flat loaded plate, the distribution of the contact pressure has a maximum value under the centre of the plate and lesser values under the sides of the plate, on some soil conditions. This tendency is more intensified under a tractor tire, and this is disclosed and certified by many researchers.

In the reports of other workers, however, the effects of the tire tread pattern on the distribution of the pressure have not been considered, and tires have been treated as if they had no treads. The effects of the tread have only been considered when traction due to the shearing reaction of the soil has been discussed. Yet the shear does not take place in the soil under the tire tread bars, but in the soil inserted between them. Moreover, in discussing traction problems, it has been assumed that the normal pressures, that is, ground contact pressures, under the tread bars were equal to the pressures on the inserted soils. But these assumptions will often bring false conclusions, because the distribution pattern varies with the soil type and soil condition. It may be asserted that a large part of the load on a tire is supported by the tire tread bars when operating on soil conditions such that the shear in the soil can substitute for the friction of the tire to the soil.

Our test was conducted to investigate contact pressure distribution under several tire treads and to show the effects of the tire treads on this distribution, as well as to determine whether a new approach to the fundamental concept of the tractor traction problem can be derived.

*Professor and Assistant Professor, respectively, in the Division of Farm Power, Faculty of Agriculture, Kyoto University, Kyoto, Japan.

TEST SOIL

1. Test Soil

The soil used in our tests was a pulverized clay which had been passed through a 4 mm sieve and packed artificially in a soil bin. The depth of the soil bin was 30 cm, the length was 90 cm and the width was 30 cm.

2. Compression Test by the Test Plates

In order to know the compressibility of the test soil, a compression test was carried out by using the two plates having different widths, according to Bekker's method. One plate was 51 mm wide and the other 24.5 mm. Both plates were 98.5 mm long and 9 mm thick. Thus the areas of the two plates were respectively 50.25 and 24.15 cm², and the weights were 416 gr. and 313 gr. The compression results are shown in Figure 1. By Bekker's theory, these curves of plates loads vs. plate sinkage should be straight lines, but in this test the curves of both

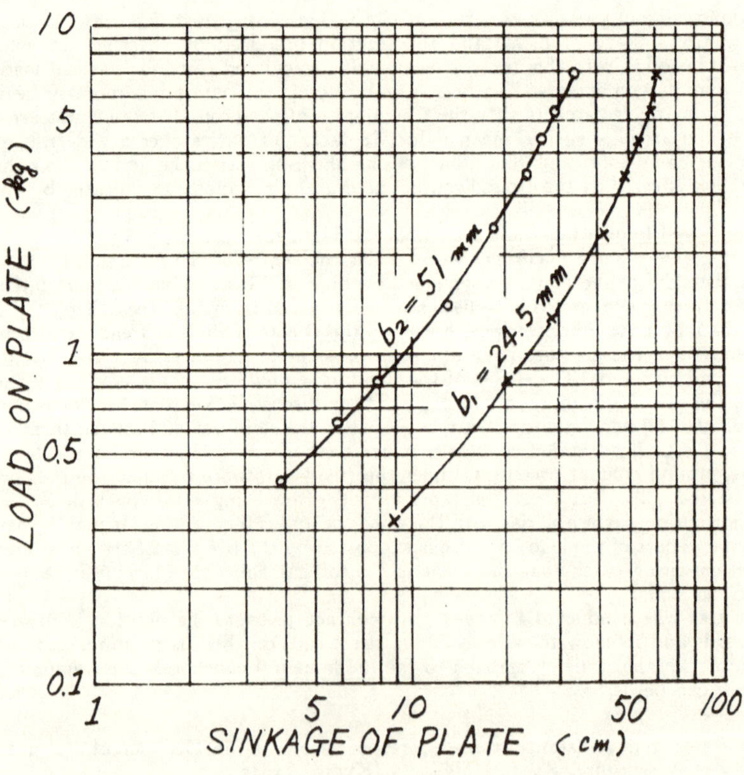

FIGURE 1

relations had a concave shape. Furthermore, in the actual field (for instance, on paddy fields or wheat lands), it will be almost impossible to obtain a linear relation, owing to the differences in the hardness of the subsoil layers. The results in Figure 1 can be regarded as almost linear for our experiments. Bekker's soil values calculated from these curves were $n = 1.23$, $kc = 12.0$ and $k\phi = 14.4$. The internal friction angle of the soil measured by a tri-axial compression test was $\phi = 36°30'$, and the cohesion was $c = 0.328$ kg/cm^2.

3. Soil Hardness under the Plate Compression Test measured by the pencil-type Cone Penetrometer

It was intended to obtain the distribution of the contact pressure from the calibration curves between the hardness of the soil and penetrating depth of a sharp cone. To obtain the calibration curve, the soil was compressed by the sinkage plate prior to the tire tests, and the hardness of the soil was measured by a small cone type penetrating device. The cone penetrating device had the sharp angle of 16.3 degrees, about 25 mm cone length and 21.75 gr weight. It was just like a pencil. The larger of the two sinkage plates was used in this test.

To check the uniformity of the distribution of contact pressure under the plate, at many contact points, the hardness of the soil was measured. But we could not find sufficient differences by the measuring magnitude of the pencil cone. Figure 2 gives the results of the test and shows the relation of penetrating depths of the cone in cm, and the mean contact pressures under the sinkage plate in kg/cm^2. The relation between these factors was represented by the straight lines on a log–log sheet.

EXPERIMENTS

1. Test Tire

The test tire was a small pneumatic tractor tire, 4.00–10, 2-ply, with a diameter of about 49 cm. The tread pattern is shown in Figure 3. Pneumatic pressure was adjusted to 0.5 kg/cm^2. Since the soil was soft, the tire deflection could not be seen within the range of the loads used in this experiment. The tire was put on the soil carefully and loaded by adding concrete blocks.

2. Measurement of the Hardness of Tire Mark

By removing the tire and load from the soil surface carefully, the mark of tire tread remained intact, as shown in Figure 3. Since the soil had no elasticity, the restoration of compression was not a problem. The hardness of this tire pattern marked on the soil was measured by the

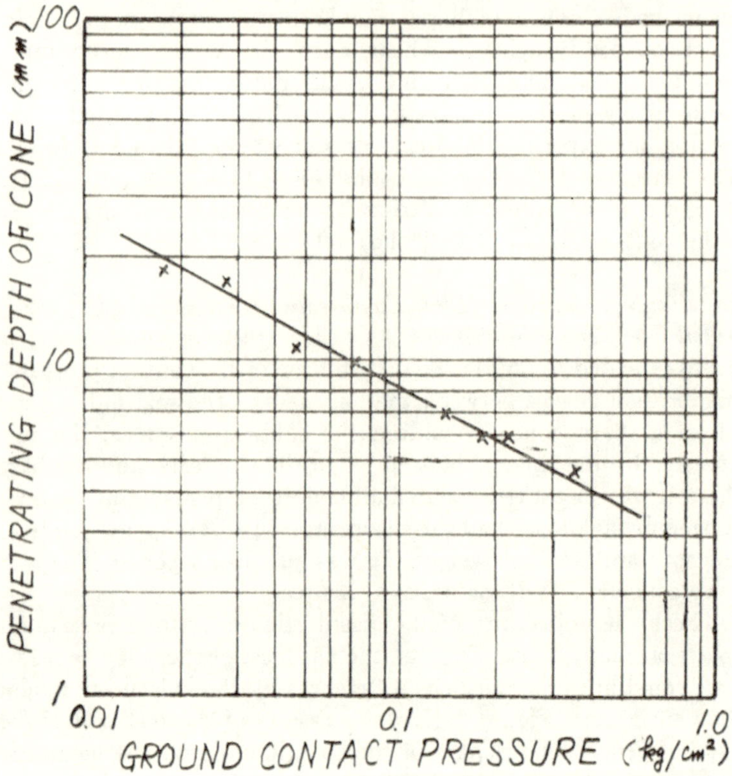

FIGURE 2

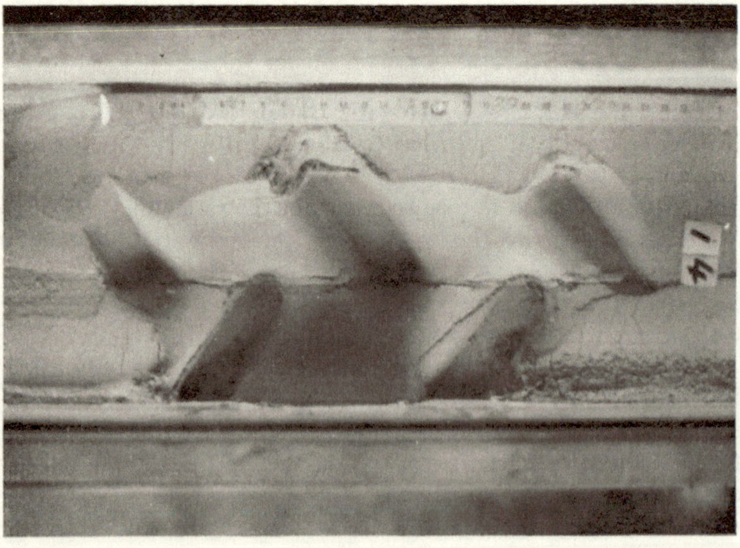

FIGURE 3A

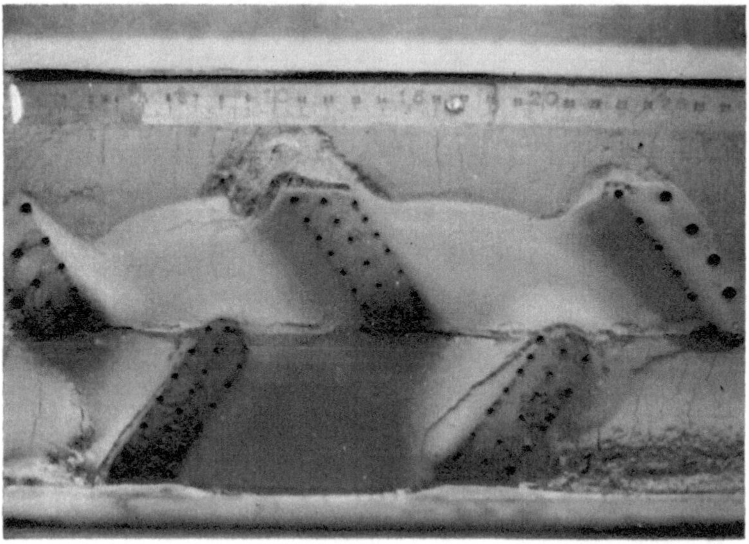

FIGURE 3B

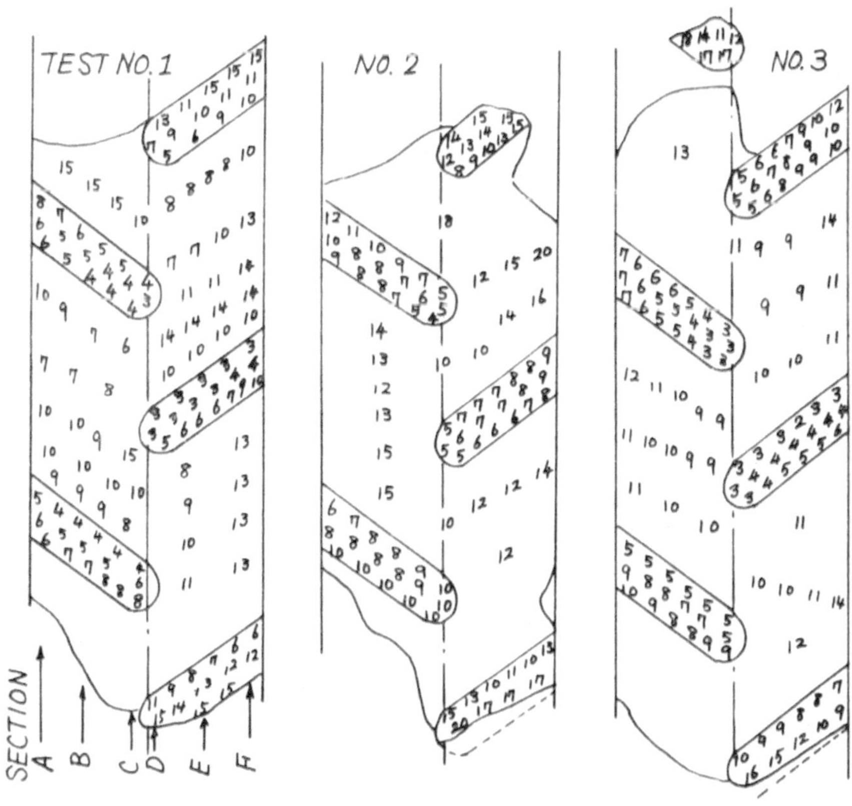

FIGURE 4

pencil cone. The small black holes in Figure 3(*b*) were the marks of this pencil cone and the differences in the diameters of these holes indicates the degree of compression. The numbers shown in Figure 4 are the penetrating depths of the pencil cone in mm at the measuring points. The test was repeated and the hardness was measured under three different loads. Details of each test—loads, contact lengths of tire, areas of contact, mean contact pressures, etc.—are shown in Table I.

TABLE I

Test	Loads	Contact Lengths	Contact Widths	Contact Areas	Contact Pressures	Contact Pressures from Figure 7	Calculated Loads from Figure 7
No.	*kg*	*cm*	*cm*	*cm*^{2}	*kg/cm*^{2}	*kg/cm*^{2}	*kg*
1	31.5	30	11.4	342	0.092	0.105	36.0
2	14.0	28	11.4	319	0.044	0.042	13.2
3	42.2	33	11.4	377	0.111	0.130	49.3

DISTRIBUTION OF CONTACT PRESSURES; DISCUSSION

1. *Distribution of Contact Pressures*

It is easily understood from the penetrating depths of the pencil cone shown on Figure 4, that the distribution of the contact pressures under the tire varied over a wide range; and compression of the soil under the tread was extremely large, when compared with the other parts. From the numbers in Figure 4, and the calibration curve in Figure 2, the contact pressures under the tire can be shown, as in Figure 5. Ground contact pressures under the deepest tread had very high values. The horizontal lines illustrated on the tire sketch in Figure 5 are the depths of tire sinkages in each test.

Under general observation, each result shows a peak value at the bottom of wheel centre, and pressures decreased along the circumference of the tire and become zero at the crossing points of the tire with the soil surface.

These distributions are the same as the results reported by many other researchers. But, as can be seen in these figures, there was a large difference between the contact pressures under the tire tread and the pressures on the soil between these treads; and we can presume that a large part of the load may be supported by the surface of the tire treads. Though it may be surmised that these results will be varied by the soil characteristics, this cannot be avoided. Observing carefully, it may be noted that the inside of the tire tread had a higher pressure

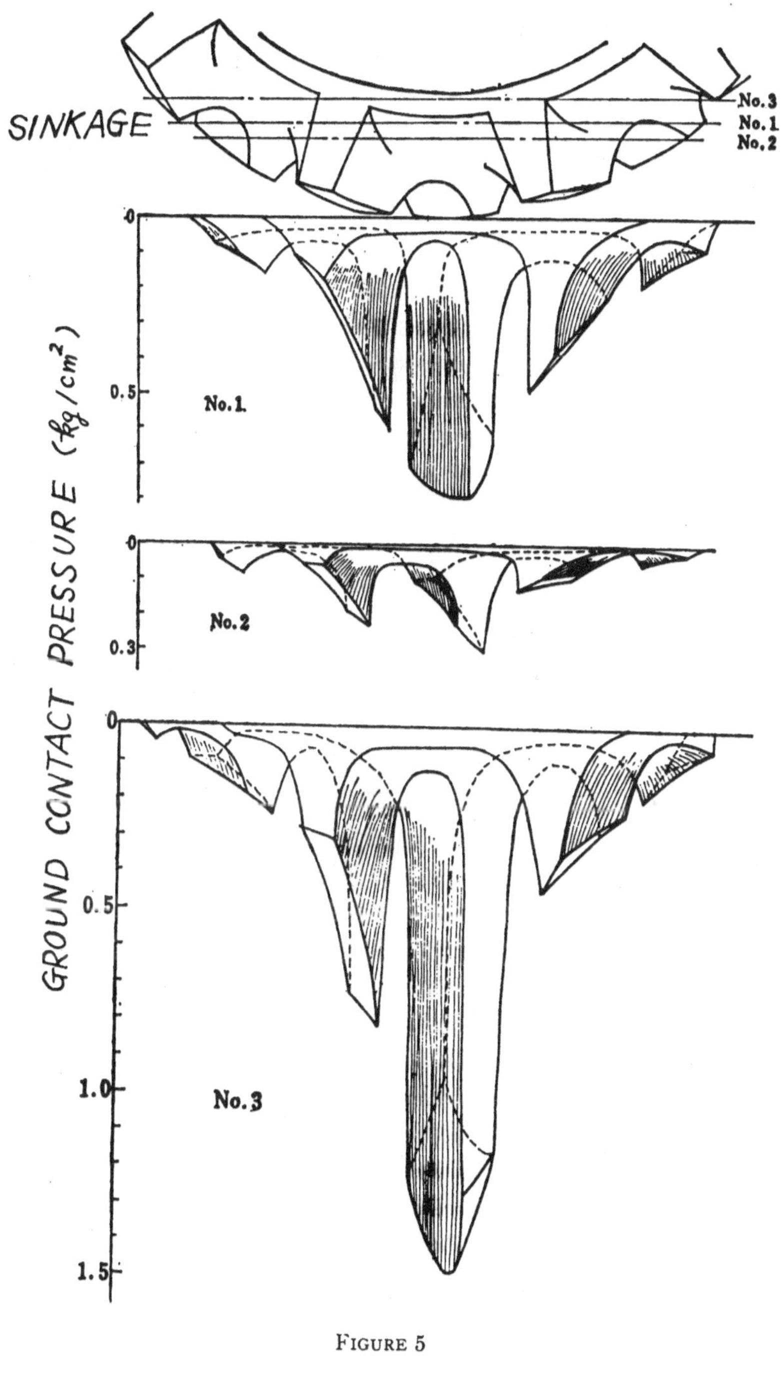
SINKAGE
No. 3
No. 1
No. 2
GROUND CONTACT PRESSURE (kg/cm²)
0
0.5
No. 1
0
0.3
No. 2
0
0.5
1.0
1.5
No. 3

FIGURE 5

than the outside of the same tread. It may be presumed that this was the result of the differences of compression at the tread surface.

2. Ground Contact Pressure and Load

To find the mean values of the contact pressures and to check the accuracy of the pressure distribution, these distributive pressures were integrated and compared with the actual loads. The pressure distributions along six planes perpendicular to the tire axle were obtained from Figure 5. The positions of these six planes were: both outsides of the tire, middle positions of each half side, and two positions near the centre of the tire, as indicated on Figure 4. As an example of the results, the distributions on the six planes of Test No. 1 are illustrated in Figure 6.

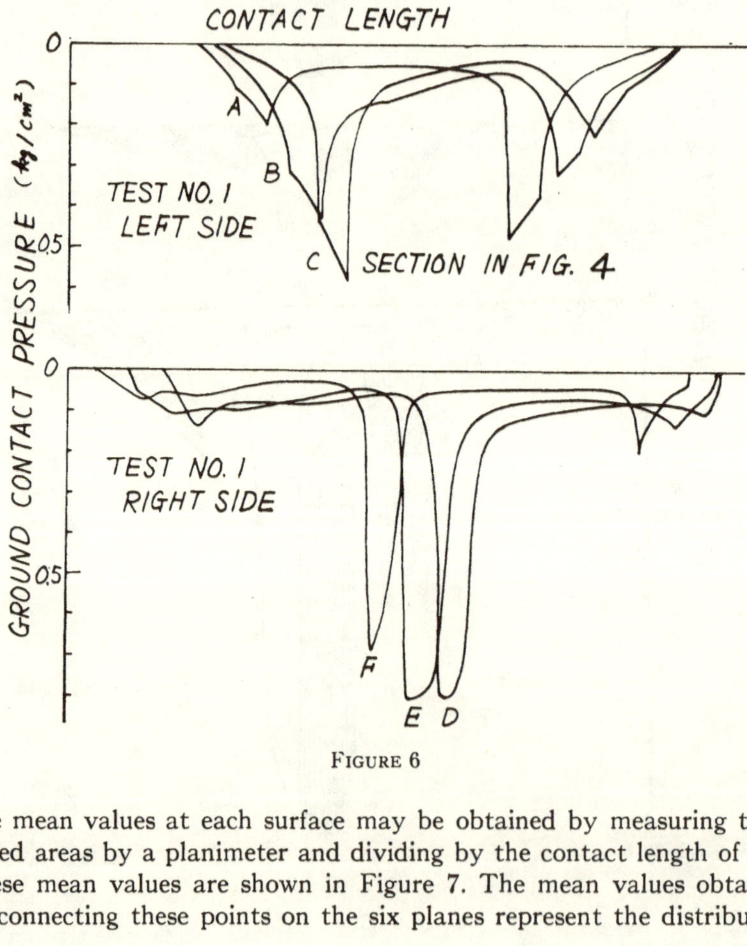

FIGURE 6

The mean values at each surface may be obtained by measuring these closed areas by a planimeter and dividing by the contact length of tire. These mean values are shown in Figure 7. The mean values obtained by connecting these points on the six planes represent the distribution

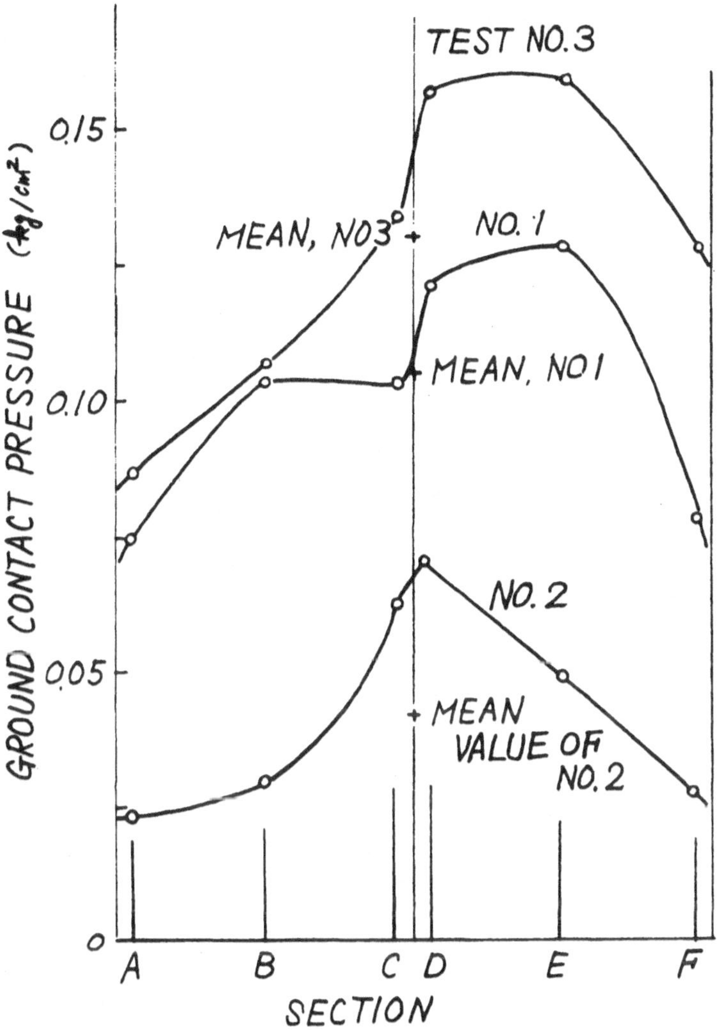

FIGURE 7

of the contact pressure along the normal sections to the moving direction of the tire. Pressure distribution of the tire having the greater load was larger than that having the smaller load. The right-hand side had larger values than the left one, and it may be presumed from Figure 5 that the deepest tread was positioned on the right-hand side of the tire. Hence, it may be considered that the high pressure at this position affected the results.

The averaged values of these curves are indicated by a mark on the centre line of Figure 7. Since these values show the mean pressures on the whole contact area of the tire, we can calculate the loads from these mean values and contact areas. These calculated loads (which are indicated in the last column of the Table I) should be equal to the actual loads. It would be proof of the accuracy of the distribution of the contact pressures if these calculated loads were found to be equal to the actual loads. In Tests No. 1 and 3, the values calculated from the ground contact pressures were slightly larger than the actual loads; and in Test No. 2, the calculated load was slightly smaller than that actual load. However, in general, they are almost in agreement with each other. It may be asserted that the contact pressures obtained by this method closely represent the actual ground contact pressures.

3. Ground Contact Pressure and the Drawbar Pull of the Tractor

The drawbar pull of a tractor is shown as the difference between the traction force and the running resistance. The traction force is the reaction of the soil shear, or friction between the soil and the tires of the tractor. The friction between the soil and the tire was considered as the most important factor when traction was first discussed, but the soil shear resistances were later introduced in the theory. Söhne discussed the traction problem by this concept. Bekker derived the equations for prediction of the traction and slippage of a vehicle by using the soil parameters. Further similar discussions have been reported by many researchers. But none of them have discussed the effects of the tire treads. Furthermore, soil shear has been considered only in the soil between the tire treads and not in the soil under the treads. Also, the two ground contact pressures have been considered as having the same value, and this concept may have brought many false conclusions.

From the results obtained in this test, it is contended that the normal pressure on the soil between tread bars is extremely small compared with the pressures under the tire treads.

It may be concluded that the fundamental concept of traction theory must be reconsidered in view of the differences in the ground contact pressures as discussed above.

Sinkage and Rolling Resistance of Wheels on Loose Sand

G. SITKEI*

ABSTRACT

The mechanics of forced and towed wheels is analyzed and the relationship between the factors influencing the rolling of forced and towed wheels are investigated. Dimensional analysis technique is used for the determination of the relative sinkage and the coefficient of rolling resistance. Based on experimental data, the criterion for the occurrence of bulldozing on loose sand is established.

THE MECHANIZATION of cultivation works on loose sand soils leads to many problems concerning cultivating implements and the traction efficiency of tractors. From the viewpoint of traction efficiency, it is possible with a suitable tracklayer to reach a satisfactory tractive effort and efficiency. However, the life-time of a track on loose sand is not more than 600–800 hours, therefore it seems to be more suitable to use a four-wheel-drive tractor equipped with pneumatic tires.

When a wheel rolls on loose sand it always sinks into the soil to a considerable extent. If the relative sinkage of the wheel surpasses a critical value then bulldozing occurs leading to a large motion resistance.

The tire-deformation on loose sand is very little, therefore it is possible to see the tired wheel as a rigid wheel.

MECHANICS OF WHEELS

1. The Forced Wheel

When a forced wheel rolls with positive slip, the instantaneous centre of motion /O′/ is over the bottom of the wheel. In a given point of the wheel rim the velocity vector has an angle γ to the radius of this point. This angle γ can be obtained from the following relationship:

$$\gamma = [(1 + \delta) - \cos \sigma]/\sin \sigma. \qquad (1)$$

Here σ = the central angle,
δ = the slip.

*Department of Agricultural Engineering, Technical University of Budapest.

The minimum value of γ is reached when $\cos \sigma_1 = (1/1 + \delta)$. The relationship between the angles γ and σ for various slippages is shown in Figure 1.

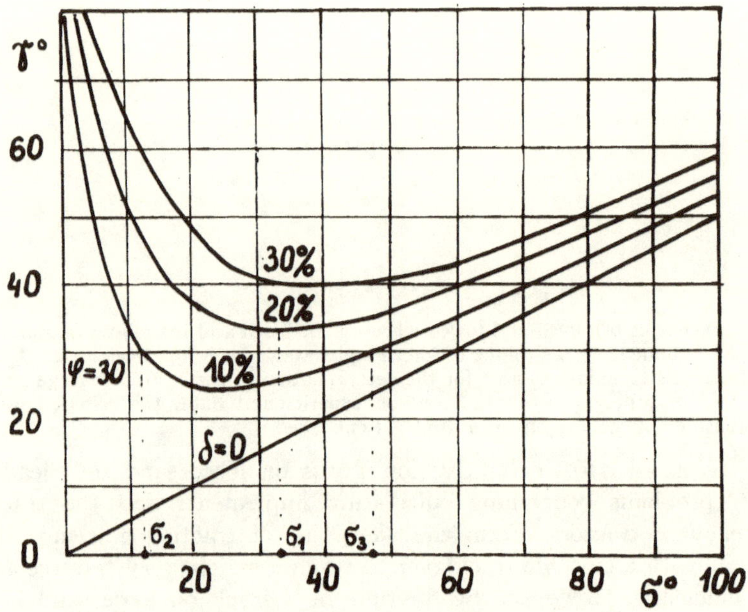

FIGURE 1. γ versus σ for various slips

For a moderate slippage, the wheel-soil interface surface may be divided into three characteristic zones:

1. $\sigma_3 < \sigma \leqslant \sigma_0$, where $\gamma > \varphi$;
2. $\sigma_2 \leqslant \sigma \leqslant \sigma_3$, where $\gamma \leqslant \varphi$;
3. $0 \leqslant \sigma < \sigma_2$, where $\gamma > \varphi$.

Here φ is the angle of friction between the rim of the wheel and the soil.

There is relative motion between the soil and the wheel in zones 1 and 3, while the soil sticks to the rim in zone 2. Elementary soil reaction (ΔP) vectors deviate from the wheel radius (or normal) by angle γ in zone 2, while in zones 1 and 3 this angle becomes φ. In practical cases the width of zone 2 is not large and therefore it may be assumed that elementary soil reaction vectors deviate from the wheel radius by angle φ in all zones.

A slow rolling forced wheel is subjected to the following forces and moments:

1. the weight of the wheel (or its vertical loading), G
2. the drawbar pull, P_v

3. the driving moment, M
4. soil reaction, $\Sigma\Delta P$

The equilibrium of the wheel is expressed by the following equations:

$$\Sigma P_x = P_k - P_f - (\Sigma bp\Delta s)_x = 0; \tag{2}$$

$$\Sigma P_y = G - (\Sigma bp\Delta s)_y = 0; \tag{3}$$

$$\Sigma M_{o'} = P_y\overline{O'C} - \Sigma M_{o'}^{OB} = 0; \tag{4}$$

$$\Sigma M_C = M - \Sigma M_C^{OB} = 0. \tag{5}$$

Here

$(\Sigma bp\Delta s)_x$ is the horizontal component of the resultant soil reaction,
$(\Sigma bp\Delta s)_y$ is the vertical component of the resultant soil reaction
$\Sigma M_{o'}^{OB}$, the moment of the resultant soil reaction taken around the instantaneous centre O'
ΣM_C^{OB}, the moment of the resultant soil reaction taken around the centre of rotation C
P_k the circumferential force (thrust)
P_f the rolling resistance.

The values of the above mentioned integrals are given by the following equations:

$$\Sigma M_{o'}^{OB} = \int^{OB} bp\zeta ds \quad \text{and} \quad \Sigma M_C^{OB} = \int^{OB} bp\zeta' ds$$

where ζ is the moment arm of an elementary soil reaction force taken around the centre of the wheel

ζ' is the moment arm of the same force elements taken around the instantaneous centre of motion.

The relative value of ζ and ζ' versus x/R are shown for various slippages and friction angles φ in Figure 2. When the zone 2 is absent or can be neglected, then the value of ζ is constant and it is determined by the angle of φ: $\zeta/D = \sin\varphi/2$. The value of ζ' varies linearly and depends on φ only to a negligible degree:

$$\zeta'/D = (\delta \sin\varphi/2) + (1-\delta)\sqrt{(1/2) - (1/2)\sqrt{1-(x/R)^2}},$$

or with a further approximation:

$$\zeta'/D = (\delta \sin\varphi/2) + (1-\delta)0.54(x/R).$$

For the sake of general validity, it is convenient to carry out the integration with respect to the relative co-ordinate x/R. The length of arc ds is: $ds = dx/\sqrt{1-(x/R)^2}$; and further:

$$dx = (D/2)d(x/R).$$

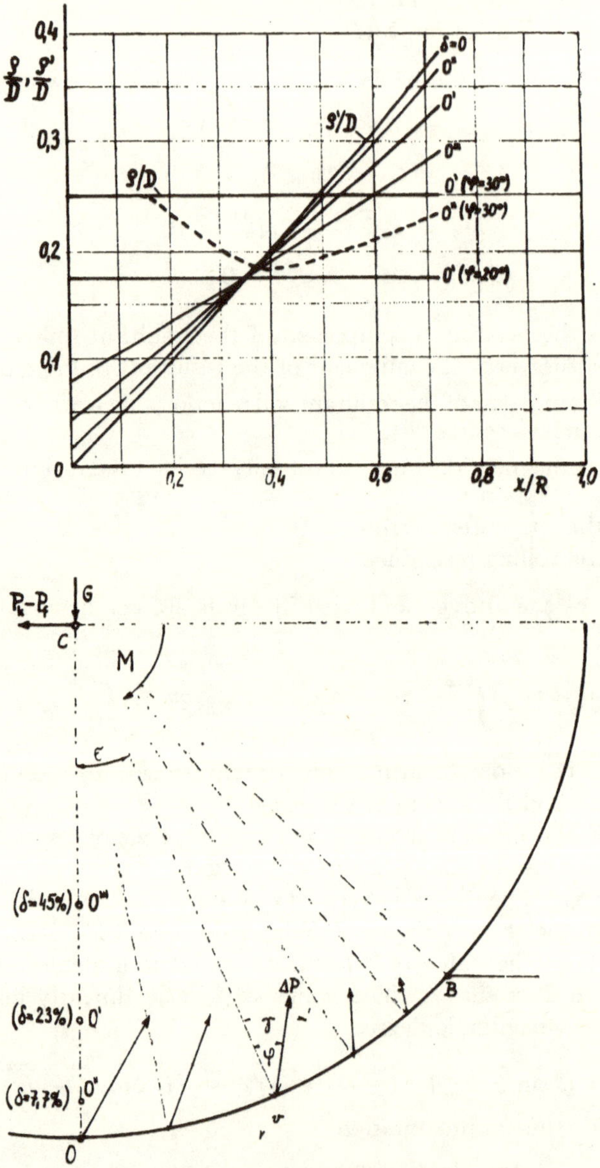

FIGURE 2. Schematics of a forced wheel. ζ and ζ' versus x/R for various slippages and friction angles φ

When the $p = kz^n$ pressure-sinkage relationship is accepted, the moments may now be rewritten as:

$$\Sigma M_C^{OB} = (bkD^3/2)\int (1/\sqrt{1-(x/R)^2})(z^n/D)(\zeta/D)d(x/R);$$

$$\Sigma M_{o'}^{OB} = (bkD^3/2)\int (1/\sqrt{1-(x/R)^2})(z^n/D)(\zeta'/D)d(x/R);$$

or in a short form:

$$\Sigma M_C^{OB} = (bkD^3/2)\Phi \tag{6}$$

$$\Sigma M_{o'}^{OB} = (bkD^3/2)\Phi'. \tag{7}$$

The integrals on the right-hand side of Equations 6 and 7 cannot be solved analytically. A graphical solution is presented in Figure 3.

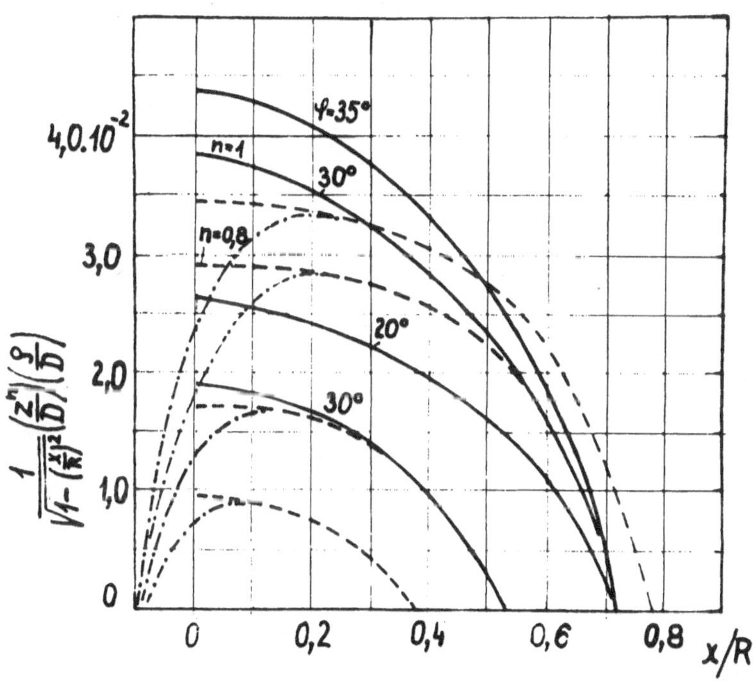

FIGURE 3. Graphical integration of the function $1/\sqrt{1-(x/R)^2}(z^n/D)(\zeta/D)$. ——— Theoretical pressure pattern ($n = 1$). - - - - - Theoretical pressure pattern ($n = 0.8$). —·—·—· Actual pressure pattern ($n = 0.8$)

In terms of Equation 5: $M = P_k(D/2) = bk(D^3/2)\Phi$; and therefore:

$$P_k/(bkD^2) = \Phi. \tag{8}$$

Equation 8 indicates that the relative value of the thrust is equal to the integral Φ. Likewise, the relative value of rolling resistance may be established according to Bekker (1):

$$P_f/(bkD^2) = z^{n+1}/[D^2(n+1)]. \tag{9}$$

On the basis of Equations 8 and 9, the drawbar-pull of the wheel may be obtained as a function of the relative sinkage (Figure 4).

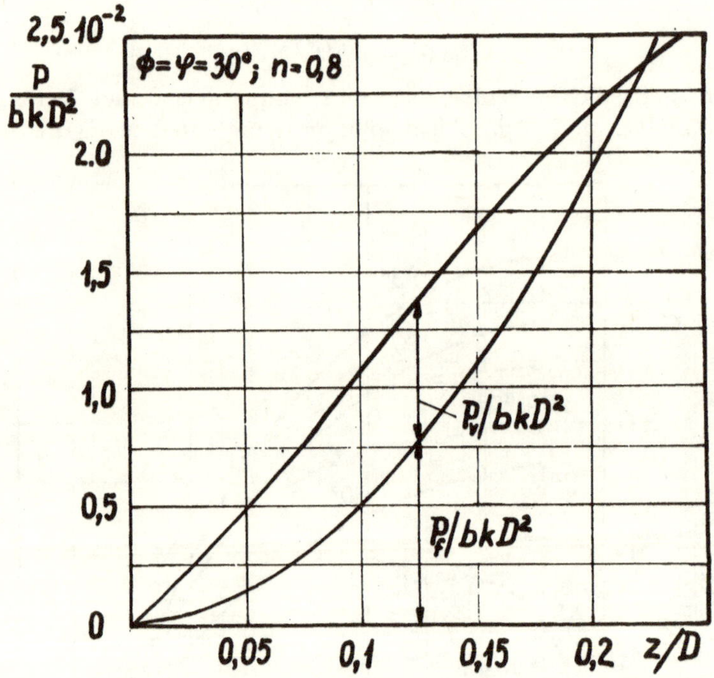

FIGURE 4. Plot of functions P/bkD^2 versus z/D

The variation of the drawbar-pull and the rolling efficiency as a function of relative sinkage are shown in Figure 5 for a loose sand. The maximum drawbar-pull is given at a considerable relative sinkage with unsuitable rolling efficiency.

2. *The Towed Wheel*

In this case the slip will be negative (skid). The instantaneous centre of motion is below the bottom of the wheel (Figure 6). The angle γ between the normal and the velocity vector is given by the following equation:

$$\gamma = \pm[\cos\sigma - (1-\delta)]/\sin\sigma. \tag{10}$$

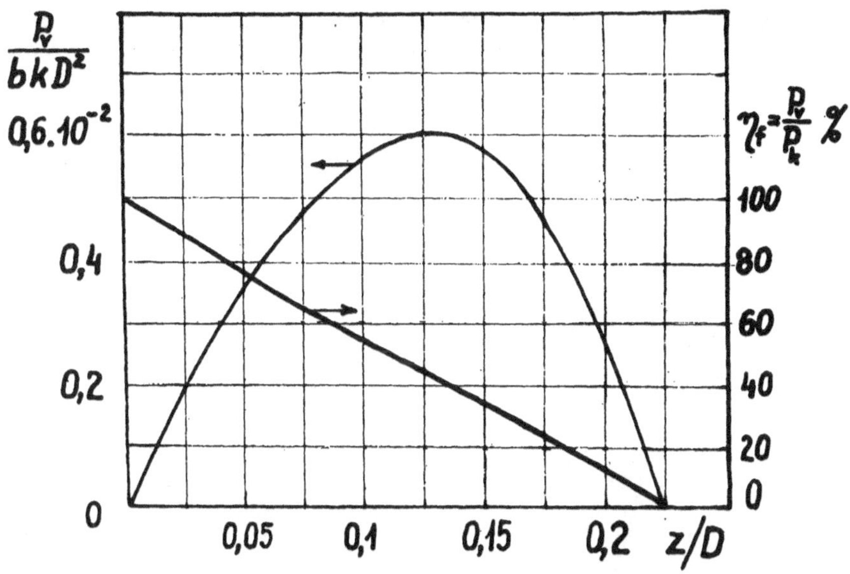

FIGURE 5. Drawbar-pull and rolling efficiency as a function of relative sinkage

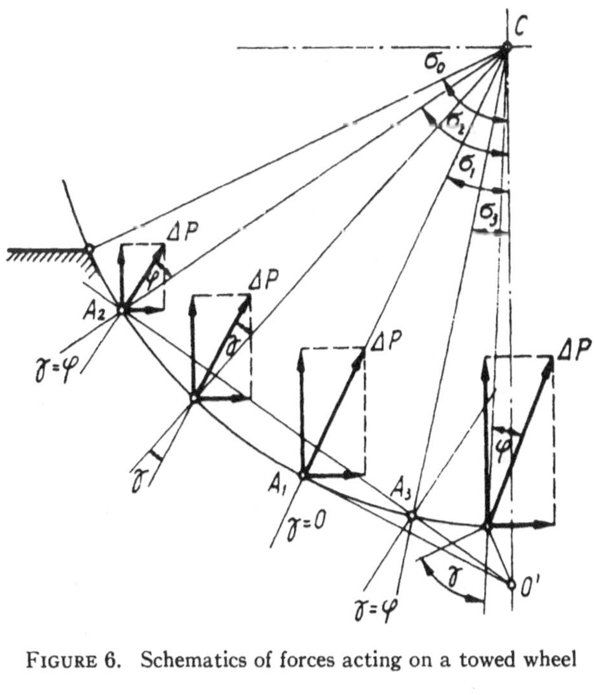

FIGURE 6. Schematics of forces acting on a towed wheel

The positive sign is valid for the region $\sigma < \sigma_1$, while the negative sign is used when $\sigma > \sigma_1$. Point A_1, at which $\gamma = 0$, is characterized by the central angle σ_1 according to the following relationship: $\cos \sigma_1 = 1 - \delta$.

Figure 7 depicts γ as a function of σ for various slippages.

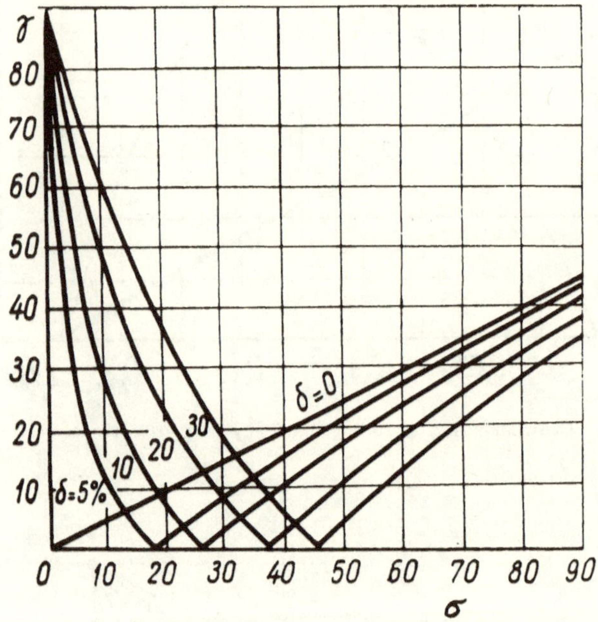

FIGURE 7. γ versus σ for various slips

In general, the wheel-soil interface surface may be divided into three characteristic zones:

1. $\sigma_2 < \sigma < \sigma_0$, where $\gamma > \varphi$
2. $\sigma_3 < \sigma < \sigma_2$, where $\gamma \leqslant \varphi$
3. $0 < \sigma < \sigma_3$, where $\gamma > \varphi$

There is a relative motion between the soil and the wheel in zones 1 and 3, while the soil sticks to the rim in zone 2.

A slow-rolling towed wheel is subjected to the following forces and moments:

1. the weight of the wheel (or its vertical loading), G
2. the towing force, P_f
3. the moment due to bearing friction, and, in some cases, the moment required by an implement, M_s, M_0
4. soil reaction $\Sigma\Delta P$.

The equilibrium of the wheel may be expressed by the following equations:

$$\Sigma P_x = P_f - (\Sigma bp\Delta s)_x = 0 \tag{11}$$

$$\Sigma P_y = G - (\Sigma bp\Delta s)_y = 0 \tag{12}$$

$$\Sigma M_{o'} = P_f \,.\, \overline{O'C} - \Sigma M_{o'}^{OB} = 0 \tag{13}$$

$$\Sigma M_C = -M_s - M_o + \Sigma M_C^{OB} = 0 \tag{14}$$

The values of moments $\Sigma M_c{}^{OB}$ and $\Sigma M_{o'}^{OB}$ are given by Equations 6 and 7 with the difference that the moment arms ς and ς' vary as shown in Figure 8. The variation of the integral Φ is shown in Figure 9 as a

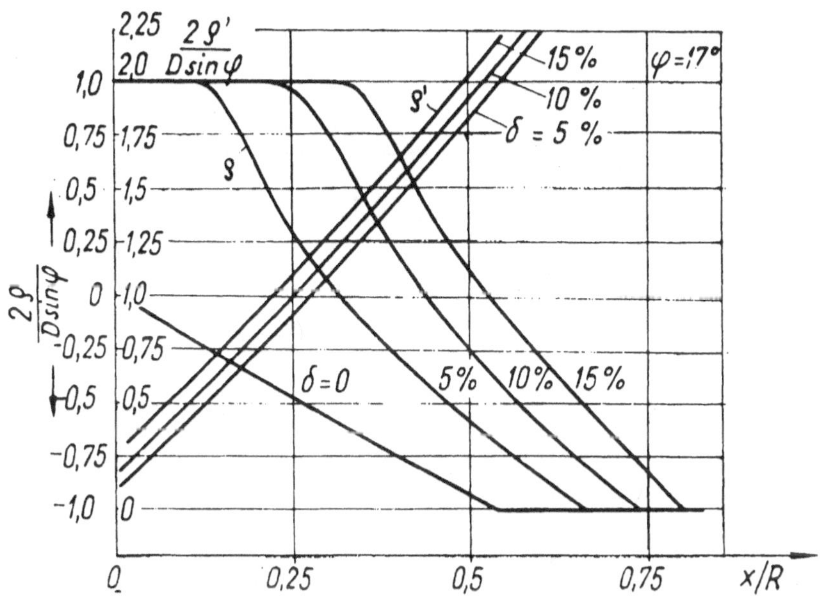

FIGURE 8. Plot of the relative moment arms ς and ς', as a function of co-ordinate x/R

function of slip for various relative sinkages. It follows from the graph that when the bearing friction is zero, and hence the resultant soil reaction vector passes through the centre of the wheel, the negative slip (skid) depends on the relative sinkage (Figure 10). When the bearing friction is not negligible, or the wheel furnishes torque for an implement, the slippage will be greater than for the case of zero moment.

The function ς' versus x/R can be accurately approximated with a linear equation and, assuming $n = 1$, the equation 7 can be integrated (2):

$$\Sigma M_{o'}^{OB} = bkDz^2/4(1 - \delta)^2. \tag{15}$$

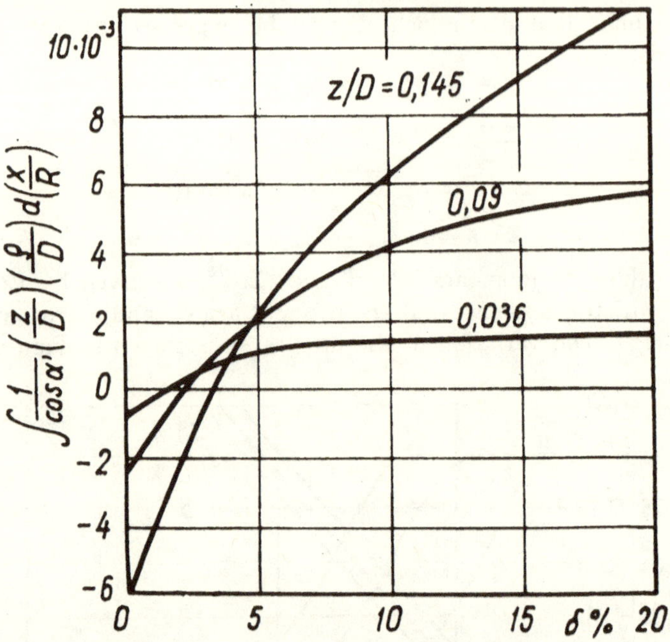

FIGURE 9. Plot of integral Φ as a function of slip

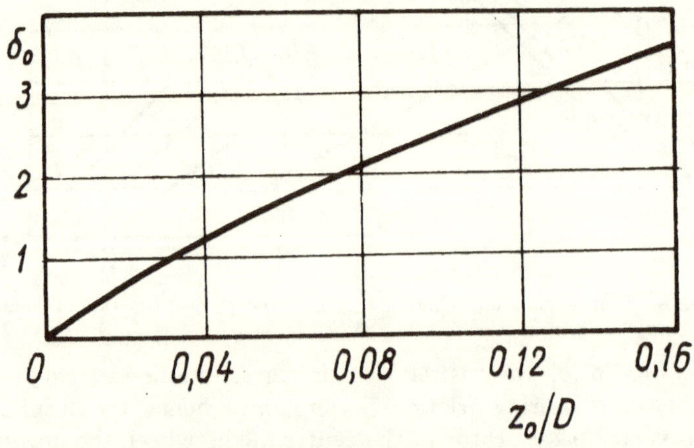

FIGURE 10. Slip of a towed wheel without bearing friction as a function of relative sinkage

The slip measured on sand is greater than that indicated by theoretical investigations. This happens because the upper layer of the sand deforms considerably in the direction of motion. In other words, the wheel is transported by the soil forward, without rolling, so that an additional slip occurs. This phenomenon may also be interpreted in the following way: only a fraction of the total slip occurs between the wheel and the soil. The rest of the slip is due to the relative motion between certain soil layers.

Figure 11 shows the motion of the soil under the wheel.

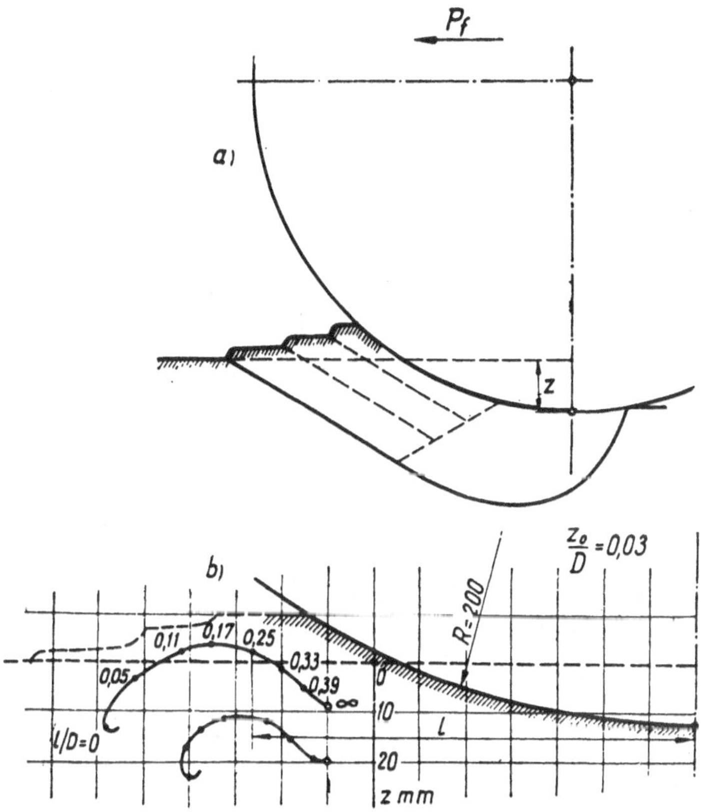

FIGURE 11. (A) zone of deformation under the wheel; (B) trajectories of particles at various depths.

THE SINKAGE OF WHEELS

The total sinkage of wheels may be determined from two components [3, 4]:

1. sinkage due to the vertical deformation of the soil
2. slip sinkage.

The existing methods for the determination of total sinkage take into account only the first part of the sinkage and therefore they can be used for the prediction of sinkage on cohesive soils, where slip sinkage is negligible.

Therefore it seems to be more suitable to use the method of dimensional analysis for determining the relationship between total sinkage and their controlling factors.

Total sinkage is a function of the variables listed below:

$$z = f(b, \gamma, k, G, D, \delta, \varphi) \tag{16}$$

Where: b is the width of the wheel, γ is the density of the soil, k is the soil-compaction-factor, G is the load on the wheel, and its weight, D is the diameter of the wheel, δ is the slip, φ is the angle of friction between the soil and the wheel. δ and φ are dimensionless quantities so that they can be taken as dimensionless criterions.

Applying the well known techniques of dimensional analysis the following dimensionless products are obtained:

$$\Pi_1 = z/D; \quad \Pi_2 = b/D; \quad \Pi_3 = \gamma D^3/G; \quad \Pi_4 = kD^3/G$$

A familiar expression is established by multiplying Π_4 by Π_2:

$$\Pi_4' = G/bkD^2.$$

Thus the controlling equation takes the following form:

$$z/D = f(b/D, G/\gamma D^3, G/bkD^2, \delta, \varphi) \tag{17}$$

According to our experimental results, the influence of the relative width of the wheel may be neglected. Further, the density of soil varies in practical cases only in a narrow region, therefore the relationship 17 may be simplified to the following equation:

$$z/D = f(G/bkD^2, \delta) \tag{17a}$$

Our experimental results for forced and towed wheels are shown in Figure 12.

ROLLING RESISTANCE

The main problem faced in the mechanization of agricultural works on sand-soils stems from the high rolling resistance encountered due to the looseness of the soil structure. Towed implements require high pulling forces, while the drawbar-pull of the tractors is small due to the high rolling resistance.

Present theories of rolling resistance are based on the assumption that rolling resistance is created solely by the vertical deformation of the soil and it is assumed that the resultant soil reaction elements pass through the centre of the wheel (5). Further, the theories do not take into account the effect of bow-wave formation (bulldozing effect).

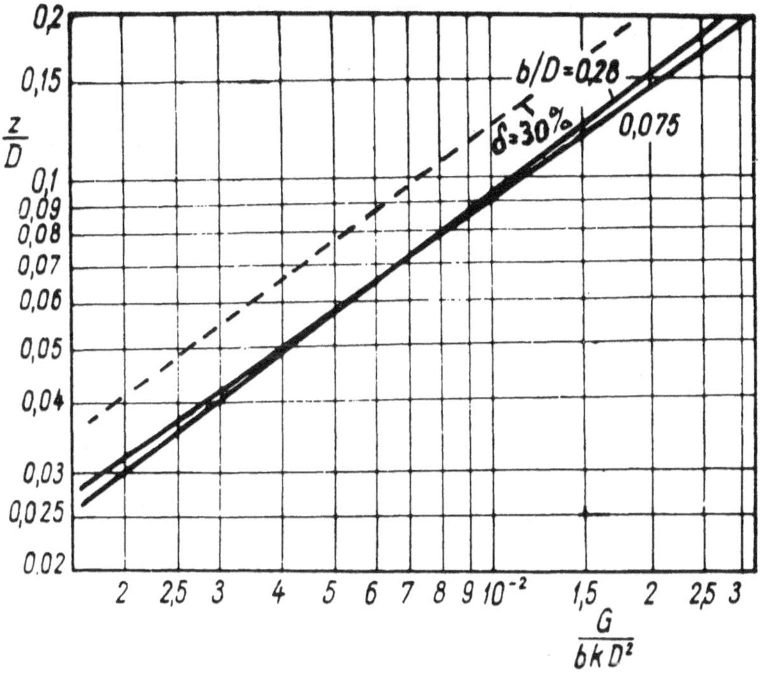

FIGURE 12. Relationship between the relative sinkage and the coefficient G/bkD^2 for forced and towed wheels

Since the quest for an accurate analytical solution is a very difficult task, the application of the dimensional analysis technique seems to be practical.

Rolling resistance is a function of these variables:

$$P_f = f(b, z, k, G, D, \delta, \varphi).$$

The following dimensionless relationship may be obtained:

$$P_f/G = f(b/D, z/D, G/bkD^2, \delta, \varphi) \tag{18}$$

Since a given relationship exists between the variables z/D and G/bkD^2, the above equation may be rewritten as:

$$P_f/G = f(b/D, G/bkD^2, \delta, \varphi). \tag{18a}$$

The experimental results for towed wheels are shown in Figure 13. The first important observation is that a change in slope of the curves occurs when the coefficient of rolling resistance is plotted for constant b/D ratios. The "break" takes place within the interval of 4.6 to $5.0.10^{-3}$. To the left of this point the sinkages are small and thus a bulldozing-free zone.

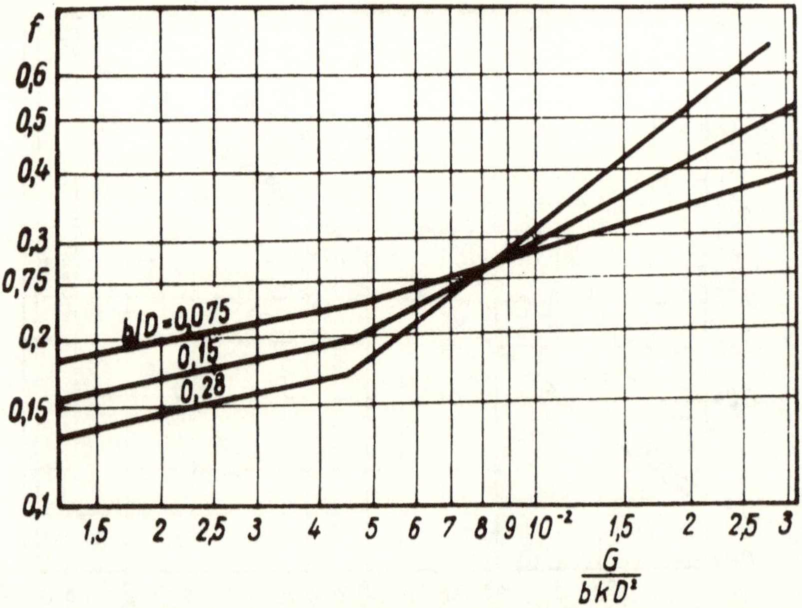

FIGURE 13. Plot of coefficient of rolling resistance as a function of dimensionless coefficient G/bkD^2

The slopes of the curves increase with increasing relative width to the right of the break-point. This happens while the soil cannot pile up considerably in front of a narrow wheel, since it is forced aside. The soil cannot flow aside, however, when the wheel is wide, therefore it piles up in front of the wheel causing an additional resistance. The relative sinkage at which the change of slope takes place is $z/D \cong 0.06$. If this criterion is not observed bulldozing occurs, which causes a large motion resistance especially for wide wheels. In the case of a forced wheel the bow-wave formation is less than in the above case. Therefore, the coefficient of rolling resistance does not increase as suddenly as in the case of a wide towed wheel.

REFERENCES

1. BEKKER, M., *Theory of Land Locomotion*. Ann Arbor, Mich.: The University of Michigan Press, 1956.
2. SITKEI, G., Investigation of the Bulldozing Resistance Encountered by Towed Rigid Wheels in Loose Sand. *J. of Terramechanics*, Vol. 3, 1966.
3. REECE, A., Principles of Soil–Vehicle Mechanics. *Proc. Inst. of Mech. Eng.*, London, 1965.
4. SOLTYNSKI, A., Slip Sinkage as one of the Performance Factors of a Model Pneumatic-tyred Vehicle. *J. of Terramechanics*, Vol. 3, 1965.
5. WILLIS, B., *et al.*, Investigation into Rolling Resistance. Theories for Towed Rigid Wheels. *J. of Terramechanics*, Vol. 1, 1965.

The Energy Loss of a Wheel

D. SCHURING*

ABSTRACT

The current concept of "motion resistance force versus thrust force" gives way to a general definition of energy loss from stress distribution in the wheel-soil interface. It is assumed that any deviations from the circularity of the wheel are small, that no forces are acting parallel to the wheel axle, that stresses in the contact area are constant across the width of the wheel, and that the interface is short.

Our general approach is exemplified by Bekker's method of stress approximation and the conventional method of coefficient of rolling resistance. A simple formula developed for the coefficient of rolling resistance for soft soils shows fairly good agreement with experimental data, except for small and very large sinkage.

Underlying assumptions and simplifications are emphasized.

NOTATIONS

- b wheel width
- c soil cohesion
- f distance (see Figure 2)
- j soil displacement in wheel-soil interface
- k Bekker's soil deformation modulus
- n exponent in Bernstein–Bekker pressure-sinkage relationship, $p = kz^n$
- p vertical stress in contact area
- $\bar{p}$ averaged vertical force per unit area of penetration plate
- r wheel radius
- s slip
- t horizontal stress in contact area
- $\bar{t}$ averaged horizontal force per unit area of shear plate
- v axle velocity
- z soil penetration
- z_0 sinkage
- E energy loss
- $E/r\theta = (E/L)(1-s)$ energy loss per unit distance traveled by point on wheel periphery
- E/L energy loss per unit distance traveled by wheel axle
- F resultant soil force
- $H = M/r$ driving torque per unit radius, or *thrust*
- K deformation modulus in Janosi–Hanamoto equation
- L distance wheel axle traveled
- L_θ distance point on wheel traveled
- M driving torque
- P vertical axle load
- $R_c \equiv R$ energy loss per unit distance traveled by slipless wheel, or *compaction resistance*
- T horizontal axle force
- ϵ coefficient of rolling resistance due to tire deflection
- θ angle of rotation
- $\rho = f/r = \rho' + \epsilon$ total coefficient of rolling resistance
- ρ' coefficient of rolling resistance due to soil friction
- σ local normal stress in contact area
- τ local tangential stress in contact area
- φ angle (see Figure 4)
- φ_1 angle (see Figure 4)
- φ_2 angle (see Figure 4)
- ϕ angle of internal soil friction
- ω angular wheel velocity

---

*GM Defense Research Laboratories, General Motors Corporation, Santa Barbara, California.

THE WHEEL AS A FREE BODY

A large portion of the original supply of energy in a vehicle is consumed by the wheels through bearing friction, internal tire friction, acceleration, and, finally, through friction at the wheel-soil interface. This paper focuses only on energy losses at the interface of a steadily moving wheel; losses due to tire deflection and bearing friction are not considered.

To study losses in this area, we think of the wheel as a free body, mentally disconnecting it from both vehicle and soil, and restoring equilibrium by forces and moments at the axle and at the contact area. (We assume throughout this study that no forces are acting parallel to the axle.)

At the axle we apply vertical force P, horizontal force T, and torque M. (Figure 1 illustrates all possible combinations of these.) All three are counteracted by the resultant soil force F acting at distance e from the axle and at angle α from the vertical line through the axle (Figure 2).

For equilibrium:

$$M - Fe = 0, \tag{1}$$

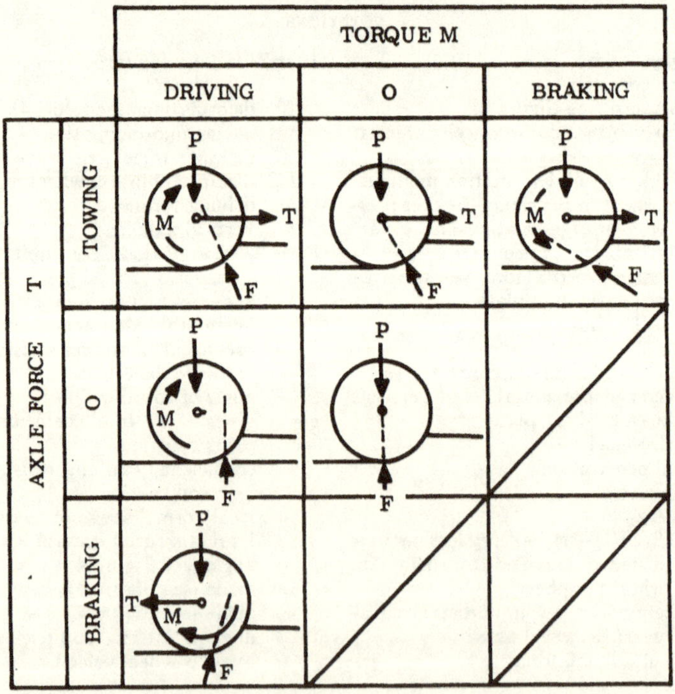

FIGURE 1. Combination of torque M and horizontal axle force T of a free-bodied wheel

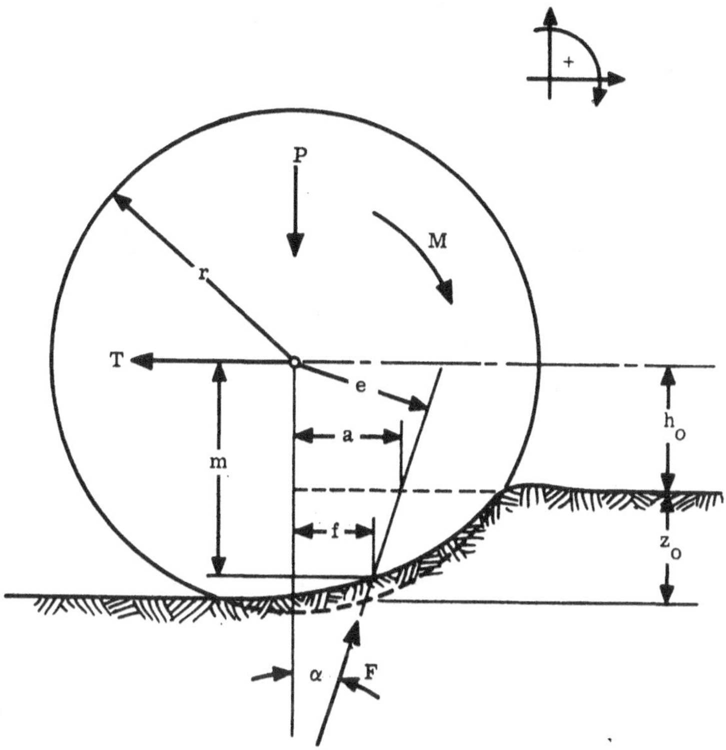

FIGURE 2. Forces, moments, and dimensions of a free-bodied wheel

$$F \sin \alpha - T = 0, \tag{2}$$

$$F \cos \alpha - P = 0. \tag{3}$$

Equations 1 to 3 are exact and independent; their form remains unchanged in spite of wheel deformation, sinkage, and slip.

The position of F is determined by angle α and distance e. Other specifications are conceivable, however; for instance, Phillips (1) suggests distances a and k_o and angle α (see Figure 2). His approach introduces a third variable, the distance k_o between the axle and the undisturbed soil surface.

Conventionally, the position of F is determined at the point where F intersects the periphery of the wheel. In this case distances f and m and angle α fix the position of F. To reduce the number of variables, it is often assumed that the wheel deformation or the sinkage of the wheel are both small and that, as a consequence, m approximates the wheel radius. In this case, Equation 1 reads:

$$M - Ff \cos \alpha - Fr \sin \alpha \approx 0. \tag{4}$$

Or, with the aid of Equations 2 and 3,

$$M - Pf - Tr \approx 0. \quad (5)$$

DEFINITION OF ENERGY LOSS

In travelling distance L, the wheel consumes energy equal to the driving torque M times the rotation angle θ (θ corresponds to distance L). Part of this energy displaces the horizontal force T by distance L, and part of its dissipates as loss E.

$$M\theta = TL + E. \quad (6)$$

This equation can be written in a different form by introducing the angular velocity of the wheel, $\omega = \theta/t$, and the horizontal velocity of the wheel axle, $v = L/t$, where t is time. Then:

$$M\omega = (T + E/L)v. \quad (6a)$$

Equations 6 or 6*a* define energy loss E, which encompasses all losses in the soil-wheel interface. Not included are losses due to friction inside the tire and friction in the wheel bearings, as stated.

Because E/L has the dimension of a force, it is sometimes thought to be an actual motion resistance force acting on the wheel in the contact area. This error would be incompatible with the conditions of wheel equilibrium stated in Equations 1 to 3.

We can write the energy-loss definitions 6 and 6*a*, which hold true for all conceivable soil-wheel combinations, in a different way by introducing the slip of the wheel. When the axle travels distance L, a point on the periphery travels distance L_o. If these distances are not equal, the wheel slips by the travel reduction $L_o - L$. Hence, a dimensionless number s, called slip, can be derived: $s = (L_o - L)/L_o$. Travelling distance L, the axle rotates by angle θ. If the wheel is rigid,

$$L_o = \theta r$$

where r is wheel radius. If the wheel is flexible, and therefore deformed at the contact area, L_o is smaller than θr (where r is the radius of the undeformed wheel). In this case the radius of the undeformed wheel may be replaced by a fictitious radius derived from the hard-surface rolling circumference; if the wheel deformation is not very great, however, the error is small. Thus for rigid and moderately flexible wheels,

$$s \approx (\theta r - L)/\theta r. \quad (7)$$

By dividing the numerator and the denominator by time interval t we obtain:

$$s = (r\omega - v)/r\omega \quad (7a)$$

where $\omega = \theta/t$ is the angular velocity of the wheel, and $v = L/t$ is the velocity of the wheel axle,

Substituting 7 into the basic equation, 6, we obtain:

$$E = r\theta[M/r - T(1 - s)] = r\theta[Ts + (M/r) - T]. \tag{8}$$

We get an equivalent expression by substituting 7*a* into the basic equation, 6*a*:

$$E = [L/(1 - s)][(M/r) - T(1 - s)] = [L/(1 - s)][Ts + (M/r) - T]. \tag{8a}$$

Both formulas, of course, contain no new information; they are only another way of writing the basic Equation 6. But they let us visualize the energy loss in graphical form (Figure 3). This figure shows schematic-

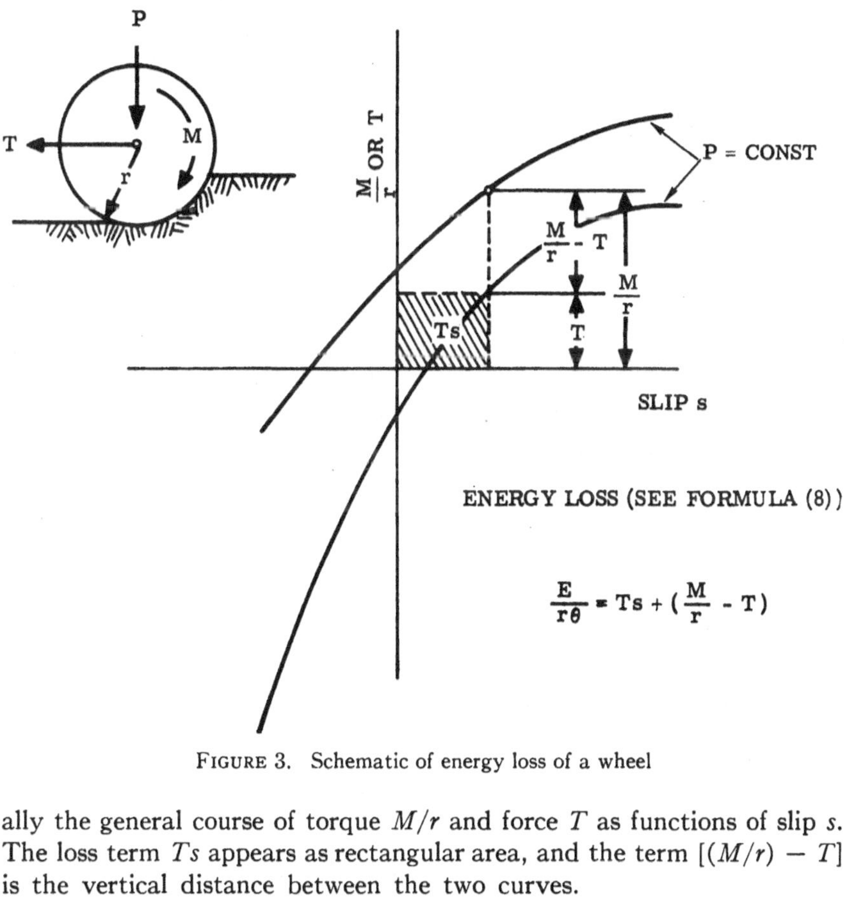

FIGURE 3. Schematic of energy loss of a wheel

ally the general course of torque M/r and force T as functions of slip s. The loss term Ts appears as rectangular area, and the term $[(M/r) - T]$ is the vertical distance between the two curves.

SOME ESTIMATES OF ENERGY LOSS

Before introducing new information by linking the external forces T and P and the external moment M with the soil-reaction forces in the contact area, we shall discuss three other approaches of estimating energy losses.

In the first approach, wheel losses are often estimated by distance f in Figure 2. Usually the ratio f/r is considered to be independent of slip, varying only with properties of the soil and of the wheel. The ratio $f/r = \rho$, called *coefficient of rolling resistance*, varies between 0.01 (highway) and 0.30 (loose sand). $\rho = f/r = \rho$ (soil, wheel).

At small wheel deflections the approximated equation of equilibrium 5 can be substituted into 8:

$$E = r\theta[P\rho + Ts]. \tag{9}$$

If the coefficient of rolling resistance ρ is considered to be independent of slip, the second term represents energy losses due to "slipless" soil deformation and wheel deformation.

If the slip of a towed wheel is considered to be negligibly small, Equations 8 and 9 can be simplified to: $E/\theta r = P\rho$ (see 9 with $s = 0$); $E/\theta r = -T$ (see 8 with $s = 0$ and $M = 0$). Hence, for a towed wheel with very little slip:

$$E/\theta r = P\rho = -T. \tag{10}$$

Occasionally, two erroneous conclusions are drawn from this relationship:

(1) Because loss can be approximated (for the special case of a pulled wheel with very little slip), by towing force T, it is confused with a "force" acting in the contact area in a direction opposite to the wheel movement. But $E/\theta r$ is energy loss per unit distance and has nothing in common with a force except for the dimension.

(2) Because the relationship $P\rho = -T$, has the appearance of Coulomb's friction law, the assumption is that Coulomb's law governs the resistance of a wheel. Actually, it is vaild for a pulled body with full slip, whereas relationship 10 holds for a wheel with *no* slip.

A second approach to determining energy loss has been attempted by Chudakow (quoted by Phillips (1)), who uses:

$$E/L = \rho_{CH}P. \tag{11}$$

This equation is similar to Coulomb's friction law, where friction force $\rho_{CH}P$ is proportional to weight P. As indicated, this approach is not meaningful. Moreover, ρ_{CH} has to be considered as being dependent on slip and is, therefore, more complex than the coefficient or rolling resistance ρ mentioned previously.

The third approach to determining energy loss has been attempted by Bekker (2), who discusses the energy loss of a towed wheel by looking at the wheel rut after the wheel has moved distance L. He contends that the rut could have been formed by a plate with length L and width b penetrating the soil to depth z_o. The work to do this is:

$$E = bL \int_o^{z_o} p(z)dz, \tag{12}$$

where $p(z)$ is the pressure underneath the plate located at distance z from the original surface. Bekker's formula is one of the first attempts at calculating the energy loss of a wheel in terms of pressure distribution along the contact area by identifying the pressure underneath a plate with the pressure underneath a wheel. We plan to elaborate on this notable advance.

ENERGY LOSS IN TERMS OF STRESS DISTRIBUTION IN THE WHEEL-SOIL INTERFACE

Figure 4 shows the dimensions of a wheel undeformed by contact with the soil. We assume that the periphery of the undeformed wheel is circular, and that any deviations caused by wheel tread are small. By contacting the ground, enveloping small obstacles, and sinking into the soil, the elastic wheel loses its circular contour. If the soil is soft, however, or the wheel is fairly stiff, the change in contour will be small, and we assume throughout that any deviations from the circular form are negligible. This assumption restricts the study to soil without obstacles sharp enough to bend the tire, and to fairly stiff wheels.

Considering now any minute portion of the contact area, we may state that the sum of all stresses acting on this portion can be resolved into two stresses: σ directed towards the wheel centre, and τ directed parallel to the undisturbed wheel circumference (see Figure 4). Moment M, horizontal force T, and vertical force P can be expressed in terms of stresses σ and τ (φ is angle along the contact area, ϕ is angle across the contact area),

$$M = \int_{\varphi_1}^{\varphi_2} \int_{\phi_1}^{\phi_2} r^3 \tau(\varphi, \phi) d\varphi d\phi \tag{13}$$

$$T = \int_{\varphi_1}^{\varphi_2} \int_{\phi_1}^{\phi_2} [\tau(\varphi, \phi) \cos \varphi - \sigma(\varphi, \phi) \sin \varphi] r^2 d\varphi d\phi \tag{14}$$

$$P = \int_{\varphi_1}^{\varphi_2} \int_{\phi_1}^{\phi_2} [\tau(\varphi, \phi) \sin \varphi + \sigma(\varphi, \phi) \cos \varphi] r^2 \, d\varphi d\phi \tag{15}$$

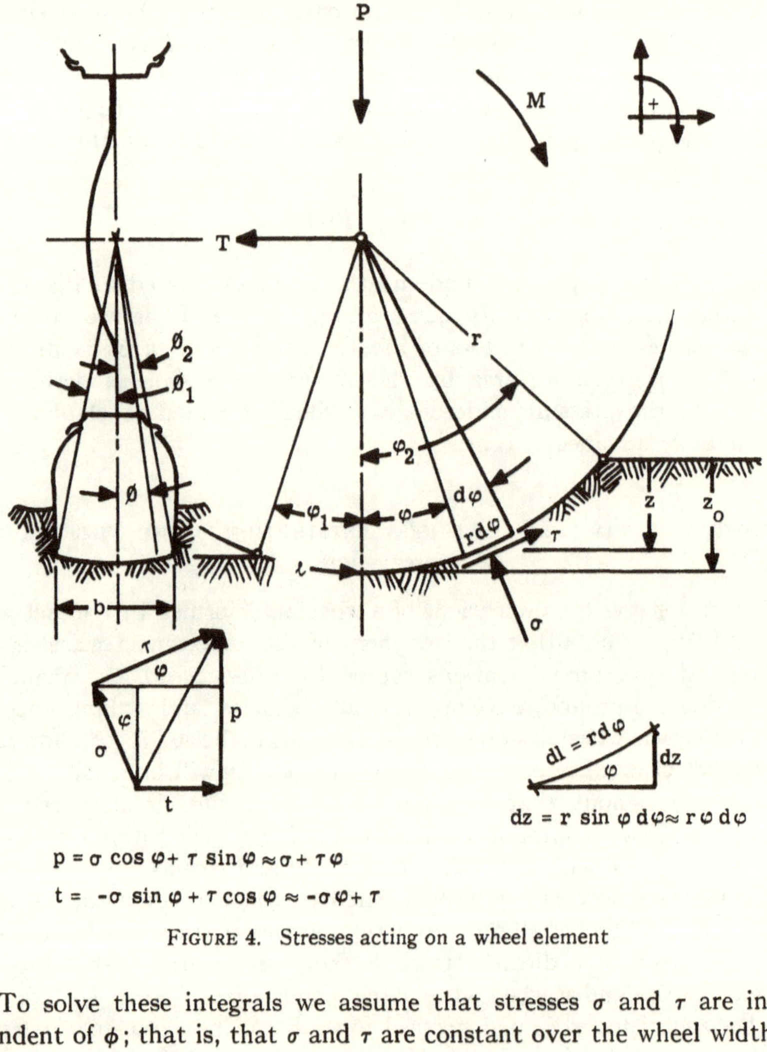

$p = \sigma \cos \varphi + \tau \sin \varphi \approx \sigma + \tau \varphi$

$t = -\sigma \sin \varphi + \tau \cos \varphi \approx -\sigma \varphi + \tau$

FIGURE 4. Stresses acting on a wheel element

To solve these integrals we assume that stresses σ and τ are independent of ϕ; that is, that σ and τ are constant over the wheel width b.

$$M = r^2 b \int_{\varphi_1}^{\varphi_2} \tau(\varphi) d\varphi \tag{16}$$

$$T = rb \int_{\varphi_1}^{\varphi_2} [\tau(\varphi) \cos \varphi - \sigma(\varphi) \sin \varphi] d\varphi \tag{17}$$

$$P = rb \int_{\varphi_1}^{\varphi_2} [\tau(\varphi) \sin \varphi + \sigma(\varphi) \cos \varphi] d\varphi \tag{18}$$

Next we introduce a drastic simplification: the interface is short. Hence,

$\sin\varphi \approx \varphi$ and $\cos\varphi \approx 1$. The error associated with this simplification depends on stress distribution. A rough estimate shows that to keep the error at about 5 to 10 per cent, angle φ_2 should not exceed 45°. Accordingly, the three integrals can be simplified:

$$M = r^2 b \int_{\varphi 1}^{\varphi 2} \tau\, d\varphi \tag{19}$$

$$T = rb \int_{\varphi 1}^{\varphi 2} (\tau - \sigma\varphi) d\varphi \tag{20}$$

$$P = rb \int_{\varphi 1}^{\varphi 2} (\tau\varphi + \sigma) d\varphi \tag{21}$$

The next step is to transform the two stresses σ and τ, oriented tangentially and radially to the wheel, into two corresponding stresses t and p oriented horizontally and vertically. According to Figure 4, then,

$$\tau = t + p\varphi \tag{22}$$

$$\sigma = -t\varphi + p \tag{23}$$

$$t = \tau - \sigma\varphi \tag{24}$$

$$p = \tau\varphi + \sigma \tag{25}$$

For this, the three integrals take the form:

$$M = br^2 \int_{\varphi 1}^{\varphi 2} (t + p\varphi) d\varphi \tag{26}$$

$$T = br \int_{\varphi 1}^{\varphi 2} t d\varphi \tag{27}$$

$$P = br \int_{\varphi 1}^{\varphi 2} p d\varphi \tag{28}$$

As a last step we substitute the two integrals 26 and 27 into the energy-loss formula 8.

$$E = \theta r^2 b \left[s \int_{\varphi 1}^{\varphi 2} t d\varphi + \int_{\varphi 1}^{\varphi 2} p\varphi d\varphi \right] = \theta r \left[Ts + rb \int_{\varphi 1}^{\varphi 2} p\varphi d\varphi \right]. \tag{29}$$

This formula is the focal point of this study. It expresses, in terms of stress distribution along the interface, the energy loss of a wheel with moderate deflection, and it can be applied to any ground that has no sharp obstacles. In the next part of the study we use Formula 29 to estimate energy loss on soft ground, but it works as well as for hard ground. Formula 29 gives only losses due to soil-wheel friction; those from internal tire friction are not considered.

By comparing the two energy-loss formulas, 8 and 29, we see that:

$$\frac{M}{r} - T = rb \int_{\varphi 1}^{\varphi 2} p\varphi d\varphi. \tag{30}$$

With the notation:

$$R = br \int_{\varphi 1}^{\varphi 2} p\varphi d\varphi, \tag{31}$$

Formula 30 takes the form:

$$(M/r) - T = R, \tag{32}$$

and the energy-loss Formula 29 reads:

$$E = \theta r(Ts + R). \tag{33}$$

By discussing several special cases, we can assess the significance of this formula.

(1) If the wheel is driven by pure torque with no horizontal force acting at the axle ($T = 0$), energy loss is reduced to:

$$E = \theta rR = \theta r^2 b \int_{\varphi 1}^{\varphi 2} p\, \varphi d\varphi. \tag{34}$$

(2) If the wheel moves without slip ($s = 0$), the same reduced formula applies. The two cases are not automatically identical, however, because pressure distribution $p(\varphi)$ is a function of slip, as Hegedus has shown (3). Figure 5 illustrates both cases: energy loss is signified by distance ($M/r - T$) between the two curves $M/r(s)$ and $T(s)$, at $T = 0$ for the first case, and at $s = 0$ for the second.

(3) If the wheel is spinning ($s = 1$), the energy-loss formula reads:

$$E = \theta r\left[T + rb \int_{\varphi 1}^{\varphi 2} p\varphi\, d\varphi\right] = \theta r[T + R]. \tag{35}$$

The loss is entirely one of slip, yet the pressure term R remains in the loss equation, as Figure 5 demonstrates. This, of course, reflects the influence of pressure on the shear stress that is causing slip loss.

(4) If the wheel is pulled and no torque is applied ($M = 0$), the energy-loss equation takes the form (see Equations 32 and 33),

$$E = \theta r(s - 1)T. \tag{36}$$

Usually a pulled wheel is assumed to develop negligible slip. This may be true where sinkage is small; but where sinkage is as much as 5 per cent of the radius, the slip of a pulled wheel cannot be neglected, as

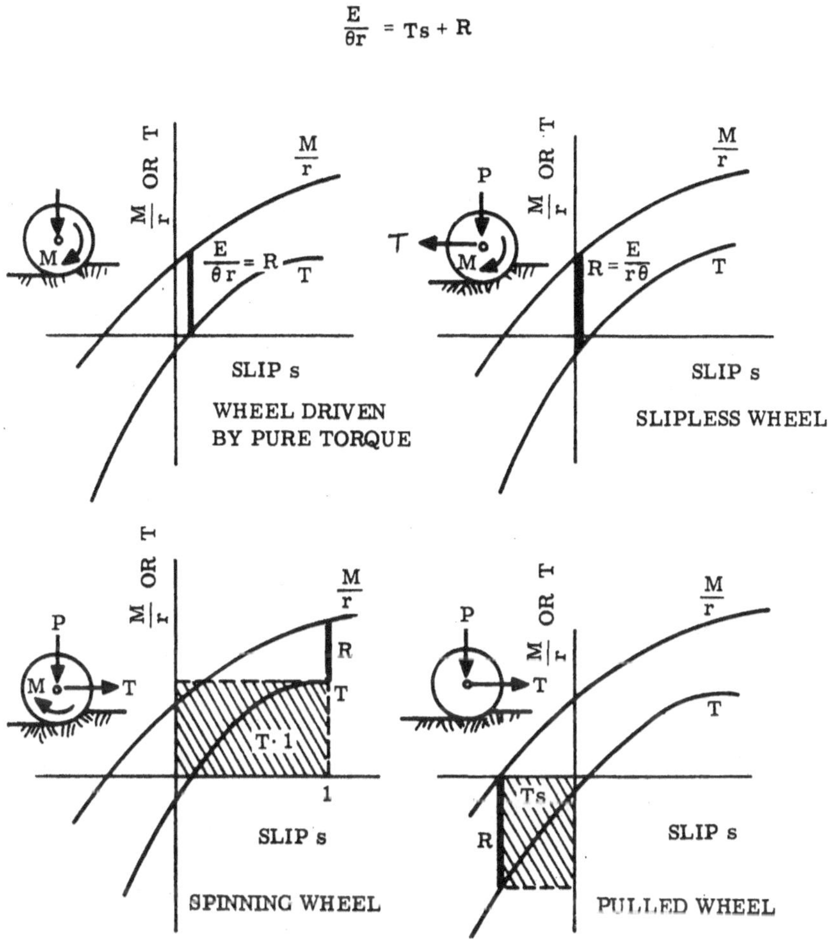

FIGURE 5. Schematic diagram of energy loss

Figure 6 demonstrates. (Figure 5 illustrates energy loss for the four cases discussed.)

COMPACTION RESISTANCE FORCE AND THRUST FORCE

Equation 20 is considered to be a fundamental relationship between vehicle and soil, and its correct interpretation is necessary for analytically predicting energy losses and the mobility of land vehicles.

$$T = rb \int_{\varphi_1}^{\varphi_2} \tau \, d\varphi - rb \int_{\varphi_1}^{\varphi_2} \sigma \phi \, d\varphi. \tag{20}$$

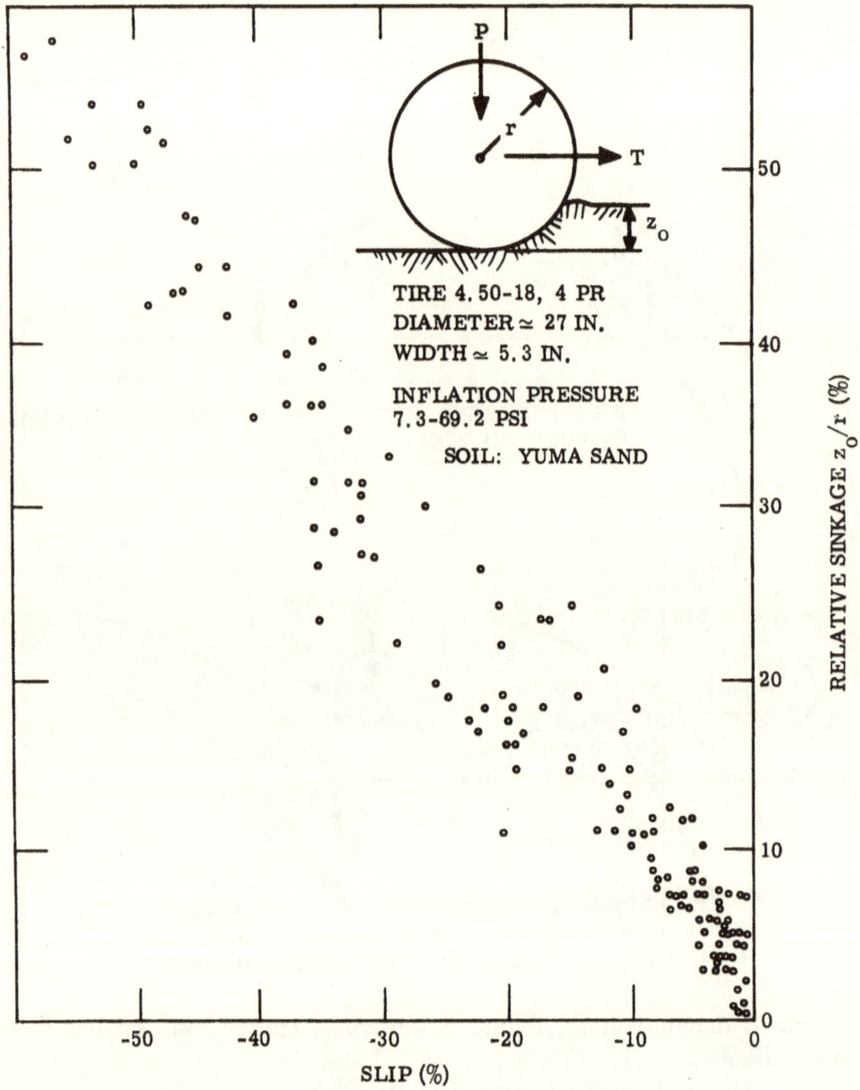

FIGURE 6. Slip *versus* sinkage of a towed wheel (from Reference 4)

This formula is identical with a relationship proposed by Bekker (2),

$$DP = b\int_o^{l_o} \tau dl - b\int_o^{z_o} \sigma dz \tag{37}$$

where l_o is length of interface and z_o is sinkage. Drawbar pull DP is identical to horizontal axle force T.

According to Figure 4 the differential penetration depth dz can be replaced by:

$$dz = r \sin \varphi \, d\varphi \simeq r\varphi \, d\varphi, \tag{38}$$

and the differential length dl of the contact area is:

$$dl = r \, d\varphi. \tag{39}$$

By substituting the two terms into 20 and letting $\varphi_1 = 0$, we arrive at Bekker's relationship, 37.

With the introduction of the notations "thrust" H and "compaction resistance" R_c, coined by Bekker (2), Formula 20 can be symbolized by:

$$DP = T = H - R_c, \tag{40}$$

where:

$$H = br \int_{\varphi_1}^{\varphi_2} \tau d\varphi \, ; \tag{41}$$

$$R_c = br \int_{\varphi_1}^{\varphi_2} \sigma \varphi d\varphi. \tag{42}$$

If we transform the stresses σ and τ into the stresses p and t by applying the stress transformations, 22 and 23, we obtain:

$$R_c = br\left[- \int_{\varphi_1}^{\varphi_2} t\varphi^2 d\varphi + \int_{\varphi_1}^{\varphi_2} p\varphi d\varphi \right];$$

$$H = br\left[\int_{\varphi_1}^{\varphi_2} t d\varphi + \int_{\varphi_1}^{\varphi_2} p\varphi d\varphi \right]. \tag{43}$$

Because only first-order terms are considered, the relationship for R_e simplifies to the following, which is identical with the term R (see Equation 31:

$$R_c = br \int_{\varphi_1}^{\varphi_2} p\varphi d\varphi = R. \tag{44}$$

Unfortunately, the second term in Equation 43 is neglected frequently, and only the truncated form is used:

$$H = br \int_{\varphi_1}^{\varphi_2} t d\varphi.$$

This is confusing because then both H and T would be calculated by the same formula (see Formula 27).

We think the notations "thrust" H and "compaction resistance" R_c should be dropped, because they give the erroneous connotation of the two terms being real forces acting upon the wheel. Actually, H and R_c are nothing more than two analytical terms determining horizontal wheel force T. If we wish to give them physical meaning, R_c can be visualized as representing energy loss per unit distance traveled by the slipless wheel (see Formula 29): $R_c = R = E/L(s = 0; \theta r = L)$; whereas H represents the torque per unit radius: $H = M/r$. Hence, we suggest the Formula 32 $M/r = T + R$, instead of 40, $H = T + R_c$. The only real forces acting on the wheel are the external vertical force P and the external horizontal force T; these, together with the external moment M, induce in the soil-wheel interface reaction forces whose distribution results in F (see Figure 1 and Equations 1 to 3). Looking at R_c and H as forces acting in the contact area is unrealistic.

Further steps in determining energy loss are possible only if the stress distributions $t(\varphi)$ and $p(\varphi)$ are investigated. But first we should discuss some other energy-loss formulas in the light of the general formulation 29.

COMPARISON OF ENERGY-LOSS FORMULAS

It is interesting to compare our energy-loss formula 29 with the approaches discussed previously, estimating energy loss in the contact area. Bekker's (2) approach applies to a towed wheel with very little slip, so slip loss can be neglected. He arrived at the expression, Equation 12,

$$E = bL \int_0^{z_0} p \, dz.$$

This is a special case of our Formula 29 with slip $s = 0$ and $\varphi_1 = 0$. Thus, according to the slip definition, $\theta r = L$. From 38, $r\varphi \, d\varphi = dz$. Hence, Formula 29 takes the form:

$$E = bL \int_0^{z_0} p \, dz.$$

This formula is identical to Bekker's, applying only to wheels with no slip. The slip of a towed wheel may not always be negligibly small, as Powell and others (4) have demonstrated; an example of their measurements was given in Figure 6, which presents high slip values at relatively small sinkage.

Hence, we think it more appropriate to use Equation 36, $E = \theta r(s-1)T$, instead of Bekker's formula, 12, to estimate energy loss in a wheel towed in soft soil.

Conventionally, losses in the interface area are estimated by Formula 9, $E = \theta r[Ts + P\rho']$. This coefficient of rolling resistance ρ' does not, however, include energy losses due to internal friction in the wheel. It reflects only losses due to friction in the contact area. Comparing this with Formula 29, we find that the coefficient of rolling resistance ρ' can be expressed as:

$$\rho' = \frac{br}{P} \int_{\varphi_1}^{\varphi_2} p \varphi d\varphi. \tag{45}$$

Or, if the integral 28 is used for the vertical force,

$$\rho' = \frac{\int_{\varphi_1}^{\varphi_2} p \varphi d\varphi}{\int_{\varphi_1}^{\varphi_2} p d\varphi} \tag{46}$$

Thus ρ' depends solely on pressure distribution in the contact area, which is a function of many variables, such as vertical load, driving torque, slip, soil properties, and wheel geometry.

If the pressure distribution in the contact area were known, the coefficient ρ', and consequently, energy loss E could be determined quantitatively. This leads to the question of stress distribution in the contact area.

STRESS DISTRIBUTION AND COEFFICIENT OF ROLLING RESISTANCE

Reliable experimental investigations into stress distribution along the contact area are not numerous. Sela (5) measured the distribution of shear and normal stresses of slipping wheels in sand by means of cells embedded in the wheel's periphery. Hegedus used the same method to measure the normal stresses in the contact area of rigid wheels operating in sand and sand-clay mixtures (3, 6). Uffelmann reports on stress measurements of a rigid wheel in clay (7). All authors recommend further research before establishing comprehnsive theoretical approach.

Pending a correct solution of the energy-loss problem by applying wheel-stress measurements, a rather crude approach has been initiated by applying plate test results to stress distribution on wheels. Plate tests are conducted exclusively in soft soils, driving a plate vertically into the soil (penetration test), or pushing it horizontally over the soil (shear test). Usually, a vertical penetration force per unit area $\bar{p}$ and a horizontal shear force per unit area $\bar{t}$ are associated with the geometrical soil-plate configuration and with physical soil values. If we apply these plate-test results to our energy-loss problem, several additional assumptions and simplifications have to be accepted:

(1) Both $\bar{p}$ and $\bar{t}$ are average stresses, whereas p and t are local stresses changing from point to point; therefore, to use $\bar{t}$ and $\bar{p}$ instead of p and t will introduce errors of unknown magnitude.

(2) The average shear stress $\bar{t}$ is a function of soil displacement. The displacement of soil particles along the contact area is thought to be a function of wheel slip, wheel geometry, stress distribution, and soil conditions, and very little is known about the interrelations of all these. Consequently, only a rough approximation can be applied.

(3) The contact area can be divided into a front part characterized by the angle φ_2 and a rear part characterized by the angle φ_1. (See Figure 4.) If the sinkage is fairly deep, the front part of the contact area will be large and the influence of the rear part on the stress distribution can be neglected. With decreasing sinkage, however, the front part shrinks and the rear part enlarges, as illustrated in Figure 7,* so that neglecting the rear part with shallow sinkage results in a large error. Since, unfortunately, plate penetration tests are not designed to take into account pressure distribution of the rear part, we must therefore either neglect the influence of the rear part or assume a certain pressure distribution.

(4) Soil deformation due to pressure is influenced by the presence of shear stresses, assumed for simplicity to be negligible so that the principle of superposition can be applied. Hegedus has shown that although this assumption would produce only minor error for a wheel in clay, it could entail substantial error in sand (3).

These difficulties prevent the solving of the stress-distribution problem, and any attempt based on plate tests to tackle the energy-loss problem must be regarded as no more than preliminary. Final results can be expected only from stress-distribution measurements on wheels, but as yet too few are available.

A partial solution to the problem can be attained, however, without knowing the exact stress distribution. If we postpone the determination of the empirical stress distributions $t(\varphi)$, we find that distribution $p(\varphi)$ is not critical to energy loss.

The different versions of energy-loss formulations found so far, all equivalent, are as follows:

$$\frac{E}{r\theta} = \frac{E}{L}(1 - s) = \begin{cases} Ts + \left(\dfrac{M}{r} - T\right) & (8) \\ Ts + R & (33) \\ Ts + P_\rho' & (10) \\ \left[rb \displaystyle\int_{\varphi_1}^{\varphi_2} t\,d\varphi\right]s + rb \displaystyle\int_{\varphi_1}^{\varphi_2} p\varphi\,d\varphi & (29) \end{cases}$$

*The data in Figure 7 are derived from plates 9 and 10 of a report by Smith and Freitag (8).

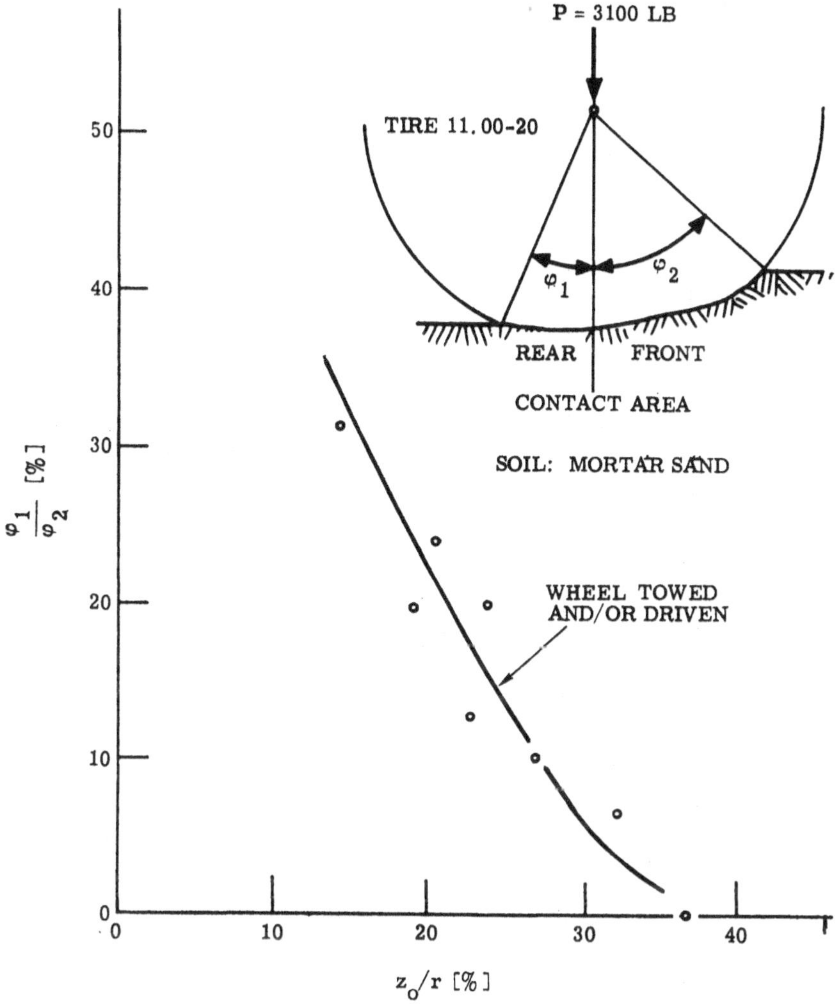

FIGURE 7. Rear and front parts of interface area as a function of sinkage

All first terms behind the bracket are identical, and all second terms are identical. Now we can concentrate on the second terms, especially on the coefficient of rolling resistance ρ', which usually encompasses both the energy loss due to soil-wheel friction and the energy loss due to internal tire friction—here, however, it represents only the loss due to soil-wheel friction. According to the listed formulas and Equation 46, ρ' can be expressed:

$$\rho' = \frac{M}{rP} - \frac{T}{P} = \frac{\int_{\varphi_1}^{\varphi_2} p\varphi d\varphi}{\int_{\varphi_1}^{\varphi_2} p d\varphi} \tag{47}$$

The integral term represents the location of the centre of gravity of the area underneath the pressure distribution $p(\varphi)$, a fortunate coincidence because with it we can make a simple estimate of the coefficient of rolling resistance for soft soils! Without checking particular pressure distributions, we can state that for all symmetrical pressure distributions the coefficient of rolling resistance must be:

$$\rho' = \tfrac{1}{2}(-|\varphi_1| + \varphi_2). \tag{48}$$

This result is significant because Hegedus has proved that a rigid wheel in cohesive soil develops a symmetrical pressure distribution for all slip values (3) (Figure 8). Consequently, we expect a wheel operating on cohesive soil to follow relationship 48.

In pure sand, however, pressure distribution does depend on slip. Although pressure distribution may be unsymmetrical in sand, as Figure 8 illustrates, we can demonstrate that the lack of symmetry does not change the centre of gravity significantly. Therefore, Formula 48 might apply to all soils, including sand.

Even a pressure distribution deviating greatly from the symmetrical does not completely invalidate Formula 48. Take, for instance, the Bernstein-Bekker distribution derived from plate penetration tests. Bernstein and Bekker (2) found that underneath a plate penetrating the soil vertically the average pressure depends on plate sinkage, as follows:

$$\bar{p} = kz^n, \tag{49}$$

where k is the soil deformation modulus, and n a soil value varying between 0.4 (soft clay) and 1.3 (coarse sand). This pressure-sinkage relationship is applied frequently to the pressure distribution of the wheel, even though this may entail significant errors. That $\bar{p}$ is not a pressure in the sense of a local stress but an averaged force per unit area of the penetration plate has already been stated. And it has been pointed out that a pressure-sinkage relationship such as the Bernstein-Bekker formula is valid only for the front part of the interface area and does not include the additional pressure of the rear.

For these reasons we should expect that to apply Formula 49 to our energy-loss problem would result in serious error. This is not quite the

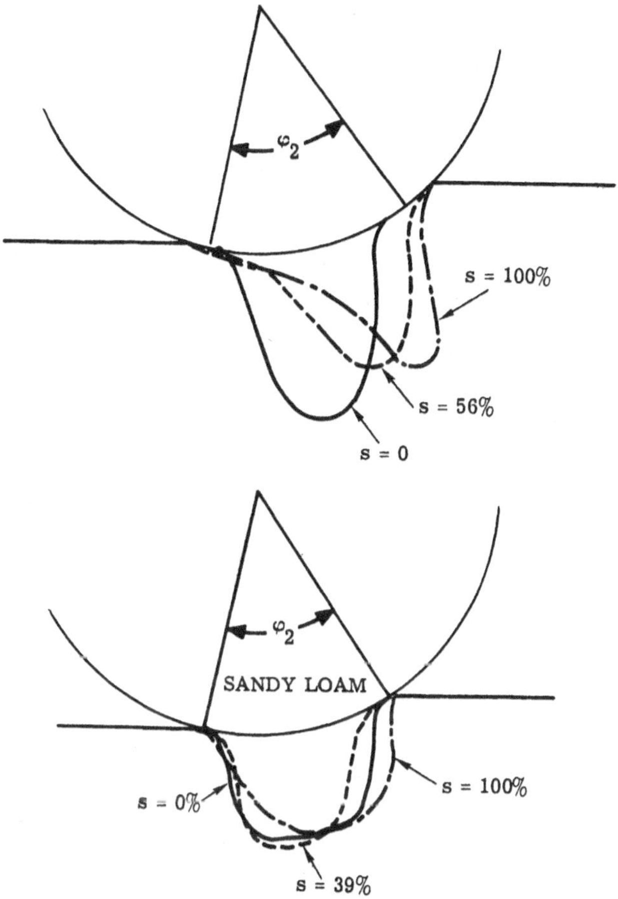

FIGURE 8. Distribution of normal stress under rigid wheel (from Reference 8)

case, however, and to prove it we first transform Formula 49 into a function of the angle φ. With the aid of the geometrical relationship (see Figure 4), $z = r(\cos \varphi - \cos \varphi_2)$, expanded into a series and truncated after the first term:

$$z \approx (r/2)(\varphi_2^2 - \varphi^2), \tag{50}$$

Formula 49 takes the form:

$$\bar{p} \approx k(r/2)^n(\varphi_2^2 - \varphi_2)^n. \tag{51}$$

The error induced by the truncation is small when the angle φ_2 does not exceed about 60 degrees.

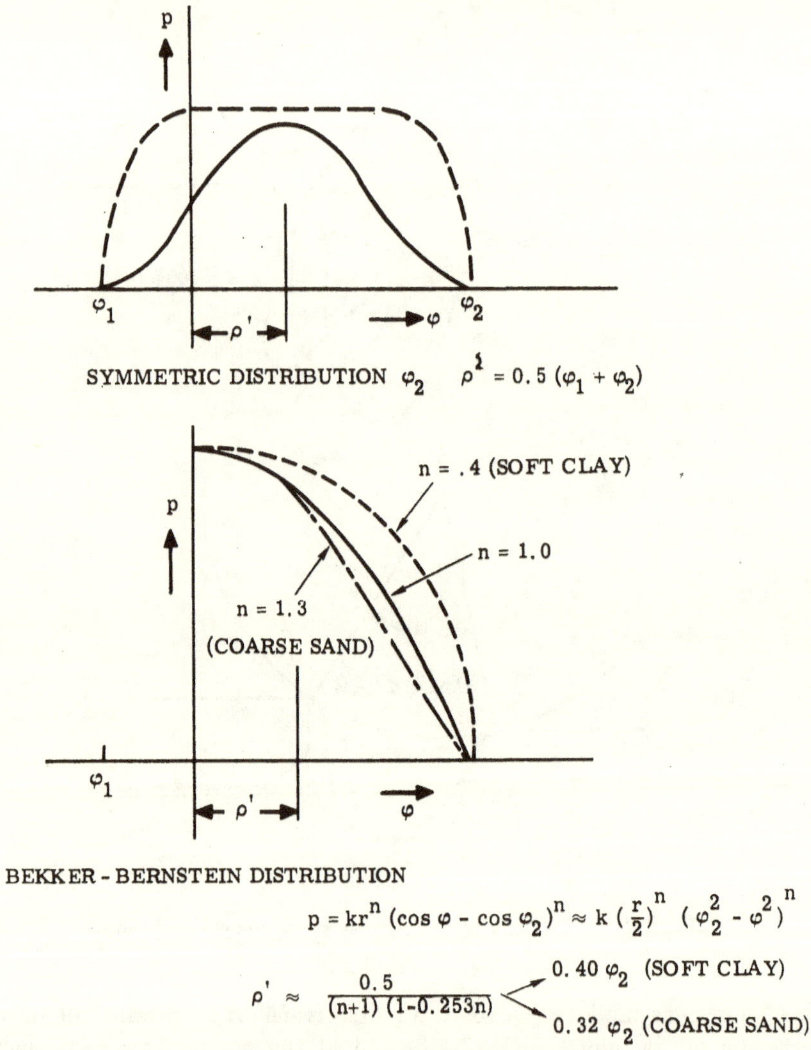

FIGURE 9. Some pressure distributions

Figure 9 shows this function for two extreme n-values: $n = 0.4$ for soft clay, and $n = 1.3$ for coarse sand.

To find the centre of gravity of this distribution, Formula 47 has to be integrated.

$$\int_{\varphi_1}^{\varphi_2} p\varphi d\varphi = k\left(\frac{r}{2}\right)^n \int_{\varphi_1}^{\varphi_2} (\varphi_2^2 - \varphi^2)^n \varphi d\varphi = k\left(\frac{r}{2}\right)^n \frac{\varphi_2^{2n+2}}{2(n+1)} \tag{52}$$

$$\int_{\varphi_1}^{\varphi_2} p d\varphi = k\left(\frac{r}{2}\right)^n \int_{\varphi_1}^{\varphi_2} (\varphi_2^2 - \varphi^2)^n d\varphi$$

$$= k\left(\frac{r}{2}\right)^n \varphi_2^{2n+1} \times \left[1 - n\sum_{m=1}^{\infty} \frac{(-1)^{m+1}}{m!(2m+1)}\right] \quad (53)$$

The expression in the brackets of Equation 53 is identical with:

$$1 - n\sum_{m=1}^{\infty} \frac{(-1)^{m+1}}{m!(2m+1)} = 1 - 0.253n$$

Accordingly, the centre of gravity of the Bernstein-Bekker distribution is located at:

$$\rho' = \varphi_2/[2(n+1)(1 - 0.253n)]. \quad (54)$$

Since the soil value n usually varies between $0.4 \leqslant n \leqslant 1.3$, the centre of gravity lies between:

$$\underset{\text{(coarse sand)}}{0.32\varphi_2} < \rho' < \underset{\text{(soft clay)}}{0.40\varphi_2}. \quad (55)$$

To compare this relationship with the ideal Formula 48, derived from symmetrical pressure distribution, we assign to φ_1 and φ_2 arbitrary angles such as $\varphi_1 = -15°$ and $\varphi_2 = 45°$. The maximum difference between the ideal value and the Bernstein-Bekker value turns out to be about 20 per cent. This relatively small deviation encourages us to ignore the effect of skewed pressure distributions in the contact area and to proceed under the assumption that Formula 48, which is based on symmetric pressure distribution, represents the relationship we need to calculate the energy loss.

It is customary to express all relationships in which the dimensions of the interface are involved in terms of sinkage z_o; therefore, we replace angle φ_2 by sinkage z_o, so that according to the geometrical relationship (50),

$$z_o = (r/2)\varphi_2^2, \quad (56)$$

and combined with 47 and 48,

$$\rho' = (M/rP) - (T/P) = 0.5(\sqrt{2z_o/r} - |\phi_1|). \quad (57)$$

This formula represents the second important relationship, leading to a more applicable version of the basic Formula 8:

$$E = r\theta[Ts + P\sqrt{z_o/2r} - |\varphi_1/2|]. \quad (58)$$

To verify this formula, all that is needed is to measure horizontal force T, vertical force P, sinkage z_o, and slip s. The angle φ_1 is usually small and may be neglected except when sinkage is small. Before elaborating on the theoretical approach, we would like to get some idea of the empirical validity of Formula 58 by turning to experimental checks that could indicate the soundness and limits of the chosen approach.

EXPERIMENTAL CHECKS

To check Formula 57, which comprises the extent of our approach so far, we use experimental data published by Powell and others (4) and by Smith (9). These authors measured the driving torque M, vertical load P, horizontal force T, and sinkage z_o for many different wheels operating on Yuma sand and fat clay. Powell tested elastic wheels with various loads and tire pressures on Yuma sand in three ways:

(1) The wheels are pulled and no torque is acting;

(2) The wheels are driven by a pure torque and no horizontal forces are acting;

(3) The wheels are driven by a torque and are pulling a horizontal load. Smith tested a rigid wheel, towed under various loads on fat clay. We plotted Powell's and Smith's data in four diagrams, together with a family of theoretical curves representing Formula 57,

$$\rho' = \sqrt{z_o/2r} + \varphi_1/2$$

for back angles of $\varphi_1 = -5°$, $-10°$, and $-15°$. Here ρ' is either T/P (pulled wheel), Mr/P (wheel driven by pure torque), or $(M/r - T)/P$ (wheel driven by torque and pulling load T). If $\varphi_1 = 0$, the Function 57 is a straight line with a slope of 2. A small angle of $\varphi_1 = -5°$ causes a slight deviation from the straight line (on log-paper the deviation is exaggerated at low sinkage); the deviation increases with increasing rear angle, as indicated in the diagrams. We could not determine analytically which of the theoretical curves reflects the actual wheel performance best, because the rear angle φ_1 was not measured, but we would expect the rear angle to increase with decreasing sinkage, as indicated in Figure 7.

When we plot experimental data for three pulled elastic wheels (Figure 10), the relationship (57) takes the form:

$$\rho' = |T|/P = \sqrt{z_o/2r} - |\varphi_1|/2 \quad \text{(see 37 with } M = 0\text{)}.$$

We recognize from Figure 10 that different wheel diameters and widths do not cause significant deviations, and this confirms that the coefficient ρ' is independent of wheel size.

We recognize also good agreement between prediction and experiment, although the data are widely scattered. At sinkages between 15 per cent and 60 per cent of the wheel radius, the data follow the straight

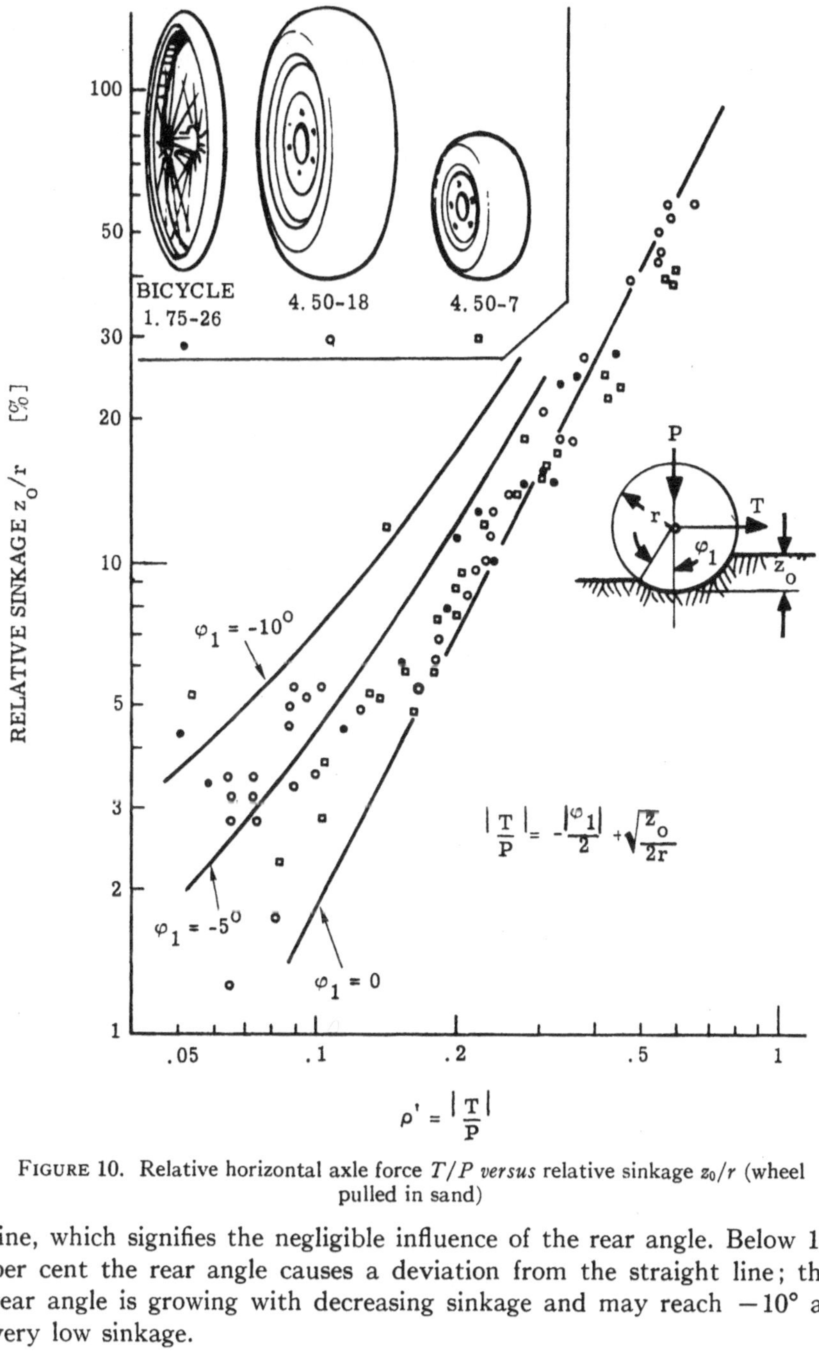

FIGURE 10. Relative horizontal axle force T/P *versus* relative sinkage z_0/r (wheel pulled in sand)

line, which signifies the negligible influence of the rear angle. Below 15 per cent the rear angle causes a deviation from the straight line; the rear angle is growing with decreasing sinkage and may reach $-10°$ at very low sinkage.

On the whole, Figure 10 can be taken as a confirmation of the approach, for it proves that an unsymmetrical pressure distribution which, according to Hegedus, may prevail in sand (3), does not cause a significant deviation from the basic relationship 57.

With small sinkage, the rear angle greatly influences the coefficient ρ'. We think that the rear angle is caused mainly by elastic rebound of the wheel carcass induced by decreasing pressure at the rear part of the contact area. Also involved may be some elastic rebound of the soil after the wheel has passed, but this is probably too slow to affect wheel performance.

Accordingly, the rear angle of an unelastic wheel should be either very small or totally absent (Figure 11). This is confirmed by Figure 12 where we plotted, in addition to the same test data on elastic wheels as in Figure 10, data on a towed rigid wheel (9). The experimental data

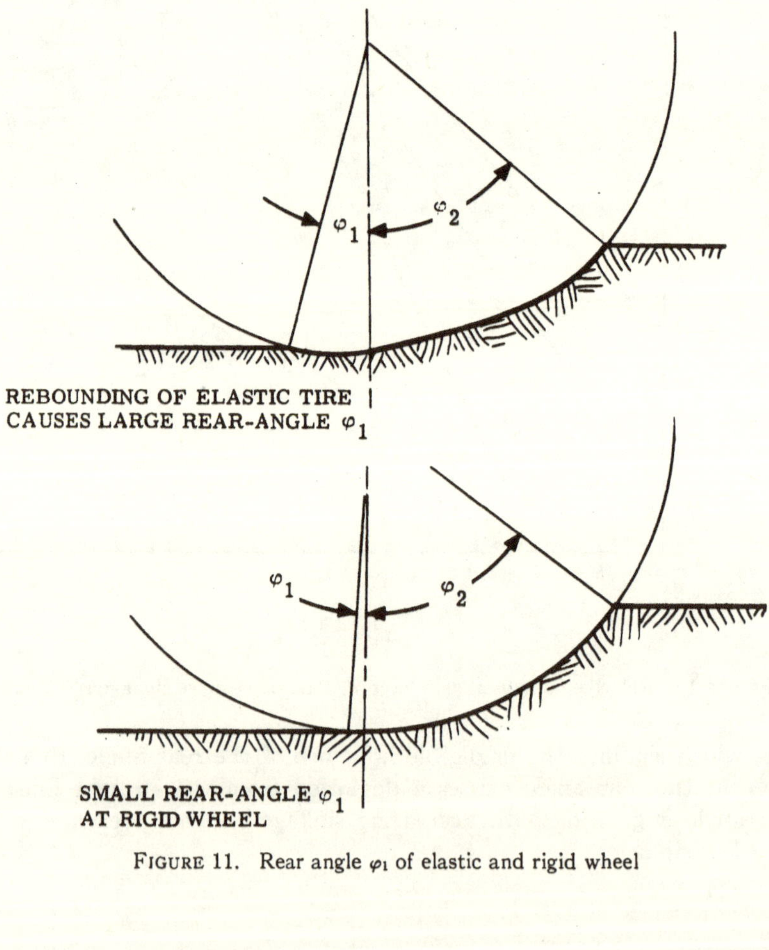

FIGURE 11. Rear angle φ_1 of elastic and rigid wheel

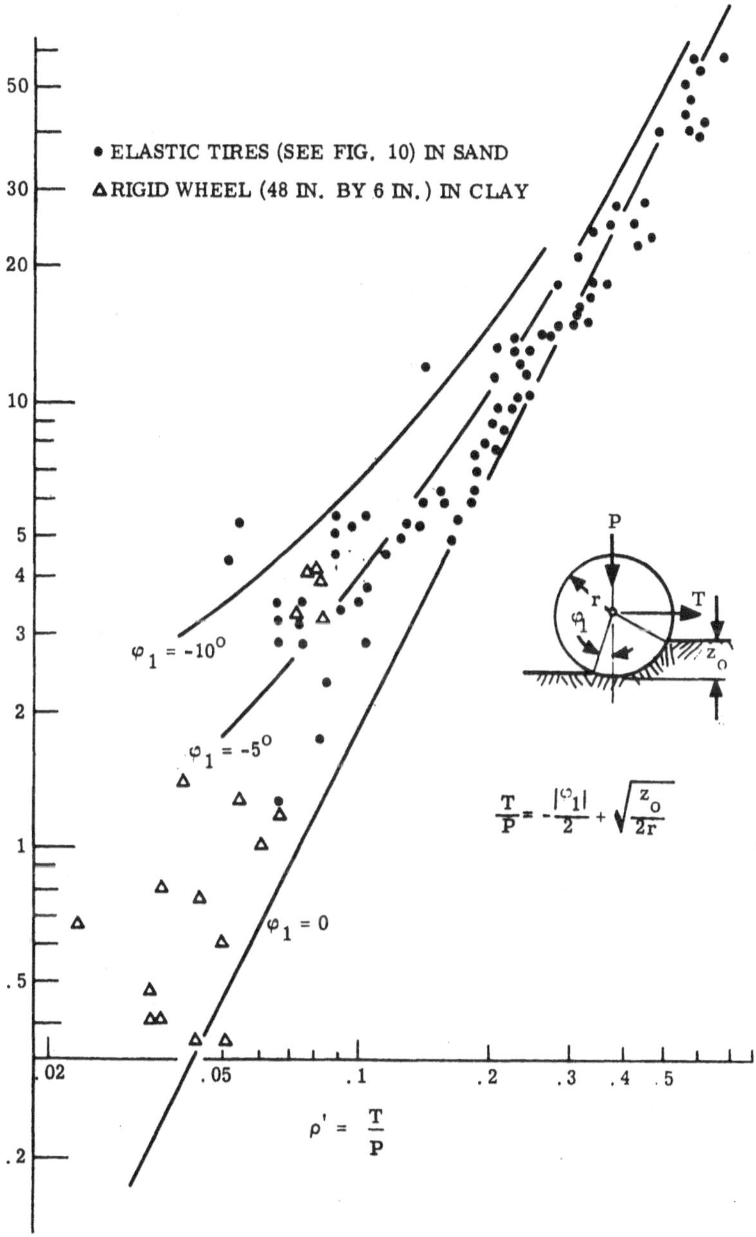

FIGURE 12. Relative horizontal force T/P vs. relative sinkage z_0/r (wheel pulled in sand and clay)

follow the predicted course down to 0.2 per cent, with indication of a very small rear angle.

Note that the data plotted in Figure 12 stem from rigid and elastic wheels of various sizes and with various inflation pressures operating on sand and clay. Despite this variety the performance of all wheels follows with satisfactory uniformity the predicted course, thus strengthening confidence in this approach. This is true also of the next two figures.

Figure 13 shows relative sinkage z_0/r as a function of the relative driving torque M/Pr for the same wheels as shown in Figure 10, except that these are driven by pure torque instead of being pulled by a horizontal force. In this case the basic Equation 57 takes the form:

$$\rho' = M/rP = \sqrt{z_0/2r} - |\varphi_1|/2 \quad \text{(see 57 with } T = 0\text{)}.$$

Again we plot a family of theoretical curves for different rear angles, together with experimental data. The scatter of experimental data is even wider than in Figures 10 and 12, but it is evident that the wheel size does not cause significant deviations from the predicted course. The rear angle, which is larger than that of the towed wheel, seems to reach $-10°$ and more.

Figure 14 shows relative sinkage z_0/r as a function of $\rho' = (M/r - T)/P$ plotted for the same three wheels and representing the total formula 57 for the driven and pulling wheel. As in Figures 10, 12, and 13, the scatter of experimental data is wide, but prediction and experiment do not deviate significantly.

STRESS DISTRIBUTION AND PLATE TESTS

Experimental checks proved the validity of Formula 58:

$$E/r\theta = Ts + P(\sqrt{z_0/2r} - |\varphi_1|/2),$$

within the limits of 5 per cent $< z_0/r <$ 60 per cent. In many instances, however, this formula does not apply, because sinkage z_0 and slip s are unknown. Both quantities could be determined analytically if we were familiar with the distributions of vertical stress p and horizontal stress t along the interface. Unfortunately, the available theory of stress distribution is crude, since it is based on plate penetration and plate shear tests. Nevertheless, because nothing better is available, we might use the "plate" approach, being always aware, however, of its shortcomings.

The current concept of stress approximation is based on two formulas derived from plate tests. One of them is the Bernstein–Bekker pressure-sinkage relationship (49), already discussed, which we transformed into a function of the angle φ (Equation 51): $\bar{p} = k(r/2)^n(\varphi_2^2 - \varphi^2)^n$. The

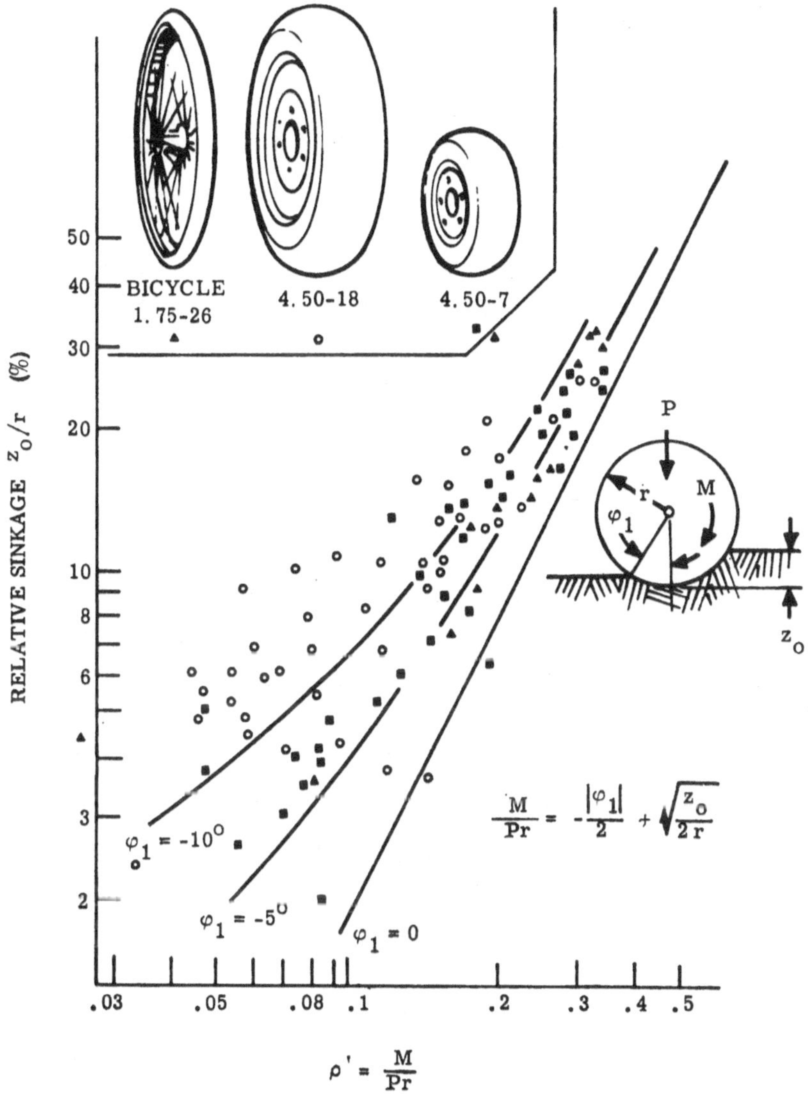

FIGURE 13. Relative driving torque M/Pr vs. relative sinkage z_0/r (wheel driven by pure torque in sand)

other, derived from horizontal shear plate tests, was first developed by Bekker (2) and later changed and simplified by Hanamoto and Janosi (10):

$$\bar{t} = (c + \bar{p} \tan \phi)(1 - e^{-j/n}). \tag{59}$$

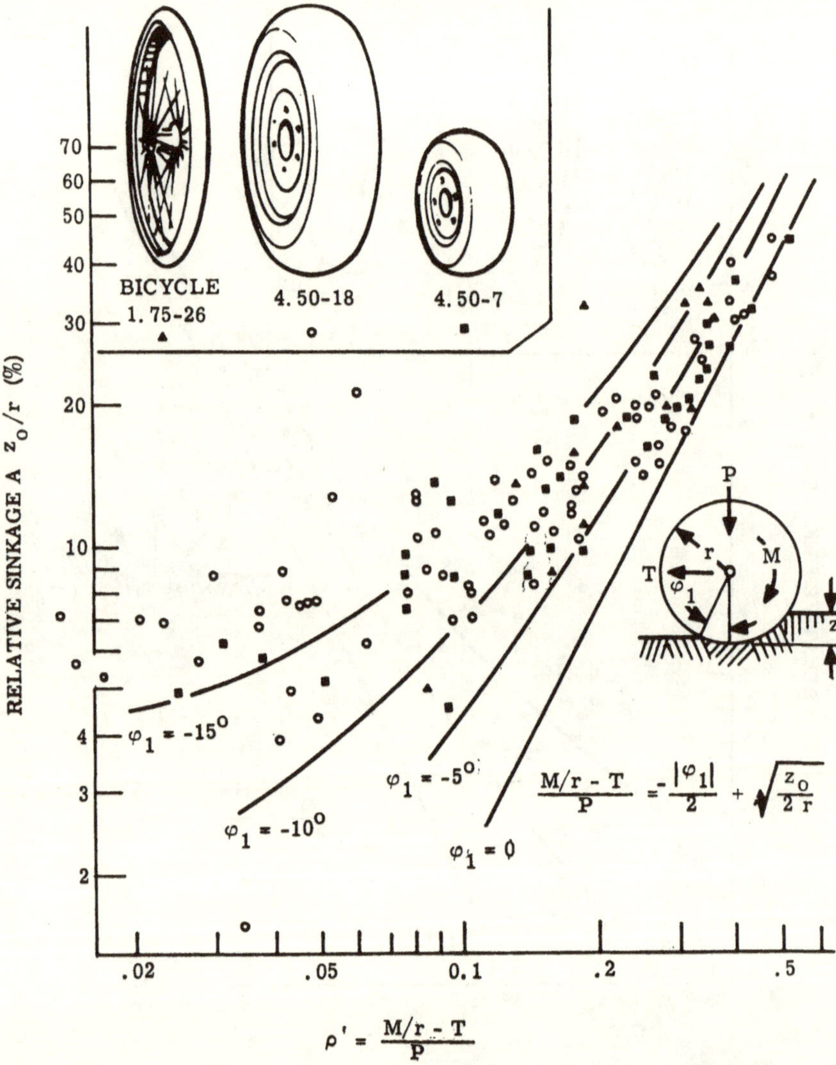

FIGURE 14. $M/r - T/P$ vs. z_0/r (wheel driven and pulled in sand)

Here, $\bar{t}$ is the average shear stress of the plate, K is the deformation modulus of the soil, j is horizontal displacement, c is soil cohesion, and ϕ is the angle of internal soil friction.

The displacement j is a function of the angle φ, but very little is known about how the soil particles are displaced along the contact area. We think that displacement at the beginning of the contact area ($\varphi = \varphi_2$) must be zero, and that at the end of the contact area ($\varphi = \varphi_1$) the wheel has slipped by the amount of $[rs(\varphi_2 - \varphi_1)]$. Furthermore, we find it conceivable to assume one-directional displacement along the interface. It is, then, only one step further to the assumption of linear displacement,

$$j = rs(\varphi_2 - \varphi). \tag{60}$$

This assumption agrees with the approach of Soehne, who maintains that displacement j of soil particles along the contact area increases linearly from the front to the rear of the wheel if the sinkage is not too large (11). With 51 and 60 substituted into the stress formula 59,

$$\bar{t} = [c + k(r/2)^n(\varphi_2^2 - \varphi^2)^n \tan \phi][1 - e^{-(rs/K)(\varphi_2 - \varphi)}]. \tag{61}$$

As a next step we equate the average stresses $\bar{p}$ and $\bar{t}$ with the local stresses p and t in the contact area, knowing that this may entail substantial errors. We are aware also that the rear angle φ_1 has to be neglected because the penetration plate test cannot take care of it. Furthermore, we accept an error of unknown magnitude by superposing stresses p and t.

Then we substitute the two stress formulas for $\bar{p}$ and $\bar{t}$ into the two wheel integrals, 27 and 28:

$$P(\varphi_2) = br\left(\frac{r}{2}\right)^n \int_0^{\varphi_2} (\varphi_2^2 - \varphi^2)^n d\varphi \, ; \tag{62}$$

$$T(\varphi_2, s) = br\int_0^{\varphi_2}\left[c + k\left(\frac{r}{2}\right)^n(\varphi_2^2 - \varphi^2)^n \tan \phi\right][1 - e^{-(rs/K)(\varphi_2 - \varphi)}]d\varphi. \tag{63}$$

Integral 62 has been solved already (see Formula 53):

$$P(\varphi_2) = kb(r^{n+1}/2^n)\varphi_2^{2n+1}(1 - 0.253n).^* \tag{64}$$

The Integral 63 can be solved either with high accuracy on a digital computer, or with low accuracy by assuming that the pressure is not variable but constant along the contact area. For instance,

$$p = P/br\varphi_2. \tag{65}$$

This leads to the well known solution (2),

$$T(\varphi_2, s) = (c/br\varphi_2 + P \tan \phi)[1 - (K/sr\varphi_2)(1 - e^{-sr\varphi_2/K})]. \tag{66}$$

One important feature of the equation for horizontal axle force T is that T is zero if slip s is zero, which means that a wheel driven by a pure torque has no slip. If this could be confirmed by actual wheel performance, the confidence in Formula 63 would be strengthened.

*This result differs from Bekker's (2). Instead of the correct value of $(1 - 0.253n)$ he arrived at $(1 - 0.333n)$, which is equivalent to truncating the series in 53 after the first term.

From the tests that Powell recently made of wheel performance on Yuma sand (4), we selected data from two tires driven by pure torque ($T = 0$) and plotted slip values (slip taken from hard-surface circumference) versus pertinent vertical loads (Figures 15 and 16).

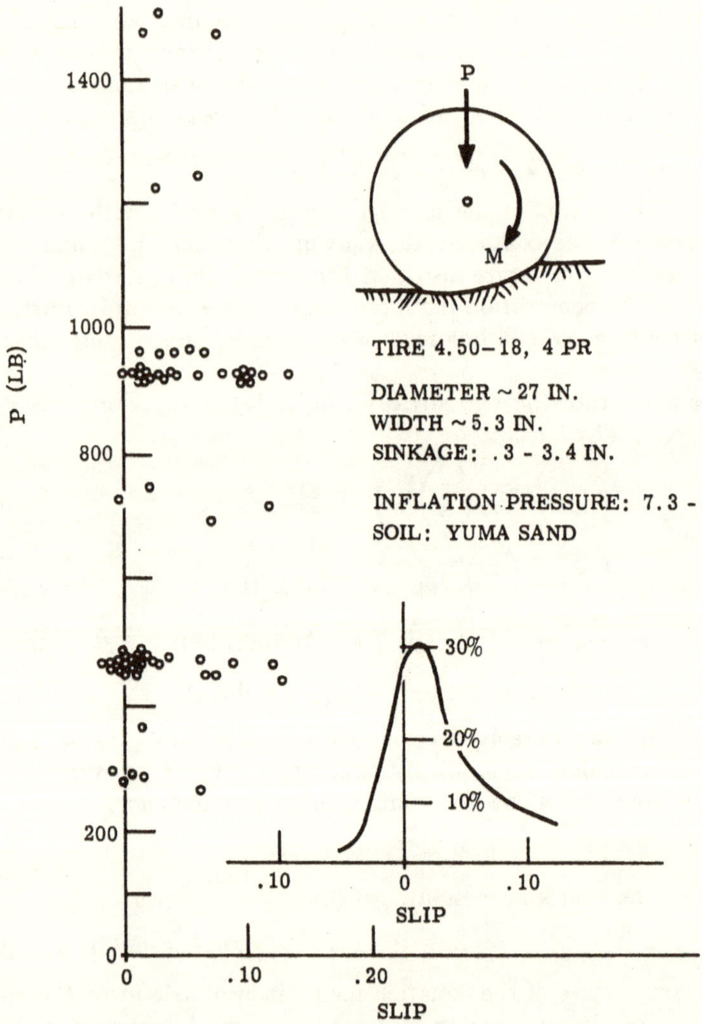

FIGURE 15. Slip of wheel driven by pure torque (tire 450-18)

Although tire loads, as well as inflation pressure, sinkage, and soil data, varied widely, it is evident from Figures 15 and 16 that the average slip of both wheels is low, thus backing up our approach. The distribu-

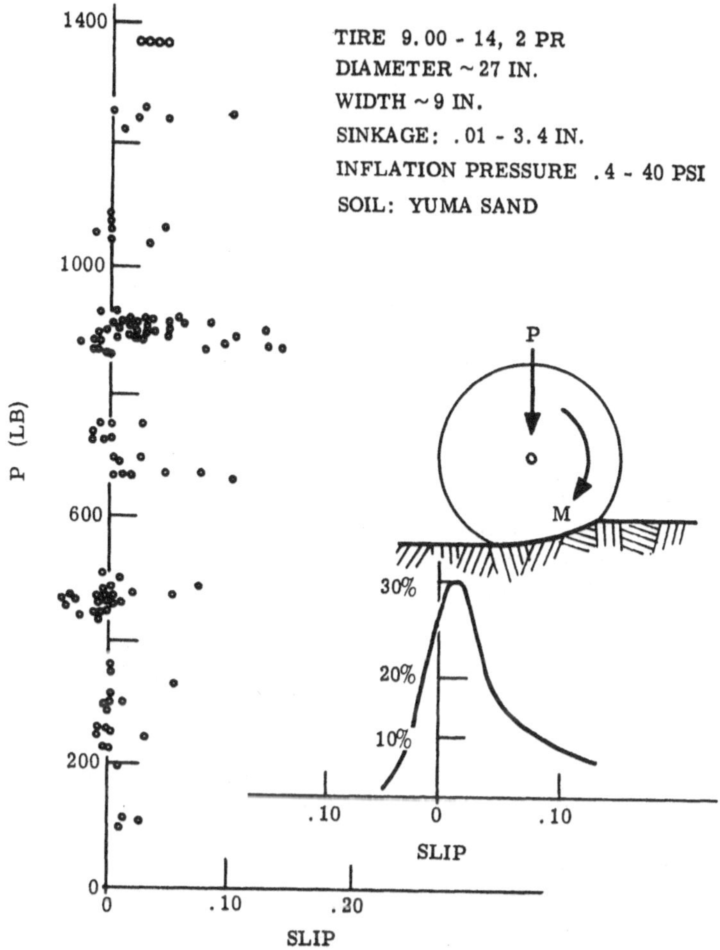

FIGURE 16. Slip of wheel driven by pure torque (tire 900-14)

tion of slip values is somewhat logarithmic, with a peak at approximately $s = 2$ per cent, as the distribution curves in each figure indicates. The distribution of slip values seems to be independent of load, at least in the plotted range. At large sinkages, however, such as at least half the wheel radius, slip could increase with load. Nonetheless, as stated, this paper deals only with small sinkages. Similar tests in clay or other soils are not known.

At this point it seems that the problem of determining energy loss in the wheel-soil interface is practically solved. In the following section we outline several ways of estimating energy loss.

SUMMARY AND CONCLUSIONS

To estimate energy loss quantitatively in the wheel-soil interface, the following assumptions and simplifications were made:

(1) Energy losses due to internal tire friction are not considered;
(2) No forces are acting parallel to the wheel axle;
(3) No forces are acting at the wheel sides; only forces at the wheel periphery are taken into account;
(4) The wheel is fairly stiff;
(5) The ground is smooth;
(6) The wheel tread is small.

Assumptions 4, 5, and 6 amount to the statement that deviations from the circular form of the wheel are negligible.

(7) Stresses in the contact area are constant over the width of the wheel;
(8) The interface is short;
(9) "Bow wave" is neglected.

Accepting these assumptions we derived Equation 58 for the energy loss due to wheel-soil friction:

$$E/r\theta = Ts + P(\sqrt{z_0/2r} - |\varphi_1|/2).$$

The term in parentheses is the coefficient of rolling resistance, Equation 57.

The coefficient of rolling resistance is used extensively to estimate energy loss of a tire on ground that is fairly solid, like streets and dry dirt roads. Numerical values of the coefficient, derived entirely from vehicle tests and listed as a function of the road conditions, vary between 0.01 and 0.15, depending on the roughness and hardness of the ground. Estimates of the coefficient of rolling resistance for soft ground have been vague, but now they can be improved by using relationship 57. Ordinarily, the coefficient ρ encompasses energy losses due to both soil friction and internal wheel friction. Formula 57 takes care only of losses due to soil friction, but by adding the term ϵ, we can establish the conventional coefficient of rolling resistance describing the total loss in the wheel contact area; ϵ lies between zero and 0.04 depending on the nature of the tire.

$$\rho = \rho' + \epsilon = \sqrt{z_0/2r} - |\varphi_1|/2 + \epsilon. \tag{57}$$

Figure 17 depicts this coefficient of rolling resistance as a function of relative sinkage for different rear angles and ϵ-values. For a rigid wheel, the rear angle φ_1 and the ϵ-value are zero; both parameters increase with increasing wheel softness.

We feel that the curves should not be extended beyond a relative sinkage of around 60 per cent, because the assumption of small angles

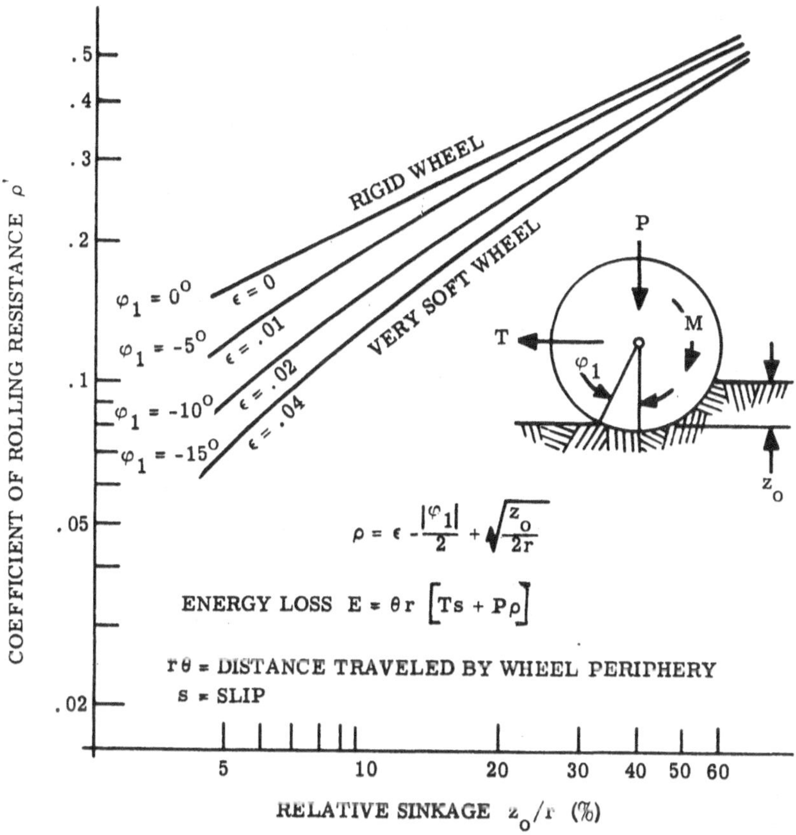

FIGURE 17. Coefficient of rolling resistance for soft soil

may not carry further without substantial error. On the other hand, approximation of energy loss may lose validity at sinkages lower than around 5 per cent, because we lack pertinent experimental data.

Although this approach to estimating energy loss is fairly accurate, it presupposes knowledge of sinkage z_0 and slip s. If both quantities are unknown, they must be estimated analytically by introducing stress distribution $p(\varphi)$ and $t(\varphi)$ along the interface. In this study we adopted two stress distributions derived from plate tests; accepting these entails further assumptions and simplifications as follows:

(1) The contact area does not reach beyond the toe of the wheel; that is, $\varphi_1 = 0$.

(2) Stresses in the two "plate" distributions are equated with local stresses in the soil-wheel interface.

(3) The two stress distributions were derived independently from two different plate tests; we assume linear superposition.

(4) Soil displacement along the contact area is proportional to slip and contact angle.

These assumptions lead to the desired determination of wheel loss as a function of slip s and sinkage expressed by the angle φ_2.

To calculate the three unknown quantities E, s, and φ_2, we use the following equations:

$$E(s, \varphi_2) = \theta r[T(s, \varphi_2)s + R(\varphi_2)]; \tag{33}$$

$$R(\varphi_2) = br \int_{\varphi_1}^{\varphi_2} p\varphi d\varphi = b(r/2)^{n+1} k \frac{\varphi_2^{2n+2}}{n+1} \quad \text{(see 31 + 52)} ;$$

$$T(s, \varphi_2) = br \int_{\varphi_1}^{\varphi_2} \left[c + k \left(\frac{r}{2}\right)^n (\varphi_2^2 - \varphi^2)^n \tan \phi \right] \times [1 - e^{-(rs/K)(\varphi_2 - \varphi)}] d\varphi ; \tag{63}$$

or:

$$T(s, \varphi_2) = [(c/br\varphi_2) + P \tan \phi][1 - (K/sr\varphi_2)(1 - e^{-sr\varphi_2/K})] ; \tag{66}$$

$$P(\varphi_2) = kb(r^{n+1}/2^n)\varphi_2^{2n+1}(1 - 0.253n). \tag{64}$$

These present a crude analytical solution to the energy loss problem. More measurements of stress distribution as a function of wheel and soil parameters and more thinking will be necessary to replace this solution with a better one.

REFERENCES

1. Phillips, J., A Discussion of Slip and Rolling Resistance. *Proc. 1st Int. Conf. Mech. Soil-Vehicle Systems.* Torino-Saint Vincent, 1961.
2. Bekker, M. G., *Theory of land locomotion.* Ann Arbor: The University of Michigan Press, 1956.
3. Hegedus, E., Pressure Distribution under Rigid Wheels. *Transactions of the ASAE,* 1965, p. 305.
4. Powell, C. J., *et al.*, Performance of Soils Under Tire Load. U.S. Army Eng. Waterways Experiment Station, Corps of Eng., Vicksburg, Miss. *Tech. Report 3-666,* 1965.
5. Sela, A. D., The Shear-to-Normal-Stress Relationship Between a Rigid Wheel and Dry Sand., U.S. Army Tank-Automotive Center, Warren, Mich. Land Locomotion Lab., *Report No. 8524,* June 1964.
6. Hegedus, E., A Preliminary Analysis of the Force System Acting on a Rigid Wheel. U.S. Army Tank-Automotive Center, Warren, Mich. Land Locomotion Lab., April 1962.
7. Uffelmann, F. L., The Performance of Rigid Cylindrical Wheels on Clay Soil. *Proc. 1st Int. Conf. Mech. Soil-Vehicle Systems.* Torino-St. Vincent, 1961.
8. Smith, M. E. and D. R. Freitag, Deflection of Moving Tires. U.S. Army Eng. Waterways Experiment Station, Corps of Eng., Vicksburg, Miss. *Tech. Report 3-516,* 1965.
9. Smith, M. E., *et al.*, Tests with Rigid Wheels: Tests in Fat Clay. U.S. Army Eng. Waterways Experiment Station, Corps of Engs., Vicksburg, Miss., *Tech. Report 3-565,* 1961.
10. Janosi, Z. and B. Hanamoto, The Analytical Determination of Drawbar Pull as a Function of Slip for Tracked Vehicles in Deformable Soils. *Proc. 1st Int. Conf. Mech. Soil-Vehicle Systems.* Torino-St. Vincent, 1961.
11. Soehne, W., Wechselbeziehung zwischen Fahrzeuglaufwerk und Boden beim Fahren auf unbefestigter Fahrbahn. *Grunlgn. Landtechn.* Heft 13, 1952, S.21–34.

Soil Failure Beneath Rigid Wheels

JO-YUNG WONG*

A. R. REECE*

ABSTRACT

Current theories of wheel performance are based on pressure sinkage laws and usually disregard the actual interaction between wheels and soil. In order to reach a better understanding of the behaviour of rigid wheels on soil, it is necessary to investigate the nature of the soil failure beneath wheels. It is shown, in this paper, that wheels displace soil partly sideways and partly by longitudinal flow. The flow depends on the slip and causes rut filling behind the wheel and a bow wave in front. Photographic evidence shows that there are normally two failure zones beneath a wheel, one forward and one backwards. These degenerate into a single backward zone at 100 per cent slips and a single forward zone for a locked wheel. The relationships between soil failure and wheel performance are also discussed.

THE PREDICTION of the performance of rigid wheels on soil is one of the basic problems in the field of soil vehicle mechanics, for two reasons. First because the high inflation pressure pneumatic tires of many cross-country vehicles are effectively rigid on soft soils. Second, because the rigid wheel is a simpler starting point for investigation than a soft tire with its complex geometry that varies with soil conditions as well as load and slip.

A considerable effort has been put into the investigation of rigid wheel performance. This has been based on the assumption that the drawbar pull at a particular slip is equal to the gross thrust which is a function of slip minus the rolling resistance which is generally assumed to be independent of slip. Most of the work that has so far been carried out has concentrated on the prediction of rolling resistance and considerable success has been achieved.

Current rolling resistance theories are based on pressure sinkage laws, an idea that originated with Bernstein and which has been developed principally by Bekker (1) for sand and Uffelmann (2) for clay. These theories assume that the reaction between vehicle wheels or tracks and soils is similar to that of a plate in a penetration test, in that the radial

*Department of Agricultural Engineering, University of Newcastle upon Tyne.

pressure acting on an element of the rim of a wheel will be the same as that beneath a plate penetrated to the same depth. Wills (3) has shown that this is basically an energy approach in which it is assumed that the rolling resistance of a wheel is simply due to the vertical work done on the soil in making a rut. These theories have recently been investigated by Wills (3) who showed that in sand Bekker's theory and in clay Uffelmann's theory both worked quite well. There is little doubt that these theories do give the right order for the rolling resistance although quite considerable error can be involved. The nature of the error has been investigated experimentally by Vincent (4), Wills (3), Hegedus (5), Onafeko (6), and other research workers. Their experiments have shown that rolling resistance must be considered as a function of slip. They also show that the pressure distribution around the wheel is not what would be expected from the plate sinkage analogy and that the actual pressure distribution is such that the rolling resistance is increased. Since this pressure distribution has to be used in calculating the gross tractive effort from the wheel it will result in a double source of error if it is not correct.

In order to reach a better understanding of the behaviour of rigid wheels on soil it is necessary first of all to find out by careful experiment the nature of the actual interaction between wheels and soil. Only when this has been done can a theoretical analysis of the problem be made.

This paper describes such an experimental study. It first considers infinitely wide wheels or rollers because they are a two-dimensional simplification of the normal three-dimensional problem. It then goes on to deal with narrow wheels of varying aspect ratio. Photographic methods have been employed to investigate the nature of soil failure and soil displacements have been compared with those beneath rigid plates.

APPARATUS AND EXPERIMENTAL METHODS

The principal requirement was an apparatus in which it was possible to drive small wheels of varying aspect ratio at varying slips close up against a glass plate. The nature of the soil failure could then be photographed through the plate. Figure 1 shows a satisfactory apparatus for this purpose. The soil box is 36 in. long, 12 in. deep and 7 in. wide; it is normally fitted with a $\frac{1}{2}$ in. plate glass side and for work with wide rollers a similar glass plate can be fitted at the back. The wheel or roller is hauled along the stationary box by means of a cable from a hydraulic cylinder. Various slips can be provided by anchoring the centre of a cable wrapped round two suitable pulleys fixed one on each side of the test wheel. The wide rollers were loaded by dead weights applied at either side as shown in the figure and the narrow wheels were operated at con-

FIGURE 1. The forced-slip apparatus, showing a roller set up for a high slip test on a fixed soil box, with two strain gauge dynamometers and six channel recorder.

stant sinkage by means of a horizontal guide for the axle. Provision for measuring drawbar pull and input torque was made using the two strain gauge dynamometers and multi-channel recorder shown in the photograph. These measurements were made in all the experiments but have not been described in this paper. The box was mounted on a carriage and it was possible to hold the wheel still and move the soil; this was used in the infinite skid experiments. The slip or skid was measured in all the experiments and is recorded on all the experimental results given in the paper. The percentage of slip is defined as: the wheel circumference, minus the distance travelled per complete revolution, divided by the circumference, all multiplied by one hundred. This definition results in a locked wheel being dragged along having an infinite skid or negative slip.

The wheels were all of $8\frac{1}{4}$ in. diameter and widths of $\frac{1}{4}$ in., $\frac{1}{2}$ in., $\frac{3}{4}$ in., and $1\frac{1}{2}$ in. were provided. When these were run close up to the glass side of the box the glass represented a plane of symmetry and the wheels behaved as if they were twice as wide. The wheel surface was made perfectly rough by sticking on the same sand as was used as the soil in the experiments. The wheels were made of wood but it was found that this caused trouble because sand penetrated between the side of the wheel and the glass. The wheels were therefore all fitted with a polished aluminium side plate. A wheel $4\frac{1}{2}$ in. wide running between two glass plates was used to represent an infinitely wide roller.

In most of the experiments the box was held stationary adn the wheel moved. The moving wheel caused failures of masses of sand within certain zones. These zones of moving sand can be clearly observed in a photograph taken by a fixed camera. In the resulting film the main mass of sand in the box shows up its individual grains in sharp zones while the moving sand in the zones influenced by the wheel appear as streaks. The changing directions and lengths of the streaks give a clear picture of the direction of soil flow.

The exposure time for the photograph must be sufficient to give adequately long streaks. If it is too long the wheel movement becomes excessive and the shape of the zones is distorted because the photographs show the zones with an appreciable movement. It was found satisfactory to move the wheel at a speed of 0.8 in. per sec. and it was necessary to use an exposure time depending on the degree of slip of the wheel. For a towed wheel the sand in the backward failing zone moves quite slowly and an exposure of $\frac{1}{2}$ sec. was required, for the driven wheels $\frac{1}{5}$ sec. was sufficient. A high level of illumination was achieved with two 500 watt flood lamps resulting in an aperture of f11 when using Ilford FP3 film (speed ASA125). A 4 × 5 precision plate camera was used.

A subsidiary experiment was carried out to determine whether or not the glass plate was capable of taking the place of a plane of symmetry. A rectangular footing with its long axis perpendicular to the glass plate was pushed into the soil and the soil failure pattern photographed. The plate was then pushed into the soil again but this time with another glass plate held firmly against one of its long edges. The soil failure pattern was again photographed and found to be half that from the first photograph. These two photographs are shown in Figure 2. A further experiment was carried out in which the footing was pressed into the soil with its long edge against the glass box. This photograph showed clearly the depth of the wedge that existed beneath the footing.

Work on another problem proceeding at the same time in the department showed that in general the friction between glass and sand could not be ignored. Measurements made by B. Witney showed that while the sand had a value of tan ϕ of about 0.9 it had a value of tan δ relative to glass of about 0.3. He showed that for large sinkages this could have a serious influence on the shape of failure zones against the surface of the glass, but that this could almost be eliminated by means of suitable lubricants. Some experiments were carried out with the wheels using lubricated glass and it was found that in these particular circumstances the sand/glass friction had a negligible effect.

The soil used was round particle, sieved, washed Leighton Buzzard sand, 95 per cent of which was retained on a 18 mesh sieve (aperture 0.034 in.) and all of which passed through a 8 mesh sieve (aperture 0.08 in.). It was used in a dense packing for which the angle of internal shearing resistance was 40.2°, measured with a 6 cm. square translational shearbox.

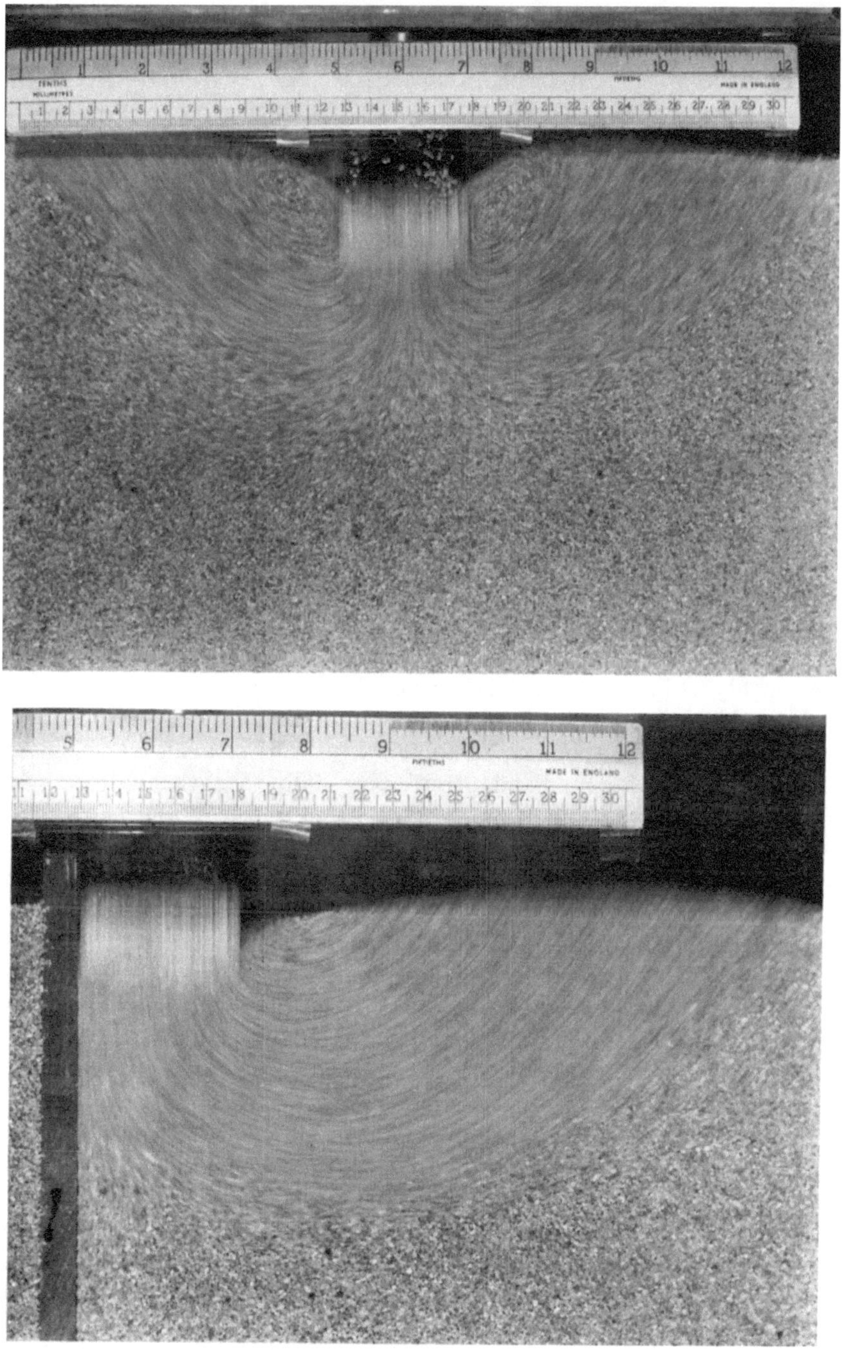

FIGURE 2. Results of the preliminary experiment designed to show that a sheet of glass can accurately represent a plane of symmetry

SOIL FAILURE UNDER WIDE WHEELS OR ROLLERS

The wide wheel ran between two glass plates which restricted the soil movement to a longitudinal plane and this, therefore, represents the case of a very wide roller. The soil was in its most dense packing and no compression was possible. It therefore follows that the soil surface has to be at the same height after the roller as before it. This requires a build up of a particular size of bow wave in front of the roller and this in turn needs a considerable forward movement of the roller before an equilibrium position is attained. This was not readily possible in the restricted length of box available and it was necessary to place the right shape of bow wave in front of the roller if equilibrium was to be obtained before the end of the box was reached. Because the soil is incompressible sinkage of the roller is only possible if it can produce a velocity of flow in the soil beneath it. This is illustrated in Figure 3. At 3 a wheel is shown rotating about an instantaneous centre I with a forward velocity of the axle v. The relative velocity of the soil to the axle at point 1 well in front of the wheel is therefore v and it is uniformly distributed down the depth of soil. At point 2 directly under the axle the stationary soil will be driven backwards by the shear stresses between the soil and the slipping wheel. The relative velocity distribution in the soil will therefore be roughly as indicated in the diagram. The line ab represents any plane below which the wheel has no influence on the soil. The mean relative velocity of the soil above ab at point 2 is v_2. The soil can be considered as an incompressible fluid and a continuity of flow equation written down for points 1 and 2. This is, $vz_1 = v_2z_2$. From which it follows that the sinkage of the roller z is given by: $z = z_1 - z_2 = z_1(1 - v/v_2)$.

From the foregoing analysis it is seen that for the case of the wide roller sinkage is impossible unless v_2 is greater than v. That is, sinkage must be a function of slip and must be zero when slip is zero. This is well borne out by Figures 4 and 5 showing the towed roller carrying varying weights with varying small degrees of skid. In each case there is no sinkage at all. It is important to notice the vital part played by the bow wave which in the case of the towed roller must be large enough to completely support the weight carried on the roller.

Figures 6 and 7 show the roller at two different slips and carrying various loads. They clearly show that the sinkage in this case increases with both load and slip. They also show that the continuity of soil flow relative to the roller is maintained and no rut is left. That is, the slipping roller imparts sufficient velocity to the soil beneath the wheel to pump the soil from the front to the back so that the rut totally recovers from the passage of the roller.

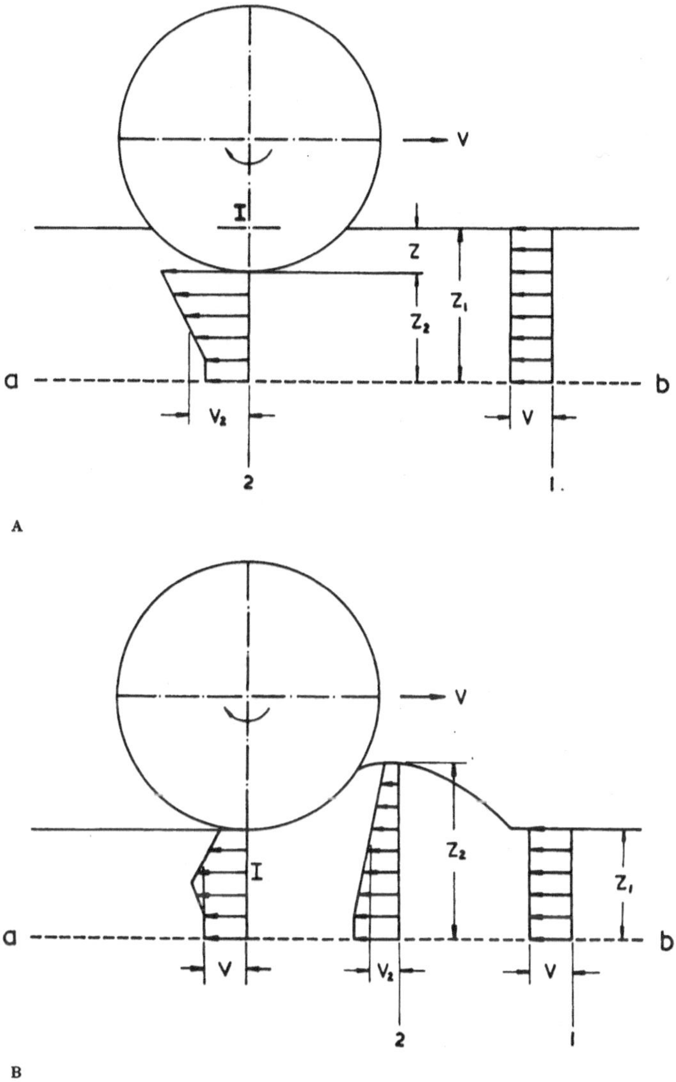

FIGURE 3. The distribution of soil velocities relative to the centre of a wide roller. A, driven, rolling with slip; B, towed, rolling with skid

Figure 7 shows that at a constant slip the sinkage increases with the load but that the depth of soil moved by the wheel remains virtually constant. This requires that the soil within this zone must have a velocity distribution which alters with increasing weight. It is interesting to observe that the increasing load is also carried out by an increasing size

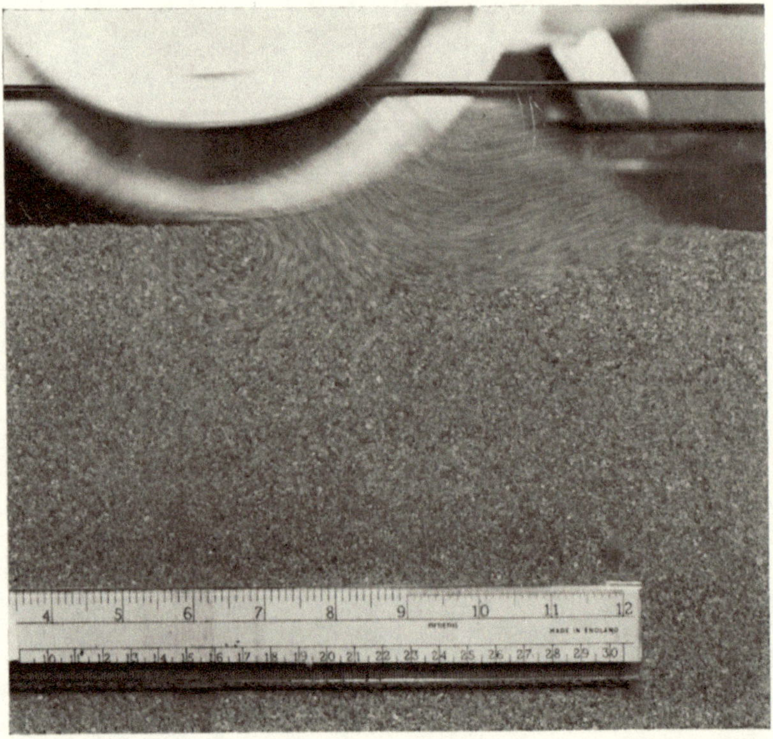

A

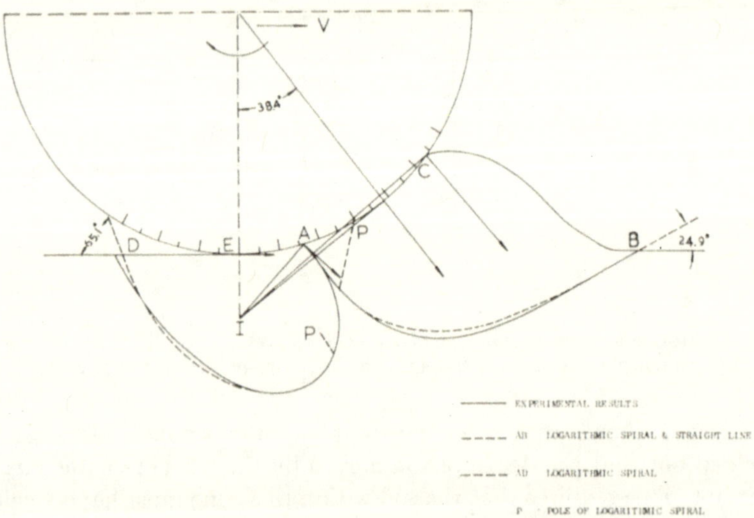

B

FIGURE 4. A, Photograph showing soil movement beneath a wide towed roller. Diameter $8\frac{1}{4}$ in. Width $4\frac{1}{2}$ in. Load 123.4 lb. Skid 28%. B, Diagram relating the failure planes in the photograph to the absolute rim velocity vectors and also comparing the actual failure surfaces with the logarithmic spirals and the straight lines of theoretical active and passive zones

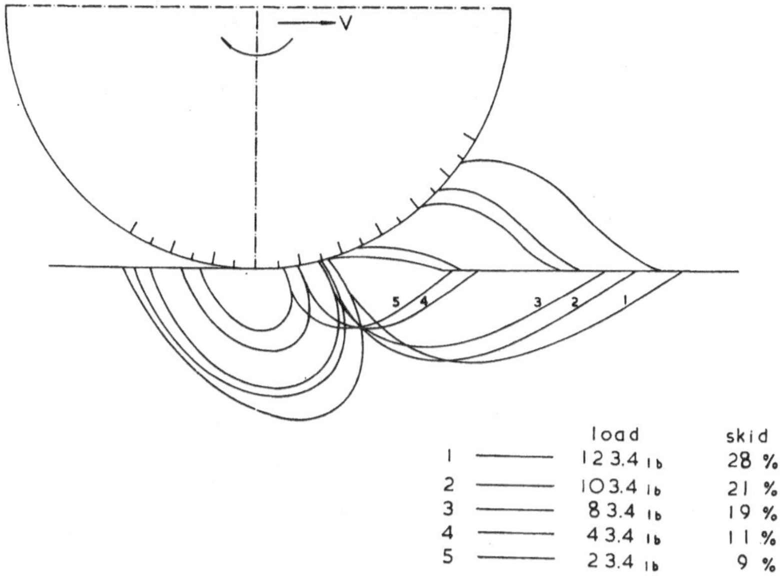

FIGURE 5. Diagram showing photographically determined failure zones and bow waves for towed rollers carrying different loads. Diameter $8\frac{1}{4}$ in. Width $4\frac{1}{2}$ in.

of bow wave. Figure 7B shows the same series of loads but at a higher slip, and it will be seen that the sinkages have increased roughly in proportion to the slip.

Figure 4 for the towed skidding wheel shows that soil beneath the bow wave and below the original surface is pushed forward by the roller. This requires a velocity distribution similar to that in Figure 3B, in order to maintain the continuity of flow at the increased depth of soil beneath the bow wave. This velocity distribution requires the mean speed of the bow wave to be less than that of the roller (this is because the forward moving part of the bow wave extends beneath the surface). This implies that the roller is continually overtaking the bow wave and pushing it down to replace the soil pushed up from beneath the bow wave.

The velocity distribution directly beneath the axle must be like that shown in Figure 3B because the zero sinkage requires that the average relative velocity at this point is v. The forward soil velocity due to the skidding rim is compensated for by the backward velocity deeper down in the soil. The circulation of soil beneath the wheel is therefore necessary in order to retain the continuity of flow.

These experiments give a very clear picture of what is almost certainly the main mechanism of slip sinkage beneath rigid wheels. Previously the only explanation for this phenomena had been presented by one of the authors in relation to tracks (7). This previous explanation depended

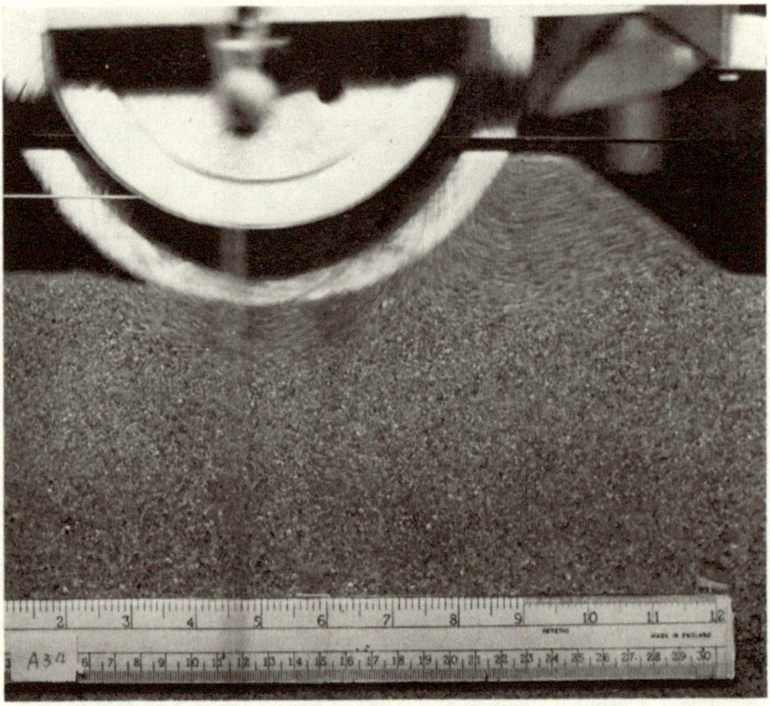

A

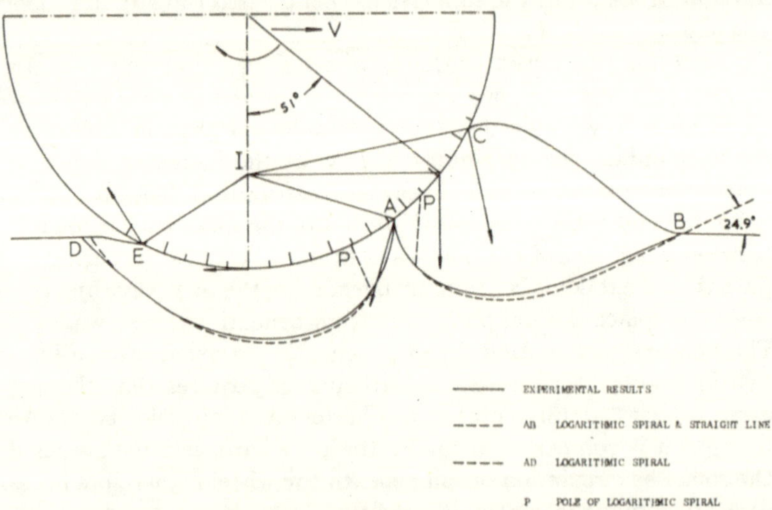

B

FIGURE 6. A, Photograph showing soil movement beneath a wide driven roller. Diameter $8\frac{1}{4}$ in. Width $4\frac{1}{2}$ in. Load 123.4 lb. Slip 37 per cent. B, Diagram relating the failure planes in the photograph to the absolute rim velocity vectors and also comparing the actual failure surfaces with the logarithmic spirals and straight lines

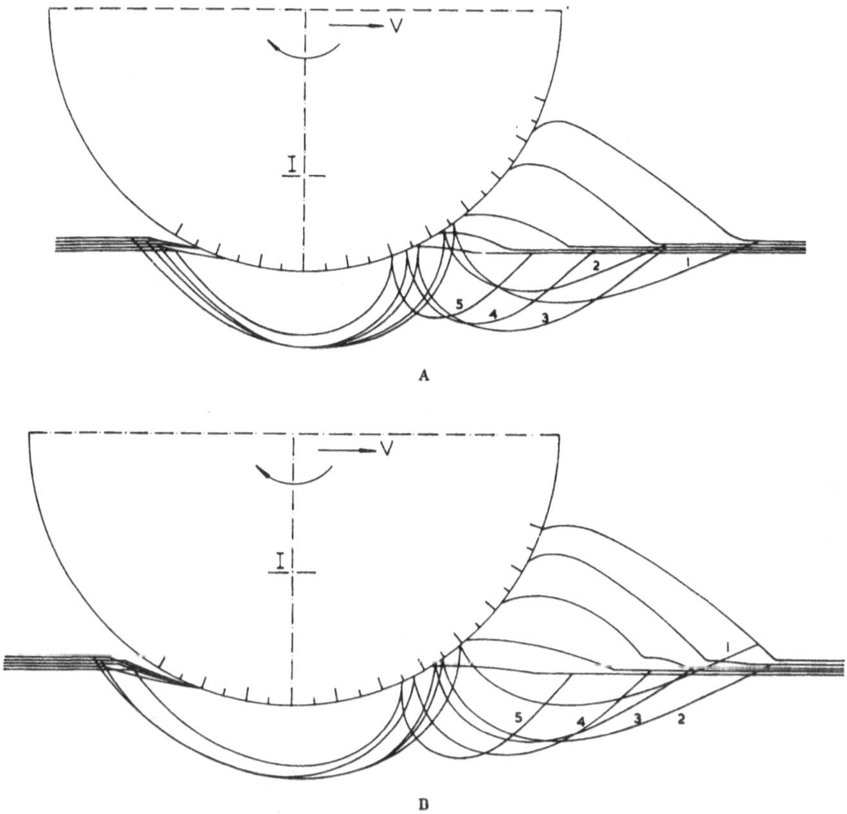

FIGURE 7. Diagram showing photographically determined failure zones and bow waves for driven rollers carrying different loads. Diameter 8¼ in. Width 4½ in. Loads: curve 1, 123.4 lb.; curve 2, 103.4 lb.; curve 3, 83.4 lb.; curve 4, 63.4 lb.; curve 5, 43.5 lb. A, Slip 37 per cent. B, Slip 48 per cent

upon a stress analysis of the soil beneath the track which showed that applying horizontal loads reduced its ability to support vertical loads. The phenomena described in the previous paragraphs is entirely different. However, it is closely akin to the phenomenon of excavation sinkage previously described by one of the authors in relation to lugged wheels and tracks (8). Excavation sinkage also has the same characteristic of being completely recovered in the rut. This phenomena of rut recovery has been pointed out by Uffelmann (9) in relation to rigid wheels operating on clay soils.

Figures 4 to 7 show that in all cases there existed two distinct soil failure zones beneath the roller. These are clearly shown in Figure 4B. The zone AC consists of soil which tries to avoid the roller by means of a forward movement in response to the general forward and downward movement of the section of the roller rim AC. AC acts on the soil in a similar way to the side of the soil wedge that forms beneath a footing when it fails. Figure 4B shows that the resulting passive failure closely conforms to the zone of radial shear bounded by a logarithmic spiral and a Rankine passive zone that would be expected beneath the footing. This soil failure forward leaves an active state in the soil behind the roller axle and the zone AD is formed, bounded by a logarithmic spiral which meets the soil surface at the correct angle for an active failure. It should be noted that the wheel rim velocity vectors in the region *EA* are forward and downward. Nevertheless the soil which moves forward when it is adjacent to the roller still finds it easiest to escape the path of the roller by moving backwards by means of a rotation approximately about the instantaneous centre of the wheel movement.

The relative size of the two zones depends on the kinematics of the roller motion which is entirely described by its slip. Figure 8B shows that at 100 per cent slip there only exists the single backward zone. Figure 8A shows on the other hand that with the infinite skid (roller locked against rotation) there is only the single forward zone. It should be noted that the roller in Figure 8A is not in equilibrium, because this photograph was taken before the importance of attaining equilibrium was appreciated. The locked roller is equivalent to a backward raked blade which can be expected to have a very large draught. The soil can arrange to reduce this draught by converting the backward raked blade into a vertical one made of soil by means of the fixed soil body clearly shown in Figure 8A. This triangular body attached to the roller effectively streamlines it.

SOIL FAILURE UNDER NARROW WHEELS

A conventional vehicle wheel differs from the roller previously discussed in that it is possible for the soil to escape from the path of the wheel by sideways movement. Existing wheel theory assumes that this is the only kind of soil movement that takes place, although reference is sometimes made to an additional bulldozing resistance at deep sinkages. It assumes that the pressures under a wheel are directly related to those beneath a narrow footing for which only wideways failure is possible. We have seen that with wide wheels the failure must be entirely longitudinal. The following part of the paper describes the second part of the investigation which was intended to find out to what extent longitudinal failure occurs with narrow wheels.

Figure 9 shows a range of wheel widths at different slips. Since the glass represents a plane of symmetry these model wheels represent wheels twice as wide operating in a normal situation without the glass. The

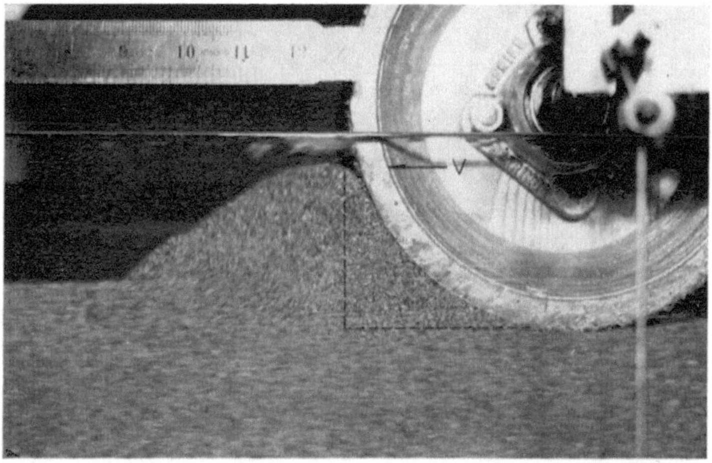

A

B

FIGURE 8. A, Photograph showing soil movement beneath a wide locked roller (infinite skid). Diameter $8\frac{1}{4}$ in. Width $4\frac{1}{2}$ in. Load 123.4 lb. B, Photograph showing soil movement beneath a wide driven roller with 100 per cent slip. Diameter $8\frac{1}{4}$ in. Width $4\frac{1}{2}$ in. Load 43.4 lb.

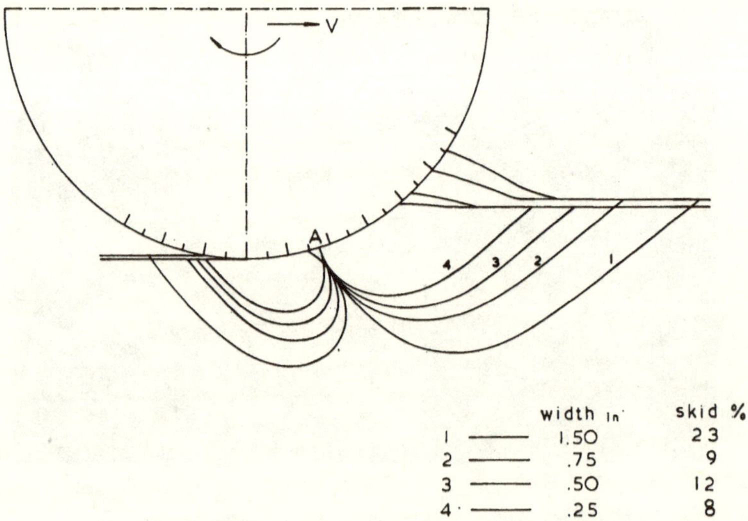
V
A
4
3
2
1
width in
skid %
1 1.50 23
2 .75 9
3 .50 12
4 .25 8

9A

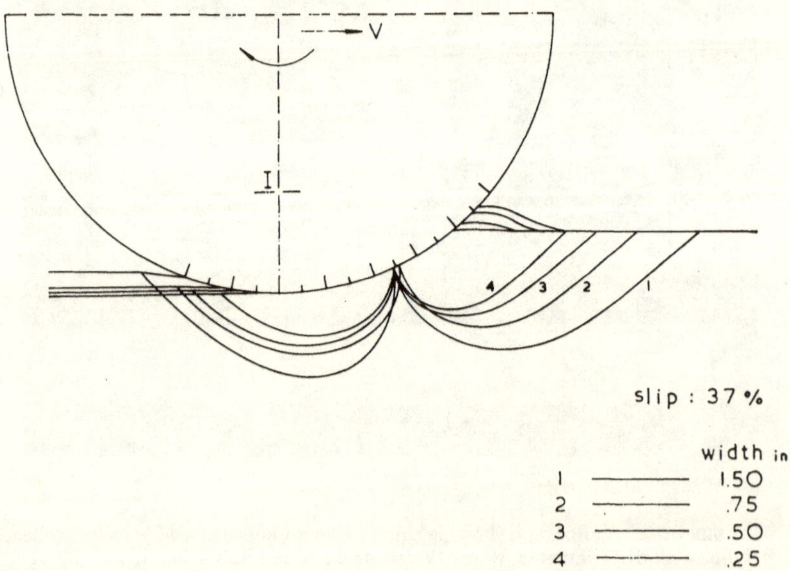
V
I
4
3
2
1
slip : 37 %
width in
1 1.50
2 .75
3 .50
4 .25

9B

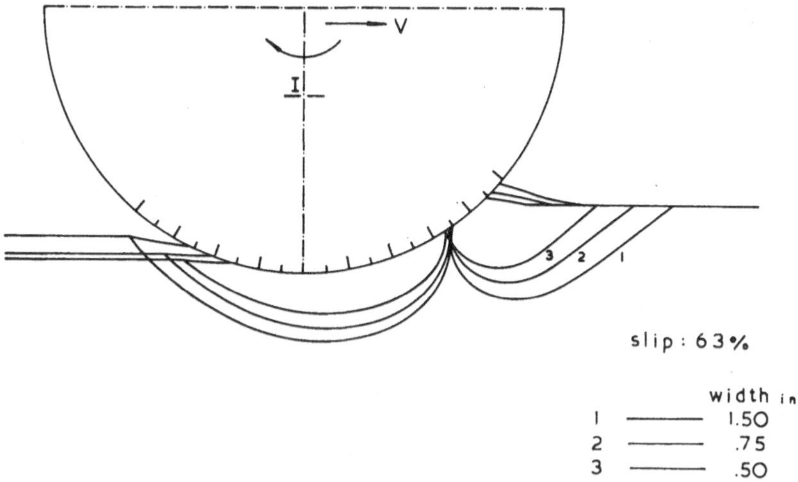

9c

FIGURE 9. Diagram showing photographically determined failure zones for narrow wheels with different widths at fixed sinkage. Diameter $8\frac{1}{4}$ in. Towed, B, Driven, slip 37 per cent. C, Driven, slip 63 per cent

narrowest $\frac{1}{4}$ in. wide wheel therefore represents a $8\frac{1}{4}$ in. $\times$ $\frac{1}{2}$ in. wheel which has about the proportions of a bicycle wheel and the widest $1\frac{1}{2}$ in. wheel represents a $8\frac{1}{4}$ in. $\times$ 3 in. wheel which corresponds to the widest type of conventional cross-country tire. These experiments were carried out at fixed sinkage, because this was more convenient than controlled load tests. However, these could be carried out using a wider box with glass on each side and using a pair of identical wheels. Since each part of Figure 9B and C represents a fixed slip and a fixed sinkage it is not surprising that the point between the two zones is fixed in each figure. However, the point does clearly shift forward with increasing slip as is the case with the roller. There is little reason to doubt that one zone would be formed at 100 per cent slip. Figure 10 shows a locked wheel with the single forward zone and the streamline wedge. The two failure patterns are seen to grow with increasing width and in no case is the bow wave of any great significance although it does increase with width of the wheel.

The phenomenon of rut recovery is clearly illustrated and is seen to grow with increasing slip and increasing wheel width. It is negligible for the towed wheel but is as large as 37 per cent for the widest wheel operated at 37 per cent slip and is 50 per cent for this wheel at 63 per cent slip.

Even the narrowest wheel exhibits longitudinal phenomena having two

FIGURE 10. Photograph showing soil movement beneath a narrow locked wheel (infinite skid) at fixed sinkage. Diameter 8¼ in. Width ¾ in.

clear zones. It is clear that both sideways and longitudinal failure occur together and that some of the soil under the edges of the wheel goes sideways while that under the centre is pumped backwards. The relative extent of the two mechanisms depends on the width of the wheel and the degree of slip. One would therefore expect a wheel to behave, as far as sideways flow is concerned, as if it was a much narrower wheel. The higher the slip the narrower the wheel. This is borne out by the measured rut profiles shown in Figure 11. It is interesting to note how the rut recovery falls off at the edges of the rut where sideways soil movement is taking place.

The observations made suggest a picture of soil deformation directly below the axle similar to that shown in Figure 12. This represents the ¾ in. wide wheel at 37 per cent slip. Directly beneath the wheel surface is a zone formed of a rectangle bounded by two triangles, that is pumped backwards by the effect of slip. The two triangular parts cause sideways failure as shown. The size of the sideways failure is drawn in correctly as would be expected from bearing capacity theory as is the failure pattern for the corresponding strip foundation. It will be seen that this assumed failure mechanism results in a rut profile very similar to that obtained in practice.

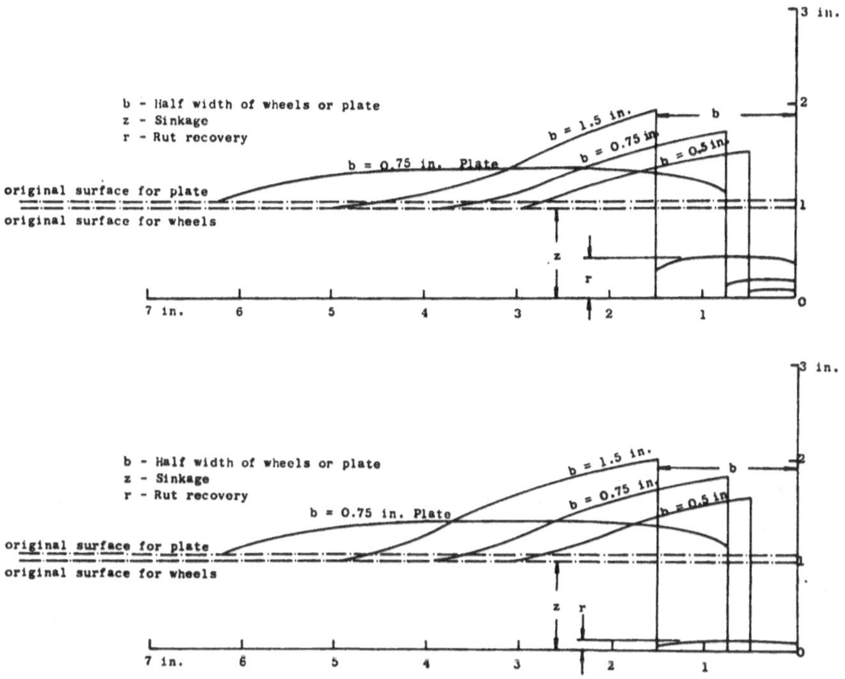

FIGURE 11. *Top*, Diagram showing the comparison between the profiles of ruts left by driven wheels at 37 per cent slip and a long narrow plate on wet coarse sand; *bottom*, Diagram showing the comparison between profiles of ruts left by towed wheels and a long narrow plate on wet coarse sand

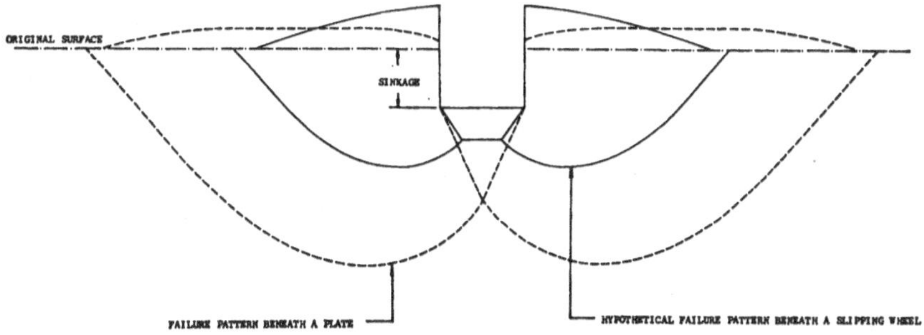

FIGURE 12. Comparison of the hypothetical soil failure mechanism beneath the axle of a slipping wheel and a narrow plate, showing the actually measured soil heave

STRESS DISTRIBUTION AROUND THE RIM

If the two failure zones beneath the wheel are considered in a conventional soil mechanics manner an approximate pressure distribution can be obtained. In a cohesive soil it would be expected that the pressures would be uniformly distributed. In a sandy soil on the other hand the pressures would be expected to grow more or less in proportion to the depth below the nearest free surface. This would give a pressure for each zone as shown in Figure 13. The pressure corresponding to the backward failure would grow steadily from E to A. The pressure in the forward

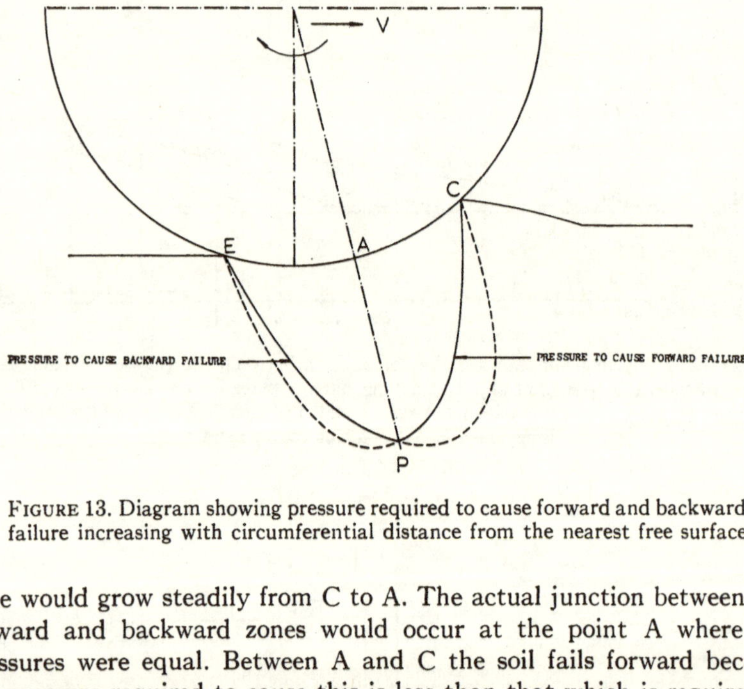

FIGURE 13. Diagram showing pressure required to cause forward and backward failure increasing with circumferential distance from the nearest free surface

zone would grow steadily from C to A. The actual junction between the forward and backward zones would occur at the point A where the pressures were equal. Between A and C the soil fails forward because the pressure required to cause this is less than that which is required to make it fail backwards. Between A and E the converse is the case. It follows then that the maximum pressure can be expected to occur at the junction between the two zones and in a sandy soil the distribution will be like that shown as CPE in Figure 13. The stresses beneath large slipping rigid wheels have been measured by O. Onafeko working in the Department of Agricultural Engineering, University of Newcastle (6). Both the large scale and the model (Figure 9B) experiments show that the position of point A does not change appreciably with skid in the region from 0 to 25 per cent. Therefore Figure 14 showing typical results for a towed wheel 6 in. wide and 49.4 in. diameter operating at 21.6 per cent skid in compact sand can be compared with the $8\frac{1}{4}$ in. $\times$ $\frac{1}{2}$ in. wheel at 12 per cent skid in the model experiments. The point A from the model experiments is seen to occur very close to the maximum pressure point on the full scale wheel. Figure 15 shows a comparison between the two wheels but this time at a considerable slip. Once again there is excellent agreement.

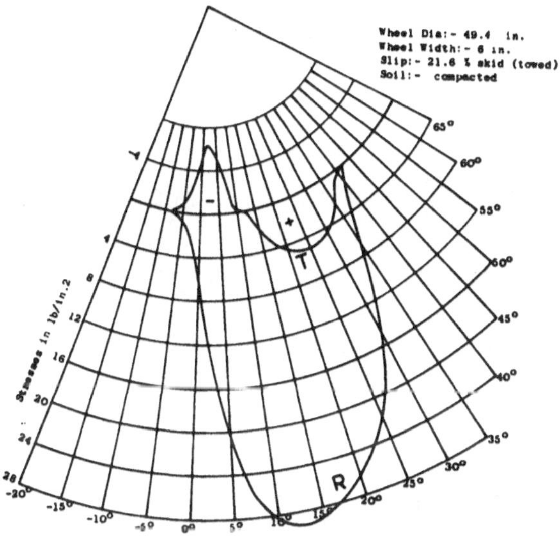

FIGURE 14. Distribution of radial and tangential stresses beneath a rigid towed steel wheel (after O. Onafeko)

Figures 4B and 6B enable some conclusions to be drawn about the tangential stresses. In the case of the towed wheel the soil in the region *ACB* must move upwards and it therefore slides along *AC* in such a way as to produce shear stresses in the direction opposite to wheel rotation. Between *E* and *A* the soil remains stationary or moves forward slowly while the wheel rim moves forward relatively fast. In this zone the shear stresses therefore act in the direction of rotation. The point *A* can be expected therefore to mark a change in sense of shear stress. This is very well borne out by Onafeko's large wheel results shown in Figure 14.

In Figure 6B it is clear that although the soil in the zone AED moves backwards it cannot move as fast as the wheel rim; therefore, for the slipping wheel the tangential stresses can be expected to point forward over the whole of the contact zone as is shown in Figure 15.

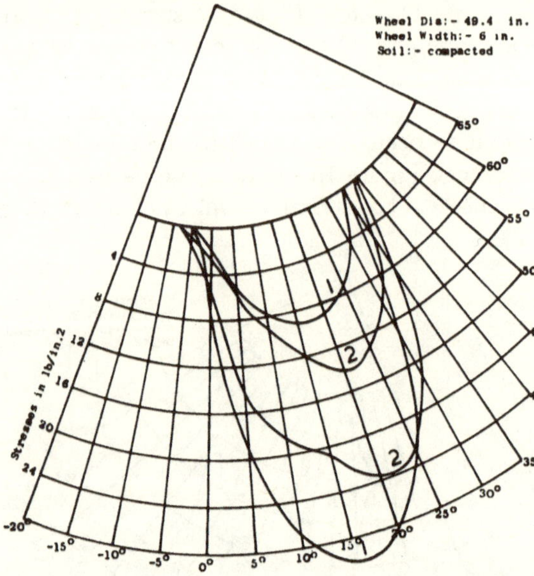

FIGURE 15. Distribution of radial and tangential stresses beneath a rigid driven steel wheel at different slips (after O. Onafeko)

CONCLUSIONS

1. The experiments described show that soil failure beneath a rigid wheel is a three dimensional phenomenon. They show that the assumptions underlying current theories of rolling resistance are inadequate in that the soil does not fail sideways in a manner akin to that beneath a strip footing.

2. Slip sinkage beneath wheels is caused mainly by a pumping action in which soil is transported longitudinally backwards along the wheel. This process also explains the phenomena of rut recovery.

3. There are normally two failure zones beneath a wheel, one forward and one backwards. These degenerate into a single backward zone at 100 per cent slip and a single forward zone for a locked wheel at infinite skid.

4. The soil fails beneath a narrow wheel in a composite way, partly longitudinal and partly sideways, the proportions of each movement depending on the wheel width and the slip.

5. Since the nature of the soil reaction to the wheel depends fundamentally upon the degree of slip, it follows that rolling resistance must also

be a function of slip. It is also clear that rolling resistance depends on soil movement in the direction of motion and is not solely due to vertical and sideways displacement.

ACKNOWLEDGEMENT

This paper is published by permission of H.M.S.O. Crown Copyright Reserved.

REFERENCES

1. BEKKER, M. G., *Off-the-road Locomotion.* Ann Arbor: University of Michigan Press, 1960.
2. UFFELMANN, F. L., The Performance of Rigid Cylindrical Wheels on Clay Soil. *Proc. 1st Int. Conf. Mech. Soil-Vehicle Systems.* Turin, 1961.
3. WILLS, B. M. D., F. M. BARRETT, and G. J. SHAW, An Investigation into Rolling Resistance Theories for Towed Rigid Wheels. *J. of Terramechanics,* Vol. 2, No. 1, 1965.
4. VINCENT, E. T., Pressure Distribution on and Flow of Sand past a Rigid Wheel. *Proc. 1st Int. Conf. Mech. Soil-Vehicle Systems.* Turin, 1961.
5. HEGEDUS, E., Pressure Distribution under Rigid Wheels. *A.S.A.E. Transactions,* Gen. Ed., Vol. 8, No. 3, 1965.
6. ONAFEKO, O., An Analysis of Soil Stresses and Deformations Beneath a Rigid Wheel. Unpublished Ph.D. Thesis of the University of Newcastle-upon-Tyne, 1965.
7. REECE, A. R., Principles of Soil-Vehicle Mechanics. *Proc. Inst. of Mechanical Engineers,* 1965.
8. REECE, Q. R., The Effects of Grousers on Off-road Vehicle Performance. *J. Agric. Engng. Res.,* Vol. 9, No. 4, 1964.
9. UFFELMANN, F. L., Discussion on Reference 3. *J. of Terramechanics,* Vol. 2, No. 4, 1965, p. 91.

Rigid Wheels In Clay

R. M. CULLEN*

G. CULLINGFORD†

B. MAYFIELD†

ABSTRACT

The paper describes some of the work carried out in the Civil Engineering Department, University of Nottingham, largely supported by the Fighting Vehicles Research and Development Establishment, Chobham, U.K. The long term objective of the research is to provide an understanding of the steering of wheeled vehicles in clays.

A description of the apparatus is given together with a résumé of the results obtained from both towed and driven single wheels. An approximate theoretical model for wheels in clay is proposed and discussed with reference to the experimental results.

NOTATION

- W_l Measured vertical reaction on the left load cell (lbf)
- W_r Measured vertical reaction on the right load cell (lbf)
- W Total vertical load on wheel (1bf)
- R_l Measured horizontal reaction on the left load cell (lbf)
- R_r Measured horizontal reaction on the right load cell (lbf)
- R_0 Rolling resistance (energy dissipated in the soil in forming unit length of rut) (lbf in./in.)
- R Drawbar pull (lbf)
- Q Measured axial reaction (lbf)
- c Soil cohesion, as measured by laboratory vane (lbf/in.^{2})
- b Wheel width (in.)
- d Wheel diameter (in.)
- θ Angle of attack (the angle between wheel plane and direction of motion (°)
- z Wheel sinkage (in.)
- V_g ground speed of wheel (in/sec)
- V_w peripheral speed of wheel (in./sec)
- S Slip per cent = $V_w - V_g/V_w \times 100$
- T Torque applied at the axle (lbf/in.)
- a Coefficient of adhesion (lbg/in.^{2})
- θ_z Sinkage angle (°)
- e Rut recovery factor
- θ_c Recovery angle = $(1+.1e)\theta_z$ (°)

APPARATUS

FIGURE 1 SHOWS a general view of the rig as used in the experiments. The main beam, approximately 12 ft. long, is supported at one end on a rotating centre post and at the other end by a bogie carried on a pair

*Department of Civil Engineering, Monash University, Clayton, Australia.
†Department of Civil Engineering, the University, Nottingham, England.

FIGURE 1. General view of apparatus

of idlers running on a part circle track. A 10 H.P. electric motor powers the drive wheel through a torque controlling electric clutch and contact between the drive wheel and the upper track has been ensured by spring loading the idlers. This motor is used for driving the automatic screed and can provide a propelling force or act as a towed load during test runs. The test wheel is attached below the main beam (Figure 2 shows the driven wheel fixed load arrangement) and mean track radii of 4 ft., 6 ft., and 8 ft. have been used. For driven wheel tests the wheel is powered by a 1 H.P. constant speed hydraulic motor.

The soil is an artificial clay composed of bentonite and glycerine and has been found to be quite stable over short periods but as it is slightly hygroscopic there is a gradual decrease in strength in the long term, amounting to about 1 p.s.i. per year under normal laboratory conditions. Undrained triaxial tests carried out on the material have shown it to have zero angle of shearing resistance, and the average cohesion over the tests reported here was about 5 p.s.i. The soil is contained within a trough 7 ins. deep and trough widths of both 2 ft. and 4 ft. have been used, the latter in preparation for model vehicle experiments.

The test wheels used to date have been of 9 ins and 12 ins diameter, with a diameter/breadth ratio of approximately 10/3. For driven wheel tests both smooth and rough surfaces have been investigated, the smooth

FIGURE 2. Driven wheel; fixed load arrangement

wheel was machined mild steel, the rough face was formed by coating this with 120 grade waterproof emery paper. Towed wheel tests were carried out with a fixed sinkage arrangement, and the angle of attack (the angle between track tangent and wheel plane) was varied between $-10°$ and $+20°$. The driven wheel tests have been run with zero angle of attack under both fixed sinkage and fixed load conditions. Fixed sinkages from 0 to $1\frac{1}{2}$ ins were possible and fixed loads above the dead weight of 160 lbs were provided by the air rams shown in Figure 2.

The test wheel axle is supported on either side of the wheel in load cells whose weigh beams measure variations in the vertical and horizontal components of load with electrical resistance strain gauges, using a system proposed by Close and Muzzey (1) (Figure 3). In addition an axial load weigh beam was used for the towed wheel tests, and for preliminary qualitative driven wheel tests.

TEST PROCEDURE

The track was reformed before each test by punning clay into the rut and screeding off the resulting surface using a cutter followed by a roller. Two passes of the screed left the surface only slightly above the required level and obvious surface defects were remedied before the final pass. This treatment produced the smooth and even finish which can be seen in Figure 4.

FIGURE 3. Wheel load cell

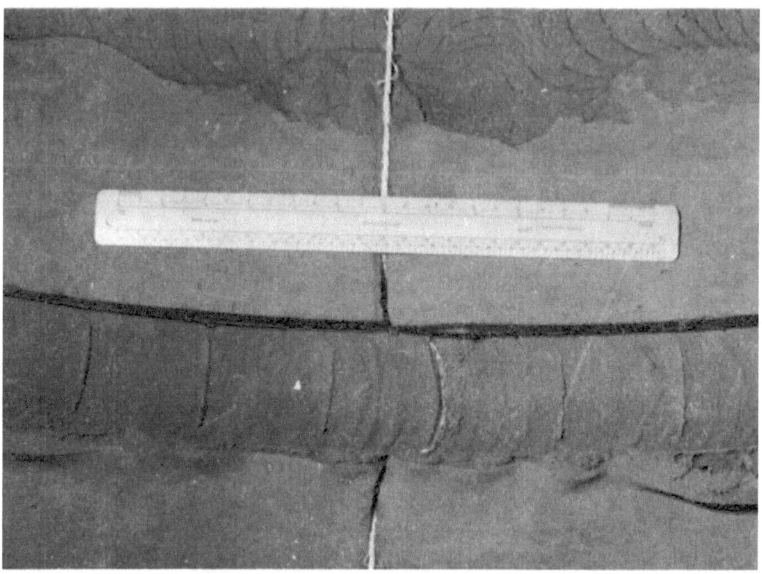

FIGURE 4. Rut formation $\theta = 0°$

Test records were obtained using a multi-channel ultra violet oscillograph. Stable running conditions were obtained over a good proportion of the track length, although more variation was experienced with the driven than with the towed wheel.

TOWED WHEEL RESULTS

The conventions used in the description of the forces and moments acting from the "chassis" to the wheel are given in Figure 5. For the towed wheel tests, which incorporated angle of attack as an independent variable

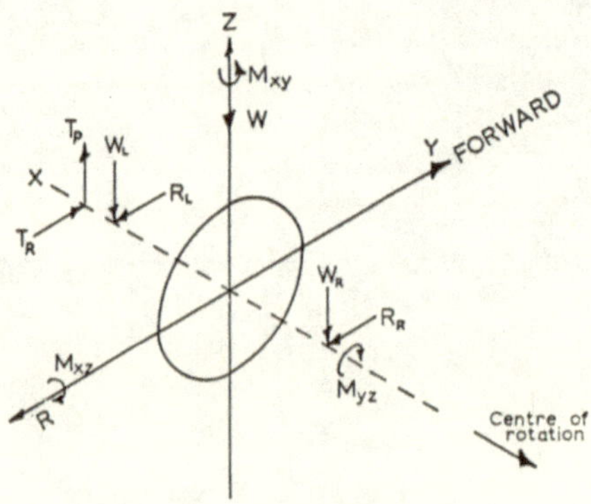

Direction of applied forces on the wheel axle.

Soil forces acting on the wheel.

$R = -(R_L + R_R - T_R)$ $\quad$ $W = -(W_R + W_L - T_p)$

$M_{xy} = -4.64(R_L - R_R - \frac{4}{3}T_R)$ $\quad$ $M_{xz} = -4.64(W_R - W_L + \frac{4}{3}T_p)$.

$M_{yz} = -$Applied Torque.

FIGURE 5. Force and moment convention

the fixed sinkage arrangement was employed and the forces measured were W_1, W_r, R_1, R_r, and Q.

Following the accepted approach, the majority of the results are given in dimensionless form. Forces have been plotted as Force/*cbd*—where *c* is the soil cohesion as measured by a laboratory vane, *b* the wheel width, and *d* its diameter. It is recognized, however, that the use of these dimensionless parameters disguises certain aspects of the behaviour of the soil. The soil used is very dependent on strain rate at high rates of shear and it is felt that the cohesion, as measured by the laboratory vane, or indeed any other method of test at a single rate does not fully describe its mechanical strength. This phenomena has also been observed in tests on natural clays and, as has been noted elsewhere (Cullen (2)) a better understanding of soil viscosity could assist the comparison of the results achieved by different mobility research workers with different clays.

Though a change in the angle of attack (θ) from $-10°$ to $+20°$ naturally showed a decrease in W_1 and an increase in W_r the sum of the two remained approximately constant, and the relationship between W/cbd and z/d observed is shown in Figure 6. This graph incorporates results for all angles, the two wheel diameters and the three radii used.

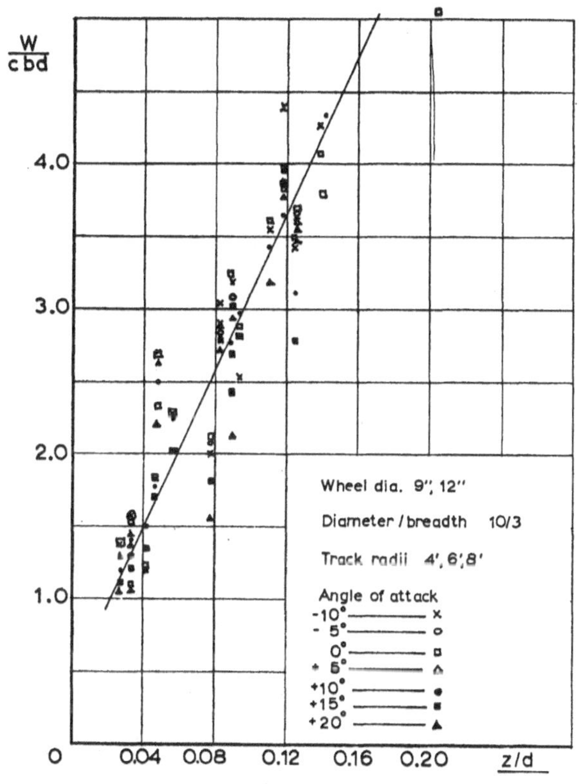

FIGURE 6. Towed wheel—W/cbd vs. z/d

Within each set of results there is a tendency for W to decrease as θ reaches $+15°/+20°$. The regression equation for the $\theta = 0°$ results is:

$$W/cbd = 27(z/d) + 0.4$$

No significant change in either R_1 or R_r is to be expected with variation of angle of attack, and none was observed. The results, again including all independent variables, are given in Figure 7 where the regression line:

$$R_0/cbd = 35(z/d)^{3/2}$$

for the $\theta = 0°$ results is shown.

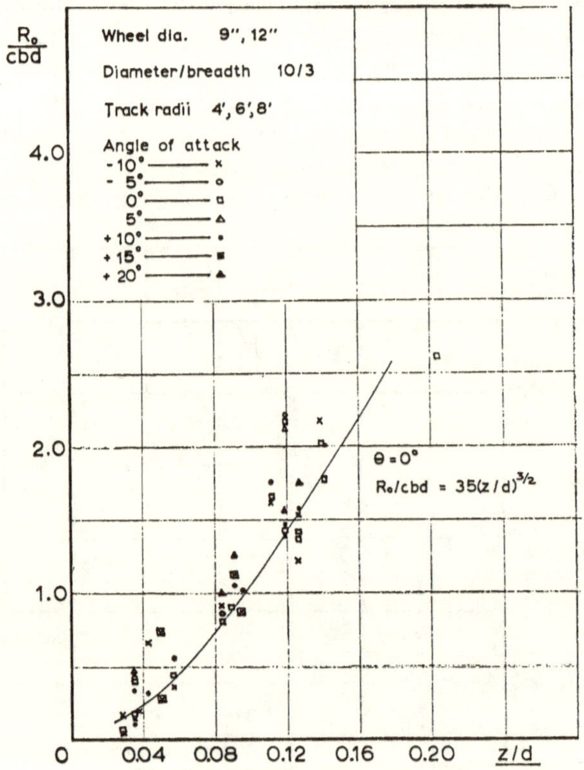

FIGURE 7. Towed wheel—R_0/cbd vs. z/d

A strong correlation with θ between both the axial force Q and the cornering force Q' (parallel to the radial arm), could be expected and was observed. Since the correlation between Q and θ was linear it was possible to plot these results in the form shown in Figure 8, the empirical line having the equation:

$$Q/cbd = 1.15\theta(z/d).$$

With the towed wheel the only two moments to be considered are M_{xy} and M_{xz}. The M_{xy} results were small and in consequence were not determined as accurately as were the forces. They did, however, change sign with the change from $-ve\ \theta$ to $+ve\ \theta$, and also increased in magnitude, for a particular value of θ, as the sinkage increased.

The M_{xz} results, on the other hand, showed a strong correlation with both angle of attack and sinkage.

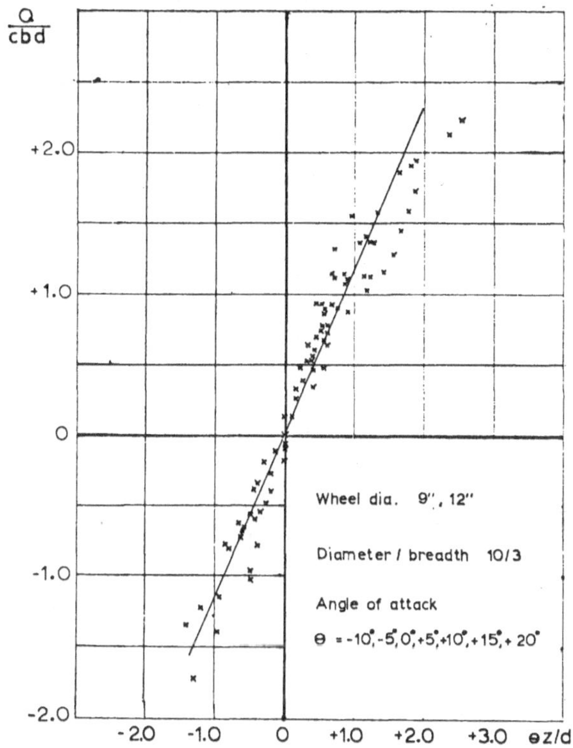

FIGURE 8. Towed wheel—Q/cbd vs. $\theta z/d$

DRIVEN WHEEL

Driven wheel results have been obtained under both fixed sinkage and fixed load conditions and a comparison of rough and smooth wheels was made at fixed sinkage. The angle of attack was kept constant at $\theta = 0°$, and a track radius of 6 ft. was used. As great a slip range as possible has been covered in the tests but since the "forced slip" principle has not been employed, results beyond the peak of the drawbar pull versus slip curve have not been obtained. In practice, because of the fluctuations in drawbar pull which occur in a test run, successful tests can be run only at slips somewhat below those for the peak drawbar pull conditions. The tests reported here were run at a constant peripheral speed of the wheel of about 6 in/sec.

FIXED SINKAGE RESULTS (ROUGH AND SMOOTH WHEELS)

Extensive tests were conducted with the smooth wheel and rather fewer with the rough wheel. Results for both wheels have been combined except in Figures 11 and 13 where they have been compared.

VERTICAL LOAD

Figure 9 shows the measured variation of vertical load with slip for sinkages of $\frac{1}{4}$ in. to $1\frac{1}{2}$ in. in $\frac{1}{4}$ in. steps on the 9 in. diameter wheel. This

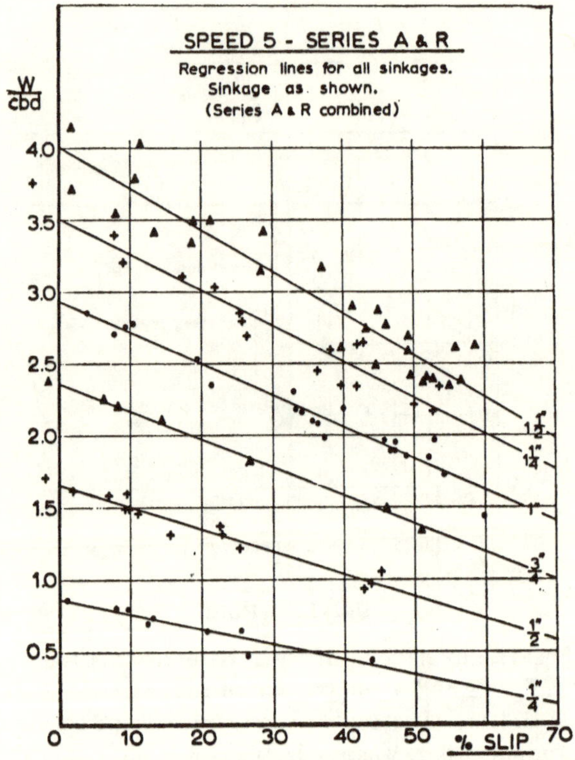

FIGURE 9. *W/cbd* vs. slip—Speed 5

represents z/d values of 0.028 to 0.167 in steps of 0.028. It can be seen that the vertical reaction at fixed sinkage decreased linearly as the slip increased. The lines shown for the various sinkages are represented by the equation:

$$W/cbd = [14.9(z/d)^{2/3} + 0.5](1 - S/175) - 1.0;$$

transposition of which gives:

$$z/d = 0.157[(W/cbd + 1.0)(175/[175 - S]) - 0.5)]^{3/2}$$

This relationship is plotted for the 9 in. diameter wheel in Figure 10, which shows the variation in sinkage with slip at given values of W/cbd which could be expected from tests run under fixed load conditions.

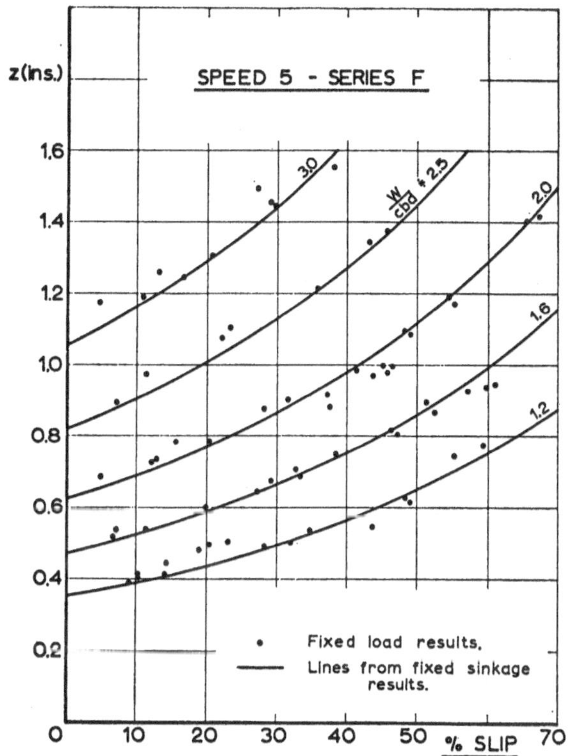

FIGURE 10. Variation of z with slip

Figure 11 compares results obtained from the rough and smooth wheels. A statistical analysis of the 1 in. sinkage results, which were taken as typical, showed the surface differences of this order to be not significant.

DRAWBAR PULL

The variation of drawbar pull with slip and sinkage was found to be as shown in Figure 12. The theoretical work indicates that the shape of these curves is strongly dependent on the sideplate action of the wheel. It is worth noting that the zero drawbar pull conditions do not occur at zero slip.

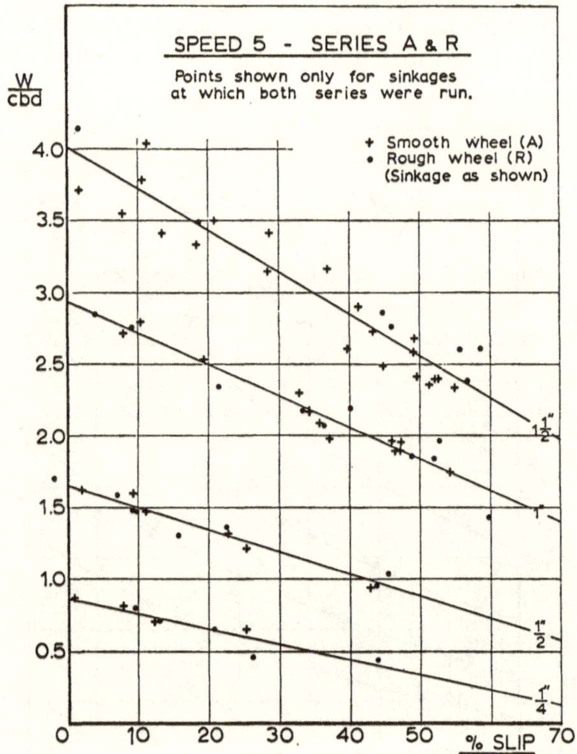

FIGURE 11. *W/cbd* vs. slip—comparison of series

ROLLING RESISTANCE

The term rolling resistance has been taken, for the purpose of this paper, to mean the total energy dissipated in the soil in the formation of unit length of rut. Characteristic values for this parameter are presented in Figure 13 which shows the results at ½ in., 1 in., and 1½ in. sinkage on the 9 in. diameter wheel. The lines drawn are described by the equation:

$$R_0/cbd = 30.1(z/d)^{1.55} + 0.0001(3 + 22.5[z/d])S^2.$$

Again, points are shown for both the smooth (Series A) and rough (Series R) wheels.

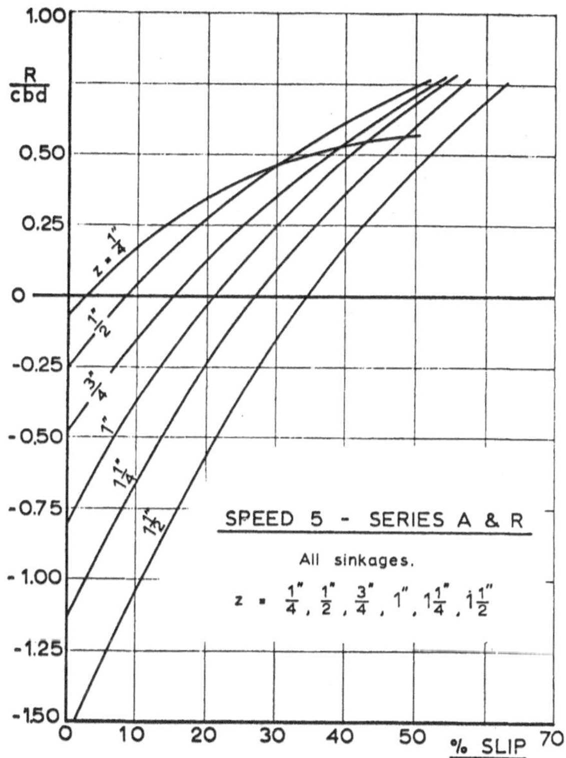

FIGURE 12. Variation of R/cbd with slip

FIXED LOAD TESTS

Using the apparatus shown in Figure 2 comparison has been made of the results measured under fixed load conditions with those predicted from the fixed sinkage results. Figure 14 shows a comparison of measured and predicted values for vertical load, and it can be seen that the agreement is good. We may thus conclude that the method of restraint of the test wheel does not affect the behaviour of the force-sinkage-slip relationships of a driven wheel.

Naturally enough the two restraint systems pose different problems and have different features in their favour. Fixed sinkage investigations make simpler comparison with most theoretical analyses since these are generally based on the assumption of a sinkage, z. They also make the study of high aspect ratio wheels (e.g. thin discs) possible even though

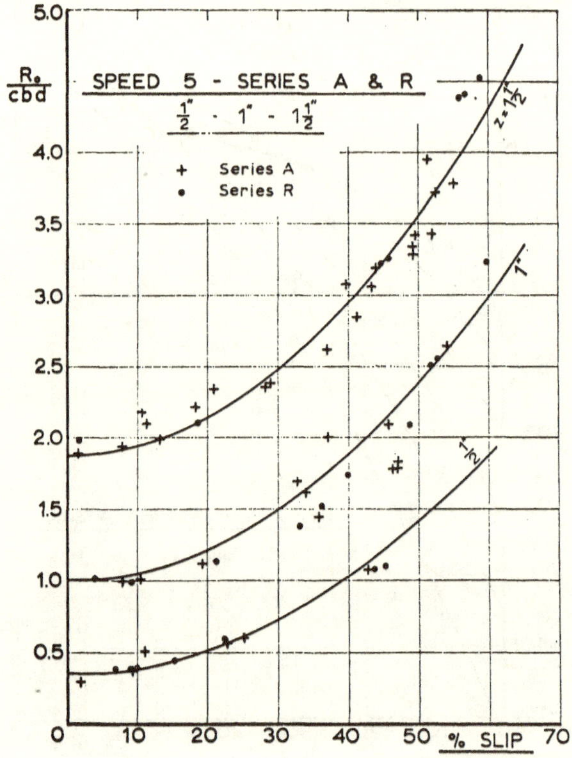

FIGURE 13. R_0/cbd vs slip—empirical lines

the vertical reactions developed are small, thus allowing investigation of sideplate effects. Fixed load single wheel systems are subject to slight variations in vertical reaction owing to frictional and inertial forces and, possibly, minor fluctuations in the applied load (e.g. in an air system), whereas fixed sinkage systems may be accurately set to a known sinkage but the load measuring is subject to experimental inaccuracy. A fixed load system represents the general condition of a real vehicle. Thus the two systems have complementary research functions.

EFFICIENCY AND WORK INDEX

Work Index may be defined as $R/W \times V_g/V_w$ which is a measure of the useful work output of a wheel at the drawbar as used by Dickson (3). Since for these tests V_w was constant, the work index at any load was

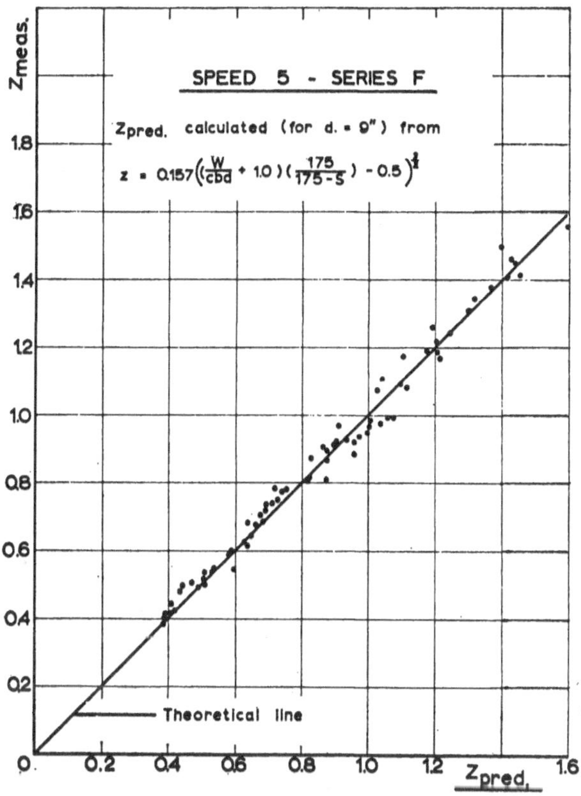

FIGURE 14. Comparison of measured and predicted sinkages

directly proportional to the work output. Efficiency of the wheel is taken as the ratio of the work output to the work input expressed as a percentage.

Figure 15 shows characteristic curves obtained for WI and E when $W/cbd = 2.0$. It will be seen that the total work output remains approximately constant over a wide slip range but that at the same time the efficiency falls rapidly as slip increases. Figure 16 shows how the maximum efficiency obtainable decreases with load. By extrapolating the line, as shown, to zero maximum efficiency, an approximate indication is obtained of the maximum load at which the wheel under test would be self propelled. A wheel under these conditions would of course be self-propelled only at the particular slip at which the drawbar pull changed from negative to zero value.

These results illustrate the important fact that in general the maximum drawbar pull, maximum work output, and maximum efficiency of a driven wheel all occur at different slips.

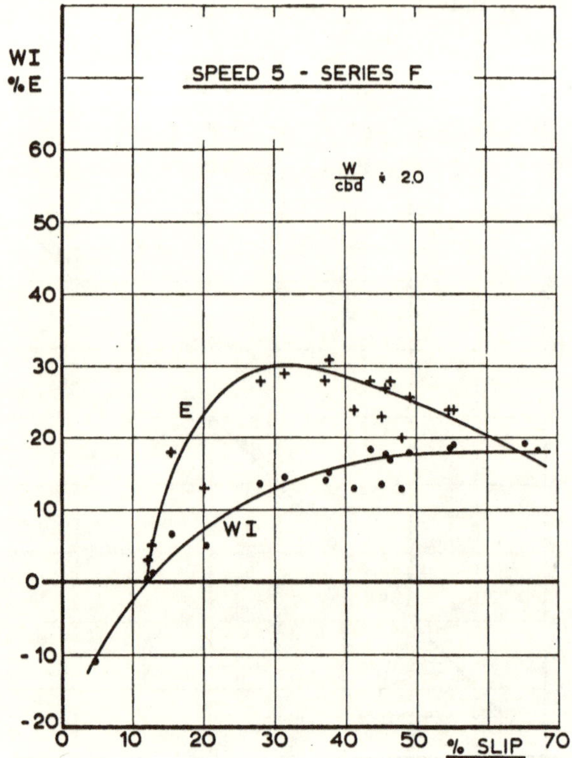

FIGURE 15. Efficiency and work index curves

RUT CHARACTERISTICS

Measurements were made of the surface displacements of the soil caused by one pass of the driven wheel operating at fixed sinkage. The nature of the fore and aft displacement of the soil can be seen in Figure 4 which shows the rut resulting from a test run at $1\frac{1}{2}$ in. sinkage and 43 per cent slip. The rearwards displacement of the string (originally radial) was found to be dependent on both sinkage and slip. With both low positive slips, that is approximately 10 per cent, and negative slips the marker was found to move forward, the bow wave being the usual visual evidence. This condition is associated with a strongly negative drawbar pull.

Rut recovery was found to be large and increased with slip as shown in Figure 17, but showed little variation with angle of attack in the towed wheel tests. The figure also shows the difference between centre recovery

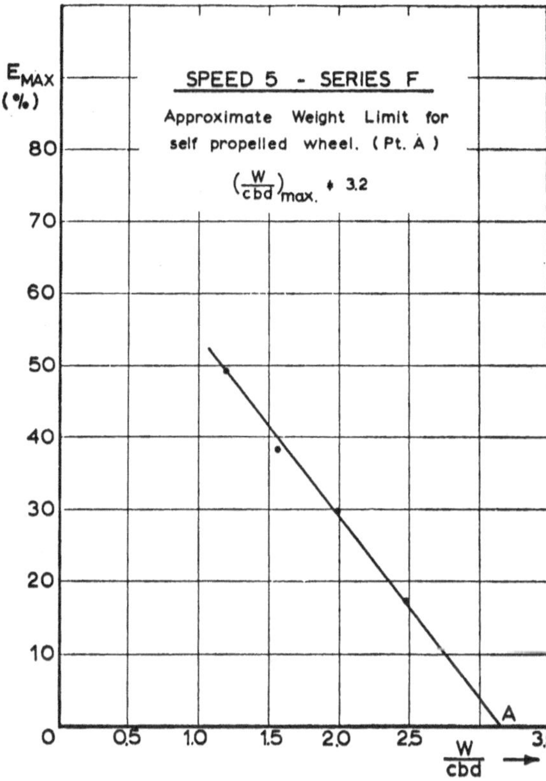

FIGURE 16. Maximum efficiency vs load

and edge recovery. This phenomenon, observed as an arch in the base of the rut, is thought to be caused by the restraint of the surrounding soil and it is unlikely that the curvature would be as marked at larger scales. This would be a scale effect not accounted for by current dimensional analysis.

DEVELOPMENT OF MODELS

It has been assumed that there is no interaction between the effects on the soil of the rim and the sideplates of a wheel, and that superposition can be used. Such an assumption is implied in most theoretical work on wheels (e.g. Uffelmann (4)) although the loads applied by the sideplates obviously alter the stress conditions on any assumed failure surfaces for the rim. Two models are therefore presented, one for the rim and the other for the sideplates; in both cases the soil is assumed to be weightless.

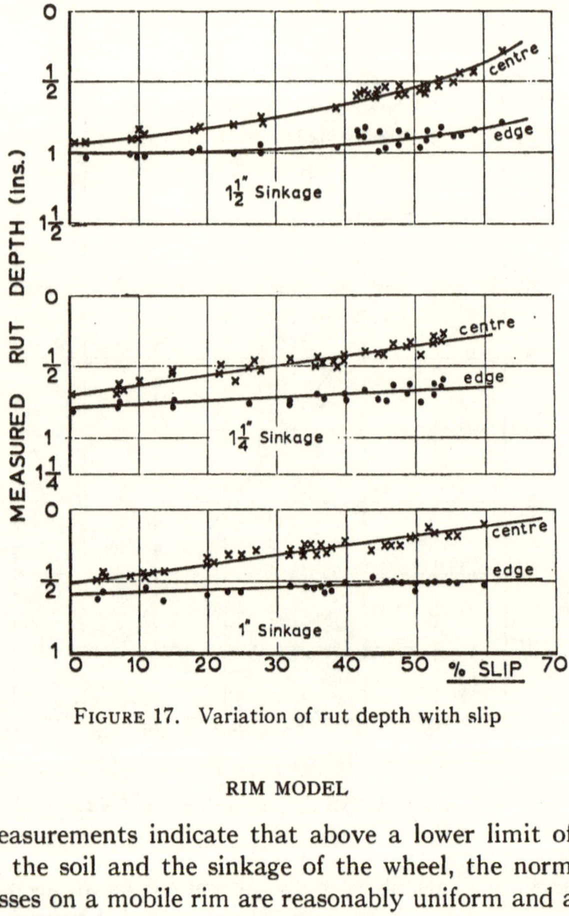

FIGURE 17. Variation of rut depth with slip

RIM MODEL

Stress measurements indicate that above a lower limit of slip which depends on the soil and the sinkage of the wheel, the normal and tangential stresses on a mobile rim are reasonably uniform and are not such as to cause failure at the soil/wheel interface unless the coefficient of adhesion is low. It was apparent from adhesion tests carried out on the soil used in the tests reported here that the adhesion of the soil to the wheel was virtually equal to the soil cohesion and, as appears reasonable from the nature of the problem, failure occurs elsewhere within the soil mass. For a wheel of finite width the problem represents a complex exercise in three dimensional plasticity even without considering viscous effects. Consideration of the related two-dimensional plasticity problem (Alexander (5), Marsden (6)) yields little that is of value in attacking the soil/wheel problem because of limitations on the boundary conditions and a more appropriate approach is required. A failure model for the two-dimensional case has been investigated; the form of the model is as shown in Figure 18, the arc ABC representing the instantaneous failure

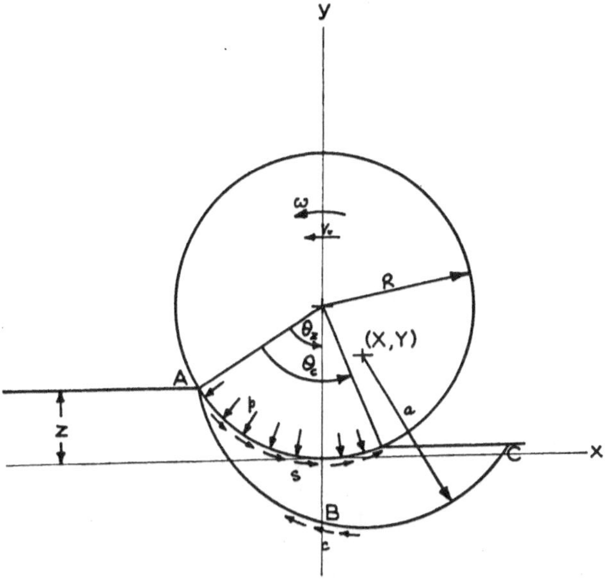

FIGURE 18. Assumed failure mode

surface. Observations of the movement of material indicated that at medium to high positive slips a considerable quantity of material was moved in the plane of the wheel while at low slips, bow wave and increased side heave indicated that if a slip circle model were appropriate, it should consist of two segments as in Figure 19, with segment *A* rotating clockwise and *B* anticlockwise. In the low slip range however it would appear that a more appropriate two dimensional model would be as suggested by Reece (7) with failure taking place on planes to either side of the wheel.

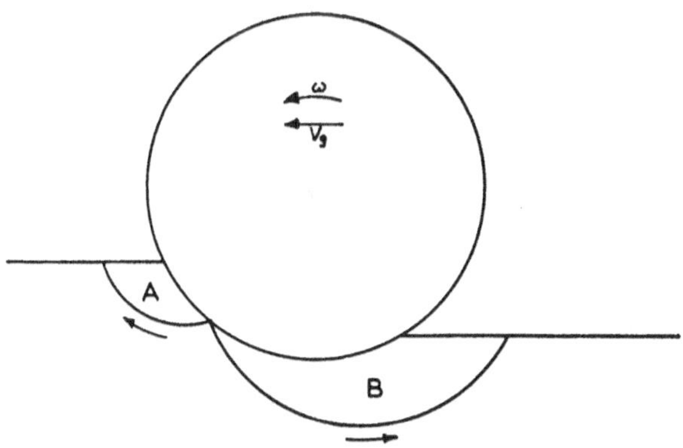

FIGURE 19. Alternative low slip failure mode

Using the model in Figure 18 for a moving wheel under applied loads *R* and *W* and torque *T*, the factor of safety of the soil mass above the arc ABC will be unity if failure is occurring on the arc, and will be greater than one if it is not. In the analysis *R*, *W*, and *T* were replaced by the uniform stress systems p and s (Figure 18) at the soil wheel interface. Taking fixed values of sinkage z, recovery angle θ_c and normal stress p, the value of s was adjusted until the minimum factor of safety of all the potential failure arcs springing from the leading point of the wheel was unity. The calculation was then repeated for other values of p until the required value of s was less than zero or greater than the cohesion of the soil (slip stall). Sinkages in $\frac{1}{4}$ in. steps from $\frac{1}{4}$ in. to $1\frac{1}{2}$ in., and ratios of recovery angle to sinkage angle between 1.0 and 1.4 (rut recovery factor, $e = 0$ to 4) have been investigated. The calculations were carried out on the Manchester University Atlas Computer.

Many of the instantaneous centres of the failure arcs were found to lie close to the vertical diameter of the wheel. As it was necessary to make some estimate of the slip required to mobilize the appropriate stress values p and s, it was assumed that the instantaneous centre of the wheel occurred at the same height as the centre of the failure circle. Figure 20 shows the vertical force results of the analysis superimposed on the experimental results, with this assumption. The range of the computed results is due to the range of recovery angles assumed for each sinkage. The experimental values of vertical reaction are not thought to be greatly influenced by the effect of the sideplates and it can be seen that although the predicted forces for $\frac{1}{4}$ in. sinkage are high, the experimental results at greater sinkages lie within the fields of prediction, with the value of recovery angle for full agreement increasing with sinkage. It can be seen that with the assumption made concerning the instantaneous centre of rotation of the wheel, the slip range covered by the calculated values is somewhat limited. As discussed earlier, observation indicates that the model is not appropriate to low-slip conditions and at very high slips it was found in the analysis that the required value of tangential stress s was tending rapidly to the cohesion of the clay. In testing it was also found that there was an effective upper limit of slip for any sinkage.

Comparison of experimental results of the prediction of *R* and *T* given by the model is complicated by the larger effects of the sideplates on the test wheel.

Inspection of the analytical results showed that different running conditions could exist at different recovery angles for a rim at different sinkages with the same values of axle load *R* and *W*. It being common experience that the horizontal and vertical loads on a wheel uniquely determine the sinkage and slip this point obviously needs consideration.

Taking the ground speed drawbar pull and vertical load of a wheel

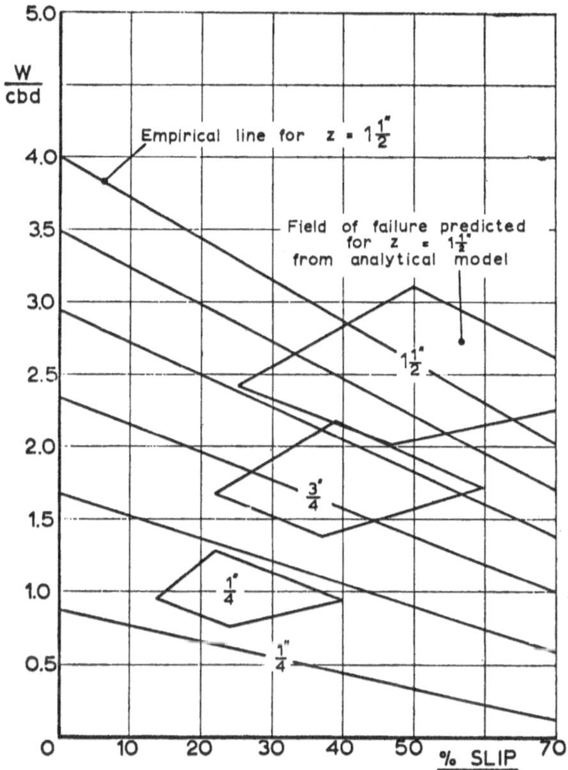

FIGURE 20. *W/cbd* vs slip—comparison of theory with experiment

as fixed it seems plausible to argue that the stable running condition of the wheel will occur when the power dissipated in the soil is a minimum. For a rim, power will be dissipated both by the local slipping at the soil wheel interface needed to mobilize the contact stresses, and also by shearing the clay along the failure arc. For a rectangular section wheel, power would also be dissipated by slip on the sideplates. For a rim it seems likely that the power dissipated at the interface will be small as no general relative movement occurs, the losses on the failure arc being the dominant factor. To estimate this effect the rate of shearing on the arc is required and an estimate of this may be obtained from a crude model of the soil transport process. Assuming unit ground speed the instantaneous rate of rotation of the soil mass must be high enough to remove a volume of soil from ahead of the wheel equal to zb in unit time. In fact only a proportion of the soil will be moved in this way in the plane of the wheel, as the recovery of the rut is usually not complete. An assumption of full rut recovery will thus overestimate the power loss as soil will move to the side of the wheel only when it requires less power to do this than to move in the plane of the wheel.

TABLE I

ENERGY DISSIPATION FACTORS

$1\frac{1}{4}''$ sinkage					$1\frac{1}{2}''$ sinkage				
7.729	8.761	9.838	10.955	12.110	7.814	8.877	9.985	11.134	12.319
7.641	8.701	9.807	10.954	12.139	7.711	8.809	9.952	11.135	12.354
7.559	8.652	9.791	10.972	12.189	7.620	8.757	9.939	11.160	12.416
7.487	8.618	9.795	11.014	12.268	7.546	8.728	9.954	11.217	12.512
7.431	8.606	9.827	11.086	12.380	7.497	8.730	10.004	11.311	12.648
7.400	8.625	9.894	11.198	12.534	7.485	8.774	10.098	11.453	12.833
7.405	8.686	10.006	11.358	12.737	7.523	8.870	10.248	11.651	13.074
7.460	8.802	10.176	11.577	12.999	7.626	9.034	10.466	11.915	13.381
7.583	8.987	10.416	11.865	13.331	7.812	9.280	10.763	12.258	13.763
7.793	9.258	10.741	12.236	13.742	8.100	9.623	11.154	12.690	14.231
8.112	9.634	11.165	12.702	14.244	8.509	10.079	11.650	13.221	14.792
8.562	10.132	11.702	13.273	14.845	9.055	10.661	12.263	13.860	15.455
9.161	10.766	12.366	13.961	15.554	9.752	11.381	13.002	14.617	16.227
9.926	11.549	13.165	14.775	16.380	10.609	12.246	13.875	15.496	17.112
10.866	12.490	14.108	15.719	17.326	11.633	13.262	14.884	16.501	18.114
11.986	13.593	15.198	16.799	18.397	12.822	14.428	16.033	17.635	19.235
13.283	14.857	16.435	18.013	19.593	14.172	15.743	17.318	18.896	20.475

Table I shows the energy dissipation factors, calculated on this basis, for circles with centres on a $\frac{1}{4}$ in. grid to $\frac{1}{2}$ in. either side of the vertical diameter for sinkages of $1\frac{1}{4}$ in. and $1\frac{1}{2}$ in., assuming full recovery. The lower row of figures is for centres on a horizontal line through the bottom dead centre of the wheel. On this basis the preferred running condition would be that having the smaller sinkage and with a centre as close as possible to the vertical diameter.

SIDEPLATE MODEL

The sideplate model is based on the assumptions:

(*a*) that all points on the sideplates below undisturbed ground level are in contact with the soil (edge recovery or corona is ignored);

(*b*) that the maximum possible shear stress (equal to the soil cohesion multiplied by the coefficient of adhesion *a*) is developed at all points of the sideplate contact area;

(*c*) that these shear stresses are mobilized in the direction of the velocity of the soil relative to the wheel;

(*d*) that the soil on the wall of the rut is stationary (i.e. the velocity of a point on the sideplates relative to the soil is given by the velocity vector for that point on the sideplate).

In support of assumption (*a*), no separation between the sideplates and the soil was noticed during the test runs although the circular form of the track would seem to promote separation in one quadrant of each sideplate. Adhesion tests, referred to earlier, indicated that the coefficient of adhesion *a* for the soil used was unity. Assumptions (*b*) and (*c*) are consistent with continuous slip on the interface which may be inferred from the small amount of corona observed on the side walls. This is not compatible with the complex particle motions which would result from full adhesion between the sideplate and the soil. That a small corona does occur indicates that assumption (*d*) is in error. Absolute vertical movements of the soil would introduce small components of vertical reaction with slip from the sideplates. Vertical movement would in addition cause an over-estimate of the tractive component from the sideplates at a given slip. Lacking data to evaluate these factors the model based on the four assumptions has been investigated. The forces and torques developed over a range of sinkages and slips have been evaluated by numerical integration using the Manchester University Atlas Computer. The basis of the integration is shown in Figure 21 and the computed torques and thrusts are shown in Figures 22 and 23.

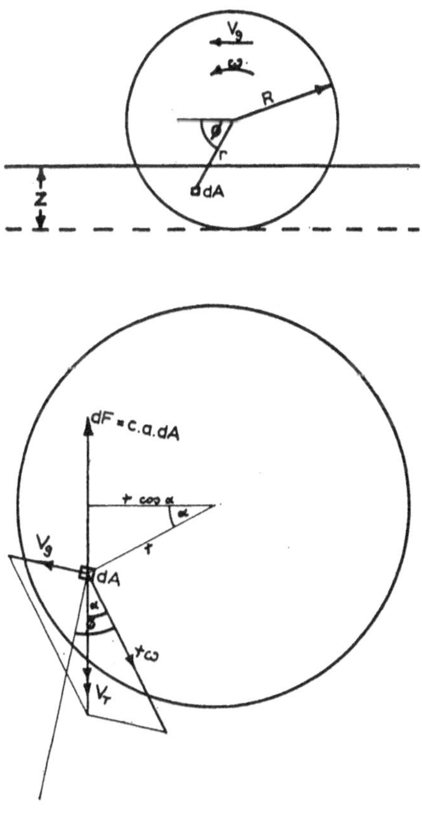

FIGURE 21. Sideplate analysis diagram

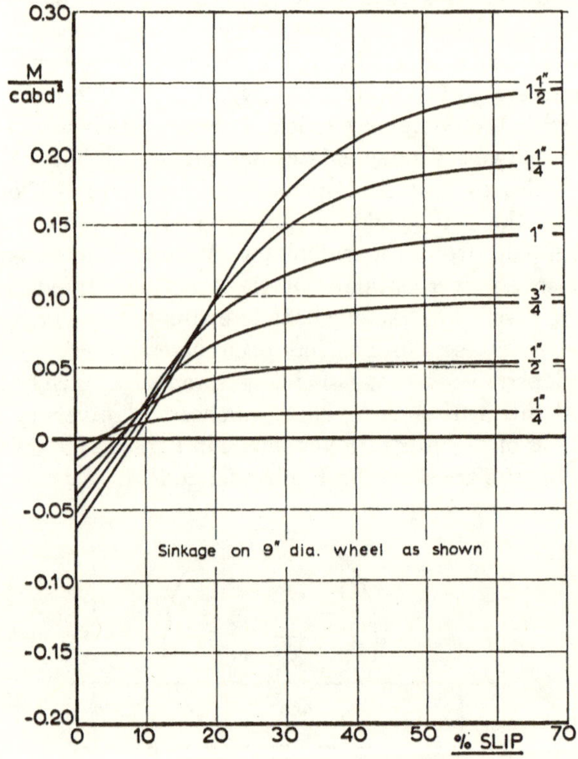

FIGURE 22. Sideplate torque vs slip—theoretical

When the computed and measured torques were compared, making allowance for the maximum torque possible on the rim, it was found that, although the form of the torque and drawbar pull variations with slip corresponded to the experimental work, the predicted torque was between one-half and one-third of the measured value at high slips, showing that the stresses mobilized in the plane of the sideplates were between two and three times the soil cohesion measured with the laboratory vane. The laboratory vane strength was found to be equal to the strength measured in the triaxial machine at strain rates of 0.016 and 0.05 per minute indicating that at low rates of strain a threshold value of soil strength was measured. When the strength was measured using the $1\frac{3}{4}$ in. diameter hand held vane, developed by F.V.R.D.E., the rate of strain was estimated to be a factor of ten greater than with the laboratory vane, and the measured strength was about twice the laboratory vane strength. Increasing the rate of shearing of the F.V.R.D.E. vane to its practical

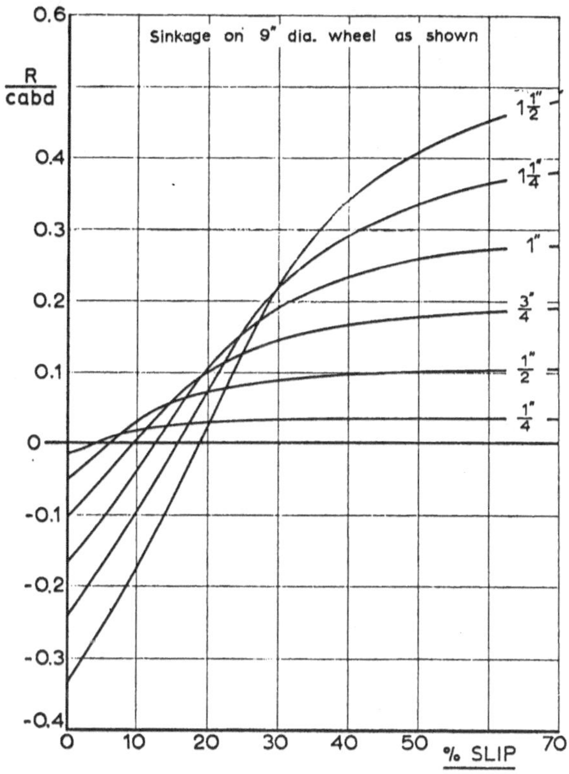

FIGURE 23. Sideplate thrust vs slip—theoretical

limit showed a further gain of 40 per cent in the measured shear strength. The discrepancy between the predicted and measured torques is thus in line with the rate dependent nature of the clay.

CONCLUSIONS

The circular track apparatus has operated successfully and both towed and driven wheel tests have produced consistent results.

The performance of a towed wheel at various sinkages has been related to its angle of attack. The dimensional similarity of results on wheels of the same aspect ratio has been demonstrated, and the non-importance (within the range covered) of track radius has been shown.

The importance of slip in the sinkage and load relationships has been demonstrated and quantitatively described. The comparability of tests carried out at fixed load and fixed sinkage has been established.

The theoretical models are attractive in their simplicity and appear to give adequate predictions of the variation of vertical load with slip over a limited slip range. The sideplate analysis is thought to be reasonably realistic and further tests on disc form wheels are being conducted to explore the problem further.

REFERENCES

1. CLOSE, W. and C. L. MUZZEY, A Device for Measuring Characteristics of Tyres on the Road. *Symposium on Research in Automobile Stability and Control and Tyre Performance.* Inst. Mech. Eng. Automobile Division, 1956, p. 311.
2. CULLEN, R. M., Forces on Rigid Wheel Moving in an Artificial Clay. PH.D. Thesis, University of Nottingham, 1966.
3. DICKSON, W. J., Ground Vehicle Mobility on Soft Terrain. *J. Soil Mechanics and Foundations Division.* A.S.C.E., August 1962.
4. UFFELMANN, F. L., The Performance of Rigid Cylindrical Wheels on Clay Soil. *1st Int. Conf. Mech. Soil–Vehicle Systems.* Turin, 1961.
5. ALEXANDER, J. M., A Slip Line Field for the Hot Rolling Process. *Proc. I. Mech. E.*, Vol. 169, 1955.
6. MARSDEN, A., Private communication, 1965.
7. REECE, A. R., Principles of Soil Vehicle Mechanics. *Proc. I. Mech. E.*, Vol. 180, Part 2A, 1965–66.

A Similitude Study of Soil Penetrometers

FRED L. SHUMAN, JR.*

W. F. BUCHELE†

W. G. LOVELY†

ABSTRACT

The principles of dimensional analysis and similitude were utilized in a study of soil penetrometers. This study was directed toward the determination of the influence of penetrometer tip size and shape on penetrometer readings and the development of an equation for determining soil shear strength from penetrometer readings. The effects of tip size were eliminated through application of dimensional analysis and similitude. Empirical equations for the prediction of shear strength were developed for all tip configurations tested.

*Assistant Professor, Agricultural Engineering Department, Mississippi State University, State College, Mississippi.

†College of Agriculture, Iowa State University, Ames, Iowa, U.S.A.

SOIL AND VEHICLE BEHAVIOUR

The Soft Ground Performance of A Vehicle when Provided with An Air Pressure Load Relief System

F. L. UFFELMANN*

ABSTRACT

The theoretical slope-climbing abilities of tracked and wheeled vehicles provided with load reductions by air pressure relief systems are discussed, and evaluated by simple approximate theories for two representative vehicles on soft to firm, purely cohesive and purely frictional soils.

The trends of the theoretical conclusions are confirmed by experiment with full scale vehicles on clay and sand of controlled characteristics.

THE HOVERCRAFT has good high-speed, level-surface, performance over water or marsh but can fail to clear vertical obstacles of a few feet in height, cross ditches of a few feet in depth, or, without the aid of powerful and uneconomical air thrust devices, to climb significant slopes; it is virtually uncontrollable in confined spaces without the addition of wheeled running gear. Conventional high pressure contact vehicles, however, have good hard ground performance and can be designed with good obstacle crossing ability, but have limitations on soft ground and in water crossing.

Various attempts have been made to combine the two types of vehicle in the hope of providing the conventional high-pressure contact vehicle with an ability to cross soft ground and water without greatly impairing its normal hard ground performance. The resulting vehicle, termed a "variable ground pressure" vehicle depends upon air pressure relief, that is, the loading of the running gear is reduced by transfer to a "plenum" chamber confined by a flexible skirt or rigid boundary wall around the perimeter of the vehicle. The hope of providing an improved soft ground performance by this means is based on the idea of reducing external rolling resistance, by reduced sinkage of the running gear, while maintaining an adequate level of tractive soil reaction. On water the plenum chamber would provide the buoyancy necessary for surface travel.

*The War Department; Fighting Vehicles Research and Development Establishment, Chertsey, Surrey, Great Britain.

In what follows it will be shown by approximate theories applied to particular conventional type vehicles, viz. the Oxford Carrier (tracked) and the Land Rover (wheeled), that an improved soft ground performance is not always obtainable by pressure relief, and generally to obtain this a lower than conventional contact pressure running gear system is necessary. Experiment at full scale with a Universal Tracked Carrier and a Land Rover confirms the trend of the theoretical conclusions.

It is interesting to note that the opposite to pressure relief, a pressure increase (or added down thrust on the running gear), can offer benefits in increasing the climb angle on firm frictional surfaces.

EQUATION OF EQUILIBRIUM OF A VEHICLE ON AN EARTH SLOPE

A simple picture of the forces acting on a variable ground pressure vehicle when climbing an earth slope can be obtained by assuming the pressure relief acts normally to the ground surface, as in Figure 1.

By resolving up the plane we have approximately:

$$A(c + \sigma \tan \phi) - R - W \sin \theta = 0, \tag{1}$$

and perpendicular to the plane:

$$W \cos \theta - F = \sigma A, \tag{2}$$

where:

A = effective area of running gear in contact with the ground
c = cohesion of the soil
ϕ = angle of soil friction
σ = stress normal to the plane
R = external rolling resistance
W = weight of vehicle
F = air thrust assumed normal to the plane
f = fraction of pressure relief = F/W.

Substituting σA of 2 in 1:

$$W \sin \theta = Ac + (W \cos \theta - F) \tan \phi - R;$$

or

$$\sin \theta = (Ac/W) + (\cos \theta - f) \tan \phi - (R/W). \tag{3}$$

We may take the rolling resistance R as a power function of the normal load and put:

$$\sin \theta = (Ac/W) + (\cos \theta - f) \tan \phi - KW^{(m-1)}(\cos \theta - f)^m, \tag{4}$$

where K and m are constants depending upon footing and soil parameters.

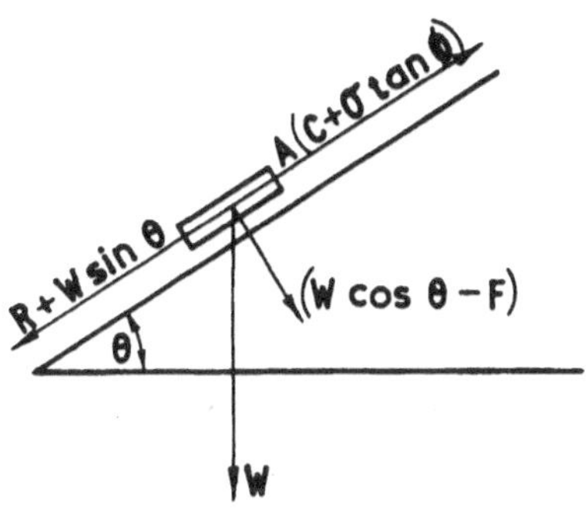

FIGURE 1. Forces acting on a ground pressure relief vehicle when in equilibrium on an earth slope

CONDITION OF ZERO NORMAL STRESS ON SOIL HAVING BOTH COHESION AND INTERNAL FRICTION

If $f = \cos\theta$ then both the frictional component of tractive soil reaction and the rolling resistance (represented per unit vehicle weight by the second and third terms of the right-hand side of the Equation 4 respectively) vanish, leaving the angle of climb, up to a maximum of $\cos^{-1}f$, dependent entirely upon the cohesion of the soil and its area of contact with the running gear.

If a slope greater than $\cos^{-1}f$ is attempted then the vehicle would be airborne and would descend to the bottom of the slope under the influence of the $W \sin\theta$ force component acting down the plane.

ZERO NORMAL STRESS CONDITION APPLIED TO A TRACKED VEHICLE (OXFORD CARRIER)

Applying the zero normal stress condition to the Oxford Carrier where $A = 1{,}878$ sq. in., $W = 17{,}920$ lb., nominal ground pressure $p_n = 9.54$ lb./in.^{2}, $f = 0.95$ (track pressure on level ground reduced to $\frac{1}{2}$ lb./in.^{2}), then $\cos\theta = 0.95$ and $\theta = 18°$.

To climb an 18° slope under these conditions the tractive soil reaction must be obtained from the first term of Equation 4 from which we deduce that the necessary soil cohesion is:

$$c = W \sin\theta / A = p_n \sin\theta = 3 \text{ lb./in.}^2$$

It has been assumed that the area (A) of running gear in contact with the soil remains unchanged by the pressure relief. This is by no means certain for the near zero normal stress condition when a reduction of contact area would be expected.

Steeper slopes can be climbed on soil having higher values of cohesion provided the pressure relief is adjusted so that the track can just remain in contact with the ground; e.g., for a 30° slope, f must be a little less than 0.87 and c greater than 4.8 lb./in.^{2}; for a 45° slope, f must be a little less than 0.71 and c greater than 6.7 lb./in.^{2}

PRESSURE RELIEF ON FRICTIONAL SLOPES—FIGURE 2

On purely frictional soil, where $c = 0$, the first term on the right-hand side of Equation 4 vanishes and the maximum slope which can now be

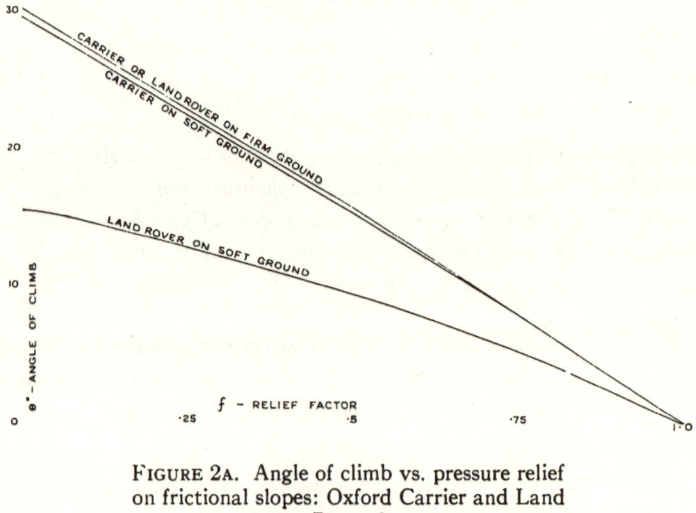

FIGURE 2A. Angle of climb vs. pressure relief on frictional slopes: Oxford Carrier and Land Rover 9

climbed by any vehicle relying on ground tractive reaction depends upon the remaining two terms. For the low rolling resistance conditions of firm frictional slopes the third term may be neglected and it can be seen that the air pressure relief factor f does not assist the angle of climb, the maximum angle being obtained without pressure relief when $\tan\theta = \tan\phi$. Zero angle of climb is obtained for $f = 1$ and only for negative values of f can the climb angle be improved. (See Figure 2B and Appendix III.)

[To climb a 90° slope with $\phi = 30°$, now regarded as angle of track-ground friction, $f = -\sqrt{3}$.]

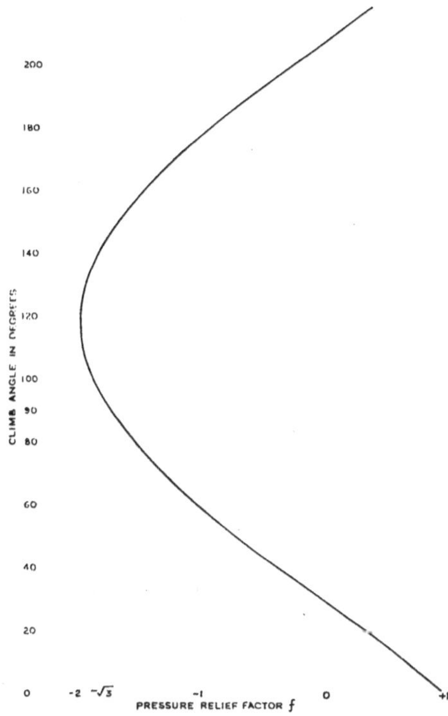

FIGURE 2B. Angle of climb vs. pressure increase on firm frictional surface

On soft frictional slopes offering greater rolling resistance there would be an advantage in the use of pressure relief only if this increases considerably the positive value of the sum of the second and third terms of Equation 4.

PERFORMANCE OF OXFORD CARRIER SOFT ON FRICTIONAL SLOPES

This is evaluated in Appendix I, showing pressure relief under these conditions to be detrimental.

The track width equivalent to pressure relief on cohesive soil is discussed in Appendix III. For the Oxford Carrier on soil of 3 lb./in.^{2} cohesion this is a 3 inches increase on 12 inches.

Wheeled vehicle performance calculations are discussed in Appendices IV and V. Values for the Land Rover 9 are given in Figures 2, 3, 4, 5, and 6.

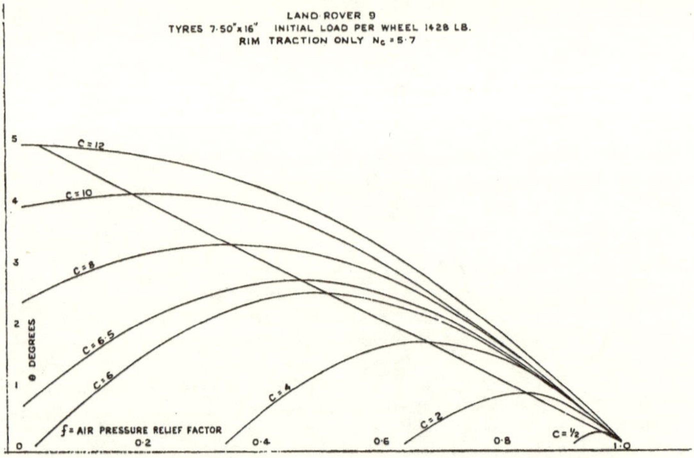

FIGURE 3. Angle of climb vs. pressure relief on cohesive soil

THEORETICAL CONCLUSIONS

1. Tracked Vehicles

(1) Air pressure relief is useful only in the case of cohesive soil where a controlled relief can reduce rolling resistance to a minimum without detracting much from the tractive soil reaction. Drawbar pull and hence angle of climb is thus increased within the limits determined by the value of soil cohesion and the remaining area of track on the ground.

(2) On hard or soft frictional surfaces the use of air pressure relief can only reduce the available drawbar pull and climb angle; these have maxima at zero pressure relief, the latter being equal to the angle of soil/running gear friction for a hard surface, and less than the angle of soil friction for a soft frictional surface. A pressure increase (negative relief) would improve the climb angle on a hard surface (see Appendix III) (Figure 2B).

(3) It can be shown (see Appendix II) that an effect equal to the beneficial air pressure relief effect calculated for cohesive soil can be achieved more simply by increasing the track width. For the Oxford Carrier a 3 in. increase (25 per cent) is adequate to achieve equality of climb performances on soil of cohesion 3 lb./in.^{2} Much higher performances than those obtainable by air pressure relief are obtainable by further increases of track width.

(4) In view of (2) and (3) above it is difficult to justify full scale development of tracked ground pressure relief vehicles as such unless they are also required to operate as hovercraft over water or on semi-liquid swamp of cohesion less than 1 lb./in.^{2}

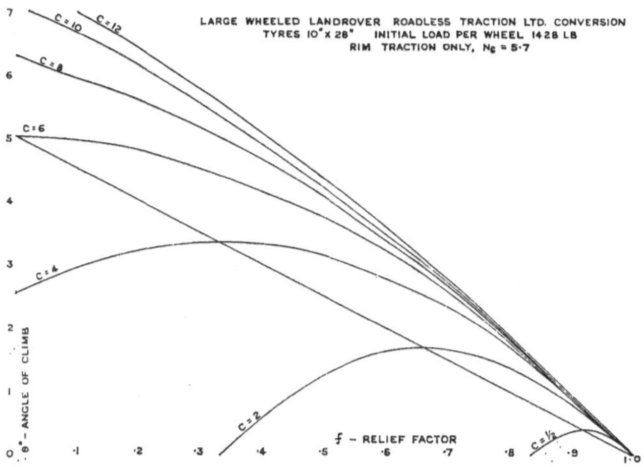

FIGURE 4. Angle of climb vs. pressure relief on cohesive soil

Ground pressure relief is not an alternative to low nominal ground pressure tracks which are vital for slope climbing of any type of ground vehicle on soft cohesive soil. Slope climbing is of course obtainable if the necessary axial thrusts are provided from the air and not the ground; this would normally mean an unacceptable increase in weight and expense.

2. Wheeled Vehicles

(1) To support the level passage of a particular wheeled vehicle with normal high pressure tires relying upon rim traction effects, non-frictional soil must have a cohesion value greater than a certain calculable minimum; below this minimum slip stall will occur (see Appendix IV). (This slip stall condition occurs at a much greater sinkage if the wheel is designed to provide a soil reaction from its side walls (see Appendix V and Figures 5 and 6).

(2) The use of pressure relief with rim traction tires allows the vehicle to pass over near level ground of lower shear strength than would otherwise be possible, provided the sinkage is maintained above a critical value dependent only upon a bearing capacity constant(see Appendix IV and Figures 3 and 4).

(3) As a maximum value of drawbar pull is obtained at an optimum sinkage, there is an optimum value of pressure relief (or pressure increase) producing this sinkage for each value of soil cohesion. The maximum drawbar pull is proportional to cohesion, however, and thus must be small for small values of cohesion (see Appendix IV, and Figures 3 and 4).

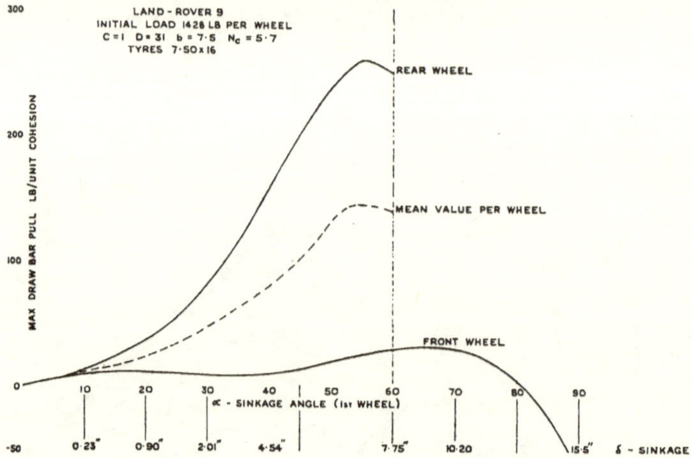

FIGURE 5. Drawbar pull vs. sinkage angle

(4) For the Land Rover 9, the maximum drawbar pull and hence climb angle on soft cohesive soil is so small for the conventional rim traction tire (climb angle 15′ to 2° 40′ for $c = \frac{1}{2}$—$6\frac{1}{2}$ lb/in^2), that the use of this type of tire with pressure relief does not seem to be a practical proposition (see Appendix IV, and Figure 3).

(5) Except for the weakest soils a much improved slope climbing ability on cohesive soil than is possible by pressure relief can be obtained by the use of special tires or wheels (as for the large wheel Land Rover,

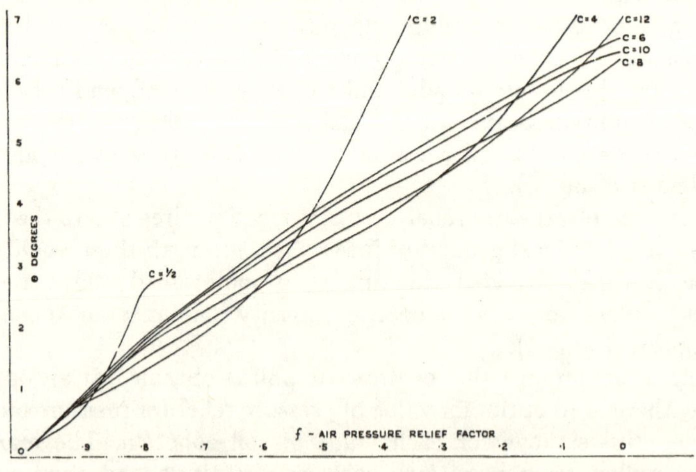

FIGURE 6. Angle of climb vs. pressure relief for Land Rover 9 provided with side wall traction tires

Figure 4). Thin walled, low pressure "terra tires" operate at high deformation on soft ground providing an effective diameter, to which the maximum drawbar pull is proportionate, greatly in excess of any practicable high pressure tire; pressure relief on such a tire would not appreciably improve performance. Improved drawbar pull can also be obtained with tires such as the Dunlop 8.20 × 15 Sand Tire (which has a circular cross section) providing some side wall traction effect.

(6) As for tracked, the effect of a pressure relief on wheeled vehicles running on firm frictional soil is to reduce performance (see Appendix IV, and Figure 2).

(7) It is calculated that slip stall failure does not occur at full loading with the Land Rover 9 on very loose dry sand, but the climb angle is limited to some 15°.* Air pressure relief reduces performance and indeed would be impracticable on this soil due to excavation which would result beneath the vehicle.

3. Tracked and Wheeled Vehicles on Soil with Both Friction and Cohesive Content

Although the full load performance of a cohesive soil would be improved by a frictional content, this improvement would in part be reduced on application of pressure relief.

EXPERIMENTAL VERIFICATION OF THEORETICAL CONCLUSIONS

In order to check the theoretical conclusions of this paper practical tests were made in the F.V.R.D.E. soil pits using specially designed pressure relief rigs (Figures 7 and 8) for the Land Rover 9 and for the Universal Carrier. The pressure relief provided was not by plenumchamber but by a transfer of the load from the vehicle running gear to rails on either side of the soil pits via Dunlop pneumatic suspension units. Two soil types were used, a London clay of low frictional content which was varied in shear strength from 0.6 to 3.8 lb/in.^{2}, and a dry round grained sand (particle size 0.024–0.055 in. with compacted density 108 lb/ft^2, uncompacted 106 lb/ft^2 and angle of friction 32°).

1. Tracked Test Rig

Figure 9 shows experimental values of drawbar pull and rolling resistance on clay for two values of soil shear strength (3.8 and 1 lb/in.^{2}). It

*This is an overestimate due to errors in the method of calculation. Subsequent experiment showed the Land Rover on 7.50 × 16 tires could only just propel itself on loose sand. 8.20 × 15 tires, however, gave a drawbar pull indicating a maximum climb angle of 18° (see Figure 12).

FIGURE 7. Modified Land Rover

FIGURE 8. Modified Universal Carrier

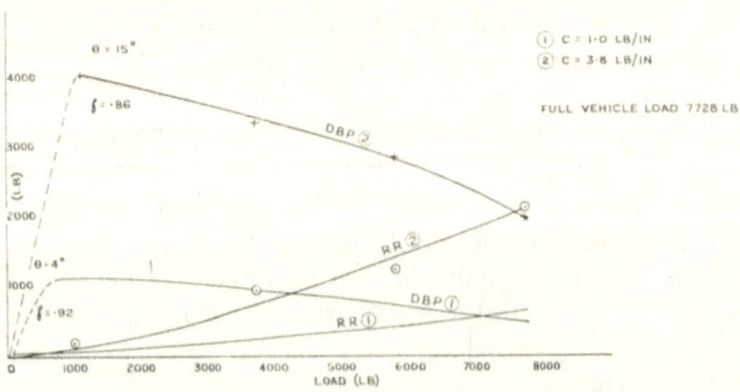

FIGURE 9. Universal tracked carrier on clay (experimental results). Soil shear strength = 3.78 p.s.i.

is seen that drawbar pull is improved by pressure relief up to maximum values at high pressure relief (around 0.9) after which a rapid drop to zero occurs due to reduction of track-soil contact area. The maximum drawbar pull is proportional to the value of soil cohesion; the equivalent climb angles being 15° and 4° for the two values of shear strength.

Figure 10 shows the experimental results on sand, the drawbar pull being at its highest for no pressure relief, the equivalent climb angle being 18°.

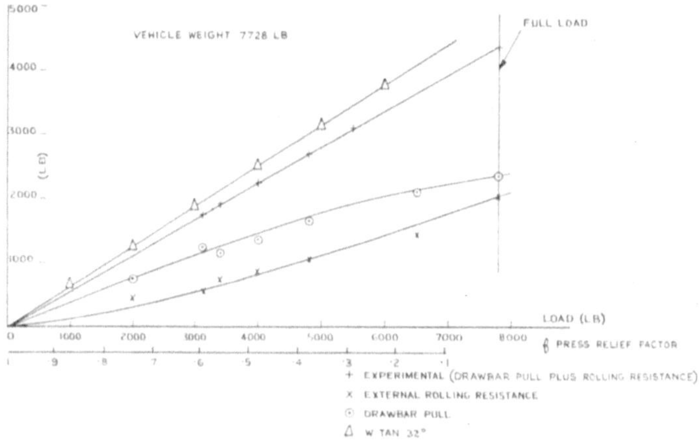

FIGURE 10. Universal carrier on sand ϕ 32 (experimental results).

2. Wheeled Test Rig

With this rig, operating in clay, side wall effects were evident (Figure 11). It was found that on soil of cohesion around 3 lb/in.^{2} torque stall occurred upon application of draught at full load, the drawbar pull at stall corresponding to a climb angle of about 13°; the application of pressure relief under these conditions resulted in slip stall at a reduced drawbar pull.

For weaker soils a maximum drawbar pull was obtainable by adjustment of pressure relief, e.g. at 2 lb/in.^{2} cohesion and relief factor of 0.3

FIGURE 11. Land Rover wheel in clay: note side-wall effect

a climb angle of 7° was obtained, while 1 lb/in.² cohesion and relief factor of 0.7 gave a climb angle of 3°.

The measured maximum in drawbar pull for particular values of shear strength occurred exactly at the loads predicted, although the calculated maxima tended progressively to be lower than the experimental with increase of shear strength. Compare Figure 12 (experimental) with

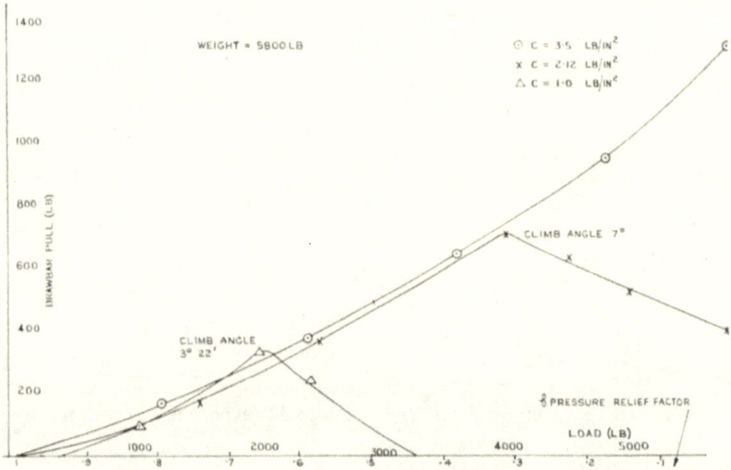

FIGURE 12. Land Rover 109 on clay—d.b.p./load (experimental results)

Figure 13 (theoretical). This effect can in part be explained by a small angle of friction developing in the higher shear strength soil ($\phi = 2°$ for $c = 3.5$ lb/in.²) and in part by the approximate nature of the theory used which, although giving a good values of rolling resistance does not take into account the full arc of contact providing tractive reaction.

On loose sand the Land Rover rig fitted with 7.50 × 16 standard road tires could only just propel itself and slip stall occurred at any attempt at pressure relief. Now the estimated climb angle using the Bekker theory of rolling resistance is 15° (see Figure 2); experimental values of towed rolling resistance (Figure 14) although in agreement at low loading (below 1,000 lb) were, however, some 70 per cent higher than that calculated at full load, this indicating a climb angle of only 6° and more in keeping with the observed performance.

To obtain a drawbar pull-pressure relation the vehicle was fitted with 8.20 × 15 Dunlop sand tires. These gave a lower rolling resistance (Figure 14), a maximum drawbar pull corresponding to a climb angle of 18° 30′, and zero drawbar pull at a load of 1,500 lb ($f = 0.74$).

It is therefore confirmed that pressure relief can only reduce performance

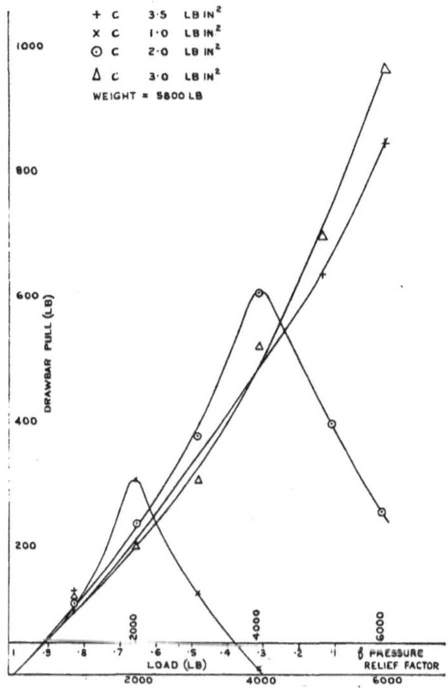

FIGURE 13. Land Rover 109 on clay: theoretical d.b.p./load

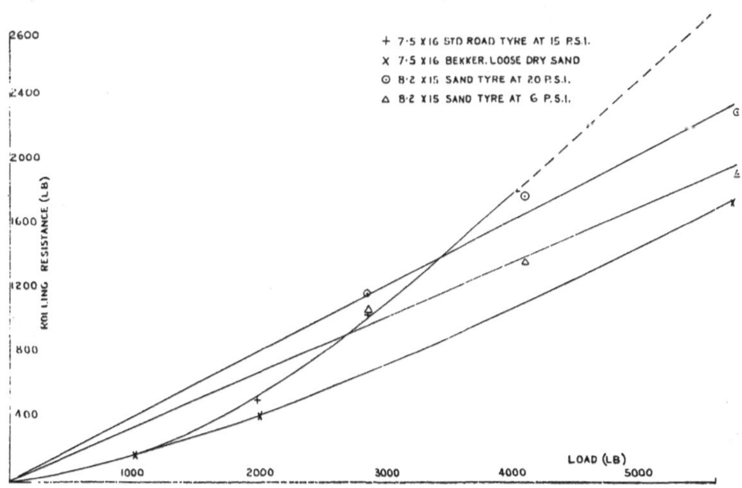

FIGURE 14. Land Rover on sand: rr/load. Experimental results compared with Bekker

of the Land Rover on loose sand, while a pressure increase may be beneficial. Special smooth wall tires of curved section are necessary and even with these a minimum load is required before drawbar pull develops.

OVER-ALL CONCLUSIONS ON THE VALUE OF PRESSURE RELIEF

1. On firm frictional slopes the angle of climb is not determined by the type of running gear (track or wheels) but is limited to the angle of contact friction with the ground; it can only be reduced by pressure relief. A pressure increase can, however, improve the climb angle to any required degree in excess of the angle of contact friction.

2. On loose dry sand the performance of the vehicles considered at normal full loading is again reduced by pressure relief but is more dependent upon the type of running gear.

3. On cohesive slopes an improved performance for tracks can result within the limits determined by the area of the track on the ground and the value of soil cohesion. An equivalent performance may generally be obtained more economically by a small increase of track area. With or without pressure relief an increased track area is necessary for improved climbing on weak soils.

4. For vehicles depending upon rim traction wheels, only small angles of climb may be negotiated on weak cohesive soils, depending upon the value of cohesion and the axle loading; this latter has an optimum value at a fixed value of sinkage, and pressure relief will assist or reduce performance depending upon the sinkage at zero relief.

Better slope performance is obtainable using side wall traction wheels at deep sinkage. Here pressure relief *reduces* climb performance except for the weakest soils, where marginal gains can be obtained and where failure due to lack of ground clearance or sinkage beyond wheel radius may be avoided by its use.

5. The added weight, size, and expense of the variable ground pressure vehicle cannot be justified for the limited benefits obtainable. Pressure relief is only valid for hover over water or marsh under conditions where a tractive reaction can be obtained from the air and is not required from the medium over which the vehicle is crossing.

APPENDIX I

CALCULATION OF PERFORMANCE OF A RIGID TRACK ON LOOSE SAND

Using the Bekker system of constants the level ground rolling resistance of a vehicle having two uniformly loaded rigid tracks is:

$$R_0 = \frac{2}{(n+1)(k_c + bk_\phi)^{1/n}} \left[\frac{W}{2l} \right]^{(n+1)/n} \tag{5}$$

In Equation 4 we have assumed:

$$R_0 = KW^m. \tag{6}$$

Thus by comparison of 5 and 6 we have:

$$K = \frac{2}{(n+1)(k_c + bk_\phi)^{1/n}} \times \frac{1}{(2l)^{(n+1)/n}}; \quad m = \frac{n+1}{n}$$

Suitable constants for very loose dry sand (1), are:

$$k_c = 0; \quad k_\phi = 4.3; \quad n = 1; \quad \phi = 30°$$

whence $K = 0.0582/bl^2$.
For the Oxford Carrier, $l = 78\frac{1}{4}$ in.; $b = 12$ in.; $W = 17{,}920$ lb;
∴ For the soil and vehicle under consideration:

$$K = 7.90 \times 10^{-7}; \; m = 2;$$

$$R_0 = 254 \text{ lb}; \quad \text{tractive reaction } T_0 = W \tan \phi = 10{,}350 \text{ lb}.$$

Substituting the above values of K, m, and ϕ in Equation 4 shows the angle of slope θ which can be climbed with $f = 0$ is about 29° 30′.*

Application of pressure relief reduces the numerical values of both tractive and resistive terms, the net result being a reduction in drawbar pull almost entirely due to the reduction in the tractive term.

Table I compares the slope climbing ability of the Oxford Carrier on firm and soft friction slopes over the full range of pressure relief.

TABLE I

SLOPE CLIMBING ABILITY OF OXFORD CARRIER ON FIRM AND SOFT FRICTIONAL SLOPES (RIGID TRACK THEORY)

	Angle of Climb		
f	Firm Slope θ_f	Soft Slope θ_s	Difference $\theta_f - \theta_s$
0	29° 59′	29° 27′	32′
.25	22° 49′	22° 28′	21′
.5	15° 31′	15° 21′	10′
.75	7° 58′	7° 56′	2′
1.0	0	0	0

*Experimental measure of d.b.p. for the Universal Carrier indicates a worse climb performance, viz. 18° (or 0.55ϕ) than given by the rigid track theory (31° 30′ for soil of $\phi = 32°$). See para. 10.1.

Appendix II

CALCULATION OF ADDITIONAL WIDTH OF TRACK NECESSARY TO GIVE A PERFORMANCE EQUAL TO THAT OF A TRACKED VARIABLE GROUND PRESSURE VEHICLE ON COHESIVE SOIL

Increase of track area both increases the tractive soil reaction on cohesive soil and reduces the rolling resistance. In general, for the reduction of rolling resistance, it is more effective to increase the length of the track rather than the width, but for the vehicle under consideration (the Oxford Carrier) a small increase of width is all that is necessary to equal the performance of the ground pressure relief version; this is demonstrated by the following calculation.

The rolling resistance of a tracked vehicle on cohesive soil may be deduced with good accuracy by Evan's theory (2), (3), (4) of sinkage, viz.:

$$(q_1 - q)/(q_1 - q_n) = 1 - e^{-5(\delta/b)}, \qquad (7)$$

For Oxford Carrier

where $q_1 = W/nbd$; or $\frac{1}{2}(WE/nbr)^{1/2}$ whichever is the smaller.

δ = sinkage
b = width of track = 12 in.
n = number of bogie stations = 8
W = weight of vehicle = 17,920 lb
d = track pitch = 3.43 in.
E = elastic modulus for soil $\sim$1,000 lb/in.^{2}
q = bearing capacity of soil = $8c$ for track at deep sinkage
q_n = nominal ground pressure of track = 9.54 lb/in.^{2}

For the Oxford Carrier climbing an 18° slope on ground of cohesion 3 lb/in.^{2} (which is the calculated performance of the ground pressure effect version at $f = 0.95$).

$$\delta = -(b/5)\log_e[(q \sec\theta - q_n)/(q_1 - q_n)] \qquad (8)$$

(N.B. Sec θ takes charge of the reduction of normal loading due to incline.)

For $b = 12$; $q = 24$; $q_n = 9.54$; $q_1 = 54.4$; $\theta = 18°$, we find $\delta = 2.525$ in. Now the rolling resistance due to sinkage is $\delta q b$ whence for $\theta = 18°$, $c = 3$ lb/in^3, $b = 12$ in., $R_0 = 1.454$ lb.

The drawbar pull necessary to climb a slope $= W \sin\theta = 5{,}530$ lb, while the available tractive soil reaction $T = cA = 3 \times 1{,}878 = 5{,}634$ lb.

∴ Tractive soil reaction necessary to climb slope:

$$\begin{aligned} T_s &= W \sin\theta + R \\ &= 5{,}530 + 1{,}454 \text{ lb} \\ &= 6{,}984 \text{ lb.} \end{aligned}$$

Neglecting the reduction of rolling resistance resulting, this value of T_s can be achieved by increasing the track area in the ratio T_s/T, i.e. to $(6{,}984/5{,}634) \times 12 = 14.9$ in. A recalculation using the new width (15 in.) shows the calculated width for the required performance is correct to the nearest inch ($b = 15$ in.; $c = 3$ lb./in.^{2}; $\theta = 18°$).

It is therefore clear that an increase of track width from 12 to 15 in. will provide the same improvement of performance on cohesive soil calculated for the pressure reduction factor $f = 0.95$. Smaller increases in width are necessary to provide performances of the air pressure relief version on firmer soils at steeper slopes. These are listed in Table II together with the performance of a 2 ft wide track. The vehicle would be unable to cross soil of cohesion less than 1 lb./in.^{2} without further widening of the track, or the use of pressure relief.

N.B. The natural slopes for cohesive soil do not exceed 11°; particularly for the weaker soils considered slopes in excess of this value are unlikely to be encountered.

TABLE II

TRACK WIDTHS NECESSARY WITHOUT PRESSURE RELIEF

θ degrees	c lb/in.^{2}	δ in.	R lb	T ($=cA$) lb	T_s ($=W\sin\theta + R$) lb	b in.
0	3	2.72	1566	5634	1566	12
18	3	2.52	1454	5634	6984	14.9
30	4.8	2.25	1297	9000	10257	13.8
45	6.7	1.47	848	12590	13508	13.0
36	3	0.74	854	11260	10406*	24
18	2	2.67	2048	7500	5425*	24
0	1	9.61	3690	3750	60*	24

*$(T - R)$ values.

APPENDIX III

SLOPE CLIMBING CAPABILITIES ON FIRM FRICTIONAL SURFACES USING PRESSURE INCREASE

Equation 4, omitting the cohesive and rolling resistance terms, simplifies to $f = -\text{cosec}\,\phi \sin(\theta - \phi)$; for $\phi = 30°$; $f = -2\sin(\theta - 30)$. Values of θ for particular values of f and ϕ are as follows:

f	ϕ	θ
$+1$	All values	0
0	All values	ϕ
$-\frac{1}{2}$	30°	{44° 26′, 195° 25′
-1	30°	{60°, 180°
$-\sqrt{3}$	30°	{90°, 150°
-2	30°	120°

The values for $\phi = 30°$ are plotted in Figure 2B. These show a maximum negative pressure relief factor of 2 is required on a 120° slope. It is suggested a "pressure increase" vehicle could be developed for cliff-scaling purposes.

Appendix IV

RIM TRACTION THEORIES FOR CALCULATING THE PERFORMANCE OF A WHEELED VEHICLE ON PURELY COHESIVE AND PURELY FRICTIONAL SOILS

1. Equilibrium Equation for Wheels on Purely Cohesive Soil

(*a*) *General Equation*

The general equation of equilibrium of a vehicle on a soil slope, Equation 3, may be written

$$W \sin \theta = T - R \tag{10}$$

where T is the tractive soil reaction.

A standard type pneumatic road tire running on a purely cohesive soil when inflated at a pressure well in excess of the surface bearing capacity may be regarded for the purpose of calculation as a rigid cylindrical wheel; the simple plastic theory of soil deformation (5) with consideration of rim traction only may then be taken to apply. Equation 10 may then be written:

$$\begin{aligned} W \sin \theta &= cbl - qb\delta \\ &= cb(1 - N_c\delta) \end{aligned}$$

Substituting $l^2 = \delta(D - \delta)$:

$$W \sin \theta = cbD[\sqrt{\Delta - \Delta^2} - N_c\Delta] \; ; \tag{11}$$

where l = effective length of footing (component of contact arc parallel to initial soil surface)

b = breadth of footing

D = wheel diameter
δ = sinkage*
$\Delta = \delta/D$
N_c = Terzaghi surface bearing capacity factor for the appropriate footing

(*b*) *Critical Sinkage Condition*

For the slip stall condition $T - R = 0$, and

$$\Delta_{crit} = 1/[N_c^2 + 1]. \tag{12}$$

It will be noted that Δ_{crit} is dependent only upon the bearing capacity factor and is independent of the value of soil cohesion. For a strip footing $N_c = 5 - 7$ and $\Delta_{crit} = 0.03$. For a circular footing $N_c = 7.4$ and $\Delta_{crit} = 0.018$. Δ may be estimated from the relationship:

$$\Delta = [W/qbD]^2, \tag{13}$$

where $q = cN_c$; W = wheel loading normal to the soil surface.

Thus to support the passage of a particular wheel the cohesion must be in excess of the value:

$$c_{min} = W/(bDN_c\sqrt{\Delta_{crit}})\ ;$$

$$c_{min} = (W/bD)[1 + 1/N_c^2]^{1/2} \tag{14}$$

For the strip footing assumption $c_{min} = 1.015W/bD$. For the circular footing assumption $c_{min} = 1.009W/bD$.

(*c*) *The Condition for Maximum Drawbar Pull*

The Formula 11 exhibits a maximum in drawbar pull with respect to Δ when:

$$d(W \sin\theta)\, d\Delta = 0;$$

i.e., when:

$$\Delta = (1/2)[1 - N_c/\sqrt{1 + N_c^2}]\ ; \tag{15a}$$

or approximately:

$$\Delta^{\frac{1}{2}} \simeq 1/(2N_c)\ ; \tag{15b}$$

whence, for the strip footing assumption, $\Delta = 0.0077$ and, for the circular footing assumption, $\Delta = 0.0046$.

Substituting for Δ in Equation 15 in terms of wheel dimensions and load, we find the maximum drawbar pull is obtained when:

$$W \simeq \tfrac{1}{2}(cbD). \tag{16}$$

*Sinkage should not be confused with rut depth which may be considerably less due to plastic recovery.

It has a value (determined from Equations 11 and 15):

$$(T - R)_{max} \simeq cbD/4N_c. \qquad (17)$$

(*d*) *Load Conditions with Pressure Relief*

Here the effective load W_e for the calculation of sinkage (Equation 13) is:

$$W_e = W_0(\cos\theta - f) \qquad (18)$$

$= W_0(1 - f)$ for horizontal or small inclination running, where W_0 = actual wheel load without pressure relief.

2. Performance of Land Rover 9 Wheel

Tire 7.50 × 16; b = 7.5 in.; D = 31 in.; maximum load per wheel W_0 = 1,428 lb.

(*a*) *The Minimum Value of Soil Cohesion to support Passage of the Wheel*

From Equation 14:

$$c_{min} = 6.14[1 + 1/N_c^2]^{1/2}$$

For the strip footing assumption, N_c = 5.7; c_{min} = 6.24 lb./in.^{2}
For the circular footing assumption, N_c = 7.4; c_{min} = 6.19 lb./in.^{2}

(*b*) *The Maximum Angle of Climb and the Effect of Pressure Relief*

The maximum angle of climb can be estimated for soil of cohesion greater than the critical value from Equation 11.

Air pressure relief will modify this maximum and allows the effective load W_e to be adjusted to the optimum value (W_{opt} = [1/2]cbD) giving the maximum drawbar pull and climb angle for the particular value of cohesion. [E.g. using N_c = 5.7; c = 6.5 lb/in.^{2} – angle of climb without pressure relief = 40′; angle of climb with optimum pressure relief (f = 0.47) = 2° 40′.] Slightly smaller values are obtained with N_c = 7.4 assumption.

The benefit of pressure relief decreases with increasing value of cohesion (see Figure 3), and at c = 12.3 lb/in.^{2} the optimum pressure relief is zero. Above this value pressure would have to be increased in order to operate at optimum load.

Pressure relief allows the wheel to operate at values of cohesion below the critical for the full load (6.25 lb/in.^{2}) but must be adjusted to near the optimum value for soil cohesion. As the maximum angle of climb is proportional to the value of cohesion the use of conventional size wheels on ground effect machines is impractical on very soft ground.

The improved performance deduced for the large wheel Land Rover is shown in Figure 4. Here a pressure increase and not relief would improve performance for soils of cohesion greater than 6 lb/in².

Much improved angles of climb are obtainable using wheels providing a side wall traction effect. This is evaluated in Appendix V, and may be appropriate to Circular Section tires operating on "sticky" clay (see Figure 11).

3. *Equilibrium Equation for Wheels on Frictional Slopes*

(*a*) *Firm Slopes*

On firm frictional slopes where both cohesive and rolling resistance terms may be neglected the equilibrium equation (Equation 4) reduces to (as for the tracked vehicle):

$$\sin \theta = [\cos \theta - f] \tan \phi \tag{19}$$

and the maximum angle of slope for positive pressure relief is when $f = 0$ and $\theta = \phi$.

(*b*) *Soft Slopes*

On soft frictional slopes the rolling resistance term of Equation 4 cannot be neglected. It may be evaluated from the Bekker's formula* for the first pass rolling resistance of a rigid towed wheel, viz.:

$$R_{\text{wheel}} = \frac{1}{(3-n)^{(2n+2)/(2n+1)}(n+1)(k_c + bk_\phi)^{1/(2n+1)}} \left[\frac{3W}{\sqrt{D}}\right]^{(2n+2)/(2n+1)}. \tag{20}$$

Any further pass in the same rut should result in some additional compaction of the soil but reduced rolling resistance.

Using the known constants for very loose dry sand: viz., $n = 1$; $k_c = 0$; $k_\phi = 4.3$; $\phi = 30°$, we obtain: $R = 0.5288[W^4/bD^2]^{1/3}$; and using the Land Rover 9 wheel parameters, $b = 7.4$ in.; $D = 31$ in.; we obtain:

$$R_{\text{wheel}} = 0.0273 W_e^{4/3};$$

and Equation 4 becomes:

$$\sin \theta = (\cos \theta - f) \tan \phi - 0.0273 W^{1/3} (\cos \theta - f)^{4/3}. \tag{21}$$

It is interesting to note that according to this equation slip stall on level ground ($\theta = 0$) does not occur until $W = 9.400$ lb.

*This is known to give low values of rolling resistance on soft soil, by as much as 50 per cent at high loadings. The climb performance can therefore be taken as inferior to that calculated.

Angles of climb calculated from the equation with the normal wheel load of $W = 1{,}428$ lb are listed against pressure relief in Table III.

TABLE III

CLIMB ANGLE ON SOFT FRICTIONAL SOIL

f	θ
0	15° 16′
.25	12° 36′
.5	9° 25′
.75	5° 25′
1.0	0

It will be seen that, as for firm frictional soil, pressure relief on the soft soil has the effect of reducing the angle of climb. Equation 20 shows the rolling resistance for loose sand is inversely proportional to the cube root of bD^2. As the tractive reaction remains unchanged, steeper slopes could be climbed on tires of larger dimensions, particularly diameter.

The use of air pressure relief would appear undesirable on this type of soil for the further reason that excavation would undoubtedly occur beneath the vehicle, resulting in failure when the artificial slope produced exceeds the (reduced) climb angle.

APPENDIX V

TRACTIVE REACTION ON PURELY COHESIVE SOIL OF A WHEEL PROVIDING SIDE WALL TRACTION

The effect of radial ribs providing a soil shear surface over the full area of contact of the side walls of the wheel with the walls of the rut is to add a component of soil reaction of maximum value T_{sw} to that provided by the rim T_r, thus for a first pass:

$$T_{sw1} = (1/2)cD^2[\sin a - \cos^2 a \log(\sec a + \tan a]) \tag{22}$$

where a is the angle of sinkage and $\sin a = 2\sqrt{\Delta - \Delta^2}$; $\cos a = (1 - 2\Delta)$.

The maximum drawbar pull of a wheel including rim reaction and rolling resistance in addition to side wall reaction is now

$$(T - R)_1 = (1/2)cDb[\sin a - N_c(1 - \cos a)] + (1/2)cD^2[\sin a - \cos^2 a \log(\sec a + \tan a)]. \tag{23}$$

A plot (Figure 5) of $(T - R)$ against a for the particular circumstances $b = 7.5$ in.; $D = 31$ in.; $N_c = 5.7$; $c = 1$ lb/in.^{2}; shows slip stall is delayed until 81° (13 in. sinkage) and two maxima in drawbar pull exist at 20° and 65° corresponding to 1 in. and 9 in. sinkage respectively.

Sinkage is incremental on cohesive soil and we now assume T_{sw2} for a second wheel following in the rut of the first as being obtained at double the determined sinkage of the first (strictly this is only correct for sinkage $>2b$). The main drawbar pull per wheel for the two wheels in tandem is then:

$$(T - R)_{\text{mean 1 and 2}} = cDb[\sqrt{\Delta - \Delta^2} - \Delta N]$$
$$+ \tfrac{1}{2}cD^2\Big[\sqrt{\Delta - \Delta^2} - \tfrac{1}{2}(1 - 2\Delta)^2 \log\left(\frac{1 + 2\sqrt{\Delta - \Delta^2}}{1 - 2\Delta}\right)$$
$$+ \sqrt{2\Delta - 4\Delta^2} - \tfrac{1}{2}(1 - 4\Delta)^2 \log\left(\frac{1 + 2\sqrt{2\Delta - 4\Delta^2}}{1 - 4\Delta}\right)\Big] \quad (24)$$

A plot of this function against a (Figure 5) shows a considerable increase in mean drawbar pull; failure is now assumed to occur at 60° sinkage for the first wheel (90° for the second wheel).

When plotted against a pressure relief factor for an initial load per wheel W_0 = 1,428 lb (appropriate to the Land Rover 9), and for various values of soil cohesion (Figure 6), it will be observed that the maximum drawbar pull per wheel in the soil cohesion range 6–12 lb/in.^{2} is at zero pressure relief, and has an approximate uniform value of 160 lb corresponding to a slope climbing ability of 6° 20′. Higher values are indicated for deeper sinkages and weaker soils; only for extremely weak soils, where $c < 1$ lb/in.^{2} would a pressure relief system appear advantageous in limiting sinkage to below the slip stall value ($\Delta = \frac{1}{4}$ or $\delta = 7\frac{3}{4}$ in. per wheel).

It would therefore appear advantageous to run two side wall traction wheels in tandem on soil of cohesion >2 lb/in.^{2}; on weaker soils or for exceptional wheel loads it would appear advantageous to run wheels on separate tracks, or to have recourse to a pressure relief system to limit sinkage Δ to $<\frac{1}{4}$.

REFERENCES

1. Janosi, Z., *Proc. 1st Int. Conf. Mech. of Soil-Vehicle Systems*, Torino-Saint Vincent, 1961, p. 769.
2. Evans, I., *British J. Applied Physics*, Vol. 4, November 1953, p. 330.
3. Evans, I., *J. of Terramechanics*, Vol. 1, No. 2, 1964, p. 33.
4. Uffelmann, F. and I. Evans, *J. of Terramechanics*, Vol. 2, No. 2, 1965, p. 33.
5. Uffelmann, F., The Performance of Rigid Cylindrical Wheels on Clay Soil, *Proc. 1st Int. Conf. of Soil-Vehicle Systems*, Torino-Saint Vincent, 1961.
6. Bekker, M. G., *Theory of Land Locomotion*, Ann Arbor: The University of Michigan Press, 1956.

Description of Wheel Performance on Dry Sand In Terms of Energy Parameters

E. M. LEFLAIVE*

ABSTRACT

This paper describes in qualitative terms some typical aspects of rolling motion of pneumatic-tired wheels in dry sand. The positive slip range, between the self-propelled condition and 100 per cent slip, is considered. In this range, three phases of the phenomenon are distinguished according to the rate of variation of torque and pull. The parameters used for the description are the following energy coefficients: torque energy coefficient, pull energy coefficient, and dissipated energy coefficient. Experimental results are given to illustrate the validity of this approach.

The first part of the paper describes the three phases of rolling motion with respect to slip and defines the energy parameters. The second part is a discussion of the observed effects of wheel load, sand strength, and tire characteristics upon (*a*) the rate of increase of torque at low slips, (*b*) the value of the maximum pull/load ratio, and (*c*) the rates of increase of torque energy and dissipated energy at medium and high slips.

This discussion illustrates the physical understanding of rolling motion in sand that can be gained from the comparative study of each phase for various tires in various load and soil conditions. Such a comparison provides means of specifying the differences in performance between different tires. The discussion also examines the agreement between observed facts and existing theories. It is found that important aspects consistently observed in tire performance are not predicted by any present theory.

The studies reported herein were conducted under the sponsorship and guidance of the Directorate of Research and Development, U.S. Army Materiel Command.

*Engineer, Mobility Section, Army Mobility Research Branch, Mobility and Environmental Division, USAE Waterways Experiment Station, Vicksburg, Miss.

The Mechanics of Tractor Operation on Tilled Soils

ARNOLD G. BERLAGE

WESLEY F. BUCHELE*

ABSTRACT

An instrumented tractor was operated on rotary tilled soil pulling a constant drawbar load. Using measured parameters, a new weight transfer equation was compared with the classical equation. The new equation predicts weight transfer with a 2 per cent error while the classical equation had a 32 per cent error at maximum drawbar pull.

*Agricultural Engineer, ARS, USADA, Wenatchee, Washington; Professor of Agricultural Engineering, Iowa State University, Ames, Iowa.

Theory of Bulldozer Action in Friable Soil

P. E. R. COOK

A. R. REECE*

ABSTRACT

The mechanism of failure of friable soil in front of a fully loaded bulldozer blade is described and used in a quantitative analysis of the forces involved. These are large due to the heavy surcharge caused by the weight and flow of the accumulated soil. The theory is corroborated by full scale experiments in the field.

NOTATION

ϕ	Angle of shearing resistance (undisturbed) (degrees)
ϕ_1	Angle of shearing resistance (disturbed) (degrees)
γ	Soil density (undisturbed) (lb. in.$^{-3}$)
γ_1	Soil density (disturbed) (lb. in.$^{-3}$)
δ	Angle of soil/metal friction (degrees)
C_a	Soil/metal adhesion (lb. in.$^{-2}$)
C	Cohesion (undisturbed) (lb. in.$^{-2}$)
C_1	Cohesion (disturbed) (lb. in.$^{-2}$)
N	Dimensionless factor
F, Q, P, S, G, V, H	Forces (lb.)
q	Equivalent surcharge (lb. in.$^{-2}$)
α	Rake angle (degrees)
z	Depth of cut (in.)
b	Breadth of blade (in.)
h	Blade height (not cutting) (in.)
m	Rupture distance (in.)
D_c	Horizontal draught force (unsurcharged) (lb.)
D_l	Additional horizontal draught force because of surcharge and sliding force (lb.)
L_f	Lifting force (lb.)

THERE CAN BE FEW machines whose basic functional form has been subjected to as little theoretical study and analysis as earthmoving machinery. On the other hand the mechanical construction of these machines has by virtue of the peculiarly arduous nature of their task reached a quite remarkable state of mechanical excellence. This has resulted in a rather odd combination of high quality complex machinery harnessed to a soil engaging form of remarkable crudity that has not changed noticeably in the last thirty years.

This situation is beginning to change and serious attention is at last being given to the possibility of changing the form of earthmoving machines in order to improve their efficiency. The incentive to take a

*Department of Agricultural Engineering, University of Newcastle upon Tyne.

more scientific view of earthmoving machinery design comes principally from the requirements of military, space, and commercial operations. The emphasis on air portability for military equipment has placed a great premium on high power to weight ratio. The army's earthmoving equipment has proved to be one of the most difficult to lighten. The possibility of lunar landings and the subsequent construction of shelters on the moon's surface has demanded some thought about the economical provision of moon-moving equipment. This immediately raises the whole question of the theory of earthmoving in relation to soil properties. Commercial pressure for more effective earthmoving machinery is growing, particularly as large scale earthmoving tackles ever more difficult terrain.

The history of the development of a quantitative analysis of the action of cutting tools in soil has been described by Hettiaratchi (1). Following the work of Payne (2) on the action of agricultural tines and Bekker's (3) success with the study of vehicle performance, Caterpillar Tractor Co. was perhaps the first to apply soil mechanics thinking to the design of more efficient earthmoving machinery. The soil mechanics group at Caterpillar analysed the forces involved in pushing soil up into a scraper bowl in a qualitative manner. They have used the ideas of classical soil mechanics in qualitative analyses, which have suggested the way in which the scraper form can be altered to give easier loading and these changes in form have been investigated using scale models (4) and (5). The results of the scale model tests have been related to full-scale scraper operations by means of dimensional analysis. These two techniques together have given rise to the important development of the slope-sided scraper and an improved apron which are described using the language of soil mechanics in two recent American patents (6) and (7). The improved apron design is now in use on all current Caterpillar scrapers and the slope-side is to be found on the latest Le Tourneau scrapers. These changes give rise to a very great reduction in loading force.

Caterpillar were followed by General Motors in this field and Emori and Schuring (8) have recently made valuable contributions in the field of dimensional analysis.

The next step should clearly be to attempt a quantitative analysis of the forces involved in earthmoving operations. If this could be done then it would permit rapid investigation of novel designs on paper. This approach has been pursued by the authors and their colleagues in the Department of Agricultural Engineering at the University of Newcastle upon Tyne over the last few years. Osman (9) showed the retaining wall theory developed by Ohde (10) could accurately predict the forces required to cut a wide range of soils. He showed that a computer programme originally developed by Lotkin and Bekker (11) could carry out the lengthy calculations quite speedily. Although this was a step forward it required that a cutting force calculation of the simplest kind could only be made by a complicated programme and the use of a digital computer. Reece (12) showed how it was possible to generalize the results of cutting force calculations by means of the following equation:

$$Q\ (\text{lb}) = \gamma z^3 N_\gamma + c z^2 N_c + c_a z^2 N_a + q z^2 N_q \tag{1}$$

This equation applies to the force required to produce failure of the soil beneath any form of loading structure.

Hettiaratchi (13) has shown that although this equation is not absolutely correct it can be applied to wide flat raked cutting blades with only a small error. For such a blade the equation reduces for the total force per unit width of blade to the following form:

$$T\ (\text{lb. in.}^{-1}) = \gamma z^2 N_\gamma + c z N_c + c_a z N_a + q z N q \tag{2}$$

The four N factors have been computed for a complete range of values of ϕ for the two extreme values of δ, that is $\delta = 0$ and $\delta = \phi$, for values of the rake angle α varying from $45° - \phi/2$ to $135 + \phi/2$. Hettiaratchi has also produced suitable interpolation formulae and graphs showing the rupture distance as a function of the soil and blade variables.

The existence of the computed N values has made it possible to develop a quantitative theory for the bulldozing of friable soils. It has been pointed out by Drees (14) that there are two distinct mechanisms of bulldozing action. In friable soils a heap builds up in front of the blade whereas in highly cohesive soils a "swiss roll" is produced and rolls along in front of the bulldozer. The following work puts forward a theory for the force required to push a fully loaded bulldozer blade with a triangular heap of soil in front of it.

THEORY

The initial cut made by a bulldozer is an excellent example of a wide flat angled blade as considered by Hettiaratchi. The N factors in Hettiaratchi's graphs refer to the total force on the cutting blade which acts at an angle of δ to the normal to the blade. It will be seen from Figure 1 that the horizontal component of the total force will be given by the total force multiplied by $\cos\theta$ where $\theta = 90 - \alpha - \delta$. The draught force D_c required to make the initial cut at a depth z with no surcharge is obtained by a straightforward application of equation 2 and is as follows:

$$D_c = b\{\gamma z^2 N_\gamma \cos\theta + c z N_c \cos\theta + c_a z (N_a \cos\theta + \cot\alpha)\} \tag{3}$$

The last term takes account of the actual force of soil adhesion acting up

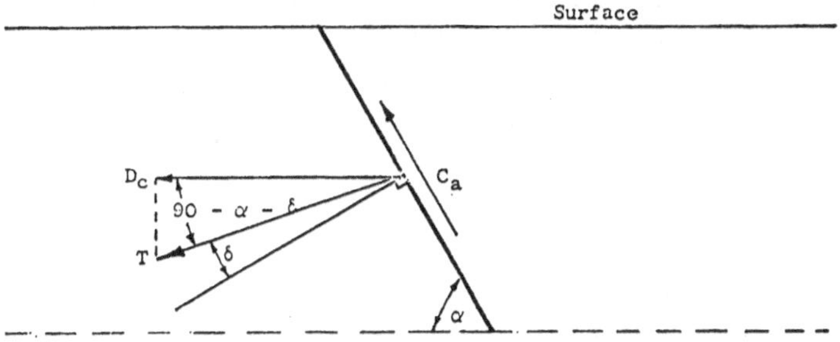

FIGURE 1. The relationship between the direction or the total cutting force F and the draught force D_c

the blade. The values ϕ, δ, c, c_a, and γ in this equation are those appropriate to the soil being cut in its dense and undisturbed state.

The soil arrangement that is initially assumed for the fully loaded bulldozer is shown in Figure 2. The triangular wedge CED is assumed to

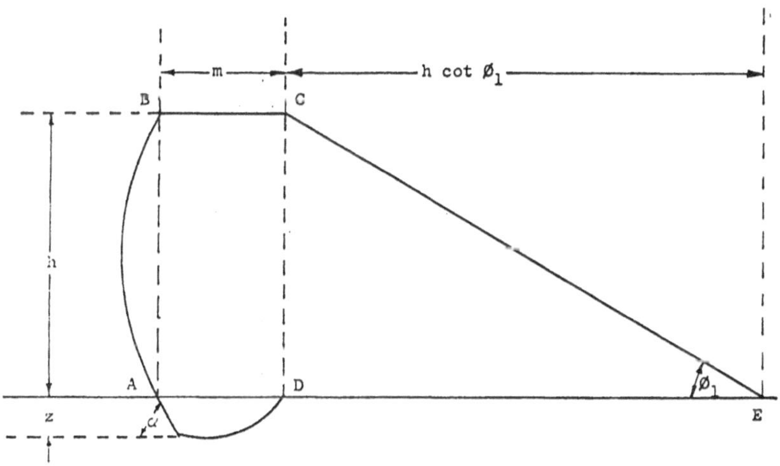

FIGURE 2. The assumed average dimensions of the soil accumulated in front of a fully loaded bulldozer blade

have the angle CED equal to ϕ_b. The triangle CDE is assumed to slide along DE, the surface left from the previous cut and the rectangular piece ABCD of width m, equal to the rupture distance, and height h is assumed to move vertically upwards. ABCD is pushed against the blade by the horizontal force required to push the front triangle along and this gives rise to vertical forces at the blade and at the interface CD. It will therefore be seen that the total cutting force is that given by equation 3 for the initial cut plus the effect of surcharging the cut and pushing along CDE. The surcharge on the cut is given by the weight of ABCD together with the sliding force along the blade and the shearing force up CD.

The force required to slide the soil up the curved surface of the blade can be approximately determined by the curved blade hypothesis, put forward by one of the authors and verified by Osman (9). This hypothesis states that the force required to slide soil along a curved surface is always more than that required to slide it along the straight chord across the curve. It therefore follows that the minimum possible force is given by that required to slide the soil on the chord assuming this to be made of the same material as the curved blade. On the other hand the maximum force is given by the straight chord made of soil. The actual force cannot be greater than this because otherwise the soil can simply fill up the blade converting it to a straight chord. Since bulldozer blades have a large radius of curvature it seems that only a slight underestimation will be made if they are taken as being straight.

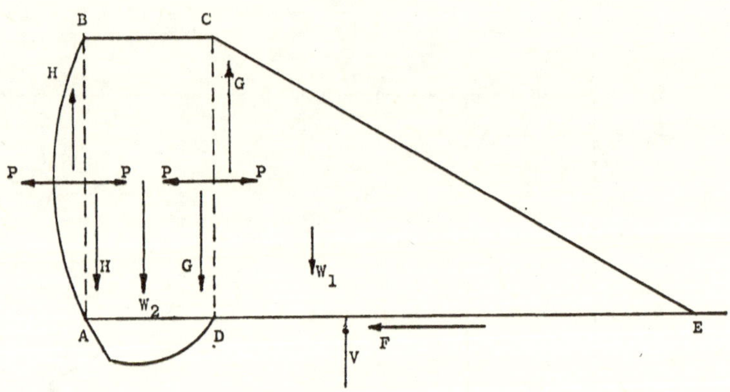

FIGURE 3. The forces acting on and within the heaped soil in front of a bulldozer blade

The soil in the heap in front of the blade is looser than that in front of the cutting edge. It is therefore necessary to modify the soil and soil/metal properties to allow for the disturbance that has occurred. It is general experience that ϕ will vary by about 7° in a sandy soil between the limits of maximum and minimum density. It is therefore suggested that in the heap ϕ_1 is taken to be 5° less than ϕ. The angle of soil to metal friction is probably unchanged. The adhesion depends upon the development of moisture tension between the blade and the soil. It is unlikely that this will change as the soil moves up the blade and therefore it is reasonable to assume adhesion similar to that on the cutting edge. The density of the soil will probably diminish by about 10 per cent and therefore $\gamma_1 = 0.90\gamma$. The cohesion will be considerably reduced and in view of the very thorough loosening action it is probably reasonable to assume that the cohesion is reduced to half what it is in the original soil, i.e. $c_1 = 0.50c$.

Considering the equilibrium of the triangle CDE for unit width of blade, the horizontal force required to slide the triangle along DE, the surface left from the previous cut, is given by: $F = V \tan \phi_1 + h \cot \phi_1 c_1$. The vertical force acting on CD will be given by $G = P \tan \phi_1 + hc_1$. Considering the vertical equilibrium of the triangle we have, since $P = F$, $W_1 = V + \{(V \tan \phi_1 + hc_1 \cot \phi_1) \tan \phi_1 + hc_1\}$, from which $V = (W_1 - 2hc_1)/(1 + \tan^2\phi_1) = (\frac{1}{2}\gamma_1 h^2 \cot \phi_1 - 2hc_1)/(1 + \tan^2\phi_1)$. Therefore the draught force F on the bottom of the triangle is given by:

$$F = \{(\tfrac{1}{2}\gamma_1 h^2 \cot \phi_1 - 2hc_1) \tan \phi_1/1 + \tan^2\phi_1\} + hc_1 \cot \phi_1. \quad (4)$$

The minimum vertical force H acting along the soil/blade interface using the curved blade hypothesis is given by $H = P \tan \delta + hc_a$. Therefore the total surcharging force on the cut is given by:

$$S = G + H + W_2 = F(\tan \phi_1 + \tan \delta) + hc_1 + hc_a + hm\gamma_1 \quad (5)$$

The draught D_t which is due to the cut soil being carried over the uncut surface of the soil by the blade, is, therefore, the sum of the additional cutting force due to the surcharge and the force required to slide the triangle CDE along the soil surface. That is:

$$D_t = b[(S/m)zN_q \cos \theta + F] \quad (6)$$

where F and S are obtained from equations 4 and 5.

For the above situation to be acceptable each part of the failure pattern must be in equilibrium and it is therefore necessary as a check to consider the rotational equilibrium of the triangle CDE. There are three forces, the location and direction of which are known. These are G, F and W_1 and they are shown as full arrows on Figure 4. The dotted arrow forces V and P cannot be accurately located. It would be reasonable to assume that if the triangle was stationary the pressure beneath the triangle would be distributed proportional to the depth of soil in it, therefore V would act along the line of W_1. A common assumption made in soil mechanics would place P at somewhere between a half and a third of h above the base line. If these positions for P and V are accepted then it will be observed that the triangle is not in equilibrium. There is a couple $P(h/3) + G(h/3) \cot \phi_1$, which is acting to tip the triangle forwards. This couple would cause V to shift forward which would reduce the couple but this would cause sheer stresses on vertical planes. It is

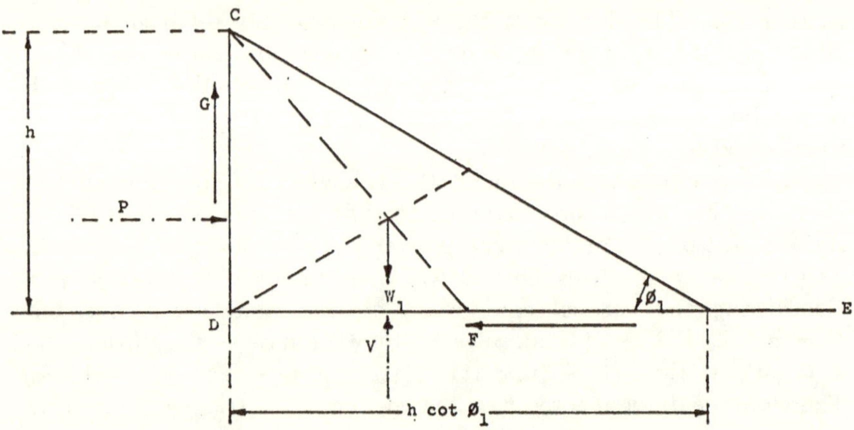

FIGURE 4. The forces involved in the equilibrium of the triangular soil mass *CDE*

most unlikely in fact that the triangle could exist under this force situation. It was therefore decided to carry out experiments designed to establish the nature of the actual failure mechanism in front of the blade.

The soil flow in front of curved and straight blades whilst bulldozing was observed through a glass-fronted box. It was found that the soil leaving the cutting zone moved up within the area ABCD the width of which is equal to the rupture distance. Thus the interface CD acts in a similar manner to a cutting blade, on which $0 > \delta > -\phi_1$, moving horizontally into the soil mass CDE. Were the soil in a rigid state then the failure surface, given by soil mechanics theory, would be an inverted log spiral, Figure 5. The upper failure surface satisfies the stress conditions on the blade for $\delta = -\phi_1$, and also the surface conditions, i.e. it is tangential to the blade surface and also to the soil surface. The lower failure surface satisfies the blade conditions for $\delta = 0$, by being tangential to a line making an angle of $45 + (\phi_1/2)$ to the blade surface, and also satisfying the surface conditions.

It was observed, Figure 6(A), that even in a loose state a similar curved failure surface occurred within the zone CDE. A section of cut soil remained stationary relative to the surface whilst the rest rotated over this section. The interface CD as it moves forward collects soil from the rear of this section, the size of which is kept constant by soil being deposited at the front. Thus the soil within the section CDE is in a continual state of rotation, not because of any curvature of the blade, but because of the negative value of δ on the interface CD, which is caused by the vertical movement of the soil ABCD.

Experiments to confirm this were carried out by arranging a negative

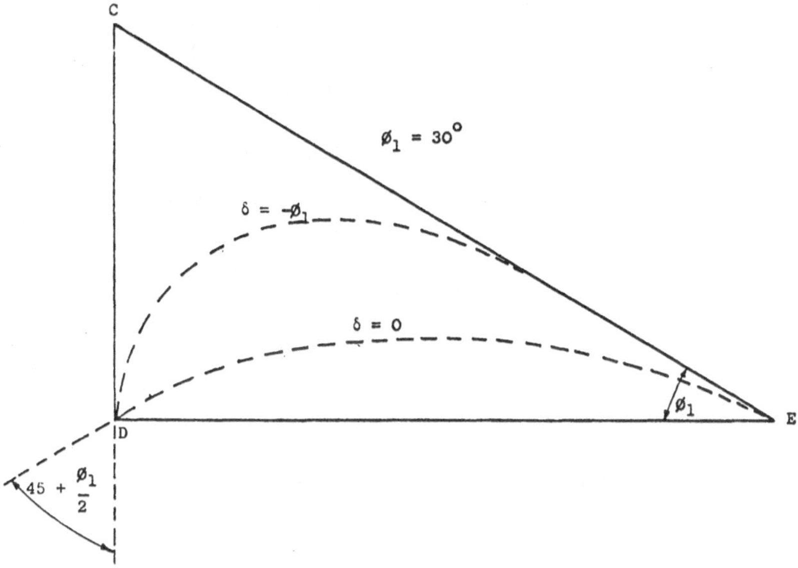

FIGURE 5. Theoretical failure surfaces between a vertically moving surface *CD* (giving negative δ) and a soil slope at the maximum angle ϕ_1 for a cohesionless soil

value of δ on a flat cutting blade. This was done by causing a continuous strip of cloth to move up the face of the blade, as it was moved forward. Figure 6(B) shows its action on a triangular heap of loose soil placed in front of the blade. This photograph, taken with the blade and camera fixed whilst the soil-box was moved, clearly shows a curved failure surface similar to that described. The soil in the stationary zone remains fixed relative to the compacted underlying soil, whilst the remainder rotates over it. However since the soil in the heap in front of a bulldozer is in a loose state, an analysis by this method of the forces involved would lead to questionable conclusions.

It seems reasonable therefore to take the previous equation for the draught force as an upper value which assumes that full shear stresses are developed on the plane CD. A lower limit would be given by the situation in which there was no shearing stresses at all on the plane CD. For this condition the following equations apply: $G = 0$; $P = F$; $V = W_1 = (1/2)h^2\gamma_1 \cot\phi_1$; $F = W_1 \tan\phi_1 + hc_1 \cot\phi_1$

$$F = (1/2)h^2\gamma_1 + hc_1 \cot\phi_1 \tag{7}$$

$$H = P \tan\delta + hc_a$$

$$S = F \tan\delta + hc_a + hm\gamma_1 \tag{8}$$

FIGURE 6A. Soil failure in front of a straight vertical blade which is bulldozing

$$D = b[(S/m)z_l\,Nq\cos\theta + F] \tag{9}$$

where F and S are determined from equations 7 and 8.

It would seem reasonable to assume that the nearest value for the cutting force would be given by the mid-point between equations 6 and 9.

EXPERIMENTAL VERIFICATION

Experimental verification of the theory was attempted using a Caterpillar D.8. tractor and blade on two sites at the M.E.X.E. experimental area near Christchurch, Hants.

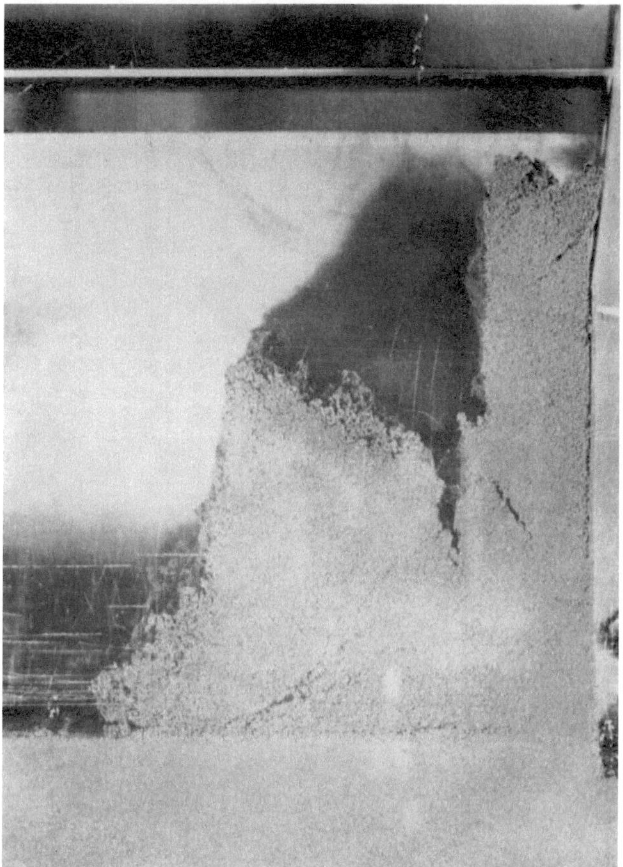

FIGURE 6B. Failure of soil in a loose state being pushed by a vertical blade with a negative δ (produced by a cloth that is moving upwards)

In order to measure the force required to push the bulldozer the trunnion pins were strain-gauged in such a way that only the horizontal force acting on them was measured. In its standard form the cables which support the blade make a small angle to the vertical and hence impose a horizontal force on the blade. For the purposes of the experiments the cable arrangement was modified to make it vertical. The strain-gauged trunnion pins, Figure 7, were calibrated on a simple rig in a large testing machine in the laboratory. They were also calibrated, mounted on the tractor, by applying dead weights over a pulley to the blade.

FIGURE 7. One of the strain gauged cantilever trunnion pins

FIGURE 8. The Caterpillar D8 tractor and blade pushing a full load of soil

A Mirror Galvanometer Recorder was used to measure the horizontal force acting on the blade. Figure 9, soil 1 and soil 2 are two traces typical of those obtained in these soils. l, the theoretical distance required to obtain a fully loaded blade starting from nothing is given by:

$$l = (mh/z) + (h^2 \cot \phi_1/2z)$$

which for soil 1 comes to 28 ft. and soil 2 to 26 ft. The experimental runs

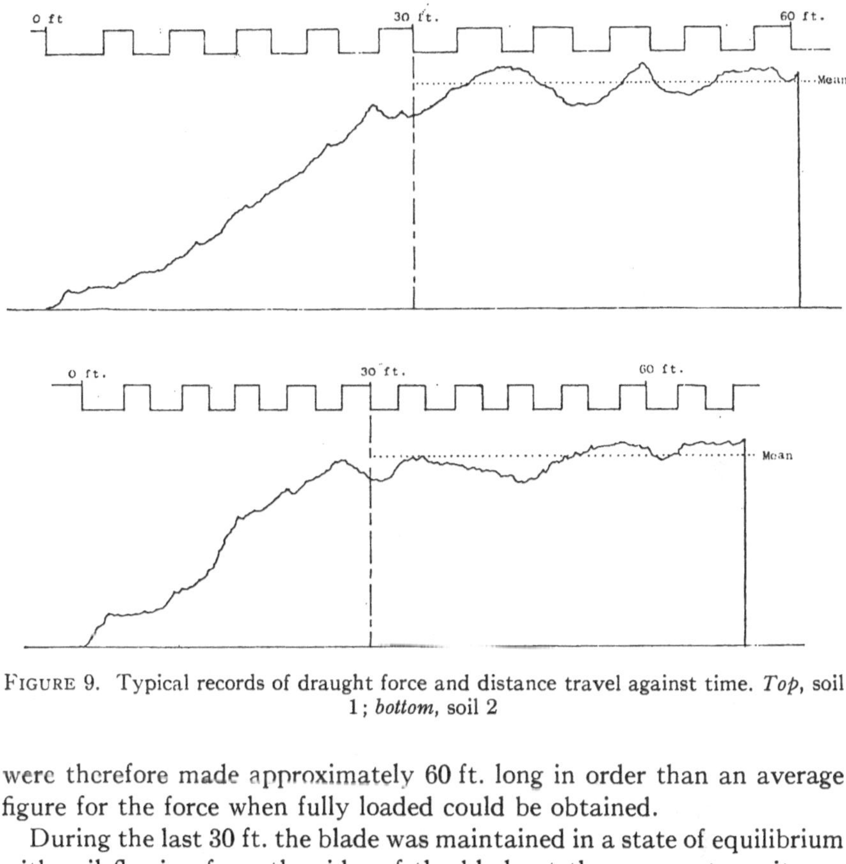

FIGURE 9. Typical records of draught force and distance travel against time. *Top*, soil 1; *bottom*, soil 2

were therefore made approximately 60 ft. long in order than an average figure for the force when fully loaded could be obtained.

During the last 30 ft. the blade was maintained in a state of equilibrium with soil flowing from the sides of the blade at the same rate as it was coming in along the cutting edge, as shown in Figure 8.

The distance travelled by the blade was measured using a tape measure attached to the back of the tractor with a manually operated event marker connected to the recorder. A change in the signal was made every 3 ft. The value of the experimental force was taken as the average force between 30 ft. and 60 ft.

A total of 9 experiments in soil 1, and 6 experiments in soil 2, were averaged to give the measured figures shown in Table II. This table also shows the calculated forces using the theory previously developed. The five main components of the cutting force are shown separately and it is instructive to compare their relative magnitudes. There is excellent agreement between the theoretical and the measured values.

The properties of the two soils used are shown in Table III. All measurements took place on the test site with the exception of the translational shear box test which was carried out in the laboratory.

TABLE I

SOIL AND BLADE DATA USED IN CALCULATING THE FORCES

Property	Blade	Soil 1	Soil 2
Breadth (b in.)	182	—	—
Cutting depth (z in.)	6	—	—
Cutting blade angle ($\alpha°$)	57	—	—
Height (not cutting) (h in.)	39	—	—
Density (undisturbed) (γ lb. in.$^{-3}$)	—	0.060	0.066
Density (disturbed) (γ_1 lb. in.$^{-3}$)	—	0.054	0.060
Shearing resistance angle (undisturbed) ($\phi°$)	—	35	32
Shearing resistance angle (disturbed) ($\phi_1°$)	—	30	27
Cohesion (undistrubed) (c lb. in.$^{-2}$)	—	0.55	1.25
Cohesion (disturbed) (c_1 lb. in.$^{-2}$)	—	0.28	0.63
Soil/Metal friction angle ($\delta°$)	—	24	23
Soil/Metal adhesion (c_a lb. in.$^{-2}$)	—	0	0
Rupture distance (m in.)	—	17.2	16.2
N factor (gravity) (N_γ)	—	1.81	1.59
N factor (cohesion) (N_c)	—	3.13	2.83
N factor (surcharge) (N_q)	—	3.62	3.17

BULLDOZING EFFICIENCY

The essential work done by a bulldozer is the cutting away and lifting of the soil from the undisturbed state. For the zone ABCD Figure 2 the soil is lifted from an average height of half the depth of cut below the soil surface to half the blade height (non cutting) above. The total height the soil is lifted is therefore $(h/2) + (z/2)$. For zone CDE the height lifted is $(h/3) + (z/2)$. The essential lifting force is thus:

$$L_t \neq \frac{\left\{W_1\left(\frac{h}{2}+\frac{z}{2}\right) + W_2\left(\frac{h}{3}+\frac{z}{2}\right)\right\}}{\text{length of cut to obtain a full load}}$$

$$\equiv \frac{b\left\{\gamma_1 mh\left(\frac{h}{2}+\frac{z}{2}\right) + \gamma_1 \tfrac{1}{2}h^2 \cot\phi_1\left(\frac{h}{3}+\frac{z}{2}\right)\right\}}{\frac{mh}{z} + \frac{h^2\cot\phi_1}{2z}}$$

The actual task of transporting the soil does not alter its energy and ideally requires no energy. The efficiency of a fully loaded bulldozer is given by the essential force divided by the actual force required for the process. That is:

$$\eta = \frac{D_c + L_t}{D_c + D_l} = \frac{D_c + \dfrac{bz\gamma_1\left\{m(h+z) + h\cot\phi_1\left(\frac{h}{3}+\frac{z}{2}\right)\right\}}{2m + h\cot\phi_1}}{D_c + D_l}$$

TABLE II

THE CALCULATED COMPONENTS OF THE DRAUGHT FORCE AND THEIR TOTALS COMPARED WITH THE MEAN MEASURED VALUES

		Force required to make initial unsurcharged cut				Additional force because of fully loaded blade				
Soil	Limits	Soil weight component $b\gamma z^2 N_\gamma \cos\theta$	Soil cohesion component $czN_c \cos\theta$	Soil adhesion component $c_a z(N_a \cos\theta + \cot\alpha)$	TOTAL D_c	Soil surcharge component $bqzN_q \cos\theta$	Sliding force component bF	TOTAL D_l	Calculated total D_c+D_l	Average measured total
1	Upper	702	1,858	0	2,560	19,950	7,260	27,210	29,770	—
	Lower	702	1,858	0	2,560	14,230	10,820	25,050	27,610	—
	Mean	702	1,858	0	2,560	17,090	9,040	26,130	28,690	29,600*
2	Upper	683	3,797	0	4,480	25,660	11,640	37,300	41,780	—
	Lower	683	3,797	0	4,480	16,300	16,960	33,260	37,730	—
	Mean	683	3,797	0	4,480	20,980	14,300	35,280	39,760	41,720*

*These figures are the average of the following mean values for each soil:
Soil 1: 29,500/30,600/25,000/24,000/25,600/30,000/37,600/29,500/35,500 (lb.).
Soil 2: 43,700/42,600/39,400/42,600/39,400/42,400 (lb.).

TABLE III

THE MEASURED SOIL AND SOIL/METAL PROPERTIES

Soil Property	Method of Determination	Results Soil—(1)	Soil—(2)
γ (lb. in.$^{-3}$)	Core samples	0.0598	0.0663
ϕ (degrees)	N.I.A.E. shear box + Hand operated	35	33
	Bevameter annulus + Torquemeter	34	30
	Traslational shear box	36	33
	Average value used in calculations	35	32
c (lb. in.$^{-2}$)	N.I.A.E. shear box + Hand operated	0.58	1.28
	Bevameter annulus + Torquemeter	0.53	1.20
	Translational shear box	0.54	1.26
	Average value used in calculations	0.55	1.25
δ (degrees)	Steel annulus and torquemeter	24	23
c_a (lb. in.$^{-2}$)	Steel annulus and torquemeter	0	0

On the two soils used for the experimental work the efficiency is given by:

$$\text{Soil 1: } \eta = 2{,}560 + 1{,}070/28{,}690 = 12.6\%$$
$$\text{Soil 2: } \eta = 4{,}480 + 1{,}172/39{,}760 = 14.2\%$$

From this it is clear that bulldozing is an inefficient way of moving soil.

FUTURE WORK

Before the theory that has been developed can be applied to the general problem of designing a more efficient bulldozer it is necessary to consider the theory for bulldozing cohesive soils. It is probable that a really weak cohesive soil, that is, a soil in which the cohesion is small relative to gravitational forces, will be covered adequately by the present theory. Strong cohesive soils, however, will require a new theory. In such a soil the bulldozer action will be that of a surcharged cut plus the force required to roll a wide soggy wheel made from the cohesive soil in front of the blade. Whilst it is realised that this force will be less than that required to slide the soil, it is not very obvious how to relate it to the soil properties. It may be that there are some interesting questions to investigate concerning the blade shape; however, the foregoing theory and the authors' observations suggest that the actual shape of the blade is not important. The fundamental rolling action of soil in front of a bulldozer blade has been shown to stem from the arrangement of forces rather than from the shape of the blade.

ACKNOWLEDGEMENTS

The authors wish to acknowledge the help of Mr. Daniel Hettiaratchi in carrying out the experiments. They would also like to express their gratitude for the assistance given by the Military Engineering Experimental Establishment, Christchurch, Hampshire, which enabled the full-scale experiments to be carried out.

REFERENCES

1. HETTIARATCHI, D. R. P., The Present State of the Theory of Soil Cutting, *J. of Terramechanics*, Vol. 2, 1, 1965.
2. PAYNE, P. C. J., The Relationship between the Mechanical Properties of Soil and the Performance of Simple Cultivation Implements, *J. of Agric. Engng. Res.*, Vol. 1, 1, 1956.
3. BEKKER, M. G., *The Theory of Land Locomotion, The Mechanics of Vehicle Mobility*, Ann Arbor: University of Michigan Press, 1956.
4. COBB, D. E., COHRON, G. T., and J. D. GENTRY, Scale Model Evaluation of Earthmoving Tools, *1st. Int. Conf. Soil-Vehicle Systems*, Turin, 1961.
5. COHRON, T., The Inception and Evolution of Earthmoving Soil Mechanics. *J. of Terramechanics*, Vol. 1, 4, 1964.
6. U.S. Patent Office: No. 3,035,361, May 1962.
7. U.S. Patent Office: No. 3,049,819, August 1962.
8. SCHURING, D. J. and R. I. EMORI, Soil Deforming Processes and Dimensional Analysis, *S.A.E. Nat. Farm Const. and Ind. Mach. Meeting*, September 1964.
9. OSMAN, M. S., The Mechanics of Soil Cutting Blades, *J. Agric. Engng. Res.*, Vol. 9, 4, 1964.
10. OHDE, J., Zur theorie des erdduckes unter besonderer berucksichtigung der erddruck verteilung (Theory of soil pressure with particular reference to its distribution) *die Bautechnick*, 16, 1938.
11. LOTKIN, M. and M. G. BEKKER, Introduction to Research on Vehicle Mobility, *Land Locomotion Lab.*, Report 1957.
12. REECE, A. R., The Fundamental Equation of Earthmoving Mechanics *Symp. Earthmoving Machinery*, 1965 (Instn. Mech. Engrs. London).
13. HETTIARATCHI, D. R. P., B. D. WITNEY, and A. R. REECE, The Calculation of Passive Pressure in Two Dimensional Soil Failure, *J. Agric. Engng. Res.*, Vol. 11, 2, 1966.
14. DREES, G., Untersuchungen uber das kraftespiel an flashbagger—schneidwerkzengen in mittlesand und schwach gindigem, sandigem schluff unter besonder berucksichtigung der planierschielde und ebenen schurf—kubelscheinden (Investigation into the play of forces on the blades of earthmoving equipment in medium sand and slightly cohesive sandy silt), *Mitteilungen Des Instituts Fur Baumaschinen Und Baubertrib*, Heft 16, Aachen, 1958.

Der Einfluss der Vibration auf den Schneidwiderstand von Planierschilden

N. PANAJOTOPOULOS*

ABSTRACT

The use of vibration on the blades of a bulldozer produces a considerable reduction on the cutting resistance of the soil. If the amplitude of vibration is small, a similar considerable reduction of the total drive power follows. The cause of this reduction is the influence that vibration exerts on the coefficient of inner and external friction. This influence is greater on coarse soils. To attain favourable results one must use an appropriate combination of drive velocity and vibration velocity. In this case the required vibrator-power is a small portion of the total drive power. The results of research up to the present stage allows the industrial production of bulldozers with vibrating blades.

IM JAHRE 1921 stellte die Firma Caterpillar die erste Planierraupe mit mechanischem Antrieb her. Damit begann die Entwicklung dieser Maschine.

Seitdem wurden zahlreiche Forschungen ausgeführt um Verbesserungen dieser Maschine zu erreichen. Dabei wurden die Kräfte, die zum Schneiden des Bodens durch den Schild nötig sind, sowie der Einfluss der verschiedenen Faktoren auf diese Kräfte, wie z. B. die Art des Schildes, der Schneidwinkel usw., untersucht.

Viele dieser Forschungen führten zu nennenswerten Ergebnissen und zu Berechnungsverfahren für die Schneidkräfte von Schilden. Fast alle dieser Arbeiten befassen sich jedoch mit starren Schilden, d. h. mit Schilden, die mit der Maschine fest verbunden sind. Nur sehr wenige Forschungsarbeiten sind über das Schneiden des Bodens mit schwingenden Werkzeugen ausgeführt worden. Diese Forschungsarbeiten sind Gegenstand der folgenden Ausführungen.

DER SCHNEIDWIDERSTAND AM PLANIERSCHILD

Viele Forscher beschäftigen sich mit dem Problem der Schneidwiderstände, die beim Bearbeiten von Böden auftreten. Sie haben alle versucht, mit der Kombination aus theoretischer Berechnung und experimentellen Versuchen zu einem Berechnungsverfahren zu gelangen, das, obwohl es

*64 Alexander Avenue, Athens, Greece.

bei den verschiedenen Forschern nicht gleich ist, viele gemeinsame Faktoren aufweist. Im allgemeinen kann man sagen, dass die Kraft F_x, bzw. die horizontale Komponente des Schneidwiderstandes (Bild 1), die

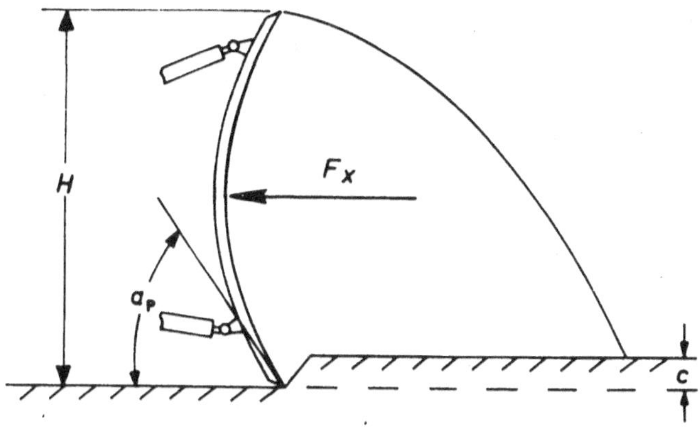

BILD 1

als Grabwiderstand eines Planierschildes bezeichnet wird, als eine Funktion verschiedener Faktoren beschrieben werden kann.

$$F_x = f(S, \alpha_p, B, H, c, a, \rho, \delta, \gamma, \gamma_p)$$

S = Formfaktor des Schildes
α_p = Schneidwinkel
B = Breite des Schildes
H = Höhe des Schildes
a = Schnittiefe
c = Kohäsionsfaktor des Bodens
ρ = innerer Reibungswinkel des Bodens
δ = äusserer Reibungswinkel des Bodens bzw. Reibungswinkel zwischen Boden und Stahl
γ = spezifisches Gewicht des ungestörten Bodens
γ_p = spezifisches Gewicht des gelockerten Bodens

Diese Funktion lässt erkennen, dass der Grabwiderstand im allgemeinen von den geometrischen Grössendes Schildes sowie von den mechanischen Eigenschaften des Bodens abhängig ist, und zwar in proportionaler Abhängigkeit.

Einfluss der Vibration

Nehman wir an, dass das Schild während des Schneidens zu vibrieren anfängt und dass es die Vibration in den Boden überträgt.

Ist die Amplitude der Schwingung im Verhältnis zur Grösse des

Schildes klein, dann können wir annehmen, dass sich die geometrischen Grössen des Schildes unter dem Einfluss der Vibration nicht verändern.

Die Vibration muss also einerseits den äusseren Reibungsbeiwert, andererseits die bodenmechanischen Faktoren, also den inneren Reibungsbeiwert und die Kohäsion, beeinflussen, um auf den Grabwiderstand zu wirken. Es ist auch möglich, dass die Vibration die Dichte des Bodens, nämlich die Faktoren γ und γ_p beeinflusst. Ausser dem vorher beschriebenen unmittelbaren Einfluss kann die Vibration einen mittelbaren Einfluss auf den Grabwiderstand ausüben. Bisher wurden die günstigsten geometrischen Grössen für ein starres Schild bestimmt. Die günstigsten Werte für das starre Schild können aber ungünstig für das vibrierende Schild sein.

Für die Betrachtung des vibrierenden Schildes sind einige zusätzliche Faktoren erforderlich, die wahrscheinlich einen Einfluss auf den Schneidwiderstand ausüben; z. B. die Frequenz, die Amplitude, die Schwingungsart, die Schwingungsrichtung, die Erregerkraft usw. Dadurch wird der Bereich zur Bestimmung der günstigsten Faktoren erweitert.

EINFLUSS DER VIBRATION AUF DEN INNEREN REIBUNGSWINKEL BZW. DIE SCHERFESTIGKEIT DES BODENS

Das Problem des Einflusses der Vibration auf die mechanischen Eigenschaften des Bodens ist von besonderem Interesse. Dieses Problem ist nicht nur mit dem Thema der vibrierenden Grabwerkzeuge, sondern auch mit anderen Themen, wie z. B. der Bodenverdichtung mit Vibrationswalzen, eng verbunden.

Man hat dieses Problem moch nicht vollkommen untersucht. Einige Forschungsarbeiten lassen jedoch schon eine allgemeine Tendenz dieses Einflusses erkennen: Die Japaner Mogami und Kubo (1) haben den Einfluss der Vibration auf die Scherfestigkeit von Sand und von bindigem Boden untersucht. Sie haben Scherversuche unter dynamischen Einfluss durchgeführt. Die Ergebnisse in Bild 2 zeigen eine starke Abnahme der Scherfestigkeit mit der Beschleunigung, aber keine klare Tendenz des Frequenzeinflusses. Es ist auch der Einfluss der Vibration auf das Trockenraumgewicht γ_{tr} dargestellt.

Der Russe Sawtschenko (2) untersuchte das Problem des Einflusses der Vibration auf den inneren Reibungsbeiwert von mittelfeinem Sand mit Hilfe einer speziellen Versuchseinrichtung, mit der er die Scherversuche mit Scherbuchse unter Vibrationseinwirkung durchgeführt hat. Die Ergebnisse zeigen, dass der innere Reibungsbeiwert bei wachsender Amplitude abnimmt (Bild 3 und 4). Denselben Einflüssen nach nimmt der innere Reibungsbeiwert bei wachsender Beschleunigung ab.

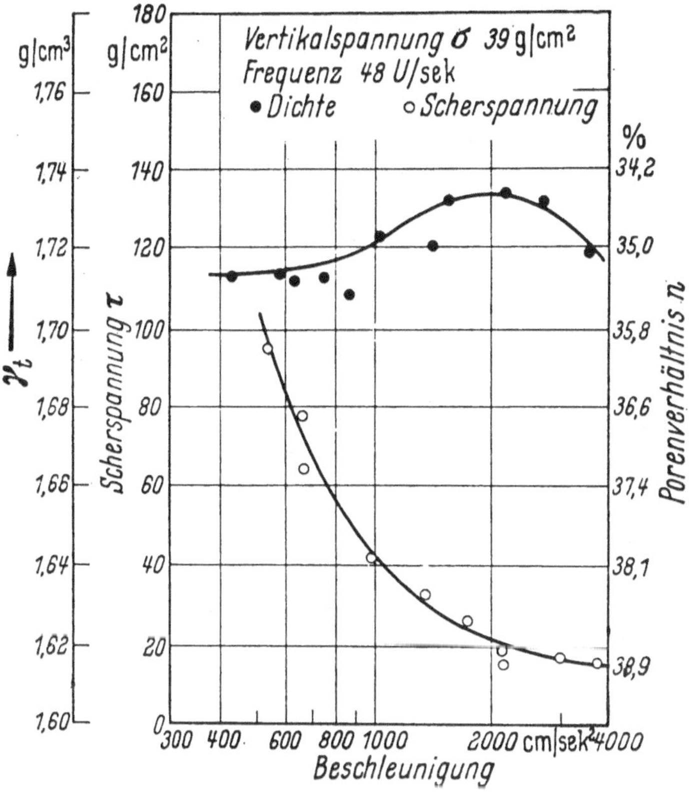

BILD 2. Einfluss der Beschleunigung auf die Lagerungsdichte und die Scherspannungen nach Mogami-Kubo

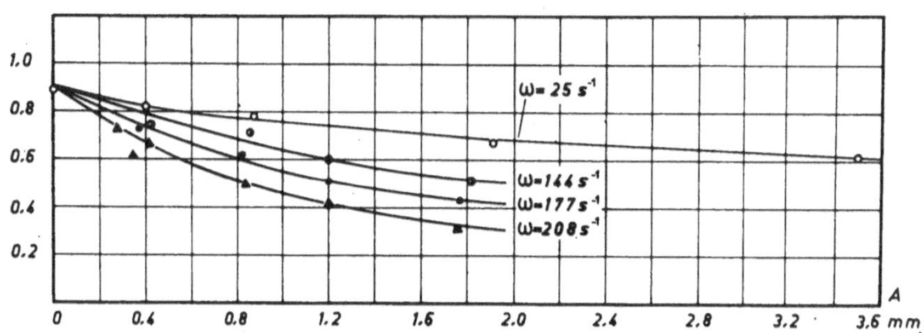

BILD 3. Abhängigkeit der inneren Reibung von der Schwingugnsamplitude bei trockenem, mittelfeinem Sand (nach Sawtschenko)

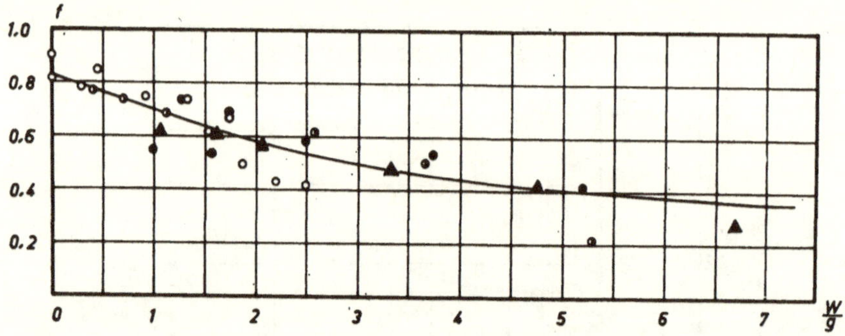

BILD 4. Abhängigkeit der inneren Reibung von der Schwingungsbeschleunigung bei mittelfeinem Sand (nach Sawtschenko)

Der Einfluss der Frequenz ist in Bild 5 dargestellt. Der innere Reibungsbeiwert nimmt auch bei wachsender Frequenz ab. Zwischen $\omega = 200\ s^{-1}$ und $\omega = 240\ s^{-1}$ findet eine deutliche Abnahme des inneren Reibungsbeiwertes statt.

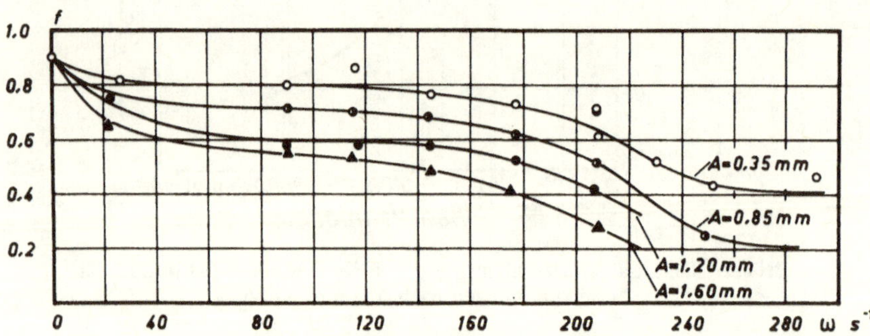

BILD 5. Abhängigkeit der inneren Reibung von der Schwingungsfrequenz bei mittelfeinem Sand (nach Sawtschenko)

Die Russen Preobraschenski und Sawtschenko (3) haben auch den Einfluss der Vibration auf den Scherwiderstand von Tonböden untersucht. Nach diesen Versuchen nimmt der Scherwiderstand bei wachsender Frequenz und wachsender Amplitude ab (Bild 6 und 7). Es gibt aber einen Frequenz- und Amplitudenbereich, in dem die Verminderung des Scherwiderstandes deutlich stärker ist.

Die Ergebnisse der obigen Forschungen zeigen, dass die Schwingungen einen starken Einfluss auf die Scherfestigkeit bzw. den inneren Reibungsbeiwert des Bodens haben. Unter dem Einfluss der Schwingungen werden die obigen Grössen immer kleiner. Dieselben Ergebnisse zeigen, welche Rolle einige Schwingungsparameter wie z. B. die Beschleunigung,

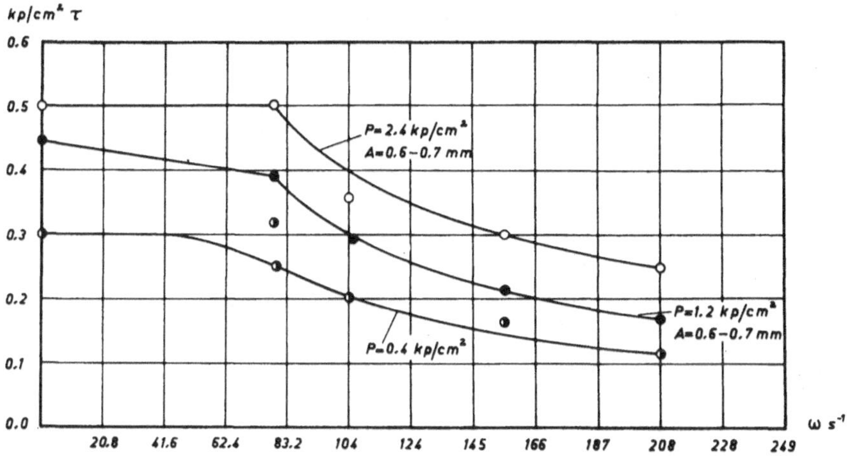

BILD 6. Abhängigkeit der Scherwiderstandes des Kudinowkaer–Salo-Tones von der Schwingungsfrequenz (nach Preobashenski und Sawtschenko)

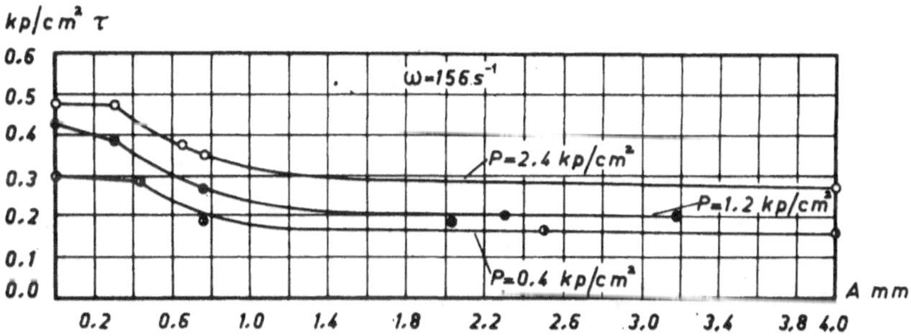

BILD 7. Abhängigkeit des Scherwiderstandes des Kudinowkaer-Salo-Tones von der Schwingungsamplitude (nach Preobrashenski und Sawtschenko)

die Amplitude, die Frequenz in dieser Verminderung der Scherfestigkeit des Bodens spielen.

Die Scherfestigkeit des Bodens entsteht wie bekannt aus den beiden Faktoren Kohäsion c und innerer Reibungswinkel ρ nach der Formel $\tau = c + \sigma tg\rho$. Wenn σ = konstant, bedeutet die Verminderung von τ eine Verminderung des Winkel ρ oder der Kohäsion c oder der beiden Faktoren.

Wir haben aber oben bemerkt, dass die Faktoren c und ρ den Grabwiderstand des Bodens beeinflussen. Man könnte also annehmen, dass tatsächlich durch die Vibration der Grabwiderstand vermindert wird.

Man kann jedoch nicht einfach die aus statischen Versuchen gewonnenen Werte für c, ρ, γ und γ_p durch die aus dynamischen Versuchten ermittelten Werte ersetzen, um so den neuen Grabwiderstand für einen vibrierenden Schild zu berechnen.

Die Umstände, unter denen der Boden in den Scherversuchsgeräten vibriert, unterscheiden sich von den Umständen, unter denen der gleiche Boden vor dem Planierschild vibriert.

Berücksichtigt man die grosse Plastizität des Bodens, so wie die starke Dämpfung, so ist es ziemlich unwahrscheinlich, dass die Schwingungsparameter, die die Schwingung des Schildes während des Bodenschneidens charakterisieren, an jeder Stelle des vor dem Schild liegenden Bodens dieselben sind. Die Tatsache, dass der vor dem Schild liegende Boden während des Schneidens sich in dauernder Bewegung befindet, und zwar mit verschiedenen Richtungen und Geschwindigkeiten von Punkt zu Punkt, stärkt die Vermutung, dass die Schwingungsparameter des Schildes nicht für den Boden gelten.

In der Praxis aber interessiert nicht so sehr, auf welche Weise der Boden vibriert, sondern nur, ob wirklich die vom Schild in den Boden übertragenen Schwingungen eine Verminderung des Grabwiderstandes verursachen.

Mit der Feststellung dieser Verminderung, sowie der Bestimmung deren Grösse, befassen sich die Forschungsarbeiten, die wie wir oben erwähnten, zwar von begrenztem Umfang sind, aber zu Ergebnissen führen, die ein allgemeines Bild des gesamten Problemes geben.

FORSCHUNGSARBEITEN ZUR FESTSTELLUNG DES VERMINDERTEN SCHNEIDWIDERSTANDES EINES PLANIERSCHILDES UNTER DEM EINFLUSS DER VIBRATION

Forschungen über das Bodenschneiden mit schwingenden Werkzeugen wurden, in erster Linie auf dem Gebiet der landwirtschaftlichen Maschinen, von Amerikanern, Russen, und Deutschen ausgeführt.

Die letzten Versuche, die im Institut für Landtechnische Grundlagenforschung der Forschungsanstalt für Landwirtschaft, Braunschweig, von Eggenmüller (4) durchgeführt wurden, zeichnen sich durch grosse Systematik und Ausführlichkeit aus und wurden sowohl in der Veruschsbahn als auch auf dem Acker durchgeführt. Diese Versuche stellen eine stakre Verminderung des Schneidwiderstandes des Bodens unter dem Einfluss der Vibration fest, die bis zu 70 Prozent beträgt. Bie diesen Forschungen wurde jedoch festgestellt, dass die erforderliche Schwingerleistung sehr gross sein muss, so dass die erforderliche Gesamtleistung für die oben genannte maximale Schneidwiderstandverminderung, bis zu 100 Prozent grösser sein muss als die Leistung, die für die starren Werkzeuge benötigt wurde. Dies ist wahrscheinlich darauf zurückzuführen, dass die angewandten Amplituden verhältnismässig gross waren.

Nach den Forschungsarbeiten von Eggenmüller muss man sich auf eine Verminderung des Schneidwiderstandes um nur 40 ± 50 Prozent beschränken, um die erforderliche Gesamtleistung kleiner oder gleich der Leistung bei Verwendung von starren Werkzeugen zu halten.

Der Russe Schkurenko (5, 6) befasste sich mit dem Problem des Einflusses von Schwingungen auf den beim Schneiden des Bodens auftretenden Widerstand in zwei Versuchsreihen in den Jahren 1952 und 1958. Die erste Versuchsreihe führte er auf Lehmboden und Sand, die zweite auf sandigem Lehmboden durch.

Beide Versuchsreihen führte er mit ebenem Schild durch, wobei die Schneidkante einmal parallel, im anderen Fall senkrecht zur Fahrichtung gestellt war. Im letzten Fall bildete der Schild einen Winkel von 45° zur Horizontalen. In beiden Fällen wurde der Schild durch Schwinger erregt. Die Schwingungsrichtung wurde senkrecht und parallel zur horizontalen Ebene und zur Fahrichtung gewählt. Die Schwingungsrichtung ist aus den Bildern 8 und 9 zu ersehen. Die Einrichtung des Bildes 8 erzeugt

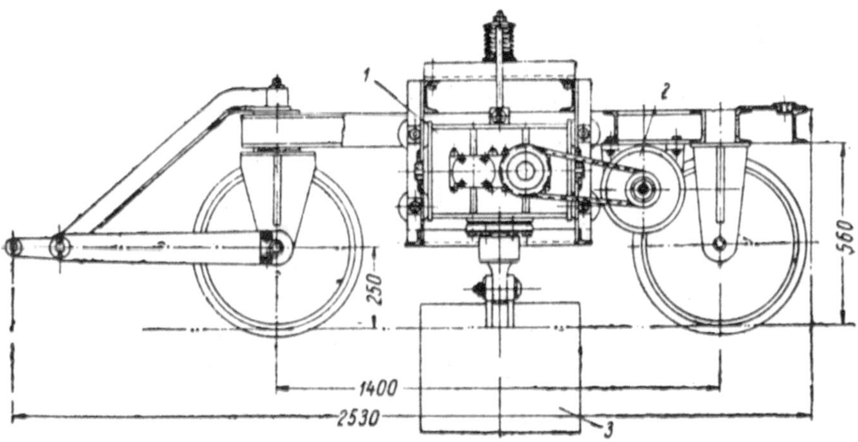

BILD 8

Schwingungen senkrecht zur Fahrichtung. Die Einrichtung des Bildes 9 erzeugt Schwingungen parallel zur Fahrichtung. Das Messer steht einmal wie das Bild 8 zeigt, längs zur Fahrichtung und einmal quer zur Fahrichtung, mit einem Neigungswinkel von 45° zur Versuchsbahn.

Das Externmoment des Vibrators war verstellbar. Die Frequenz mit der Drehzahl des Motors geändert. Für die Beschreibung der Ergebnisse wurde die Grösse $\delta = (F_s - F_d)/F_s$ angewandt, worin F_s = der statische Schneidwiderstand, F_d = der Schneidwiderstand bei Vibration ist.

BILD 9

Die Versuchsergebnisse

Während der ersten Versuchsreihe wurde eine Verminderung des Schneidwiderstandes proportional der Frequenz sowie der Schwingungsgeschwindigkeit festgestellt (Bild 10 und 11). Im Bild 12 ist auch die

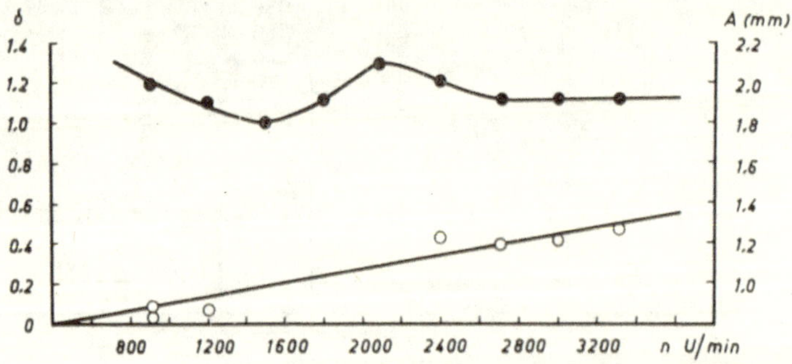

BILD 10. Abhängigkeit des Vibrationseffekts von der Schwingungsfrequenz. (Das Messer ist senkrecht zur Bewegungsrichtung gerichtet, Boden-Lehmboden, Schwingungensenkrecht.) Rechts unten: U/min. (nach Schkurenko)

Abhängigkeit von δ und der Schneidwiderstände von der Schneidtiefe zu ersehen, wenn die Schneide parallel zur Fahrichtung steht.

Während des zweiten Versuchsreihe wurde folgendes festgestellt:

1. Der Schneidwiderstand nimmt für konstante Frequenz mit der Amplitude ab (Bild 13).

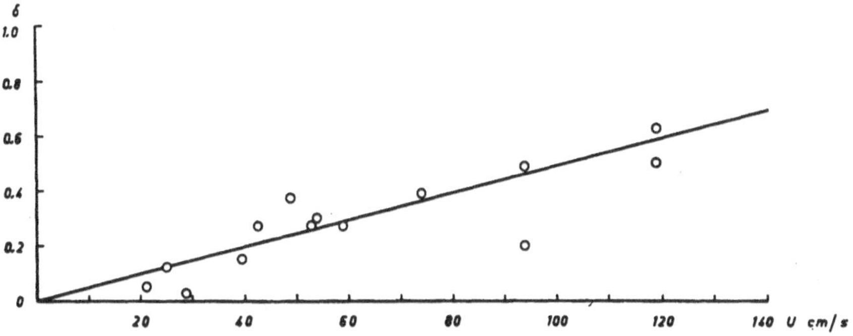

BILD. 11. Abhängigkeit des Vibrationseffekts von der Schwingungsgeschwindigkeit. (Das Messer ist parallel zur Bewegungsrichtung gerichtet, Boden-Sandboden, Schwingungen-senkrecht.) Unten rechts: cm/s. (nach Schkurenko)

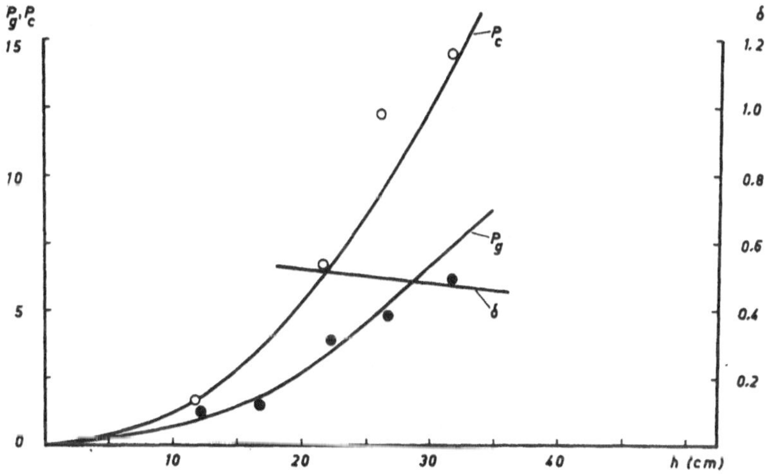

BILD 12. Abhängigkeit des Vibrationseffekts und der Schneiderkräfte von der Schneidetiefe. (Das Messer steht parallel zur Bewegungsrichtung, Boden-Sandboden, Schwingungensenkrecht, Exzentermoment 51.4 kp cm, Umdrehungszahl i.d. Min. 2100–2200; Schwingungsamplitude 3.1–3.5 mm.) (nach Schkurenko)

2. Der Schneidwiderstand nimmt auch mit wachsender Schwingungsgeschwindigkeit ab (Bild 14).

3. Die Wirkung der Vibration vermindert sich bei wachsender Fahrgeschwindigkeit (Bild 15).

4. Die Schneidwiderstandsminderung wird grösser bei horizontaler Schwingungsrichtung und senkrechter Stellung der Schneidkante zur Bewegungsrichtung.

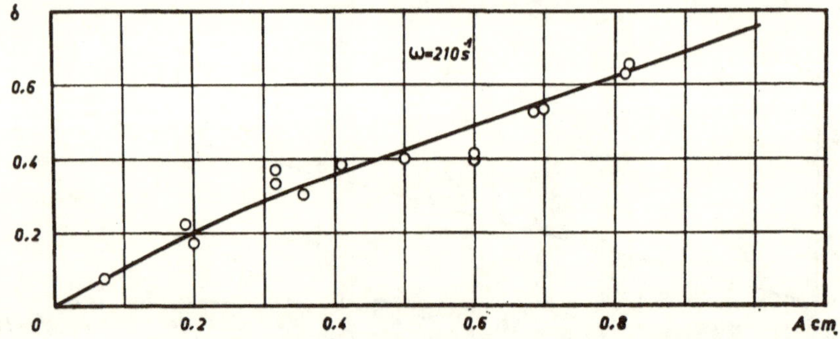

BILD 13. Abhängigkeit der Schwingunswirkung von der Schwingungsamplitude (nach Schkurenko)

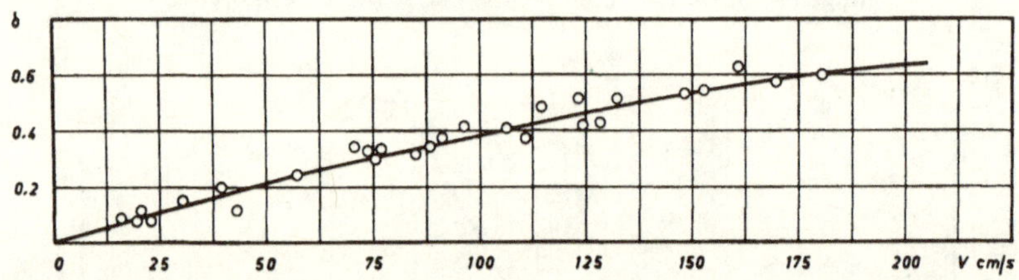

BILD 14. Abhängigkeit der Schwingungswirkung von der Geschwindigkeit der Schwingung. Vertikale Schwingungen Schneidende Messerkante senkrecht zur Bewegungsrichtung (nach Schkurenko)

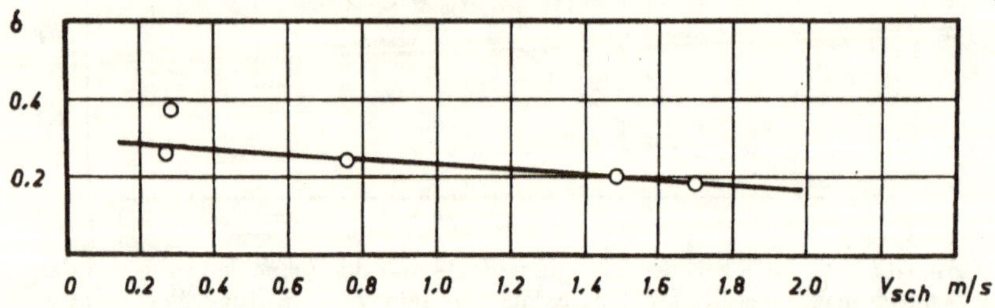

BILD 15. Abhängigkeit der Schwingungswirkung von der Geschwindigkeit des schneidenden Teiles (nach Schkurenko)

5. Wie auf den Diagrammen dargestellt ist, hat Schkurenko eine Schneidwiderstandsverminderung bis zu 60 Prozent festgestellt.

Die Ergebnisse der Versuche von Schkurenko sind sehr interessant, obwohl diese Versuche nur mit ebenem Schild und mit zwei Schwingungsrichtungen durchgeführt wurden. Es blieben aber einige Fragen offen, die in der Veröffentlichung von Schkurenko nicht ganz klar beantwortet worden sind.

Erstens ist nicht ganz klar, wie und an welcher Stelle Schkurenko die Schneidwiderstände gemessen hat. Wenn man die Abbildung der Einrichtung und ein Kraftdiagramm genau betrachtet, kann man annehmen, dass die Schneidwiderstände nicht auf dem Schild sondern am Zughaken des Versuchswagens gemessen wurden. Diese Messanordnung ist aber ungenau, da die Zugkraft bzw. die Reibungskraft des Versuchswagens nicht konstant bliebt, sondern sich bei vibrierendem Schild ändert. Das wurde bei Versuchen im Institut für Baumaschinen und Baubetrieb der R.W.T.H. Aachen festgestellt.

Die zweite offene Frage ist, wie und an welcher Stelle die Schwingungsamplitude gemessen wurde.

Weiter versucht Schkurenko zu beweisen, dass die lineare Schwingungstheorie bei der Berechnung der Schwingungen des schwingenden Schildes anzuwenden ist; das bedeutet, dass für den schwingenden und gleichzeitig schneidenden Schild die Gleichung gilt: $m\ddot{s} + \alpha\dot{s} + cs = \hat{F} \sin \omega t$. Hierbei ist: m = die Gesamtmasse des schwingenden Systems Schild + Boden ($kp\ cm^{-1}\ sec^2$); α = der Dämpfungsfaktor des Systems ($kp\ cm^{-1}\ sec$); c = die Gesamtfederkonstante des Systems ($kp\ cm^{-1}$); $\hat{F}$ = die maximale Erregerkraft (kp); ω = die Kreisfrequenz (sec^{-1}); t = die Zeit (sec).

Wenn diese Gleichung für alle Frequenzen gilt, dann ist folgendes festzustellen:

1. Der schwingende Schild kann durch ein Ersatzbild ersetzt werden (Bild 16).

2. Die Dämpfungskonstante des Bodens α_B bleibt konstant, bzw. die Dämpfung des Bodens ist proportional der Schwingungsgeschwindigkeit.

3. Die schwingende Masse des Bodens m'_B sowie die Federkonstante des Bodens c'_B bleiben konstant.

Unter diesen Voraussetzungen wäre die Berechung des schwingenden Schildes leicht.

Der Beweis von Schkurenko aber lässt aus folgenden Gründen einige Zweifel:

1. Schkurenko nimmt an, dass die Dämpfung des Bodens nur im Resonanzbereich von Bedeutung ist. Im Ueberresonanzbereich setzt er sie null. Das aber scheint sehr zweifelhaft, besonders da der Boden vor dem Schild mehr plastisch als elastisch ist.

2. Schkurenko nimmt an, dass die schwingende Masse des Bodens m'_B null ist. Es ist aber leicht zu erkennen, dass diese Annahme unzutreffend ist, besonders dann, wenn ein geneigter Schild vertikal erregt wird. Schliessblich bleibt die Frage offen, die Anwendung der linearen Schwingungstheorie zur Berechnung der Schwingungen des schwingenden

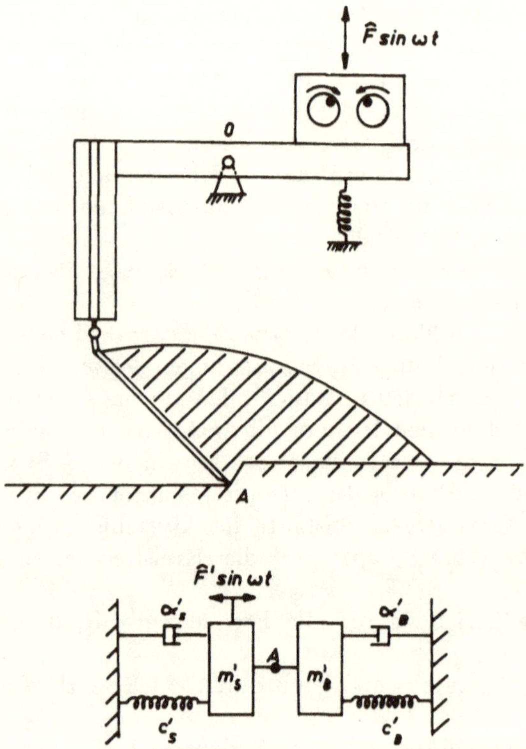

BILD 16

Schildes richtig ist. Die Beantwortung dieser Frage erfordert weitere Untersuchungen.

Bei den Versuchen von Schkurenko wurde die Schwingerleistung nicht gemessen. Daher bleibt inbekannt, ob gleichzeitig mit der Schneider-wider standsverminderung eine Verminderung oder eine Vermehrung der Gesamtleistung entsteht. Ausser Schkurenko hat auch Drees (7) im Institut für Baumaschinen und Baubetrieb der R.W.T.H. Aachen, im Jahre 1958, einige Versuche mit schwingendem Schild in Mittelsand im Rahmen einer umfangreichen Arbeit durchgeführt. Mit dieser Arbeit wurde versucht, die Schneidwiderstände auf die starren Schilde der Flachbagger und die Abhängigkeit dieser Kräfte von der Schildart, dem Neigungswinkel, der Bodenart, der Schnittiefe u. a. Faktoren zu bestimmen.

Die Versuche wurden mit einem Parabelschild durchgeführt, das mit einem Kreisschwinger erregt wurde. Diese Versuche wurden mit zwei Frequenzen, 22 und 36 Hz, und verschiedenen Neigungswinkelndes Schildes durchgeführt.

Aus diesen Versuchen ergab sich eine Schneidwiderstandsverminderung von 17 Prozent für eine Frequenz von 36 Hz und einem Schneidwinkel von 45°. Dagegen für eine Frequenz von 22 Hz und einen Schneidwinkel von 60° wurde keine deutliche Verminderung des Schneidwiderstandes festgestellt.

WEITERE UNTERSUCHUNGEN IM INSTITUT FÜR BAUMASCHINEN UND BAUBETRIEB DER R.W.T.H.–AACHEN

Seit 1963 wurden im Institut für Baumaschinen und Baubetrieb der R.W.T.H. Aachen umfangreiche Modellversuche bzw. Versuche mit einer umgebauten Planierraupe durchgeführt zur Untersuchung des Einflusses der Schwingungen auf die Kräfte, die während des Schneidvorganges auf das Planierschild wirken.

1. Modellversuche

Die Modellversuche wurden in leicht schluffigem Sand sowie auf Kiessand durchgeführt. Es wurde ein Schild von Parabelprofil eingesetzt, das 500 mm lang und 225 mm hoch war. Die Versuchseinrichtung war so aufgebaut, dass die unmittelbare Messung der horizontalen Komponente (Grabwiderstand) des auf das Schild wirkenden Schneidwiderstandes mit Hilfe von Dehnungsmessstreifen möglich war (Bild 17).

Es wurde auch die Schwingungsbeschleunigung mit Hilfe von Beschleunigungsaufnehmern auf dem Schild gemessen. Der Schwingungserreger wurde hinter dem Schild auf dem das Schild tragenden Rahmen festgeschraubt (Bild 18).

Die Lage des Rüttlers auf dem Schild war verstellbar, so dass die jeweils erwünschte Schwingungsrichtung erreichbar war. Das Schild war in den Neigungswinkeln − 15°, 0° und + 8° verstellbar, die den Schneidwinkeln von 45°, 60° und 68° entsprechen. Die ganze Einrichtung wurde auf einem auf Schienen fahrbaren Wagen festgebunden (Bild 19). Eine Winde zog den Wagen mit einer jeweils konstant bleibenden Geschwindigkeit.

Zur Darstellung der Ergebnisse wurde die Grösse δ verwendet: $\delta = (F_s - F_v)/F_s$.

Darin ist: δ = Verminderungsfaktor des Grabwiderstandes; F_s = Grabwiderstand bei starrem Schild; F_v = Grabwiderstand bei vibrierendem Schild.

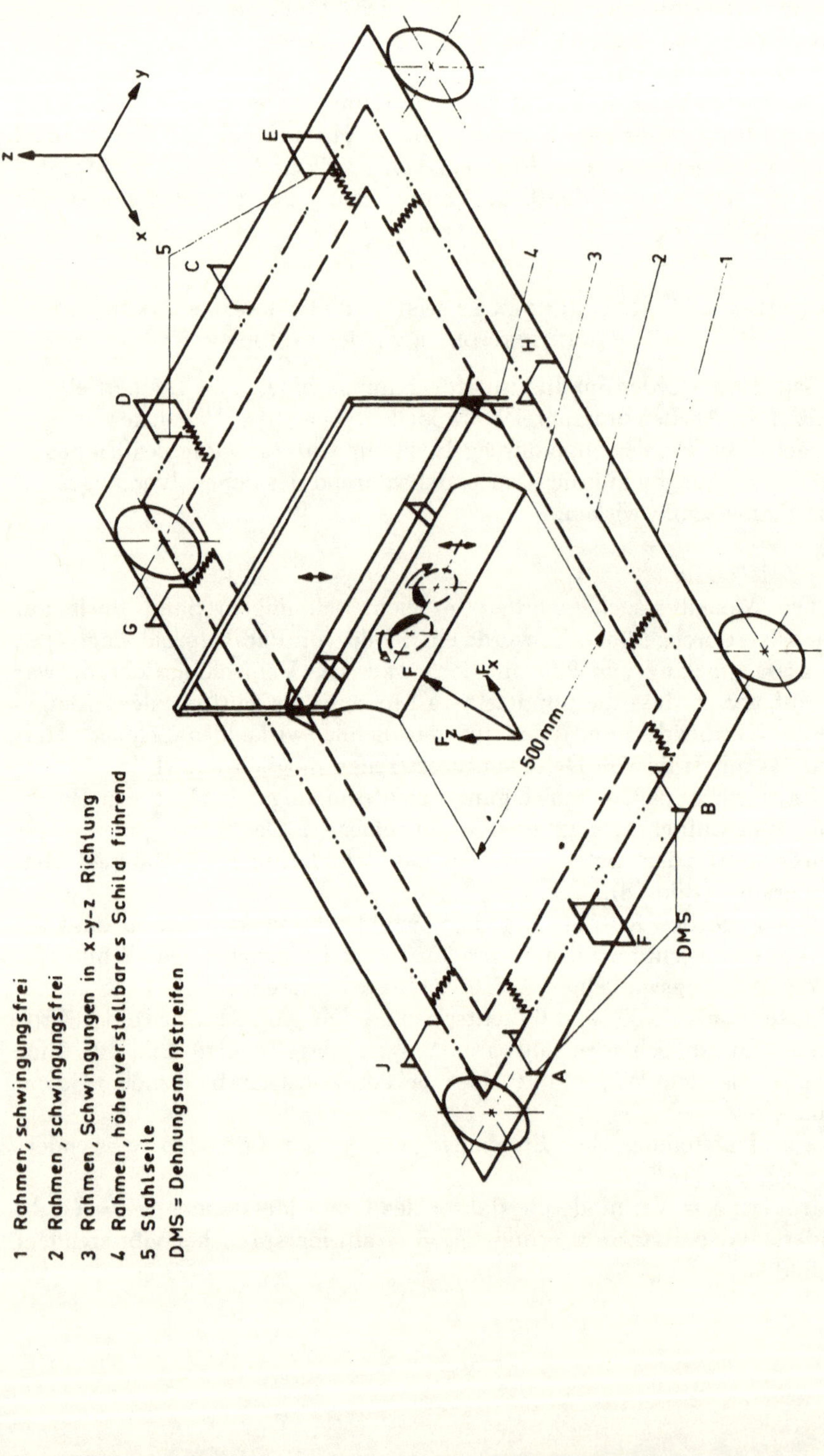

BILD 17. Schematische Darstellung der Versuchseinrichtung für die Modellversuche

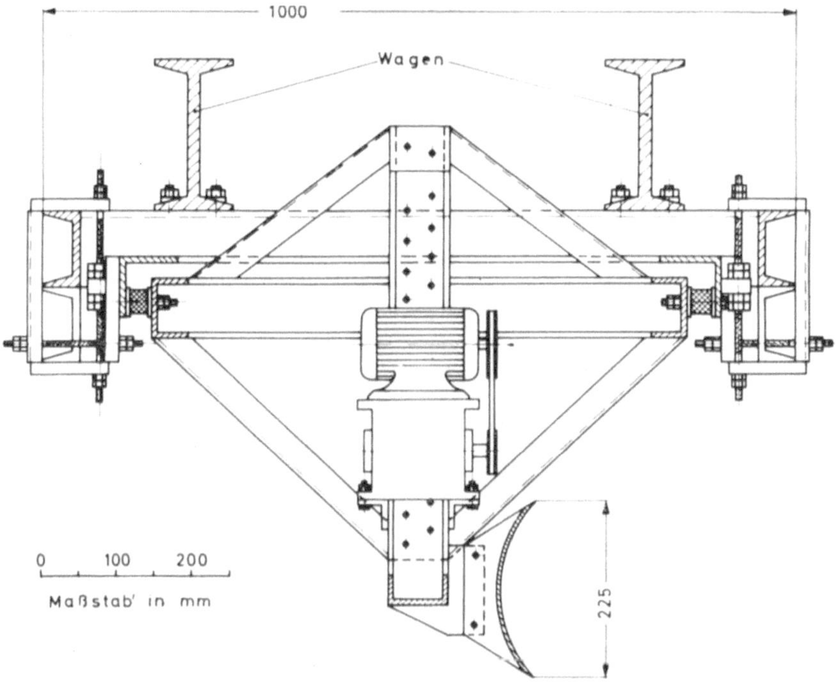

BILD 18. Die Versuchseinrichtung für die Modellversuche

BILD 19. Der Versuchswagen

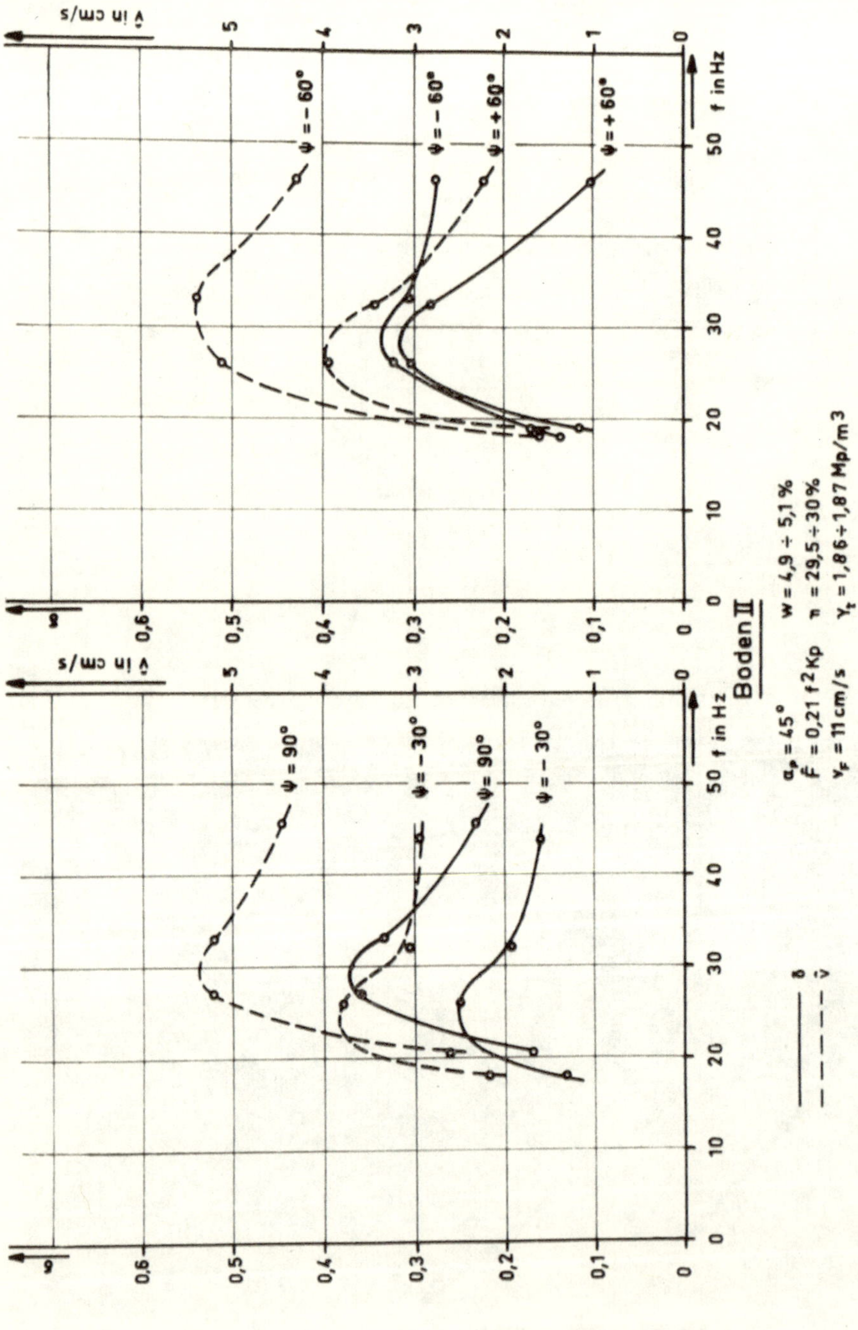

BILD 20. Einfluss der Frequenz f und der Schwingungsgeschwindigkeit v auf den Verminderungsfaktor δ des Grabwiderstandes

Ergebnisse der Modellversuche

Durch die Modellversuche wurde folgendes festgestellt:

(*a*) Die Verwendung von Schwingungen bei einem Planierschild vermindert den Grabwiderstand des Bodens. Die Verminderung ist von der Bodenart abhängig. Sie ist stärker (bis zu 40 Prozent) bie Kiessand als bei leicht schluffigem Sand (bis zu 30 Prozent).

(*b*) Bei allen Versuchen wurde ein grosser Einfluss der Schwingungsgeschwindigkeit $\hat{V}$ auf die Verminderung des Grabwiderstandes festgestellt (Bild 20).

(*c*) Die Schwingungsrichtung in Verbindung mit der Schwingungsgeschwindigkeit spielt ebenso eine wichtige Rolle auf die Verminderung des Grabwiderstandes (Bild 21).

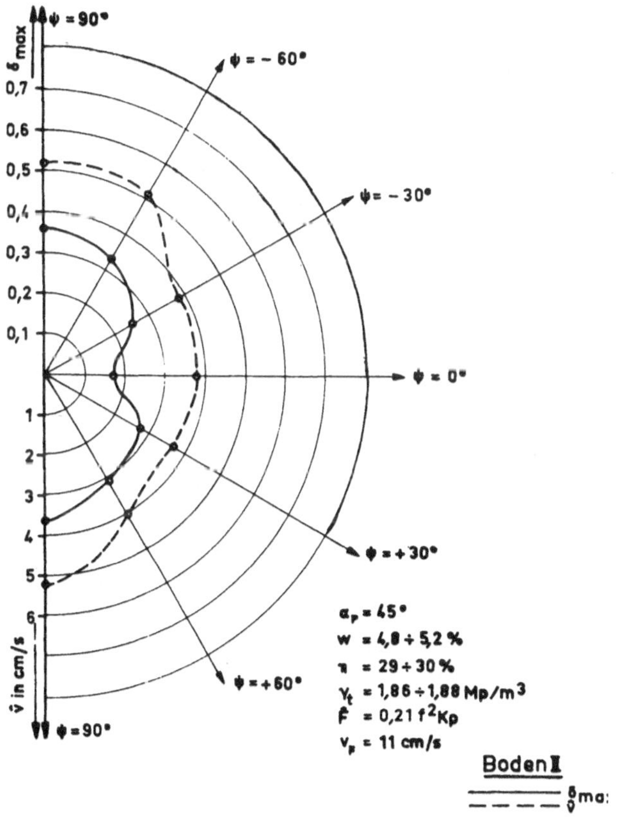

BILD 21. Einfluss der Schwingungsrichtung auf den Verminderungsfaktor δ des Grabwiderstandes. ψ = der Winkel den die Schwingungsrichtung zur Horizontalen bildet

(*d*) Der kleinste Schneidwinkel $\alpha_p = 45°$ erwies sich bei den Versuchen als günstiger (Bild 22).

(*e*) Der Wassergehalt ω des Bodens übt auf δ bis zu einem bestimmten Wert keinen Einfluss aus. Weitere Steigung von ω aber verursacht eine Verminderung von δ (Bild 23).

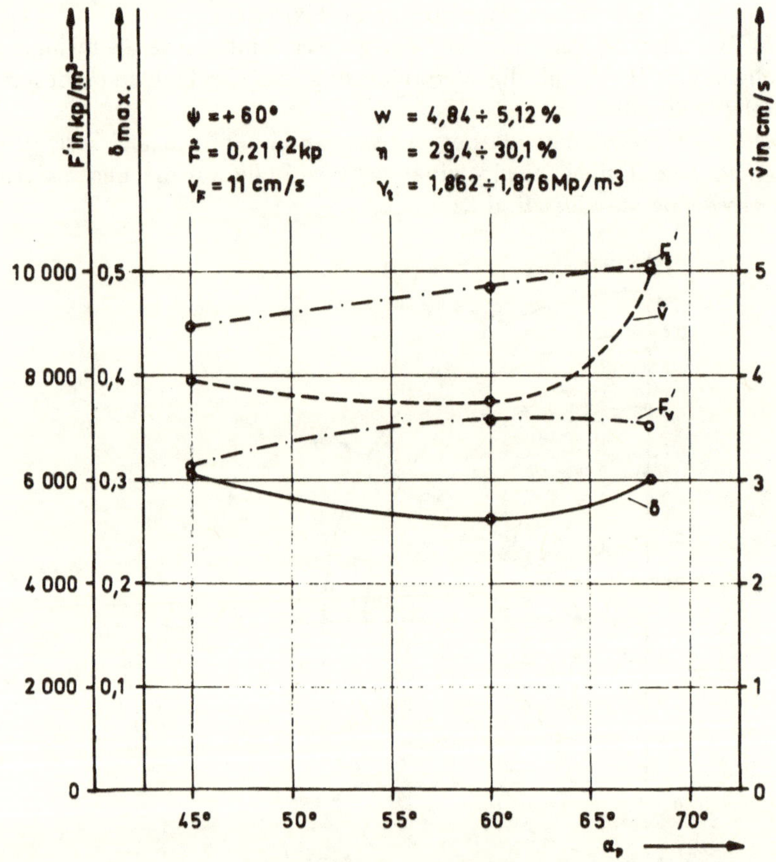

Bild 22. Einfluss des Schneidwinkels α_p auf den Verminderunsfaktor δ des Grabwiderstandes

(*f*) Die Fahrgeschwindigkeit v_F apielt eine bedeutende Rolle auf die Verminderung des Grabwiderstandes. Bei grösseren Fahrgeschwindigkeiten wird der Wert δ geringer (Bild 24). Es ist aber bemerkenswert, dass, wie weiter aus dem Bild 24 zu ersehen ist, das Verhältnis $\hat{V}_{v_F}$ zum Wert δ proportional bleibt. Diese Erkenntnis ist von besonderer Bedeutung, da sie uns die Möglichkeit gibt, den ungünstigen Einfluss von v_F

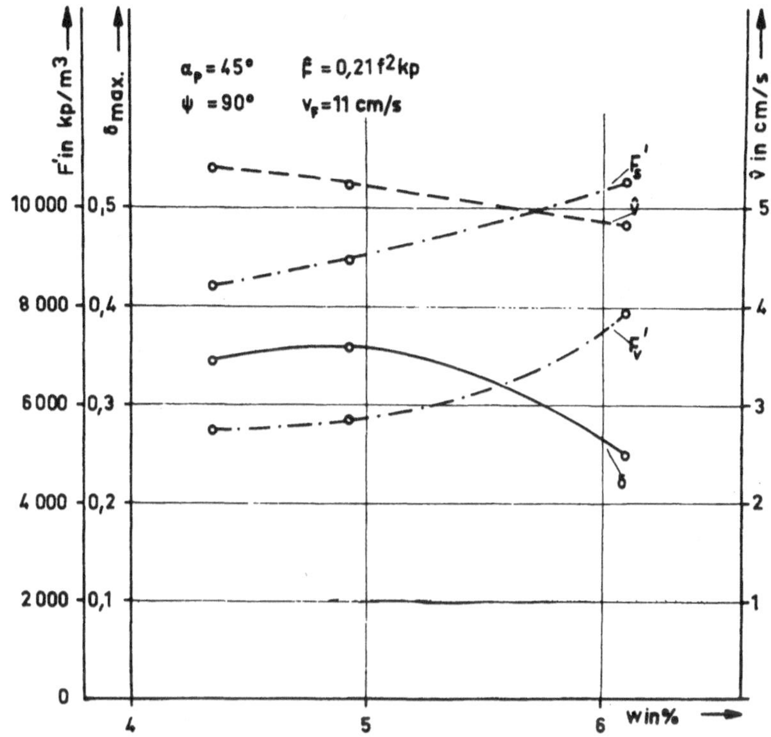

BILD 23. Einfluss des Wassergehaltes ω auf den Verminderungsfaktor des Grabwiderstandes

auf den Grabwiderstand des Bodens durch eine entsprechende Vergrösserung von $\hat{V}$ auszugleichen.

2. Grossversuche

Die Grossversuche wurden auf Kiessand durchgeführt, also auf einem der Böden, auf denen die Modellversuche durchgeführt wurden. Für die Versuche wurde eine Planierraupe Fiat Typ 30. C.A. verwendet. Sie trug ein Schild mit Evolventenprofil, das 1600 mm breit und 640 mm hoch war. An der obigen Planierraupe wurden einige Veränderungen vorgenommen, die eine elastische Bewegung des Schildes um eine mit dem Gerät verbundene, feststehende Achse ermöglichten (Bild 25).

Der das Schild tragende Rahmen wurde mit einem schweren Wagen fest verbunden, dessen Räder über die beiden seitlich der Versuchsbahn angebrachten Schienen lief. Auf diese Weise blieb die Schnittiefe während des Versuches konstant. Der Rüttler zur Erzeugung der Schwingungen wurde auf dem Schild angebracht.

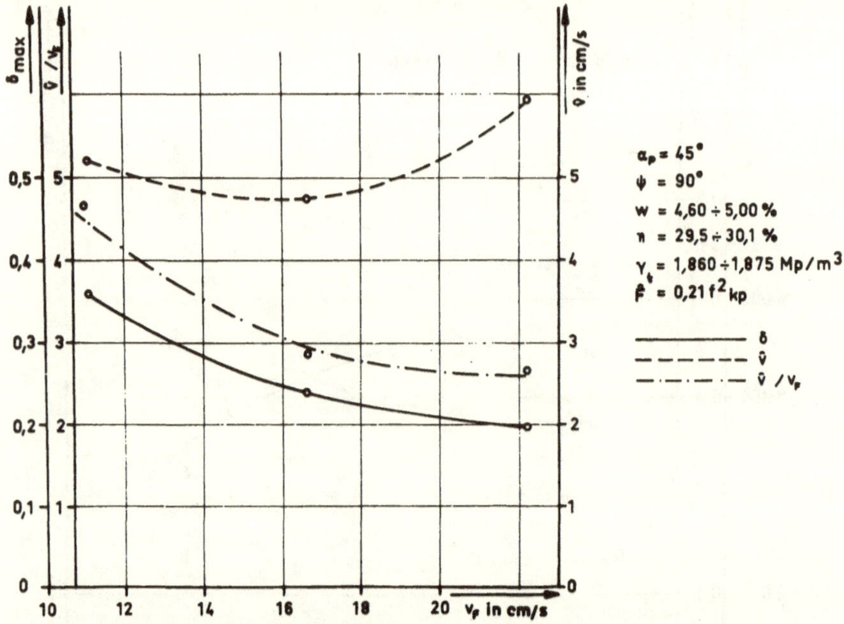

BILD 24. Einfluss der Fahrgeschwindigkeit v_F und des Verhältnisses $\hat{v}_F/v$ auf den Verminderungsfaktor des Grabwiderstandes

Die auf das Schild wirkenden Kräfte wurden in der Verbindungsstelle des Schildes mit dem Rahmen mit Hilfe von Dehnungsmessstreifen gemessen (Bild 26).

Auf diese Weise wurden die zwei Komponenten F_x (Grabwiderstand) und F_z (vertikale Kraft) des Schneidwiderstandes bestimmt. Es wurde auch die jeweilige Schwingungsgeschwindigkeit $\hat{v}$, die Schwingungsbeschleunigung $\hat{a}$ und der Phasenwinkel φ_2 zwischen Erregerkraft und Schwingungsgeschwindigkeit des Schildes sowie die Fahrgeschwindigkeit der Raupe gemessen.

Nach diesen Messungen wurde folgendes festgestellt.

(i) Die Schwingerleistung P_R
$P_R = 1/2\hat{F}\hat{v}\cos\varphi_2$ oder $P_R = 6.66\ (10^{-5}\hat{F}\hat{v}\cos\varphi_2)\ (PS)$
wobei $\hat{F}$ = der maximale Wert der Erregerkraft (*kp*); $\hat{v}$ = der maximale Wert der Schwingungsgeschwindigkeit ($cm\ s^{-1}$); φ_2 = der Phasenwinkel zwischen $\hat{F}$ und $\hat{v}$ ist.

(ii) Die Fahrleistung des vibrierenden Schildes P_{vs}
$P_{vs} = F_{xv}v_F$ oder $P_{vs} = 1.33 . 10^{-4} F_{xv} . v_F\ (PS)$
wobei F_{xv} = der Grabwiderstand bei vibrierendem Schild (*kp*); v_F = die Fahrgeschwindigkeit der Raupe ($cm\ s^{-1}$) ist.

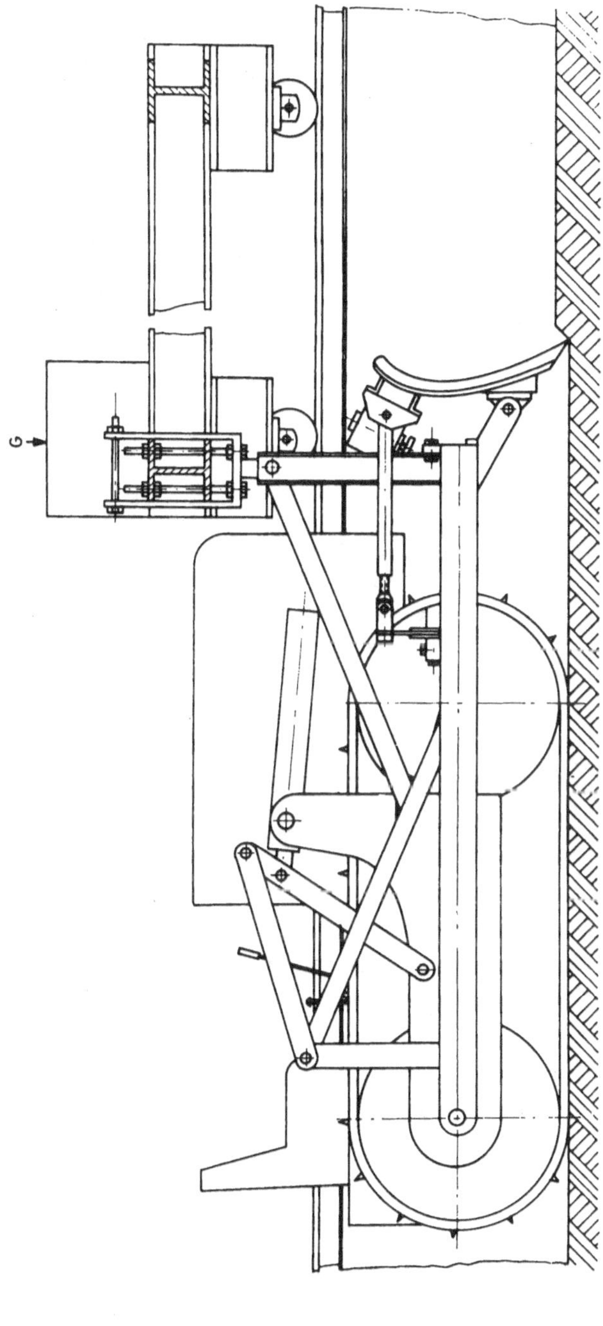

BILD 25. Die Versuchseinrichtung für die Grosseversuche

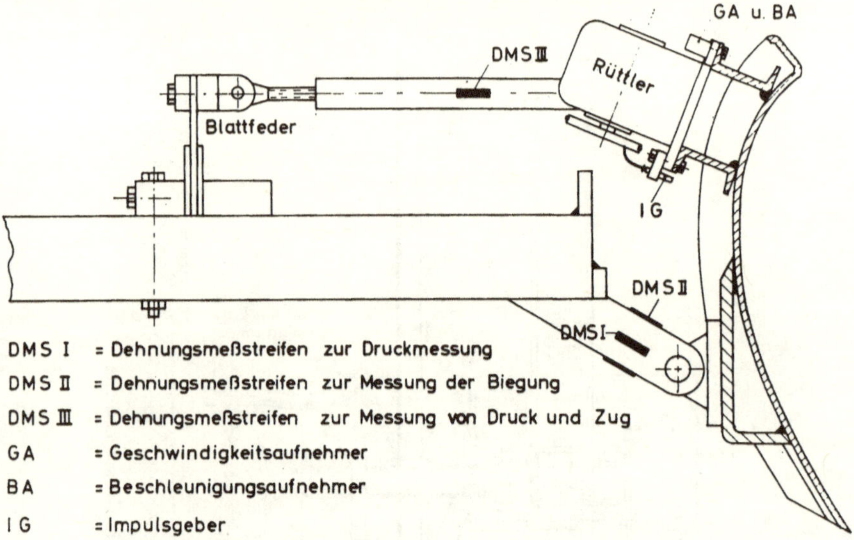

BILD 26. Die Messeinrichtung für die Grossversuche

(iii) Die Fahrleistung des starren Schildes P_s

$P_s = F_{xs} v_F$ oder $P_s = 1.33 . 10^{-4} F_{xs} . v_F$ (PS)

wobei F_{xs} = der Grabwiderstand bei starrem Schild (kp) ist.

(iv) Die Gesamtleistung P_G

$P_G = P_R + P_{vs}$

(v) Der Verminderungsfaktor δ des Grabwiderstandes F_x

$\delta = (F_{xs} - F_{xv})/F_{xs}$

(vi) Der Verminderungsfaktor δ' der vertikalen Kraft F_z

$\delta' = (F_{zs} - F_{zv})/F_{zs}$

wobei F_{zs} = die vertikale Kraft bei starrem Schild (kp); F_{zv} = die vertikale Kraft bei vibrierendem Schild (kp) ist.

(vii) Der Verminderungsfaktor λ der Gesamtleistung P_G

$\lambda = (P_s - P_G)/P_G$

Ergebnisse der Grossversuche

Nach den Grossversuchen ergaben sich die folgenden Ergebnisse.

(*a*) Die Schwingungen des Schildes verursachen für kleine Amplituden sowohl eine starke Verminderung des Grabwiderstandes F_x als auch der vertikalen Kraft F_z. Infolge der kleinen Amplitude der Schwingung tritt gleichzeitig mit der Verminderung des Grabwiderstandes eine fast ebenso starke Verminderung der Gesamtleistung P_G auf (Bild 27).

(*b*) Die Schwingungsgeschwindigkeit ϑ übt den stärksten Einfluss auf den Grabwiderstand F_x und auf die vertikale Kraft F_z aus, und zwar

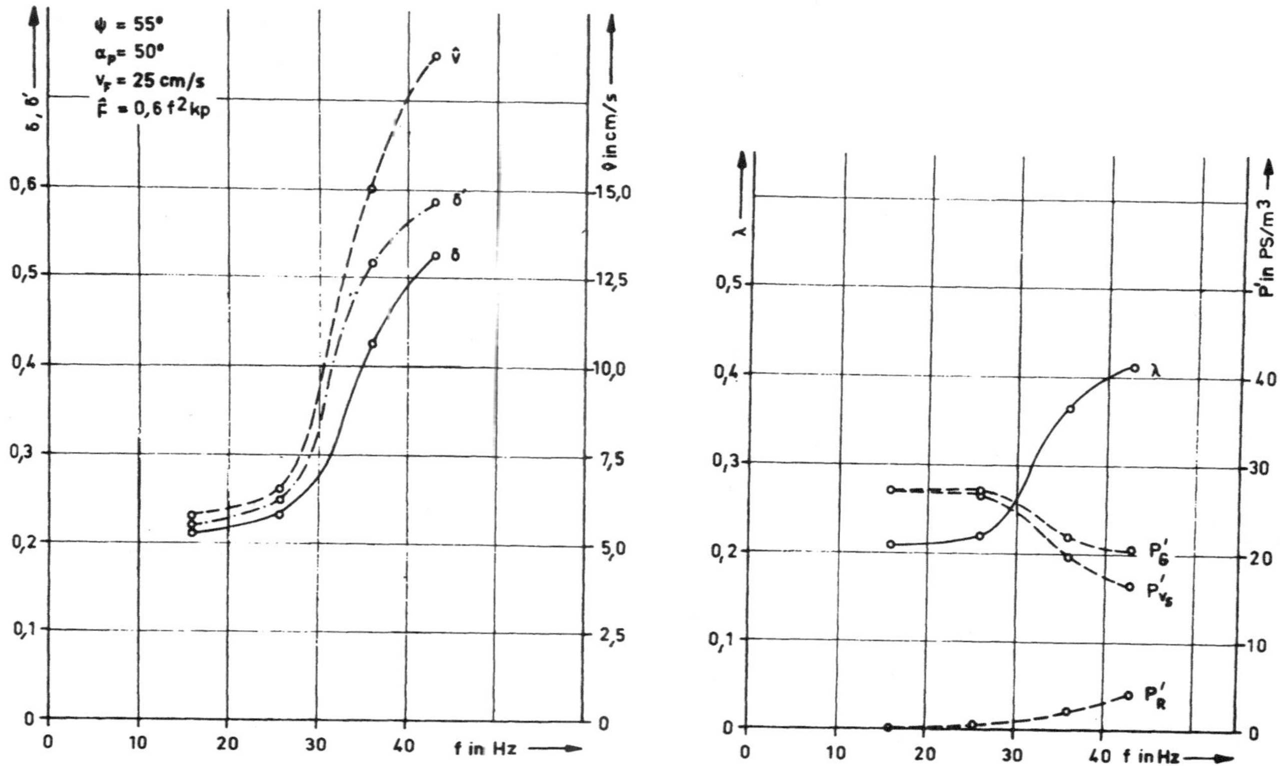

BILD 27. Einfluss der Frequenz f und der Schwingungsgeschwindigkeit $\hat{v}$ auf die Verminderungsfaktoren δ, δ', und λ

ergibt sich bei wachsendem $\hat{\vartheta}$ eine Verminderung von F_x und F_z (Bild 27).

(*c*) Es gibt einen günstigsten Wert für die Erregerkraft, bei dem sich die grösste Gesamtleistungsverminderung ergibt (Bild 28).

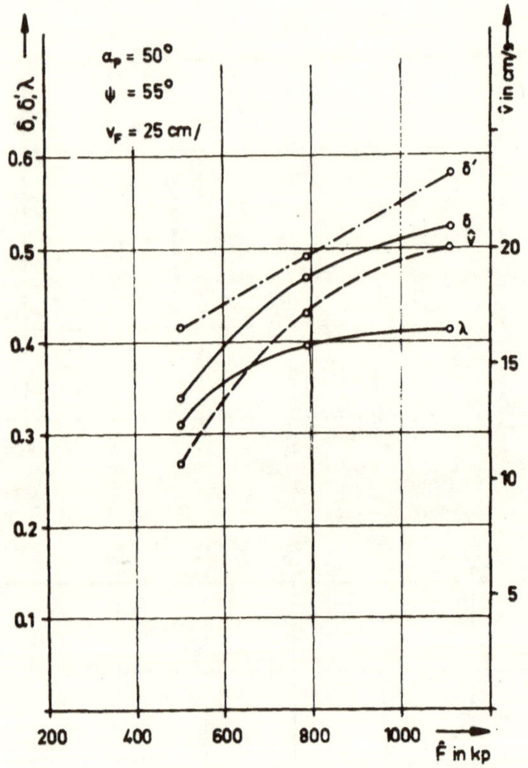

BILD 28. Einfluss der Erregerkraft $\hat{F}$ auf die Verminderungsfaktoren δ, δ', und λ

(*d*) Die Erhöhung der Fahrgeschwindigkeit v_F vergrossert bei gleichbleibender Schwingungsgeschwindigkeit $\hat{\vartheta}$ den Grabwiderstand (Bild 29). Vergrössert sich aber gleichzeitig mit der Fahrgeschwindigkeit auch die Schwingungsgeschwindigkeit derart, dass $\hat{\vartheta}/v_F$ konstant bleibt, so bleibt der Grabwiderstand F_x unverändert.

(*e*) Die Verminderung der vertikalen Kraft F_z bleibt von der Fahrgeschwindigkeit und der Schwingungsrichtung unbeeinflusst, sondern hängt nur von der Schwingungsgeschwindigkeit ab (Bild 29).

(*f*) Bei dem schwingenden und gleichzeitig schneidenden Planierschild ist die lineare Schwingungstheorie nicht anwendbar.

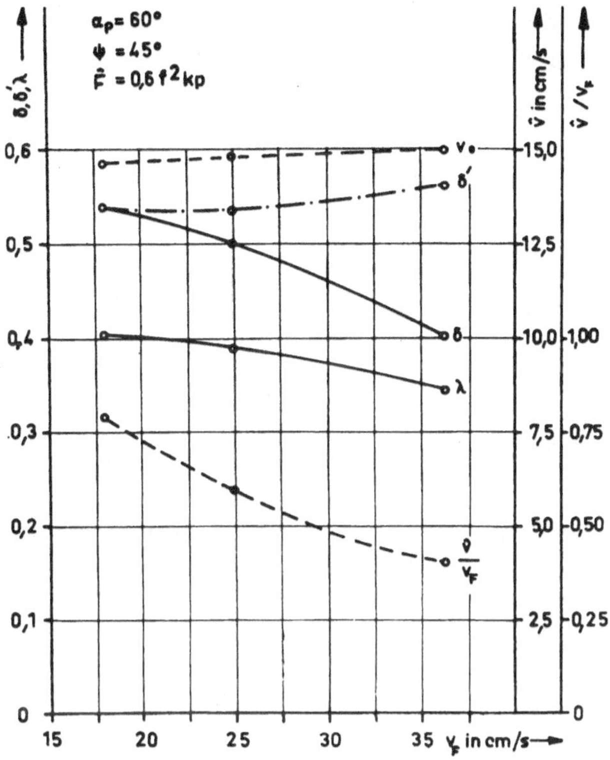

BILD 29. Einfluss der Fahrgeschwindigkeit $v/_F$ und des Verhältnisses $\hat{v}/v_F$ auf die Verminderungskaktoren δ, δ', und λ

SCHLUSSFOLGERUNGEN

Aus den Ergebnissen der oben dargestellten Untersuchungen kann man die folgenden, allgemeinen Schlüsse ziehen.

1. Durch die Schwingungen werden die Eigenschaften des Bodens beeinflusst, und zwar vermindert sich dadurch die Scherfestigkeit bzw. der innere Reibungswinkel des Bodens beträchtlich. Aus dem gleichen Grunde nimmt auch der Reibungsbeiwert zwischen Stahl und Boden ab. Die Ausnutzung dieser Tatsachen hatte bei der Anwendung von Schwingungen auf ein Planierschild eine starke Verminderung des Grabwiderstandes zur Folge, die bei Grossversuchen auf Kiesboden bis zu 50 Prozent betrug.

2. Diese hohe Verminderung des Grabwiderstandes ist schon mit Schwingungen von kleiner Amplitude zu erreichen, wodurch die zusätzlich aufzuwendende Schwingungsleistung sehr gering gehalten werden

kann, so dass die zum Bewegen des schwingenden Schildes erforderliche Gesamtleistung immer noch bedeutend kleiner ist als die Leistung, die zur Bewegung eines starren Schildes benötigt wird. Diese Leistungsersparnis gegenüber dem starren Schild betrug bei den Grossversuchen bis zu 40 Prozent.

3. Die Verminderung des Grabwiderstandes, d. h. der horizontalen Komponente F_x des Schneidwiderstandes F, ist proportional dem Verhältnis der Schwingungsgeschwindigkeit zur Fahrgeschwindigkeit $\hat{v}/v_F$. Demgegenüber ist die Verminderung der vertikalen Kraft F_z nur einfach proportional zur Schwingungsgeschwindigkeit, ohne von der Fahrgeschwindigkeit beeinflusst zu werden.

Nach den Forschungsergebnissen von Balownew (9) hängt F_x hauptsächlich von der Kohäsion c und dem Reibungswinkel des Bodens ρ, und F_z hauptsächlich vom äusseren Reibungswinkel ab. Diese beiden Erkenntnisse erlauben die Formulierung der folgenden Annahmen:

Die vom Schild auf den Boden übertragenen Schwingungen erfahren im Boden eine starke Dämpfung. Dadurch wird der Einfluss der Schwingung auf den vor dem Schild befindlichen Boden mit zunehmender Entfernung vom Schild rasch vermindert. Eine Vergrösserung der Schwingungsgeschwindigkeit dehnt den Wirkungsbereich der Schwingungen aus, wogegen eine zunehmende Fahrgeschwindigkeit ihn einschränkt, da die Menge des senkundlich hinzukommenden unbeeinflussten Bodens sich vergrössert. Die Verminderung des Grabwiderstandes ergibt sich also aus der Aufsummierung der Verminderung von c und ρ an allen von den Schwingungen beeinflussten Raumpunkten des vor dem Schild befindlichen Bodens. Die Vergrösserung des Schwingungswirkungsradius durch eine Erhöhung der Schwingungsgeschwindigkeit, welche eine Verminderung von c und ρ innerhalb eines grösseren Bereiches vor dem Schild zur Folge hat, bewirkt damit eine grössere Verminderung des Grabwiderstandes. Da die Reibungskraft zwischen Schild und Boden nur an der stets gleichbleibenden Kontaktfläche wirkt, wird ihre Verminderung zwar von der Schwingungsgeschwindigkeit, nicht aber von der Fahrgeschwindigkeit beeinflusst.

Diese Annahmen und der allgemeine Einfluss von Schwingungen auf den vor dem Schild liegenden Boden müssen durch weitere Forschungen untersucht werden. Dabei kann durch Anwendung einer geeigneten Einrichtung, z.B. die von Selig und Nelson (10) verwendeten, die Bewegung der einzelnen Bodenschichten vor einem schwingenden Schild untersucht werden. Durch den Vergleich der Ergebnisse einer solchen Untersuchung mit den Verhältnissen des starren Schildes müsste es möglich sein, zur Formulierung einer klaren Theorie zu gelangen, die Aufschluss darüber gibt, von welcher Art der Einfluss des schwingenden

Schildes auf den vor ihm liegenden Boden ist. Diese Theorie würde auch die Umwandlung der für den Grabwiderstand des starren Schildes vorhandenen Bestimmungsformeln in der Art ermöglichen, dass man damit auch das schwingende Schild erfassen könnte.

4. Auch die Schwingungsrichtung spielt eine Rolle bei der Verminderung des Grabwinderstandes beim schwingenden Schild. Es existiert eine günstigste Schwingungsrichtung, die unabhängig von der Bodenart zu sein scheint; denn in den Modellversuchen ergaben sich mit beiden Böden in bezug auf die Schwingungsrichtung fast dieselben Ergebnisse. Auch in diesem Punkt sind weitere Untersuchungen mit einer wirklichen Planierraupe, mit verschiedenen Böden und mit einer speziellen Einrichtung zum Aufhängen des Schlides nötig.

5. Die Erregerkraft weist für jedes Schild einen günstigsten Wert auf. Leigt die Erregerkraft unterhalb dieses optimalen Punktes, so wird dadurch eine Verminderung der Schwingungsgeschwindigkeit bewirkt, was eine Vergrösserung des Grabwiderstandes und damit auch eine Erhöhung der Gesamtleistung zur Folge hat. Liegt die Erregerkraft über dem Optimalwert, so wird dadurch zwar eine Vergrösserung der Schwingungsgeschwindigkeit und damit über die Verminderung des Grabwiderstandes eine Verringerung der Fahrleistung bewirkt, aber diese Einsparung an Leistung wird überdeckt von der zusätzlich aufzubringenden Rüttlerleistung, so dass insgesamt eine Vergrosserung der Gesamtleistung eintritt.

6. Die Auswertung der Modell- und Grossversuche lässt keinen klaren Einfluss der Frequenz auf den Grabwiderstand erkennen. Die grösse Verminderung des Grabwiderstandes trat genau bei der Frequenz auf, bei welcher auch die grösste Schwingungsgeschwindigkeit herrschte. Diese Frequenz ist bei beiden Böden nicht die gleiche. Eine genaue Untersuchung mit mehreren Böden könnte das Problem des Frequenzeinflusses aufhellen.

7. Die Schwingungsprobleme des vibrierenden Planierschildes sind nicht durch die Anwendung der linearen Schwingungstheorie zu lösen.

8. Durch die Grossversuche sind auf jeden Fall Voraussetzungen erbracht worden, die die industrielle Herstellung einer Planierraupe mit schwingendem Schild ermöglichen. Besonders günstig für den Bau der Maschine wirkt sich die Tatsache aus, dass die Verminderung des Grabwiderstandes durch verhältnismässig kleine Erregerkräfte und durch Schwingungen mit kleiner Amplitude erreichbar ist. Dadurch wird es möglich sein, zur Schwingungserzeugung einen hydraulischen Motor zu verwenden, obwohl der Wirkungsgrad des hydraulischen Antriebes klein ist. Bei der Konstruktion des Schildes und dessen Einrichtungen muss natürlich die dynamische Belastung dieser Teile berücksichtigt werden.

Eswerden jedoch in dieser Beziehung keine Schwierigkeiten auftreten können, da die Verwendung einer kleinen Erregerkraft die Konstruktionsarbeit wesentlich erleichtert. Schliesslich muss auch darauf hingewiesen werden, dass beim Bau einer Planierraupe mit schwingendem Schild die wegen der Schwingungseinrichtung notwendigen Zusatzkosten auf Grund der günstigen Ergebnisse, wie etwa Gewichtsersparnis, schwächerer Motor usw., weit überdeckt werden.

ZUSAMMENFASSUNG

Die Verwendung von Schwingungen auf den Planierschilden verursacht eine beträchtliche Verminderung des Schneidwiderstandes des Bodens. Falls die Schwingungen von kleiner Amplitude sind, folgt eine ebenso starke Verminderung der Gesamtantriebsleistung.

Der Grund dieser Verminderung ist der Einfluss, den die Schwingungen auf die Scherfestigkeit bzw. den inneren Reibungsbeiwert wie auch auf den äusseren Reibungsbeiwert des Bodens ausüben. Dieser Einfluss ist stärker auf grob- als auf feinkörnigen Böden. Um die günstigsten Ergebnisse zu erzielen, muss man eine geeignete Kombination von Fahr- und Schwingungsgescbwindigkeit verwenden. In diesem Fall ist die erforderliche Schwingerleistung ein kleiner Anteil der Gesamtleistung.

Die Ergebnisse der bisherigen Untersuchungen befürworten eine industrielle Entwicklung einer Planierraupe mit schwingendem Schild.

LITERATURVERZEICHNIS

1. MOGAMI, T. und K. KUBO, The Behaviour of Soil during Vibration. *Proc. 3rd Int. Conf. on Soil Mech. and Found. Eng.*, Zürich, 1953 I, S. 152–155.
2. SAWTSCHENKO, I. A., Der Einfluss der Vibration auf die innere Reibung von Sanden. Uebers. *aus Dinamika Grundow* Verlag Gostroicsdat, Moskau 1958 (S. 83–88). Deutsche Gesellschaft für Erd- und Grundbau e.V. Essen.
3. PREOBRASCHENSKI, N. A. und I. A. SAWTSCHENKO, Der Einfluss der Vibration auf den Scherwiderstand von Tonböden. *Uebers aus Dinamika Grundow*, Verlag Gostroicsdat, Moskau 1958, S. 89–92.
4. EGGENMÜLLER, A., Untersuchungen an schwingenden Bodenbearbeitungswerkzeugen. Sonderdruck aus *Grundlagen der Landtechnik* (1958), Heft 10, S. 55–95 und 143–150, VDI-Verlag GmbH, Düsseldorf.
5. SKURENKO, N., Versuchsdaten über den Einfluss von Vibration auf den Schneidwiderstand der Böden. *Zurnal techniceskoj fiziki* Bd. 23, Nr. 8, S. 1430–1434.
6. SKURENKO, N., Ergebnisse von Versuchen über den Einfluss von Schwingungen auf den beim Schneiden des Bodens auftretenden Widerstand. *Uebers aus Dinamika Grundow* (*Baugrund-dynamik*), Verlag Gostroicsdat, Moskau 1958, S. 93–103, Deutsche Gesellschaft für Erd- und Grundbau e.V., Essen.
7. DREES, G., *Untersuchungen über das Kräftespiel an Flachbagger-Schneidwerkzeugen in Mittelsand und schwach bindigem sandigem Schluff unter besonderer Berücksichtigung der Planierschilde und ebenen Schürfkübelschneiden.* Herausgeber Prof. Dr. G. Garbotz, Institut für Baumaschinen und Baubetrieb R.W.T.H. Aachen, Heft 16, 1958 (S. 24/25).

8. LORENZ, H., *Grundbau-Dynamik*. Springer-Verlag, Berlin, 1960 (S. 258/662).
9. BALOWNEW, V., Analytische Berechnung der optimalen Parameter der Graborgane und Fragen der Automatisierung des Schneid-prozesses von Böden mit Planierraupen und Schrappern. *Wissenschaftl. Zeitschr., der T.U.*, Dresden, 12 (1963), Heft 5.
10. SELIG, E. und R. NELSON, Observations of Soil Cutting with Blades. *Journal of Terramechanics*, Vol. 1, No. 3 (1964).
11. PANAJOTOPOULOS, N., Der Einfluss der Vibration auf den Grabwiderstand von Planierschilden. *Mitteilungen des Instituts für Baumaschinen und Baubetrieb der R.W.T.H.—Aachen*, Heft VII, 1965.

Performance of Soil Compaction Machinery

E. T. SELIG*

ABSTRACT

A full-scale field study was undertaken to investigate the effectiveness of various methods of compacting soil for highway construction and to determine the resulting soil properties using rapid field measurement of moisture, density, strength, and stiffness. Twelve commercial rollers representing most of the basic types available were used to compact lifts of soil varying from 6 to 18 in. thick at a range of moisture content. The soil types included silt, sand, clay, crushed stone, and gravel. Approximately 140 lifts were tested and over 10,000 soil measurements were made in the course of the study. Changes in the soil properties with passes of the rollers were determined. The paper will compare the results obtained with each type of roller including such factors as the uniformity of compaction and the rate of change with each pass.

---

*Senior Research Engineer, Soil Mechanics, IIT Research Institute, Chicago, Illinois.

PERFORMANCE AND PREDICTION

On the Theoretical Rolling Resistance Of Wheels of Different Widths Running in Tandem

LOUIS E. LEVITICUS

J. B. LILJEDAHL*

ABSTRACT

The article uses the well known sinkage equation $p = kz^n$, where the Bekker parameters may or may not be substituted for k. A relation is sought and found which will calculate the rolling resistance of wheels of different sizes running one in the track of the other as is the case for agricultural tractors.

A second equation based on similar assumptions combines the weights on the wheels into an equation of the weight distribution between front and rear wheels.

When soil parameters are known the two equations combined will make it possible to determine weight distribution for optimizing rolling resistance or will enable determination of rolling resistance of a given weight distribution.

*Graduate Research Assistant and Professor, respectively; Agricultural Engineering Department, Purdue University, Lafayette, Indiana.

Multi-pass Behaviour of A Rigid Wheel

R. LISTON

L. MARTIN*

ABSTRACT

An analysis of the behavior of a rigid wheel during multi-passes is presented. The relationship between the pressure-sinkage curve obtained by means of a circular plate is compared to the pressure distribution acting against the face of a wheel. The normal pressure distribution acting against a wheel was measured by means of load cells developed by Michigan State University. An attempt is made to justify the difference between predicted and actual pressure-distribution on the wheel.

*Land Locomotion Laboratory, Detroit Arsenal, Detroit, Michigan.

The Influence of Front Wheel Path Upon the Performance of A Following Wheel

P. H. SOUTHWELL*

M. E. MARWOOD*

ABSTRACT

Since the performance of a driving wheel is dependent upon soil condition, it may be expected that when one wheel follows behind another: (*a*) the performance of a rear wheel when operating on the path created by a front wheel will not be the same as that obtained when operating outside the front wheel track, and (*b*) the performance of a rear wheel when operating on the path of a front wheel will be influenced by the form of the path created by the front wheel. These points are, of course, important in connection with the design of multi-axle tractive machines.

An investigation was therefore made, on two soil types, in which the rear wheels of a tractor were operated both in and out of the front wheel path, and in which various forms of front wheel path were produced by both driven and non-driven front wheels. Pairs of front and rear wheels were employed, using solid axles, so as to double the differences and facilitate accurate measurement of forces. The rear axle was driven mechanically, using the highest gear ratio available in the tractor transmission, and the front axle was driven through an hydrostatic system so as to provide a range of torque loadings from negative (2-wheel-drive) to positive (4-wheel-drive).

THE REAR TIRES used were 13.6–28 and the front tires were all 5.50–16 having a contact width of one-half that of the rear tires. Thus a front-wheel-path width equal to that of the rear wheels was obtained by using dual front tires. Two front-wheel-path widths (50 per cent and 100 per cent of the rear-wheel width), two front-tire tread designs (smooth and lugged), and two front-wheel ground pressures were employed. In addition, the front wheels were (*a*) non-driven [D_1], (*b*) driven at zero thrust and zero slip [D_2], (that is, the applied torque was only that required to overcome rolling resistance, and the peripheral speed was equal to that of the rear wheels), and (*c*) driven so as to exert a net thrust and appreciable positive slip [D_3].

*School of Agricultural Engineering, University of Guelph, Guelph, Ontario, Canada.

FIGURE 1. The two front-tire tread designs, and arrangement for two path widths (dual tires equal to rear-wheel width)

The front axle was mounted on the tractor frame in such a manner as to permit measurement of the torque input and of the vertical and horizontal forces on the axle. Thus, in each experiment, the dynamic weight distribution and the energy input to the front axle and to the rear axle were recorded. Similarly, the net thrust output and the corresponding slip of the front wheels and of the rear wheels were also recorded. Hence the influence of the variations of front wheel path could be isolated in terms of differences in rear wheel performance and, in comparing the latter, the performance on undisturbed soil (outside the front wheel path) constituted a datum.

Incidental to the purpose of these experiments, the apparatus also permitted some useful measurements to be made of front-wheel rolling resistance and of weight transfer due to drawbar load.

Apparatus

The front axle was supported on the main frame by two hydraulic cylinders (double-acting) in the horizontal plane and two hydraulic cylinders (single-acting) in the vertical plane, so that the oil pressures in the pairs of matched cylinders were proportional to the forces acting on the axle. Attachment of the cylinder rods to the axle was through two roller "cage" assemblies designed to eliminate any interaction between the horizontal and vertical force measurements, which were recorded on calibrated Bourdon gauges.

The front axle propulsion system consisted of a variable-displacement hydraulic pump, of the swash-plate type, in a closed circuit with a fixed-displacement piston motor. The latter was mounted in the manner

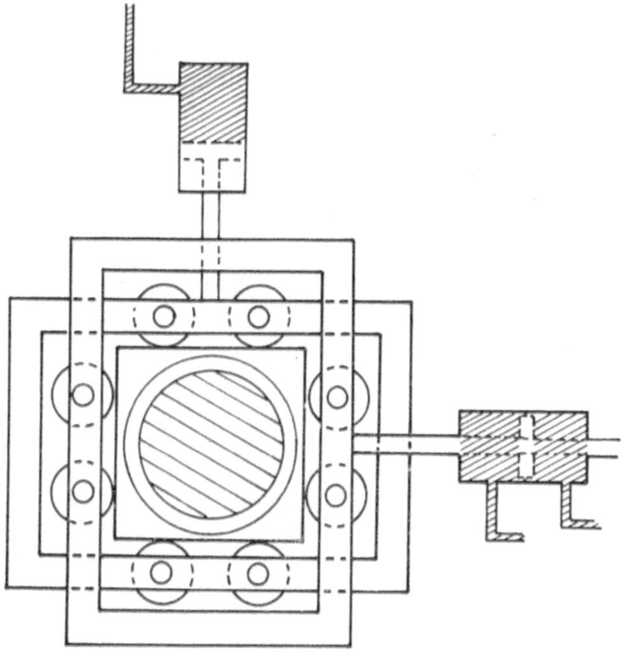

A

B

FIGURE 2. One "cage" and hydraulic cylinder assembly for front axle force measurement. A, schematic design; B, actual design of a swinging-frame dynamometer so that the torque input to the axle could be recorded; the drive to the axle was through a pair of Hooke's joints in order to ensure that only a pure couple was applied and, therefore, that the torque input did not affect the horizontal and vertical force measurements.

The power output of the engine supplying the mechanical transmission to the rear axle as well as the hydraulic transmission to the front axle was calibrated in terms of intake manifold depression at a constant speed so that, knowing the two transmission efficiencies and the output of the hydrostatic system, the input to the rear wheels could be determined by difference.

A ciné film record of the gauges indicating engine power, hydraulic motor torque, and vertical and horizontal front axle forces was made in each test. This was then projected at slow speed for analysis at leisure.

As shown in Figure 3, the complete apparatus was made rather untidy by the large water ballast tank used to permit variation of front wheel path compaction, and which also served as a structural member.

FIGURE 3. The complete apparatus

Load was applied to the complete machine by means of a tracked vehicle capable of precise regulation, and was recorded by an hydraulic dynamometer. Manual measurements of forward speed, and of front and rear wheel slip, thus permitted the total output of the whole machine in terms of drawbar pull and power to be determined. The output of the front wheels (positive or negative) was calculated from the horizontal transducer records, and the output of the rear wheels was thus obtained by difference. An excessive power supply was made available in order to ensure that wheel performance was not limited by engine torque.

Soil Conditions

Comparative experiments were made on two soils, both of which were cultivated to a depth of about 8 inches and were uniformly dry and loose. Pertinent data for the test areas of these light and medium loam soils is given in Table I.

TABLE I

Soil type	Percentage according to separates (USDA system) Sand	Silt	Clay	Organic matter content by weight
(*a*) Fox Sandy Loam	84	11	5	2
(*b*) Guelph Loam	35	48	17	5

The rolling resistance forces were greater on the loam soil than on the sandy loam, and the maximum pulls (thrust minus rolling resistance) were also greater by from 13 to 40 per cent. However, the maximum conversion efficiencies (wheel output ÷ wheel input) on the loam soil were generally somewhat lower than in the case of the sandy loam, but were derived at appreciably higher drawbar pull values.

Front-Wheel Rolling Resistance

In connection with rolling resistance, it may be of interest to note that on the sandy loam soil neither an increase in ground pressure nor a change in tire tread design had any appreciable influence on the front-wheel rolling resistance force, but that doubling the width of the front wheels increased the rolling resistance by 22 per cent. On the loam soil, on the other hand, neither the wheel width nor the tread design had any measurable influence, but an increase in ground contact pressure effected an increase in rolling resistance force by approximately 1 lb. for each 6.3 lb. of added weight. This major influence of ground pressure (sinkage) in the case of the loam soil would indicate that a decrease in rolling resistance might have been anticipated when wheel width was doubled and hence ground pressure halved. Thus it may be surmised that increased path width did effect an increase in rolling resistance force, but that this was offset by the lesser ground pressure. On the sandy loam, however, the effect of double the wheel width was more evident because ground pressure had only a minor influence on rolling resistance.

The front-wheel rolling resistance data was determined from the measurements of horizontal force on the front axle, and confirmed by the rear wheel slip—total drawbar pull curves obtained in each test on each soil. A typical example of these performance curves is shown in Figure 4, and the rolling resistance of the front wheels is represented by

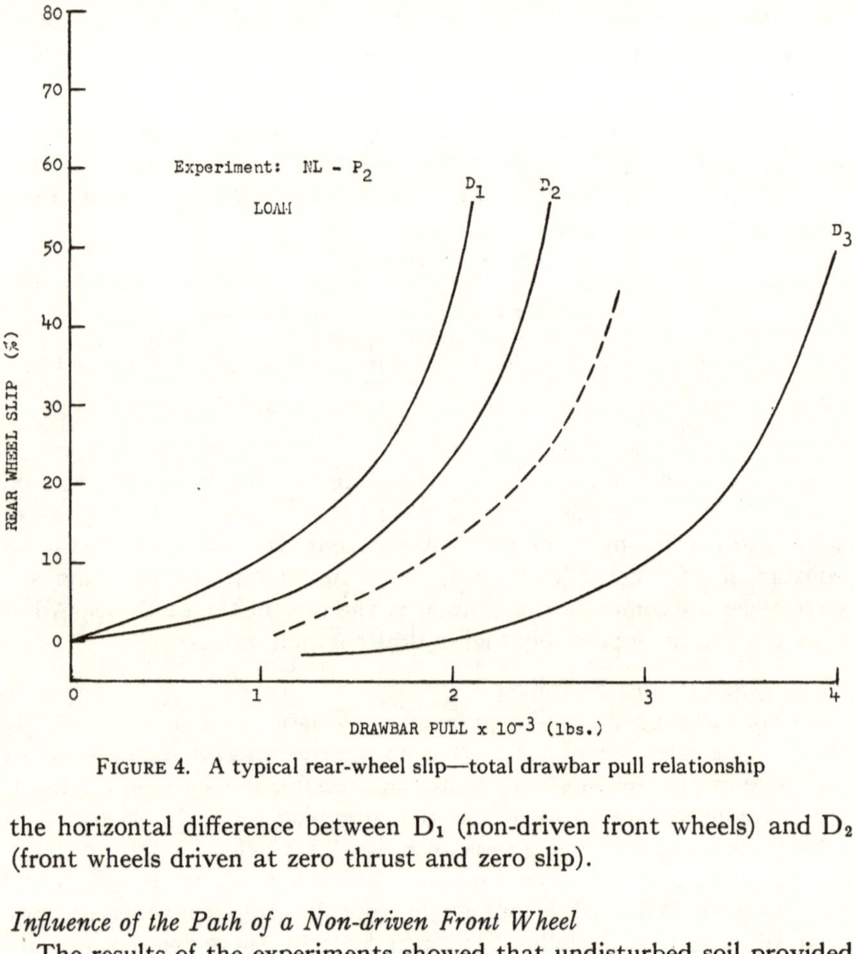

FIGURE 4. A typical rear-wheel slip—total drawbar pull relationship

the horizontal difference between D_1 (non-driven front wheels) and D_2 (front wheels driven at zero thrust and zero slip).

Influence of the Path of a Non-driven Front Wheel

The results of the experiments showed that undisturbed soil provided a slightly more favourable condition for operation of the rear wheels than did the paths created by non-driven front wheels. It is evident from the typical curves shown on Figures 5 and 6 that both higher rear-wheel pulls and higher pull/weight ratios were obtained, at any slip value, when the rear wheels were not operating on the front wheel path.

Although these differences in performance were small, they were confirmed in replications and also by the conversion efficiency values. Operation on the path of a non-driven wheel was thus clearly not advantageous.

Influence of the Path of a Driven Wheel Exerting a Thrust

The front wheels were propelled in D_3 with a constant torque which was greater than that required to overcome their rolling resistance

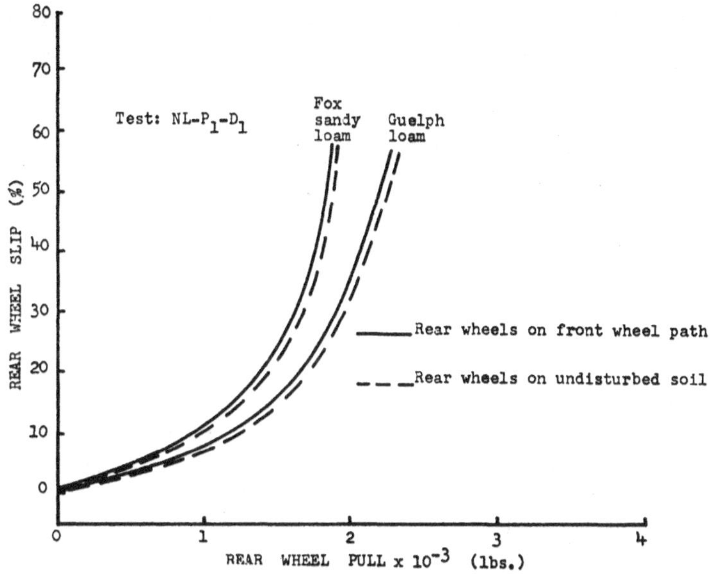

FIGURE 5

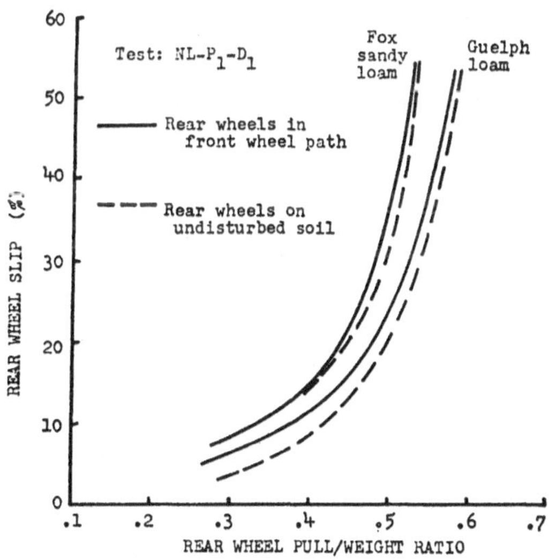

FIGURE 6

Figures 5 and 6 are typical comparisons of rear-wheel performance on undisturbed soil and on the path of a non-driven wheel.

(i.e. greater than in D_2) and with a peripheral speed which was greater than that of the rear wheels, so as to produce a positive horizontal force and a positive slip.

The horizontal difference between the D_3 and D_2 slip-pull curves on the typical data shown on Figure 4 therefore theoretically represents the magnitude of the thrust exerted by the front wheels when driven in D_3. That is, the differences on the graph should equal the recorded values of positive force on the front axle. This was not, however, found to be the case. The increase in total drawbar pull ($D_3 - D_2$) was in each experiment considerably greater than the thrust on the front axle.

The broken line on Figure 4 represents the total drawbar pull minus the horizontal force on the front axle (that is, the slip–pull relationship of the rear wheels only) and should therefore superimpose upon the D_2 curve. Hence the difference between the broken curve and the D_2 curve demonstrates an appreciable increase in performance of the rear wheels, the average increase in total pull being from 128.5 per cent to 151.0 per cent of the pull exerted by the front wheels.

The increased performance of the rear wheels when operating on the path of the front wheels driven in D_3 may be attributed to: (1) an increased dynamic weight on the rear wheels due to the thrust exerted by the front wheels, plus (2) an increase in rear wheel efficiency due to the path created by the driven front wheels. Analysis of the dynamic weight distribution data permitted calculation of the increase in performance due to (1), and thus determination of the increase in performance due to (2). It was shown that an increase in pull/weight ratio had been derived in each experiment—over the range shown by the examples given in Table II—and that the effect of the path of the driven front wheel constituted 4.5 to 12.5 per cent of the total increase in rear wheel output.

TABLE II

(S = sandy loam soil; L = loam soil)

Front wheel path (lugged tires)	Average pull exerted by front wheels (lb.)		Average increase in total pull (D_3-D_2) as % of front wheel pull		Average increase in P/W ratio (D_3-D_2) as % of P/W ratio in D_2	
	S	L	S	L	S	L
narrow; compaction P_1	766	872	128.5	143.5	4.4	5.8
narrow; compaction P_2	885	1100	135.5	139.0	4.7	10.2
wide; compaction P_1	774	930	136.0	151.0	7.2	13.5

It will be noted that this effect was opposite to that of the non-driven wheels, and may be attributed to changes in soil structure effected by the front wheels when exerting an appreciable thrust and having more than about 25 per cent slip.

It will also be noted from Table II that the influence of the driven front wheel path was more pronounced on the loam soil than on the sandy loam. Furthermore, an increase in path compaction from P_1 to P_2 almost doubled the percentage increase in P/W ratios on the loam soil, but caused only a slight increase on the sandy loam. Similarly, doubling the path width (and thus halving the ground pressure) more than doubled the percentage increase in P/W ratios on the loam soil, but increased those on the sandy loam to a lesser extent.

Influence of Variations of the Path created by a Non-Driven Wheel

Investigations were made, with D_2 propulsion of the front wheels, to determine the influence of the type of tire tread (ribbed and lugged), the ground pressure (nominal values of 6.5 and 8.0 p.s.i.), and the path width (50 per cent and 100 per cent of the rear-wheel width).

Apart from the points previously mentioned, the differences recorded were insignificant and in many cases were within the limits of experimental error. It may be said, however, that the data indicates a slightly better rear wheel performance with the narrow front wheel path on the sandy loam soil, with the ribbed front tire tread design on the loam soil, and with the lesser front wheel path compaction on both soils.

Over-all Performance of the Machine

Since an increase in rear-wheel performance is of little value if obtained at the expense of, for example, an equal reduction in the performance of driven front wheels or an increase in the power required to overcome the rolling resistance of non-driven front wheels, it was necessary to also consider the over-all performance of the complete machine.

TABLE III

(S = sandy loam soil; L = loam soil)

Front wheel path	Soil	Maximum overall efficiency and (corresponding pull)		
		D_1	D_2	D_3
narrow; ribbed; compaction P_1	S	45 (1450)	44 (1450)	—
	L	44 (1750)	45 (1650)	—
narrow; lugged; compaction P_1	S	43 (900)	43 (900)	40 (2300)
	L	45 (1500)	45 (1800)	45 (2650)
narrow; lugged; compaction P_2	S	42 (1400)	41 (1650)	40 (2250)
	L	41 (1750)	42 (1850)	42 (2600)
wide; lugged; compaction P_1	S	44 (1200)	43 (1550)	43 (2300)
	L	40 (1600)	42 (1700)	45 (2600)

Table III gives the maximum over-all efficiencies obtained in 22 tests together with, in parentheses, the corresponding total drawbar pulls.

It will be noted that the maximum over-all efficiencies were all between 40 and 45 per cent but occurred at a higher pull with D_3 than with D_2, and generally at a higher pull with D_2 than with D_1. In some cases with D_3 the maxima was also maintained over a wider range of drawbar pull, as illustrated by the example shown on Figure 7. It will also be noted that the maximum over-all efficiencies were somewhat lower with the higher front wheel ground pressure (P_2).

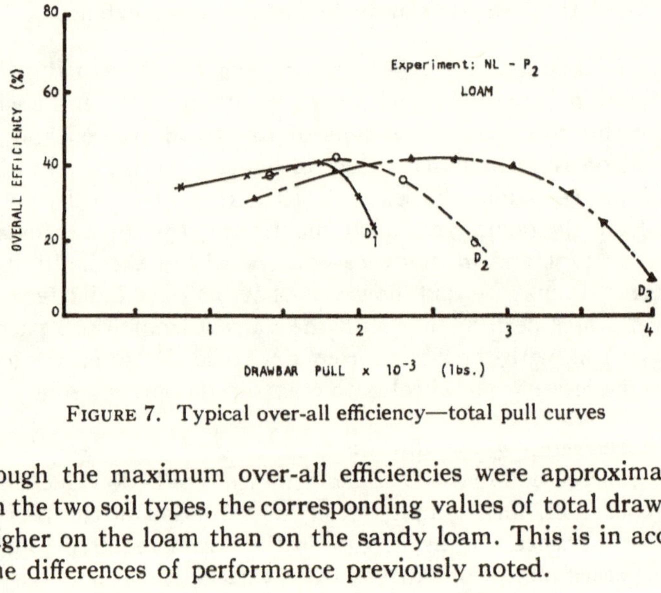

FIGURE 7. Typical over-all efficiency—total pull curves

Although the maximum over-all efficiencies were approximately the same on the two soil types, the corresponding values of total drawbar pull were higher on the loam than on the sandy loam. This is in accordance with the differences of performance previously noted.

CONCLUSIONS

The conclusions arrived at from the investigation were as follows:

1. The variations of the front-wheel path width, front-tire tread design, and front-wheel path compaction which were employed in the experiments, had no pronounced influence upon rear-wheel performance.

2. Operation of the rear wheels on the paths created by the non-driven front wheels definitely did not, as compared to operation on undisturbed soil, effect any improvement in rear-wheel performance; in fact, there was evidence of a somewhat reduced performance.

3. Operation of the rear wheels on the paths created by the driven front wheels exerting a thrust, as compared to operation on the paths

of non-driven front wheels, did effect an appreciable increase in rear-wheel performance, particularly on the loam soil. Although the data showed that this increase was predominantly due to a greater dynamic weight on the rear wheels, it was also proven that an increase in rear-wheel efficiency had in fact been derived.

4. The differences in rear wheel performance did not result in corresponding changes in over-all efficiency, but were apparent as differences in the drawbar pulls at which maximum over-all efficiencies occurred.

Dimensional Analysis of An Unpowered Pneumatic Tire

VICTOR C. PIERROT, III

WESLEY F. BUCHELE*

ABSTRACT

The factors affecting the rolling resistance and sinkage of a pneumatic tire were arranged into seven dimensionless pi-terms.

An instrument dynamometer was used to hold the loaded model and prototype tire while a soil bin containing soil was pulled under the tire. Load, state of compaction, and speed were varied.

It was found that the behaviour of the prototype tire could be predicted from a study of the model.

*Agricultural Engineer, Engineering Research, Deere & Co., Moline, Illinois; Professor, Agricultural Engineering Department, Iowa State University, Ames, Iowa.

Traction of the Tractor Based on Soil Parameters

SHOZO MASUDA

TAKASHI TANAKA

ABSTRACT

In order to find the relationship between the soil parameters and the thrust on actual paddy fields, several tractors having various running gear were tested on different fields, and the relation between the following items was investigated:

(*a*) Slippage (i) and sinkage (z)
(*b*) Running resistances (R) and sinkage (z)
(*c*) Drawbar pull (H) and ground contact areas (A)
(*d*) Drawbar pull (H) and ground contact pressure (σ)
(*e*) Unit soil thrust (T/A) and contact pressure (σ) or shear strength of soil (s)
(*f*) Unit soil thrust (T/A) and slippage (i)

ACCORDING TO THE recent development of agricultural mechanization in Japan, it has been getting more important to secure tractor trafficability and traction performance by measuring the soil conditions, especially in paddy fields.

This report is the summarized results of experiments which were carried out to respond to this requirement and to investigate the relations between the traction of tractors and the soil values in actual paddy fields. M. G. Bekker has formulated systematically the traction for other performance of the vehicle by introducing the various soil parameters, and resulting formulas have been confirmed by many researchers.

In order to investigate the effects of these soil parameters, drawbar pull tests were carried out on various fields.

TEST TRACTORS AND TEST FIELDS

1. *Test Tractors*

The dimensions of the tractors used in the tests are shown in Table I. As shown in Table I, A to I were wheel-type tractors with half tracks,

*Professor and Assistant Professor, respectively, in the Division of Farm Power, Faculty of Agriculture, Kyoto University, Kyoto, Japan.

TABLE I

TEST TRACTORS

Tractor	Output of engine (PS)	Weight (kg)	Kinds of running gears used in the test: Kinds	Size	Test field
A	18	1035	Tire	8–24	1, 2, 3, 7
			Half Track	38/115 cm²	1, 2, 3
B	27	1400	Tire	11–28	1, 2
			Floating Aid		1, 2
C	15	750	Tire	7–24	2
			Floating Aid		2
D	35	1570	Trie	11–28	1, 2
			Half Track	45/145 cm²	1, 2
E	18	950	Tire	8–24	3
			Half Track	30/120 cm²	3
			Girdle		3
			Dual Tire	(8–24) × 2	3
F	37	1430	Tire	11–28	3
			Half Track	45/145 cm²	3
G	46	1570	Tire	11–28	4
			Half Track	45/145 cm²	4
			Floating Aid		4
H	17	860	Tire	8–24	5
I	12	650	Tire	7–20	5, 6, 7
J	37	2500	Track	43/165 cm²	8
K	35	2100	Track	40/155 cm²	4
			Track	50–155 cm²	4
L	40	2330	Track	50/160 cm²	4

Japanese floating aids for paddy fields, girdles, and dual tires—J to L were crawler-type tractors. The size of the running gears given in Table I are the rear-tire sizes, or the width/ground contact length of the tracks.

2. Test Fields

Seven fields were selected for the tests. The kinds of soil, moisture content, surface hardness, internal friction angles of the surface soil and the surface conditions are shown in Table II.

The hardness of the soil was measured by a cone penetrometer having an angle of 30° and area of 0.5 square inches, and hardness values at 10 or 15 cm depths are indicated on Table II. The internal friction angles were measured by a shear vane having a diameter of 46.5 mm and an effective height of 10 mm. On the test fields 5, 6, and 7, these angles were measured by the direct shear test device in the laboratory. The values of cohesion of the soil fall within the limits of 0.05 to 0.22 kg/cm².

SLIPPAGE AND SINKAGE

The prevention of sinkage is one of the important problems when tractor performance is discussed, because sinkage increases running

TABLE II

TEST FIELDS

Test field	Soil*	Moisture content	Hardness of soil (kg)	Internal friction angles of soil (degrees)	Surface conditions of soil
1: a, b, c	LiC	Much	10.0	23–24	Paddy field after harvesting rice plant
2: a, b	CL	Medium	8.0	13–14	Paddy field after harvesting Chinese Milk Vetch
3: a–d	CL	Medium	16.0	34–36	Fallowing paddy rice field
e–f	CL	Medium	13.5	23–25	"
h	CL	Medium	7.5	14	"
4: a	Mud	Much	5.0	25	Reclaimed land under water level
b	Mud	Much	4.0	23	"
c	Mud	Much	3.5	17	"
d	Mud	Much	3.0	15	"
5: a, b	SL	Less	16.0	32	Paddy field after harvesting rice plant
6	SL	Less	22.0	25	"
7	SL	Less	16.0	28	"

*LiC: light clay; CL: clay loam; SL: sandy loam.

FIGURE 1A

FIGURE 1B

FIGURE 1C

FIGURE 1D

FIGURE 1E

FIGURE 1F

resistance and reduces the traction of the tractor. Two different types of sinkage may occur when the tractor moves on a soft surface. One of them is the static sinkage caused by the ground contact pressure and soil hardness; and the other is the slip sinkage caused by insufficient thrust. These sinkages may be quite different. The static sinkage has been discussed by many researchers; but for slip sinkage, the effects of the soil parameters have been scarcely discussed or investigated.

Figure 2 shows one of the results obtained from the drawbar pull test in this experiment, and there is a similarity between the curves of slippage vs. drawbar pull and sinkage vs. drawbar pull. The downward broken curves of the drawbar pull depended on the soil sticking with the weed fibres to the tire or to the half tracks, and also similar on both curves. Then, we can present the curves of slippage vs. sinkage by the linear relations shown in Figure 3 for all tests. Accordingly, it may be presumed that the sinkage of the tractor (z) can be expressed by the linear function of the slippage (i), that is, $z = f(i)$ or $z = Ai + B$, where A and B are constants. A is defined by the soil hardness and it may be larger on the soft surface than on the hard surface. B is defined by the static sinkage of the tractor, which is defined by Bekker's equation.

RUNNING RESISTANCE AND SINKAGE

From the results of the test on field 4, the relation between the sinkage and running resistances of the crawler type tractor on mud could be illustrated, as in Figure 4. The slippage on the ordinate is rearranged from the linear relationship between the sinkage and slippage in our test. From this Figure, it may be assumed that the running resistance B can be defined as a function of the sinkage and it may be $R = Cz + D$, where C and D are constants. Since the sinkage has a linear relationship with the slippage, we can define the running resistances as a linear function of the slippage, also. That is, $R = C'i + D'$, where C' and D' are also constants.

The difference between the soil thrust and the running resistances may be regarded as the drawbar pull. Hence, with an increase of running resistance, the drawbar pull will be reduced. When the running resistance, becomes equal to the soil thrust, the tractor will have lost its moving ability. The relation between the sinkage and the dimensionless values of the running resistance vs. soil thrust can be shown by the curves on Figure 5. The tractors lost their moving ability at the values of 0.7 in this test; and sinkage of the tractors under these circumstances was about 17 cm. This data was confirmed by many repeated tests.

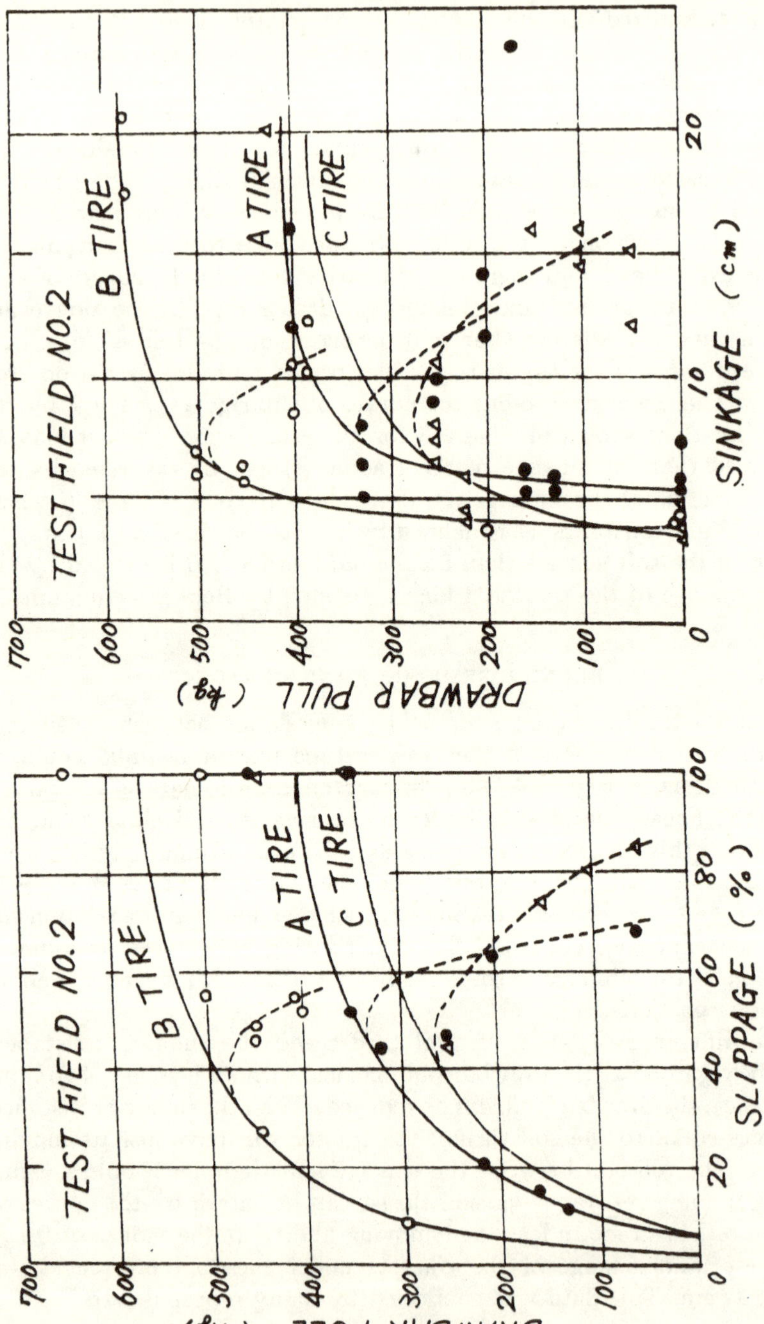
TEST FIELD NO.2
B TIRE
A TIRE
C TIRE
DRAWBAR PULL (kg)
SINKAGE (cm)
TEST FIELD NO.2
B TIRE
A TIRE
C TIRE
DRAWBAR PULL (kg)
SLIPPAGE (%)

FIGURE 2A

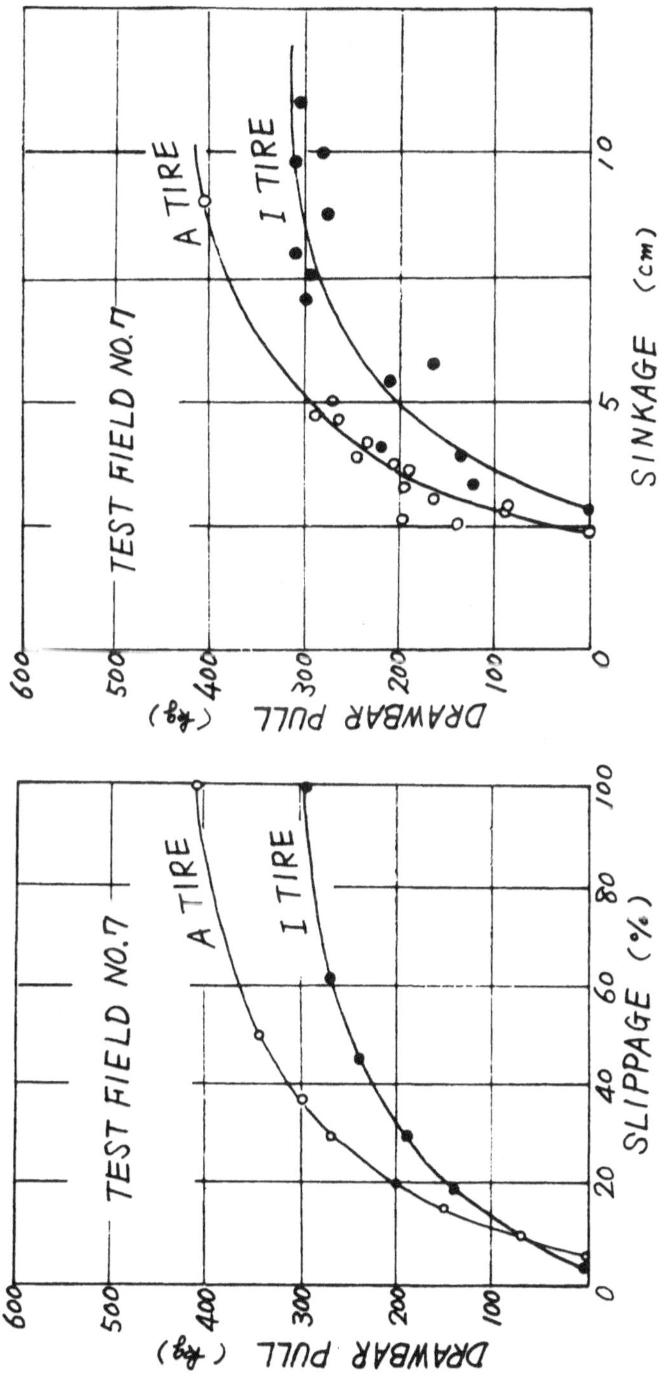
TEST FIELD NO.7
A TIRE
I TIRE
DRAWBAR PULL (kg)
0
100
200
300
400
500
600
SINKAGE (cm)
5
10
TEST FIELD NO.7
A TIRE
I TIRE
DRAWBAR PULL (kg)
0
100
200
300
400
500
600
SLIPPAGE (%)
20
40
60
80
100

FIGURE 2B

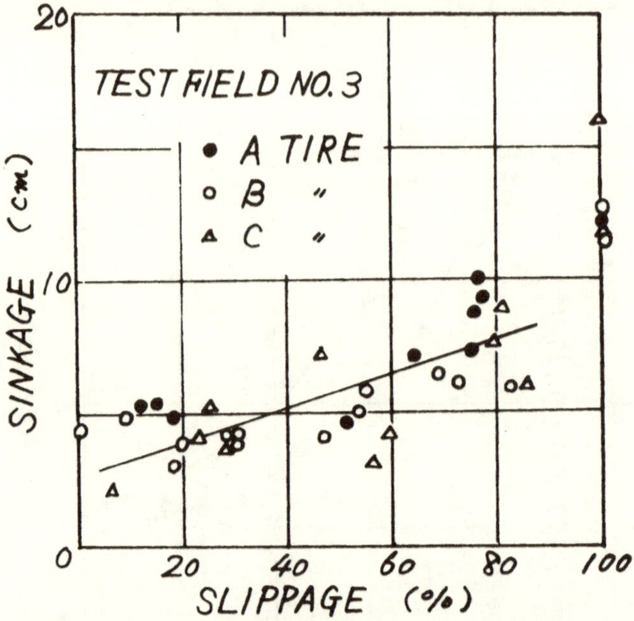

FIGURE 3A

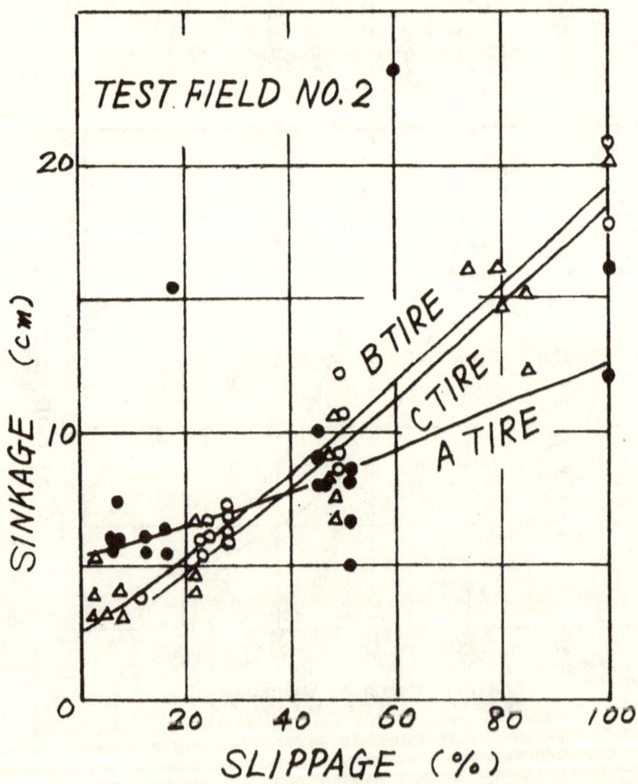

FIGURE 3B

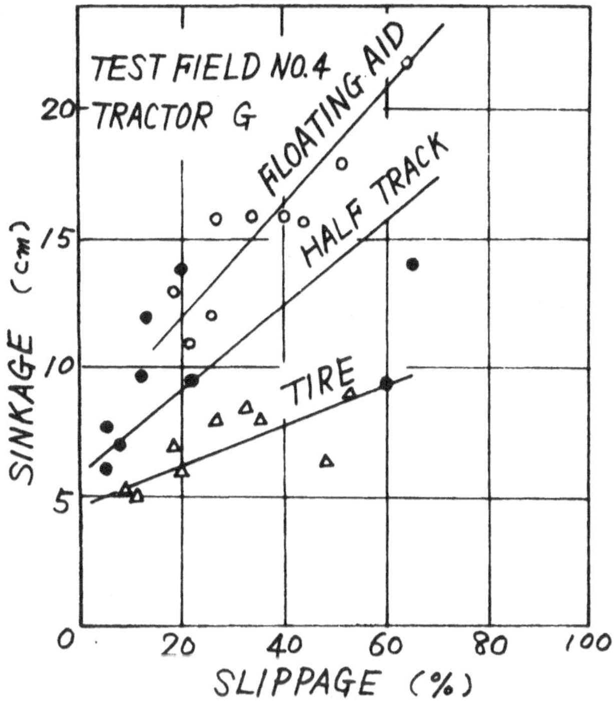

FIGURE 3c

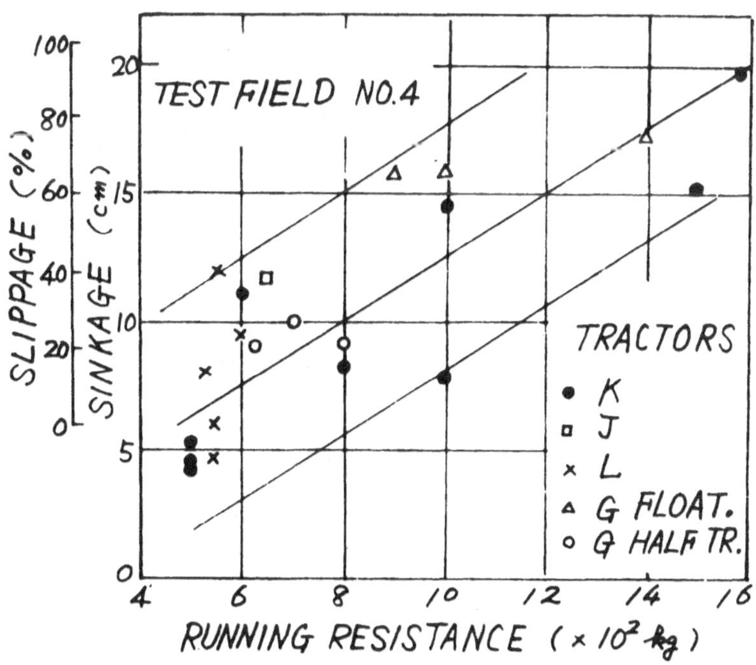

FIGURE 4

From these curves on Figure 5, the dimensionless value of running resistance R vs. thrust T can be indicated as the nth order function of the sinkage, and it may be $R/T = Ez^n + F$, where E and F are constants and n is the exponent defined by the soil conditions.

DRAWBAR PULL AND GROUND CONTACT AREAS AND PRESSURES

Since the sinkage of the tire or other running gear increased with the drawbar pull and the slippage also increased simultaneously, the ground contact area of the tire was increased in accordance with the drawbar pull or slippage. The increases in contact areas which were obtained are illustrated by the curves on Figure 6. With the increases of the ground contact areas of the tires, the ground contact pressures were reduced. This is presented on Figure 7.

UNIT SOIL THRUST AND MEAN GROUND CONTACT PRESSURES

If it be assumed that traction is a result of the reaction of the soil to shear under the tire, then the relationship between slippage and traction must be indicated by the curves of shear resistance vs. deformation of

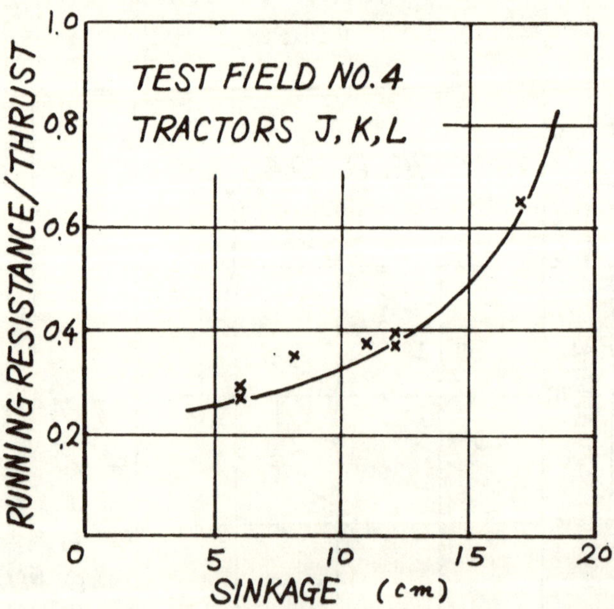

FIGURE 5

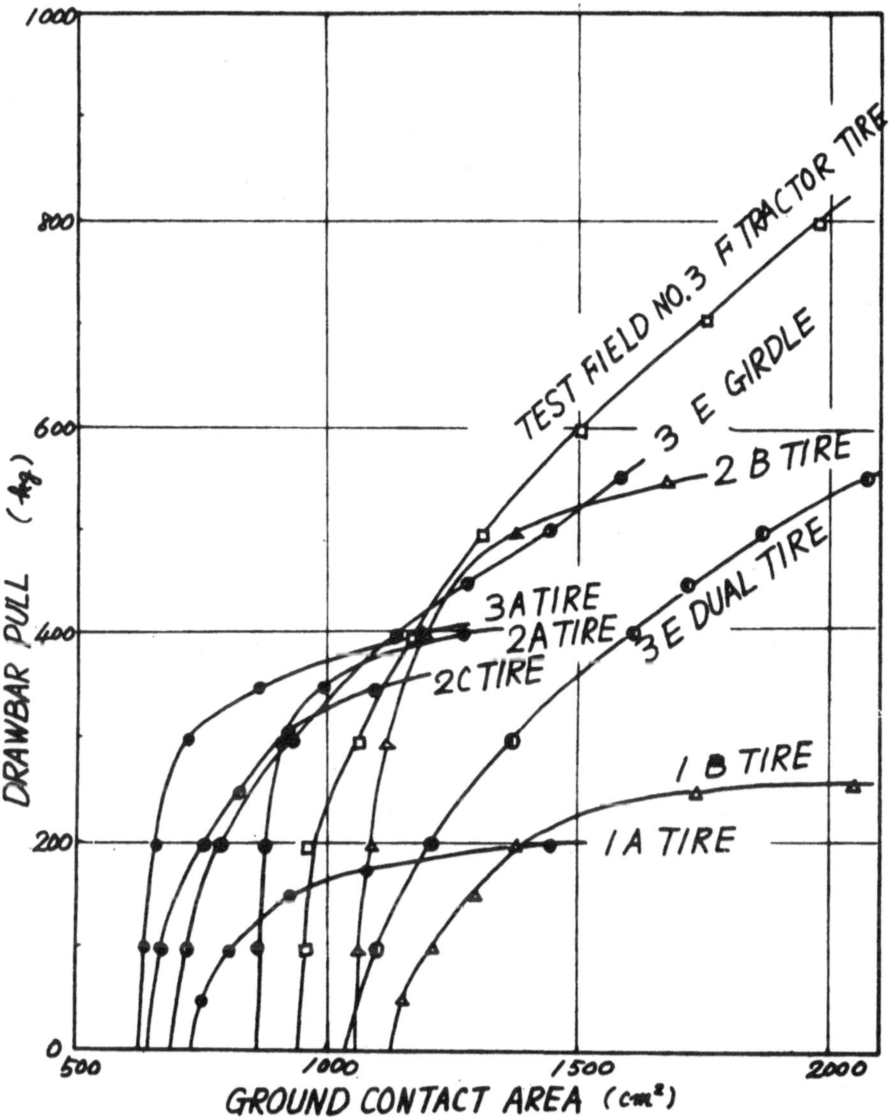

FIGURE 6

the soil; and the maximum thrust will be a value defined by Coulomb's equation at the maximum shear. Since the soil shear resistance s may be expressed by Coulomb's equation $s = c + \sigma \tan \phi$, then the maximum soil thrust per unit ground contact area ($t_{\max}$) must be

$$s = t_{\max}$$

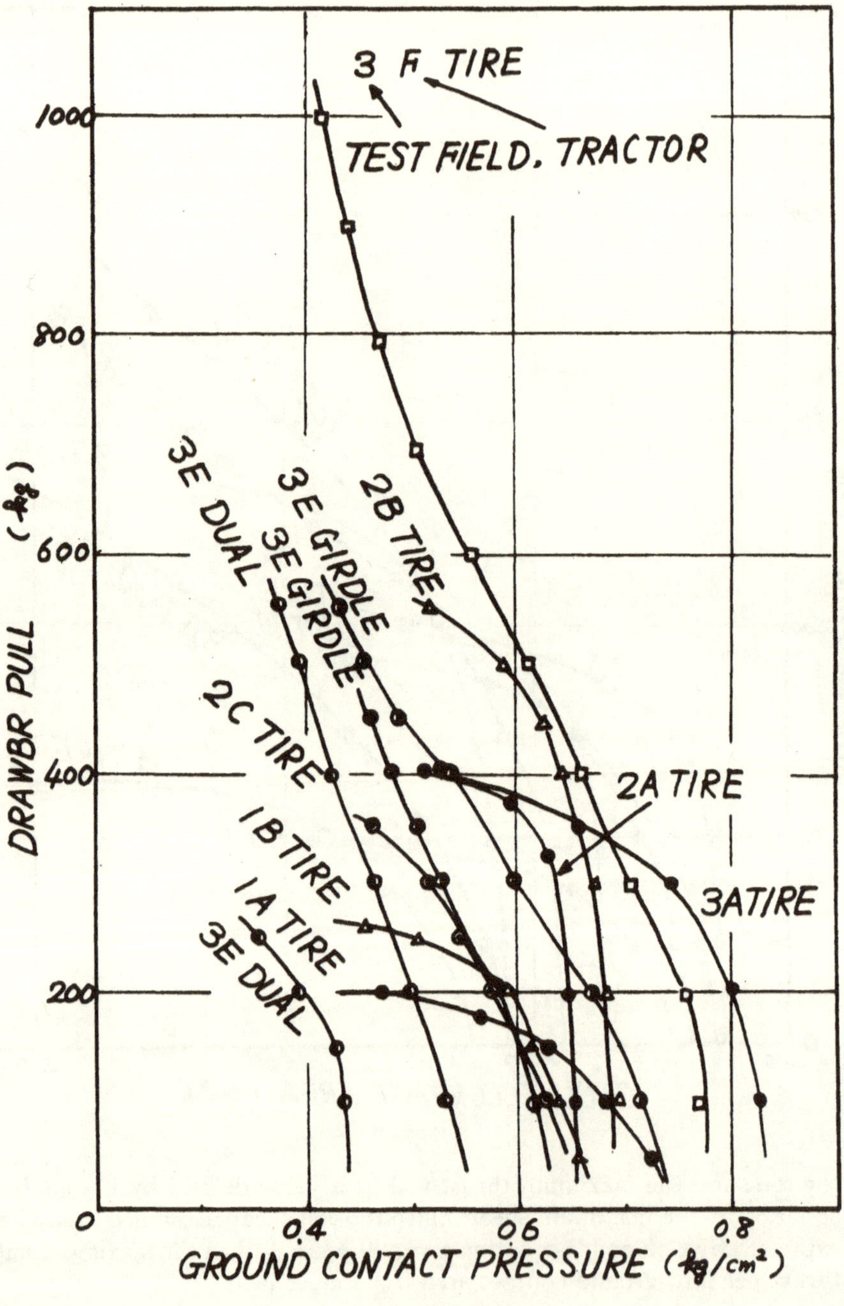
3 F TIRE
TEST FIELD. TRACTOR
1000
800
600
400
200
0
DRAWBR PULL (kg)
3E DUAL
3E GIRDLE
3E GIRDLE
2B TIRE
2C TIRE
1B TIRE
1A TIRE
3E DUAL
2A TIRE
3ATIRE
0.4
0.6
0.8
GROUND CONTACT PRESSURE (kg/cm²)

FIGURE 7

and maximum soil thrust under the tractor tire (t_{max}) may be written:

$$t_{max} = A'(c + \sigma \tan \phi) = A'c + W' \tan \phi$$

where A' is the effective ground contact area, and W' is the weight on the driven wheel. Figure 8 shows the relationship between the ground contact pressure σ and the unit soil thrust (t_{max}), calculated from the results of the drawbar pull tests. The relations when the tractor had less drawbar pull may be indicated by the right lower portions of these curves.

Until the tractor has the maximum traction, Coulomb's equation can not be applied to the relation of the unit thrust vs. ground contact pressure. But after the soil has been sheared off perfectly, the traction under the tire depends only upon the soil shear reaction, that is, the soil shear strength. Since the maximum unit thrust may be assumed as point A in Figure 8, then the left hand side from point A may be regarded as the results of the soil shear or friction, and Coulomb's equation can be applied. The soil strengths defined by Coulomb's equation, which were measured from the direct shear tests on each soil, are represented by the straight lines on the same Figure. Accordingly, maximum unit thrust shown by the point A in Figure 8 must approach these lines at each ground contact pressure. But, as can be seen, the maximum unit soil thrust (t_{max}) is always smaller than the soil strength (s). Table III gives the

TABLE III

MAXIMUM UNIT SOIL THRUST AND SHEAR STRENGTH OF SOIL

Test field	Tractor used	Max. unit thrust t_{max} kg/cm²	Shear strength of soil(s) kg/cm²	t_{max}/s
5(*a*)	D	0.28	0.45	62.2
	I	0.39	0.50	78.0
5(*b*)	D	0.32	0.52	61.5
	I	0.41	0.58	70.7
6	I	0.33	0.61	54.2
	C	0.37	0.69	53.6
7	I	0.41	0.45	91.2
	A	0.35	0.46	76.2

values of maximum unit thrust and soil strength corresponding to points B and A, respectively, on Figure 8, and ratios of those values are presented.

The value of (t_{max}) of the tractor (I) on the test field 7 was almost equal to (s), but with the other fields or tractors there were large differences between t_{max} and s. Though it may be assumed that t_{max} must be equal to s theoretically, the results obtained in these tests could not satisfy this relation.

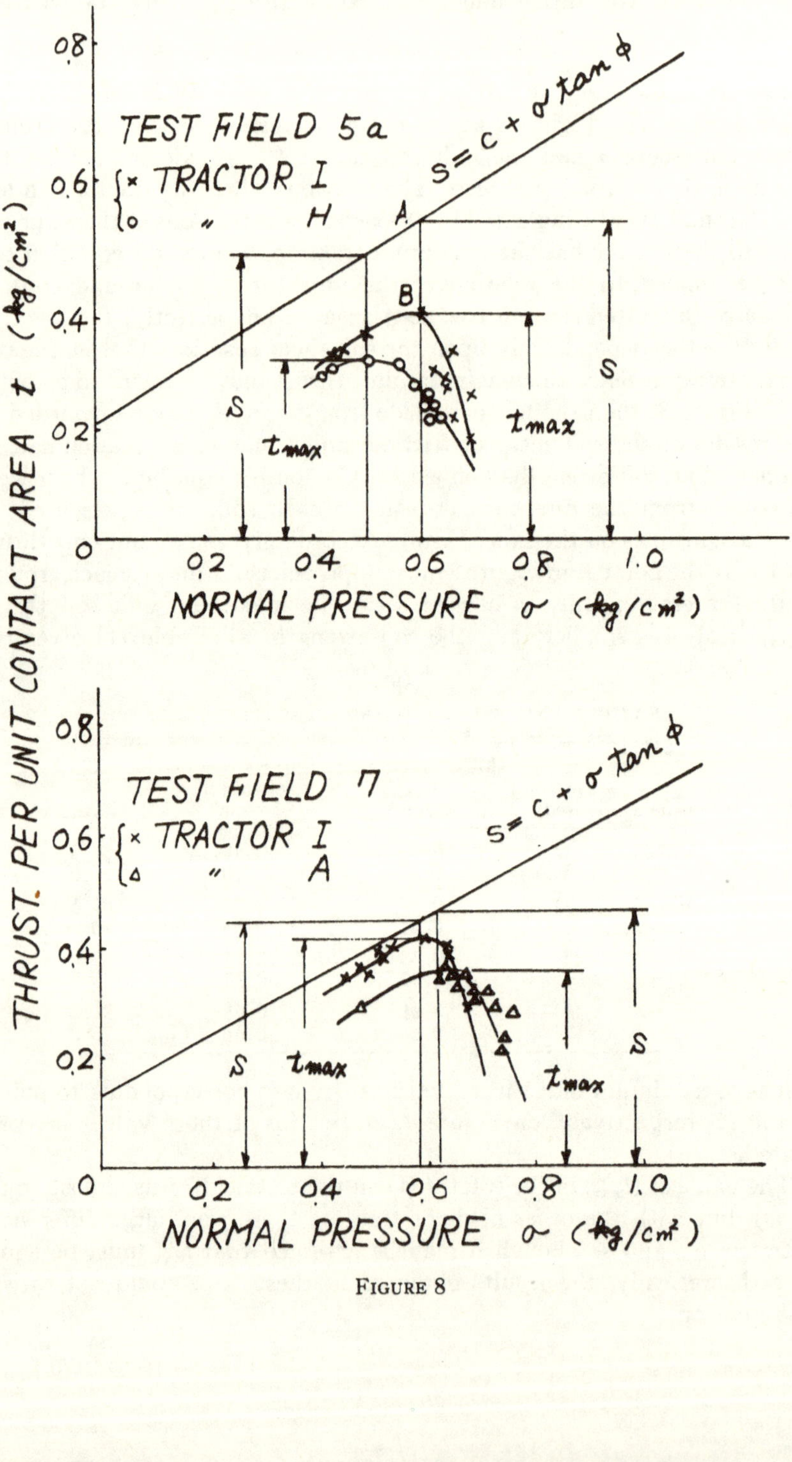
TEST FIELD 5a
× TRACTOR I
○ " H
S = c + σ tan φ
A
B
S
t max
THRUST PER UNIT CONTACT AREA t (kg/cm²)
NORMAL PRESSURE σ (kg/cm²)
0
0.2
0.4
0.6
0.8
1.0
TEST FIELD 7
× TRACTOR I
△ " A
S = c + σ tan φ
S
t max
NORMAL PRESSURE σ (kg/cm²)

FIGURE 8

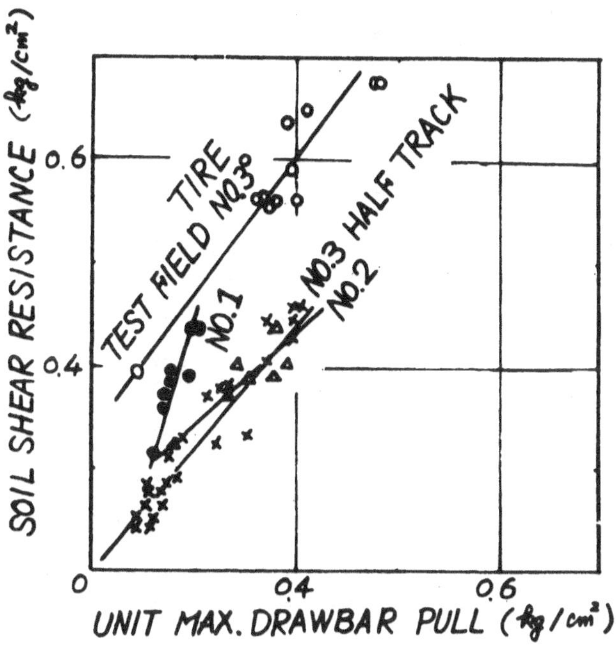

FIGURE 9

On the former tests on the fields 1 to 4, similar relationships were obtained. Figure 9 shows the comparison of the maximum unit tractions with the soil strengths under the same pressure conditions in these tests. Maximum unit tractions are always smaller than the soil strengths, as can be seen from this Figure, but linear relations of both factors can be found on each soil. If the traction is effected by only the soil shear resistance, then these linear relations would be straight lines passing through the origin and having a 45 degree slope. The only line in agreement with this hypothesis is that of the half track, as shown in Figure 9.

Referring to Figure 8, the curves of the unit thrust (t) have a parallel relation with the lines of soil shear strength ($s = c + \sigma \tan \phi$), at the left hand side from point A. Owing to a large slippage, the soil has been perfectly sheared off on these areas, and it may be assumed that the soil thrust under the tire will be a function of the ground contact pressure and will be shown by lines parallel to those of soil strength.

In the test of the crawler tractor J to L on the field 4, which was reclaimed land below the water level and an extremely soft and muddy field, the equation $t_{max} = Ac + W \tan \phi$ was almost satisfied. Table IV shows both the actual results and the calculated values from this equation.

TABLE IV

TEST RESULTS ON MUDDY FIELD

Test field	Cohesion (c) (kg/cm²)	Int. Fric. Angle (deg.)	Tractor Weight (kg)	Tractor Cont.Area (cm²)	Calculated thrust (kg)	Measured Max. DP (kg)	Measured Slippage (%)	Measured Run resis. (kg)	Measured Thrust (kg)	Calculated thrust / Measured thrust
4(a)	0.12	25	2330	8000	2050	1400	20	585	1985	1.03
(b)	0.11	23	2100	7750	1740	1250	20	825	2075	0.84
(b)	0.11	23	2330	8000	1870	1300	20	825	2125	0.89
(c)	0.09	17	2330	8000	1440	700	10	760	1460	0.99
(d)	0.08	15	2100	7750	1190	275	20	1025	1300	0.92
(a)	0.12	25	2100	6200	1720	1200	10	502	1702	1.01
(b)	0.12	25	2100	6200	1720	1000	13	555	1555	1.10
(b)	0.11	23	2500	7180	1850	1350	23	815	2165	0.86

In conclusion, it may be suggested that the maximum thrust of a crawler tractor or half-track type tractor was almost in agreement with the soil thrust calculated from Coulomb's equation. But in the case of tires, this relation cannot be satisfied. It may be considered that this difference was due to the variation of pressure distribution under the tire, and the effects of the tire friction to the soil and so on. If these effects are not taken into account in the calculation of thrust under a tire, false results may be obtained.

THRUST PER UNIT CONTACT AREA AND SLIPPAGE

Until the thrust reaches the maximum value, the relation of slippage to the thrust per unit contact area (that is, unit thrust) may be represented by the relation of the soil shear stress-strain curves, assuming thrust is the result of the soil shear reaction. These relations are therefore discussed on the basis of the shear stress-strain curves of the soil on each test field.

1. Shear Stress-Strain Curves of the Soil

An example of the results of the direct shear tests of each soil are shown in Figure 10. The normal pressures to the soil were kept constant

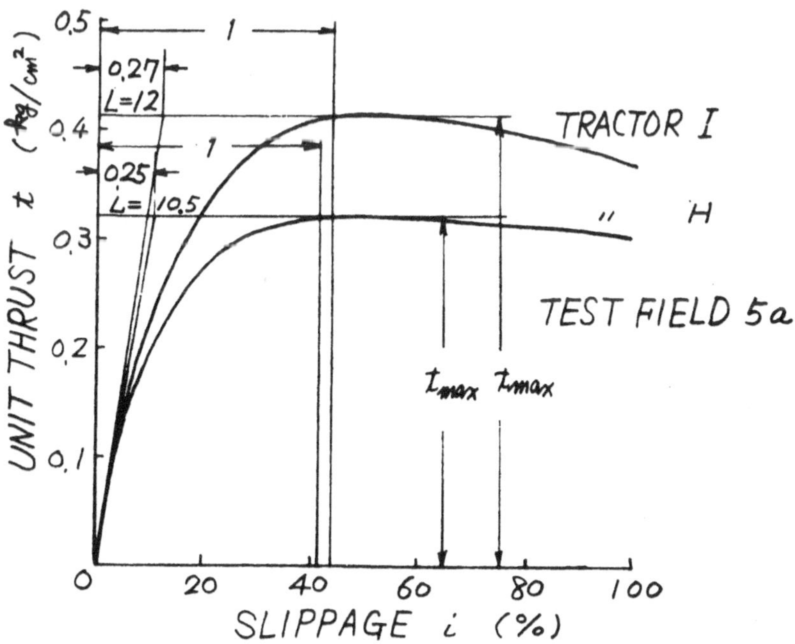

FIGURE 10

while the soil was sheared, and were equal to the mean value of the ground contact pressures of the tractor tires which were tested on each soil.

These curves may be formalized as the functions of the soil deformation (j) and the modulus of the soil deformation (K) as follows: $\tau = (c + \sigma \tan \phi)(1 - e^{-j/K})$. The derivative of this equation with respect to (j) gives $(d\tau/dj) = (1/K)(c + \sigma \tan \phi) - e^{-j/K}$. Since this equation shows the slope of the curves of stress–strain of the soil shear, the values of this derivative when $j = 0$ represent the slope of the curve at the origin. It may be written as: $(d\tau/dj)|_{j=0} = (c+\sigma \tan \phi)/K$, where K may be given by the abscissa of the cross point of both tangent lines to the curves at the origin and maximum values as can be seen on Figure 10.

On the other hand, the curves of the unit thrust vs. slippage are given by Figure 11. The values of L in this Figure are shown by the intersection of the tangent line to the curve at the origin and the parallel line to the abscissa passing through the maximum point of this curve. Maximum unit thrusts were produced at about 40 per cent slippage in these tests. Table V presents these slippage values in each test.

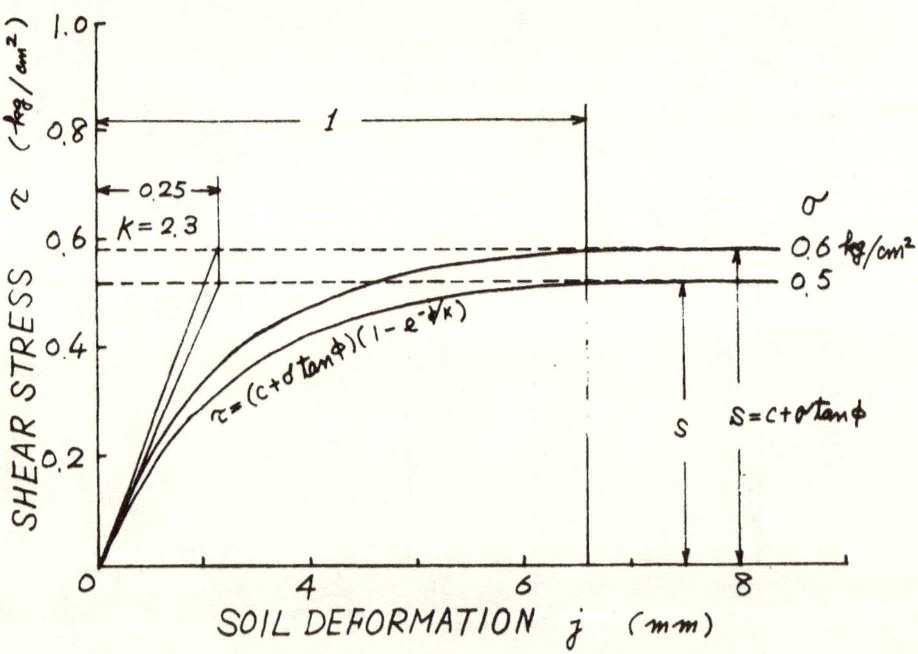

FIGURE 11

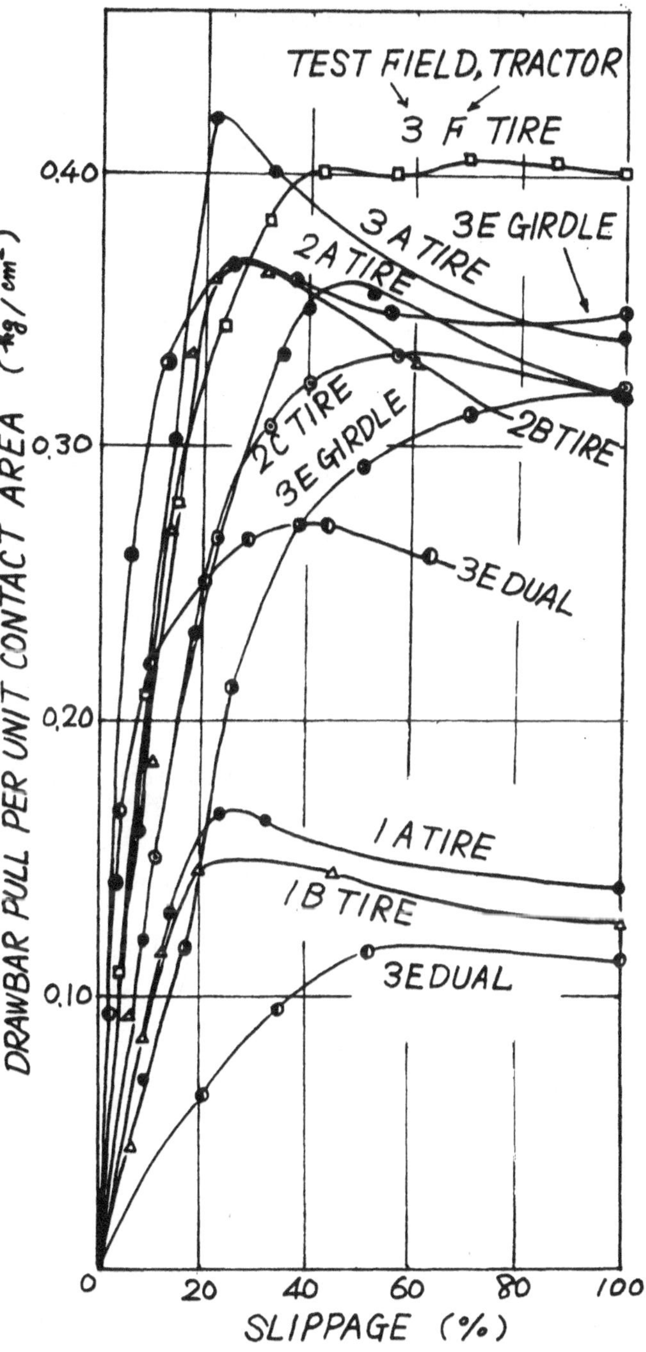
TEST FIELD, TRACTOR
3 F TIRE
3E GIRDLE
3 A TIRE
2A TIRE
2C TIRE
3E GIRDLE
2B TIRE
3E DUAL
1 A TIRE
1B TIRE
3E DUAL
DRAWBAR PULL PER UNIT CONTACT AREA (kg/cm²)
0.40
0.30
0.20
0.10
0
20
40
60
80
100
SLIPPAGE (%)

FIGURE 12

Comparing L with these slippage values, we can obtain one factor which indicates the slope of the curves at the origin. This value, (L/i), is given in the last column of Table V.

In the same manner, we can obtain (K/j) on Figure 10. In this case, maximum shear can not be obtained if the deformation (j) is not infinity; hence the values of (j) when the shear strength of the soil was 98 per cent of the maximum values ($\tau = c + \sigma \tan \phi$) were used. The resulting K/j values are given on Table VI.

TABLE V

Test field	Tractor	Slippage i (%)	L/i
5	H	42	0.25
	I	44	0.27
6	I	40	0.25
	C	30	0.27
7	I	35	0.26
	A	36	0.25

TABLE VI

Test field	Tractor	K/j
5	H	0.25
	I	0.25
6	I	0.25
	C	0.23
7	I	0.25
	A	0.25

Comparing the factor (L/i) with (K/j), we find that they are almost equal, and the curves are similar. It may be presumed that the unit thrust is given by the soil shear stress–strain curves as follows:

$$t = t_{max}(1 - e^{-i/L}).$$

Similar curves of the unit thrust vs. slippage were obtained from the tests using tractors A to F, as shown on Figure 12. From these results, it may be deduced that these curves are quite similar to the stress-strain curves obtained from the shear test of the soil.

Ein Beitrag zur Theorie der Vibrationsverdichtung von Böden

WALTER JURECKA*

ZUSAMMENFASSUNG

Theorie der geschlossenen Modelle—Energieverteilung zwischen Erreger und Boden—Materialkennwerte—Zur Theorie offener Modelle—Schwingungen am isotropen Halbraum—Vergleich der Ergebnisse aus Theorie und Praxis.

ABSTRACT: A CONTRIBUTION TO THE THEORY OF VIBRATORY COMPACTION OF SOILS

Theory of closed models; distribution of energy between vibrator and soil; coefficients of material; theory of open models; vibrations on the isotropic continuum; comparison of results of theory and practice.

ZUR ERHÖHUNG der Lagerungsdichte verschiedener Materialien werden seit mehreren Jahrzehnten Schwingungsbewegungen des lose gelarageten Material verwendet. Obwohl im Bauwesen dieses Verfahren zur Verdichtung von eingebauten Erdstoffen sowie zur Verdichtung des hinter die Schalungsformen verbrachten unverdichteten Frischbetons in grossem Umfang angewandt und hierfür eine Unzahl verschiedener Geräte gebaut und verwendet wird, ist eine theoretische Durchleuchtung des Vorganges bisher nicht möglich gewesen. Zwar wurden in Versuchsserien umfangreicher Art eine Anzahl der dabei auftretenden verwickelten Zusammenhänge empirisch abgeklärt, welche Untersuchungen auch in sehr beschränktem Umfang Voraussagen über das voraussichtliche Verhalten eines neu entwickelten Gerätes erlauben, sofern die Arbeitsweise im empirisch erfassten Bereich der Einflussgrössen liegt. Aber auch hier bleibt die Aussage vielfach problematisch, weil die Wirkung möglicher Einflussfaktoren im einzelnen noch nicht restlos erkannt und der komplexe Zusammenhang einzelner Einflussfaktoren noch nicht restlos erfasst ist. Eine theoretische Durchleuchtung des Verdichtungsvorganges scheint daher dringend notwendig.

Es ist offensichtlich, dass bei der Vibrationsverdichtung die periodisch in das zu verdichtende Material eingeleitete Bewegung die Verdichtung

*Direktor des Instituts für Baumaschinen und Baubetrieb, der Rheinisch-Westfälischen Technischen Hochschule Aachen, Aachen, Deutschland.

hervorruft. Es wird daher in erster Linie die Bewegung zu analysieren sein, und zwar sowohl diejenige des Schwingungserzeugers als auch diejenige des angeregten Materials. Beide können voneinander nicht unabhängig sein sondern beeinflussen sich wechselseitig. Dabei wird im allgemeinen die Bewegung des schwingungserzeugenden Gerätes genügend genau durch die Beschreibung der Bewegung seines Massenmittelpunktes oder beim Mehrmassenschwinger seiner verschiedenen Massenmittelpunkte erfolgen können, während die Erfahrung lehrt, dass Schwingungen im zu verdichtenden Material mit der Entfernung vom Erreger abklingen und dementsprechend ortsabhängig beschrieben werden müssen. Die Bewegungen werden zweckmässig durch die Bewegungsamplitude x, die Bewegungsgeschwindigkeit $\dot{x}$, und die Beschleunigung $\ddot{x}$ beschrieben, wobei diese Werte die auftretenden Maxima darstellen und vorausgesetztwerden soll, dass noch geprüft werden muss, ob der Bewegungsverlauf harmonisch erfolgt. Gegebenenfalls wäre die Bewegung hinsichtlich des Verlaufes zu beschreiben. Zur Beschreibung der Bewegung ist ferner die Angabe des Phasenwinkels φ erforderlich.

Erst wenn über die Bewegungsabläufe Klarheit herrscht, kann an die Beantwortung der Frage mit Aussicht auf Erfolg herangegangen werden, welcher der Bewegungsfaktoren für die Kornumlagerung von lockere in dichte Lagerung von Einfluss ist. Es wäre auch denkbar, dass diese weniger Bewegungs- als Belastungszustände sind, aber auch diese folgen letzten Endes den Bewegungsvorgängen.

DER SCHWINGUNGSERREGER

Sieht man von der zwar gebräuchlichen, aber für die Verdichtung von Material wegen der starren Frequenzabhängigkeit unbedeutenden Magneterregung ab, so erfolgt die Schwingungserregung bei der Massenkrafterregung durch rotierende Unwuchten, und zwar mit einer einfachen Unwucht durch eine rotierende Zentrifugalkraft oder mit zwei gegenläufigen Unwuchten, deren Querkomponenten der Zentrifugalkraft sich gegenseitig aufheben und deren Längskomponenten sich zu einer gerichteten Erregerkraft addieren. Dieses Resultat liefert auch der Pendelrütter, der schwenkbar gelagert ist und bei dem die Zentrifugalkraftkomponenten in einer Richtung durch Federn kompensiert werden, während senkrecht dazu die Federkraftkomponente frei zur Wirkung kommt. Bezeichnet e die Exzentrizität des Schwerpunktes der Unwuchtmassen m_0 von der Drehachse und Ω die Kreisfrequenz, so ist die Erregerkraft definiert durch $P_0 = m_0 e \Omega^2$ und $P = P_0 e^{j\Omega t}$ Bei konstanter Exzentrizität und konstanten Massen m_0 steigt die Erregerkraft quadratisch mit der Drehgeschwindigkeit und man spricht von quadratischer Erregung zum Unterschied von konstanter Erregung bei proportionaler Änderung von m_0 oder e mit Ω^2 und dementsprechend konstanter Erregerkraft P_0. Bei der Geschwindigkeitserregung erfolgt der Antrieb hingegen durch Kurbeltrieb mit vorgegebener Bewegungsgeschwindigkeit. In beiden Fällen ist die vektoriell anzuschreibende Kraft P mit der ebenfalls vektoriellen Geschwindigkeit durch die komplexe Grösse der Impedanz verbunden. Definitionsgemäss stellt sie zu jedem Zeitpunkt den Quotienten aus Kraft und Geschwindigkeit an einem Schwingsystem dar ($Z = P/\dot{x}$). Bei der Krafterregung ist P durch Rotation der Unwuchten vorgegeben und $\dot{x}$ durch $\dot{x} = P/Z$ bestimmt, bei der Geschwindigkeitserregung ist $\dot{x}$ vorgegeben durch die Bewegung der Exzenterwelle und P bestimmt durch $P = Z\dot{x}$. In beiden Fällen ist die Leistungsaufnahme aus $(1/2)P\dot{x} = P^2/(2Z) = 1/2(Z\dot{x}^2)$ definiert.

Bei den massenkrafterregten Schwingern stellen die Lager der Unwuchtwellen das die Leistung des Schwingers begrenzende Bauteil dar, wenn man von der möglichen Leistungsabgabe des Antriebsmotors absieht. Bei den als Lager üblicherweise infrage kommenden Wälzlagern ist einmal die zulässige Drehzahl nach oben begrenzt und ferner ist für jede kleinere Drehzahl eine maximale Lagerkraft zugelassen, die mit geringer werdender Drehzahl ansteigt. Den prinzipiellen Zusammenhang zeigt Bild 1 in Verbindung mit dem Zusammenhang von Drehzahl und Erregerkraft bei quadratischer und bei konstanter Erregung. Da die

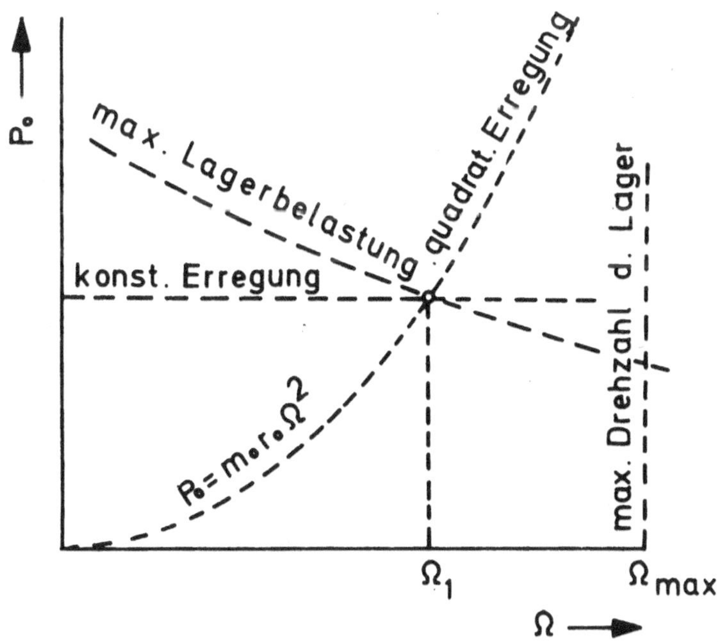

BILD 1. Begrenzung der Erregerkfrat bei Unwuchtrüttlern

konstante Erregung dem dargestellten maximal möglichen Zusammenhang noch am besten entspricht, sollen alle weiteren Betrachtungen auf diesen Fall abgestimmt werden.

SPRUNG- UND AUFLASTRÜTTLER [1], [2]

Nacht der Art der Bewegung lassen sich Vibrationsverdichtungsgeräte in Sprung- und Auflastrüttler einteilen. Grundsätzliches Unterscheidungsmerkmal ist das gegenseitige Bewegungsverhalten von Erreger und zu verdichtendem Material. Bleibt der Schwingungserzeuger ständig im Kontakt mit dem Material und machen beide stets die gleiche Bewegung mit, so spricht man vom Auflastrüttler, trennen sich beide Teile und führen sie ihre Bewegungen zeitweise getrennt und unabhängig voneinander aus, spricht man vom Sprungrüttler. Die Bezeichnung stammt von Erdbaugeräten, doch ist auch bei der Betonverdichtung, bei der im allgemeinen Kontakt zwischen Schwingungserzeuger und Material gehalten wird, ein dem Sprungrüttler gleichwertiges Verhalten festzustellen, nämlich dann, wenn sich um den Flaschenrüttler in besonders steifem und wasserarmem Beton ein Hohlraum bildet. Ob es sich bei einem Vibrationsgerät um einen Sprung- oder Auflastrüttler handelt, ist keine Geräteeigenschaft und nicht nur abhängig von den Gerätekennwerten, sondern eine Eigenschaft des verbundenen Systems aus Schwingungserreger und schwingungserregtem Material.

DER SCHWINGUNGSERREGER

In vielen Fällen ist der Schwingungserreger die von den Unwuchtkräften bewegte Masse des Gerätes. In den meisten Fällen wird jedoch die erregte Masse mittels Federn an einer nicht erregten Masse abgestützt. In diesem Falle spricht man vom Zweimassenschwinger (Bild 2). Da

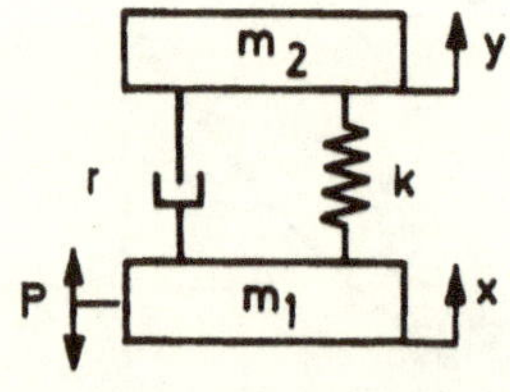

BILD 2. Zweimassen-Schwinger

kein Bewegungsvorgang ungedämpft abläuft, ist zumindest theoretisch auch ein Dämpfer vorzusehen, von dem vorerst einmal angenommen sein soll, dass die in ihm entwickelte Dämpfungskraft proportional der

Schwinggeschwindigkeit $\dot{x}$ sein mag. Inwieweit die im Schwingungserreger in den meist verwendeten Schwingmetallelementen vorhandene Dämpfung überhaupt von Bedeutung ist und inwieweit sie obiger Annahme entspricht, ist noch weitgehend ungeklärt.

Die Beschreibung des Bewegungsvorganges erfolgt beispielsweise für den Einmassenschwinger durch eine lineare, inhomogene Differentialgleichung 1. Ordnung der Form

$$m\ddot{x} + r\dot{x} + kx = P_0 e^{j\Omega t} \tag{1}$$

Sie kann als Gleichgewichtsbedingung zwischen der Erregerkraft $P = P_0 e^{j\Omega t}$, der Massenkraft $M = m\ddot{x}$, der Dämpfungskraft $D = r\dot{x}$, und der Federkraft $F = kx$ aufgestellt werden (Bild 3). Aehnliche

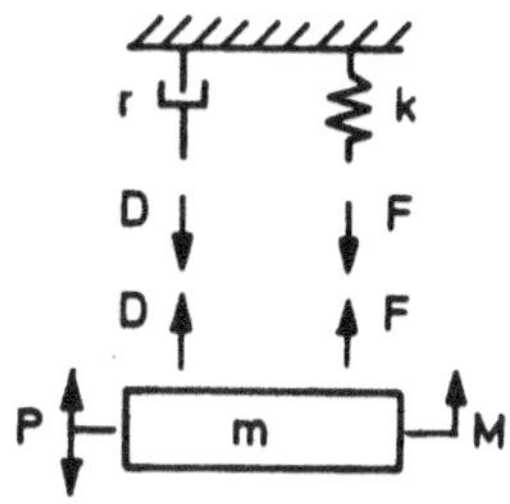

BILD 3. Schnittkräfte Einmassen-Schwinger

Gleichgewichtsbedingungen und damit definierende Differentialgleichungen lassen sich für anders aufgebaute Systeme, z. B. auch Mehrmassensysteme, aufstellen. Die Lösung der Gleichung (1) oder die Lösungen anderer inhomogenen Schwingungsgleichungen bestehen aus:

(*a*) dem vollständigen Integral der homogenen Gleichung, das zwei Integrationskonstanten beeinhaltet und die freie Schwingung darstellt. Aus ihm lässt sich die Eigenfrequenz $\omega = \sqrt{k/m}$ ermitteln.

(*b*) aus einem partikulären Integral der inhomogenen Gleichung, das mit dem vollständigen Integral additiv verbunden ist und das die durch die harmonische Erregerkraft $P = P_0 e^{j\Omega t}$ erzwungene Schwingung mit der Frequenz Ω darstellt.

Die Lösung der inhomogenen Gleichung führt unter Vernachlässigung des den Einschwingvorgang darstellenden vollständigen Integrals der homogenen Gleichung zu:

$$x = \frac{P_0}{(k - m\Omega^2) + jr\Omega} e^{j\Omega t} \tag{2}$$

und noch einmaliger Differentiation zur Definition der Impedanz

$$Z = P/\dot{x} = r + j(m\Omega - k/\Omega)(kp \text{ cm}^{-1} \text{ sec}). \tag{3}$$

Sie ist eine komplexe Grösse mit dem Absolutwert:

$$|Z| = \sqrt{r^2 + (m\Omega - k/\Omega)^2}$$

und dem Phasenwinkel: $\varphi = \text{arc tan } (r\Omega/k - m\Omega^2)$. Die Impedanz des einfachen Schwingers setzt sich aus den Impedanzanteilen der Masse, der Dämpfung, und der Feder mit den Werten:

$$Z_m = jm\Omega; \quad Z_r = r; \quad Z_k = -j(k/\Omega) \tag{4}$$

zusammen. Die reziproken Werte der Impedanzen heissen Beweglichkeiten und lauten:

$$W_m = -j(1/m\Omega); \quad W_r = 1/r; \quad W_k = (\Omega/k). \tag{5}$$

Bei Serienschaltung verschiedener Elemente addieren sich die Beweglichkeiten, bei Parallelschaltung die Impedanzen. Damit lassen sich die Impedanzen eines einfachen oder mehrfachen Schwingers ohne Ansatz der Gleichgewichsbedingung und ohme Lösung der Differentialgleichung anschreiben. Grundsätzlich ist jede Impedanz eine komplexe Zahl, die prinzipiell mit Real- und Imaginärteil als:

$$Z = X + jY \tag{6}$$

angeschrieben werden kann.

Die von einem einfachen Schwinger aufgenommene Leistung $N = (P/2)\dot{x}$ ist ebenfalls eine komplexe Grösse.

$$N = (P/2)\dot{x} = N_w + jN_p. \tag{7}$$

In Anlehnung an die Elektrotechnik wird ihr Realteil Wirkleistung, ihr Imaginärteil Blindleistung, oder besser Pendelleistung, und die absolute Grösse des Leistungsvektors $|N| = |N_s|$ Scheinleistung genannt.

Bezieht man die beim Schwingungsvorgang auftretende maximale Amplitude auf die statische Einsenkung der Feder (k) unter der maximalen Erregerkraft (P_0) und die maximal auftretende Beschleunigung auf diejenige der Masse (m) unter der gleichen maximalen Erregerkraft (P_0), so erhält man das Amplituden- (ξ) bzw. Beschleunigungsverhältnis (β) oder, wie sie in der Literatur manchmal genannt werden, die Vergrösserungsfunktion von Amplitude und Beschleunigung zu:

$$\xi = k/\Omega|z|; \quad \beta = m\Omega/|z|.$$

Betrachtet man diese Ausdrücke als Funktionen von Ω, bildet die Differentialquotienten ($d\xi/d\Omega$) bzw. $d\beta/d\Omega$, so erhält man nach Nullsetzen die Maximalstellen dieser Funktionen zu:

$$\Omega = \sqrt{(k/m) - (r^2/2m)}; \quad \Omega = \sqrt{(2k^2/2mk - r^2)}. \tag{9}$$

Sie geben diejenigen Frequenzen an, bei denen Amplitude und Beschleunigung ein Maximum werden. Hingegen wird die Schwinggeschwindigkeit unabhängig vom Dämpfungsglied (r) immer bei der Eigenfrequenz des Systems $\omega = \Omega = \sqrt{k/m}$ zum Maximum. Für diese Frequenz wird auch der Phasenwinkel $\varphi = \text{arc tan } \infty = \pi/2$. Es gibt also im gedämpften Schwingsystem getrennte Amplituden-, Geschwindigkeits- und Beschleunigungsresonanz, was bei der Angabe von Resonanzfrequenzen zu berücksichtigen wäre.

VERBUNDENE SCHWINGSYSTEME [4]

Solange der Schwinger mit dem zu verdichtenden Material in dauerndem Kontakt bleibt, liegt ein verbundenes Schwingsystem vor, denn auch das zu verdichtende Material weist elastische Eigenschaften auf. Diese Eigenschaften werden sich zwar im Verlauf des Verdichtungsvorganges ändern, was allerdings prinzipiell keinen Einfluss auf die obige Aussage hat. Wie dieser Uebergang von lockerer zu dichter Lagerung in seinem zeitlichen Verlauf erfolgt, ist weitgehend unbekannt. Ein linearer Uebergang ist unwahrscheinlich, vielmehr ist ein zeitlicher Verlauf etwa nach der Form einer e-Funktion denkbar. Für das Schwingverhalten des verbundenen Systems Erreger—Material ist die Frage jedoch nicht von Bedeutung, wenn das Bewegungsverhalten nur für den stationären Endzustand untersucht werden soll.

Um den Energifluss zwischen Erreger und angeregtem Material prinzipiell kennenzuleren, soll zuerst einmal ein verbundenes Schwingsystem, bestehens aus zwei einfachen Schwingern mit den Impedanzen $Z_1 = X_1 + jY_1 = r_1 + (m_1\Omega - (k_1/\Omega)$ und $Z_2 = X_2 + jY_2 = r_2 + (m_2\Omega - k_2/\Omega)$ untersucht werden (Bild 4), von denen das System 1 den Erreger darstellt, an dem die periodische Erregerkraft $P = P_0e^{j\Omega t}$ angreift und von der vorausgesetzt werden soll, dass P_0 konstant,

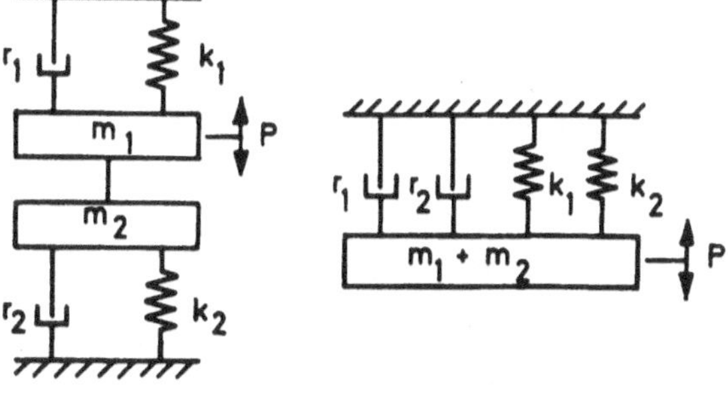

BILD 4. Verbundene Schwingsysteme

insbesondere also von Ω unabhängig ist (konstante Erregung). Da es sich hierbei um Parallelschaltung von zwei Schwingern handelt, nämlich des Erregersystems 1 mit dem erregten System 2, ist die Impedanz des Gesamtsystems:

$$Z = Z_1 + Z_2 = (X_1 + X_2) + j(Y_1 + Y_2) = X + jY. \tag{10}$$

Aus der Bedingung, dass beide Schwinger die gleiche Bewegung ausführen müssen, insbesondre dass ihre Schwinggeschwindigkeit $\dot{x} = \dot{x}_1 = \dot{x}_2$ gleich sein muss, lässt sich die Erregerkraft P bzw. ihr Maximalwert P_0 in zwei Teile P_1 und P_2 aufspalten, die jede an dem ihr zugeordneten System die gleiche Bewegung hervorrufen. Damit erhält man die in den beiden Systemen umgesetzten Energien zu:

$$\left.\begin{aligned}
N_W &= \frac{P_0^2}{2}\frac{X_1 + X_2}{(X_1+X_2)^2 + (Y_1+Y_2)^2}\\
N_{W_1} &= \frac{P_0^2}{2}\frac{X_1[(X_1+X_2)^2 - (Y_1+Y_2)^2] + 2(X_1+X_2)(Y_1+Y_2)Y_1}{[(X_1+X_2)^2 + (Y_1+Y_2)^2]^2}\\
N_{W_2} &= \frac{P_0^2}{2}\frac{X_2[(X_1+X_2)^2 - (Y_1+Y_2)^2] + 2(X_1+X_2)(Y_1+Y_2)Y_2}{[(X_1+X_2)^2 + (Y_1+Y_2)^2]^2}\\
N_P &= \frac{P_0^2}{2}\frac{-(Y_1+Y_2)}{(X_1+X_2)^2 + (Y_1+Y_2)^2}\\
N_{P_1} &= \frac{P_0^2}{2}\frac{Y_1[(X_1+X_2)^2 - (Y_1+Y_2)^2] - 2(X_1+X_2)(Y_1+Y_2)X_1}{[(X_1+X_2)^2 + (Y_1+Y_2)^2]^2}\\
N_{P_2} &= \frac{P_0^2}{2}\frac{Y_2[(X_1+X_2) - (Y_1+Y_2)^2] - 2(X_1+X_2)(Y_1+Y_2)X_2}{[(X_1+X_2)^2 + (Y_1+Y_2)^2]^2}\\
N_s &= \frac{P_0^2}{2}\frac{1}{\sqrt{(X_1+X_2)^2 + (Y_1+Y_2)^2}}\\
N_{s1} &= \frac{P_0^2}{2}\frac{\sqrt{X_1^2 + Y_1^2}}{(X_1+X_2)^2 + (Y_1+Y_2)^2}\\
N_{s2} &= \frac{P_0^2}{2}\frac{\sqrt{X_2^2 + Y_2^2}}{(X_1+X_2)^2 + (Y_1+Y_2)^2}
\end{aligned}\right\} \tag{11}$$

Die Auswertung für ein bestimmtes Zahlenbeispiel ist in Bild 5 als Zeitschaubild mit gleichzeitiger Eintragung von x, $\dot{x}$, und $\ddot{x}$ dargestellt. Die auftretenden Maxima der Amplitude, der Schwinggeschwindigkeit

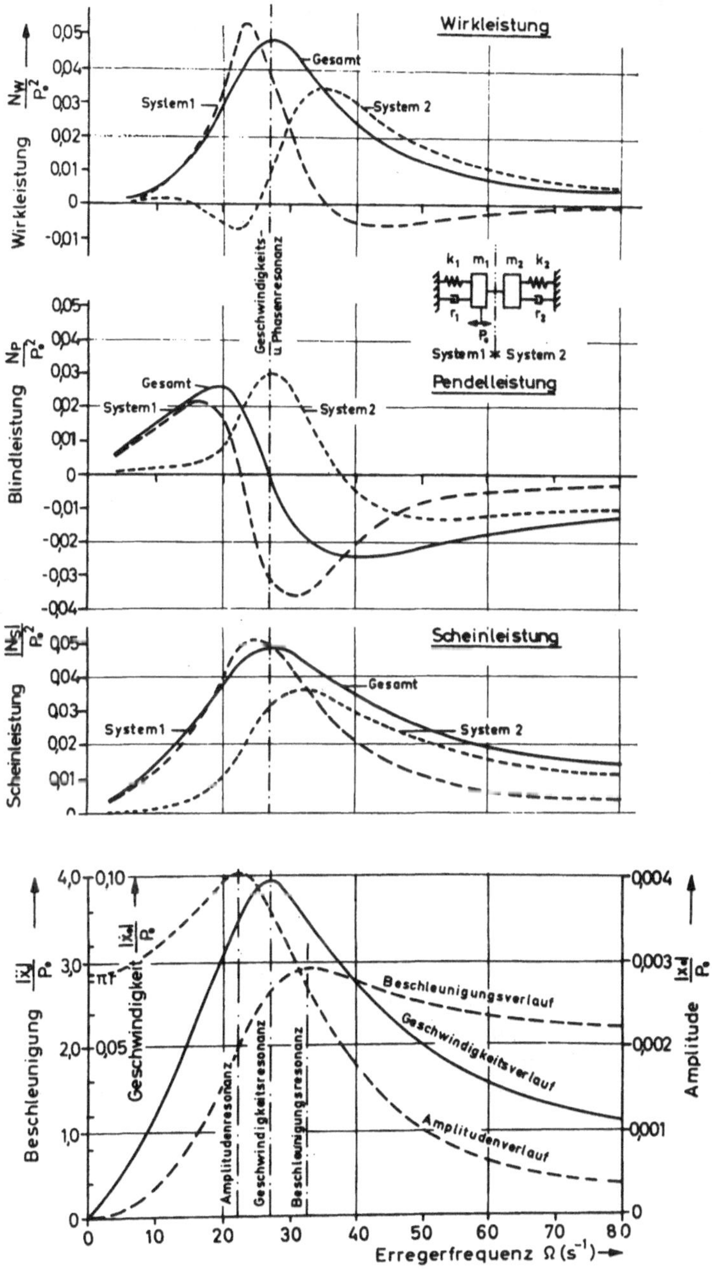

Wirkleistung
Gesamt
System 1
System 2
Wirkleistung
Geschwindigkeits- u. Phasenresonanz
k1 m1 m2 k2
r1 r2
P0
System 1 * System 2
Pendelleistung
Blindleistung
Scheinleistung
Beschleunigung
Geschwindigkeit
Amplitude
Amplitudenresonanz
Geschwindigkeitsresonanz
Beschleunigungsresonanz
Beschleunigungsverlauf
Geschwindigkeitsverlauf
Amplitudenverlauf
Erregerfrequenz Ω (s⁻¹)

Bild 5

und der Beschleunigung entsprechen den rechnerisch nach Formel 9 ermittelten Frequenzwerten. Bild 6 zeigt die Leistungsverteilung in vektorieller Darstellung.

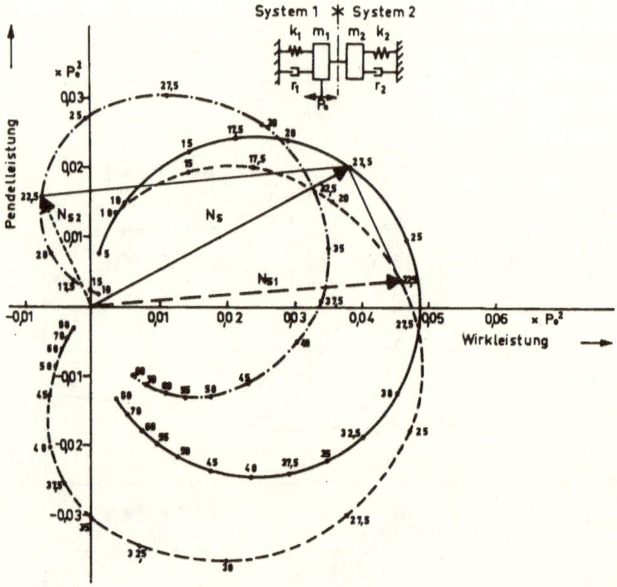

BILD 6

Es würde durchaus logisch erscheinen, wenn der Verdichtungserfolg mit der Einleitung einer gewissen Verformungsarbeit in das zu verdichtende Material verbunden wäre. Ob ein solcher Zusammenhang besteht, müsste allerdings erst durch umfangreiche Versuche abgeklärt werden, und zwar ebenso wie die Frage, welche der Leistungskomponenten von Einfluss ist. Da die Extremwerte der Leistungskomponenten in der Lage auf der Ω-Achse von dem Verhältnis der vier Impedanzbestandteile X_1, X_2, Y_1, und Y_2 bestimmt werden, ist a priori keine Aussage über das Frequenzoptimum im Hinblick auf die Resonanzfrequenz des Erregers, des zu verdichtenden Materials oder des gemeinsamen (verbundenen) Schwingsystems möglich.

DARSTELLUNG DES ZU VERDICHTENDEN MATERIALS ALS GESCHLOSSENES MODELL [1] [2]

Schon frühzeitig wurde versucht, das zu verdichtende Material in seinen Schwingungseigenschaften als geschlossenes Modell darzustellen, das heisst durch einen Schwingkreis, aus dem keine Energieabstrahlung stattfindet. Wo man glaubte, eine solche Energieabstrahlung berücksichtigen zu müssen, wurde ein Dämpfungsglied angebracht, das die Energieabstrahlung durch eine geschwindigkeitsproportionale Dämpfung ersetzen sollte. Ob diese Annahme zusammen mit der der Viskosität (Geschwindigkeitsproportionalität) des Dämpfers den tatsächlichen Verhältnissen entspricht, wurde niemals genauer untersucht.

Eine Anzahl der in früheren Untersuchungen verwendeten oder zur Verwendung vorgesehenen Bodenmodelle zeigt Abbildung 7. Die meisten

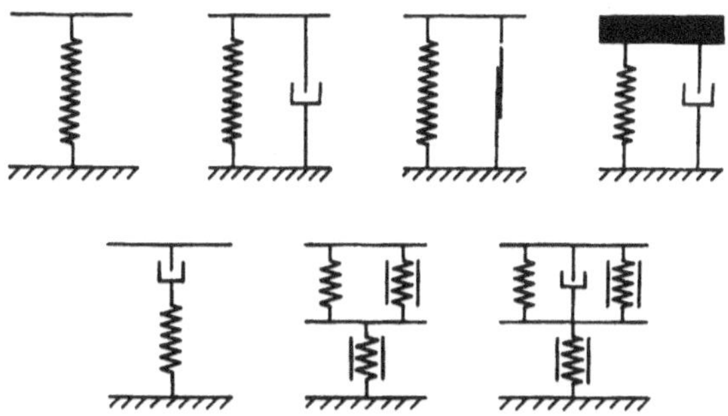

BILD 7. Verschiedene Bodenmodelle

von ihnen enthalten kein Massenglied, obwohl die Wirkung von Massenfräften kaum verneint werden könnte. Andere Modelle enthalten einrastende Federn, die die Volumsverminderung infolge des Verdichtungsvorganges darstellen. Nun bringt aber eine Materialverdichtung durch Vibration üblicherweise einige Zentimeter Einsenkung in etwa der doppelten Anzahl von Sekunden, also etwa 5 mm pro Sekunde, und bei einer Frequenz von 50 Hz etwa 0.1 mm pro Schwingung. Die Lageänderung dürfte daher auf den Bewegungsvorgang nur geringen Einfluss haben, jedoch wird die fortschreitende Verdichtung recht erhebliche Aenderungen der elastischen Eigenschaften hervorrufen, die die hinsichtlich ihrer Steifigkeit konstant bleibende einrastende Feder nicht darstellt. Hinsichtlich der Dämpfung wäre zu bemerken, dass eine sicher vorhandene Reibungsdämpfung der Schwingbewegung im Material bestimmt nicht viskos (geschweindigkeitsproportional) sondern eher wegproportional ist (Coulomb'sche Reibung). Insofern würde ein Modell mit einem solchen Dämpfungsglied die tatsächlichen Verhältnisse sicher besser darstellen. Ein geschlossenes Modell muss aber die Energieabstrahlung durch ein Dämpferglied berücksichtigen, so dass auch dieses Modell die tatsächlichen Verhältnisse nicht zweckentsprechend darstellen kann. Es ist daher ganz allgemein sehr stark zu bezweifeln, ob jemals ein geschlossenes Modell das Verhalten des zu verdichtenden Materials überhaupt auch nur annähernd richtig darstellen würde.

DIE SCHWINGENDE SÄULE [3]

Die Materialverdichtung auf dem Rütteltisch, wie sie im Betonbau zur Herstellung von Fertigteilen zur Anwendung kommt, kann durch die schwingende Materialsäule sicher besser dargestellt werden. In diesem Falle ist eine Energieabstrahlung nur an die den Rütteltisch und das Formstück umgebende Luft denkbar und diese wird verhältnismässig sehr klein bleiben. Setzt man in einer schwingenden Materialsäule (Bild 8) den zur Schwingungsrichtung parallelen Druck $p\ (t, y)$ und die

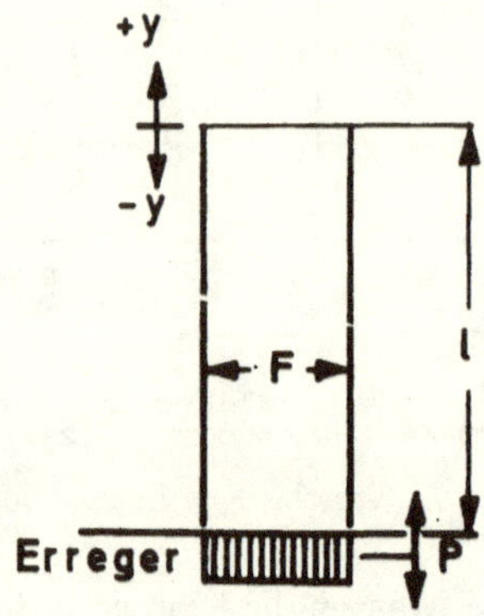

BILD 8. Schwingende Materialsäule

Geschwindigkeit $v(t, y)$ des Materialteilchens in gleicher Richtung ebenfalls zeit- und ortsabhängig an, so lässt sich für das Massenteilchen mit der Querschittsfläche F, der Dicke dy und der Masse $\rho\, F\, dy$ ein Kraftansatz anschreiben:

$$F.p(t, y) = [\partial v(t, y)/\partial t]\rho F\, dy + v(t, y) r\, F\, dy + F.p(t, y) - F[\partial p(t, y)/\partial y]dy; \quad (12)$$

sowie eine Kontinuitätsbedingung ansetzen:

$$F.v(t, y) = (1/E)F[\partial p(t, y)/\partial t]dy + F.v(t, y) - F[\partial v(t, y)/\partial y]dy. \quad (13)$$

Unter der Voraussetzung, dass nur harmonische Schwingungen auftreten, können:

$$\partial v(t, y)/\partial t = j\Omega v \quad \text{und} \quad \partial p(t, y)/\partial t = j\Omega p; \quad (14)$$

eingesetzt werden und man erhält die definierende Differentialgleichung des Schwingungsvorganges in der Säule:

$$(j\rho\Omega + r)j(\Omega/E) = d^2v/dy^2. \quad (15)$$

NachErweiterung der Gleichung mit F und der Substitution:

$$\gamma^2 = -(\rho\Omega^2F/E) + j(r\Omega F/E) = (\alpha - j\beta)^2 = \alpha^2 - \beta^2 - 2\alpha\beta j;$$

kann die Lösung der Differentialgleichung 15 mit:

$$v(y) = C_1e^{\gamma y l/\sqrt{F}} + C_2e^{-\gamma y l/\sqrt{F}}, \tag{16a}$$

angeschrieben werden und man erhält nach Differentiation und Einsetzen in 13

$$p(y) = -j\frac{E}{\Omega}\frac{\gamma}{\sqrt{F}}(C_1e^{\gamma y l/\sqrt{F}} - C_2e^{-\gamma y l/\sqrt{F}}). \tag{16b}$$

Die Integrationskonstanten lassen sich durch die Randbedingungen an der Erregerstelle für $y = -1$ mit $p(y) = p_0$ and $v(y) = v_0$ eliminieren und man erhält unter Verwendung von:

$$Z_i = -j(E/\Omega)(\gamma/\sqrt{F}) = -(E/\Omega\sqrt{F})(\beta + j\alpha),$$

die Lösungsfunktionen $v(y)$ und $p(y)$ zu:

$$v(y) = v_0\cos(\gamma/\sqrt{F})(1 + y) + (p_0/Z_i)\sin(\gamma/\sqrt{F})(1 + y); \tag{17a}$$

$$p(y) = v_0Z_i\sin(\gamma/\sqrt{F})(1 + y) + p_0\cos(\gamma/\sqrt{F})(1 + y). \tag{17b}$$

Sie entsprechen den Vierpolgleichungen der Elektrotechnik. Die Impedanz der schwingenden Säule $Z_F = p_0/v_0$ bestimmt sich aus den Randbedingungen an demjenigen Ende der Säule, das dem Erreger abgekehrt ist. Da bei der Tischrüttlung dieses End frei schwingen kann, sagt die Randbedingung für $y = 0$ aus, dass keine Kraft und damit kein Druck übertragen werden kann, d. h. es muss $p(y) = 0$ sein. Damit folgt aus 17b mit der Rücksubstitution $\gamma = \alpha - j\beta$

$$Z_F = \frac{p_0}{v_0} = \frac{E}{\Omega\sqrt{F}}(\beta + j\alpha)\frac{\sin\frac{\alpha}{\sqrt{F}}l\cos\frac{\beta}{\sqrt{F}}l - j\cos\frac{\alpha}{\sqrt{F}}l\sin\frac{\beta}{\sqrt{F}}l}{\cos\frac{\alpha}{\sqrt{F}}l\cos\frac{\beta}{\sqrt{F}}l - j\sin\frac{\alpha}{\sqrt{F}}l\sin\frac{\beta}{\sqrt{F}}l}$$

oder nach Einsetzen von $n = \Omega 1\sqrt{\rho/E}$ als Frequenzverhältnis, $\delta = r1/\sqrt{\rho E}$ als Dämpfungsverhältnis und

$$\alpha_1^2 = 4\alpha^2 1^2/F = 2n(\sqrt{n^2 + \delta^2} - n) \text{ und}$$

$$\beta_1^2 = 4\beta^2 1^2/F = 2n(\sqrt{n^2 + \delta^2} + n),$$

und Umformung und Berücksichtigung der Ankopplungsfläche F

$$Z_F = \left[\frac{\beta_1\sin\alpha_1 + \alpha_1\sin\beta_1}{2n(\cos\alpha_1 + \cos\beta_1)} + j\frac{\alpha_1\sin\alpha_1 - \beta_1\cos\beta_1}{2n(\cos\alpha_1 + \cos\beta_1)}\right] \mp \sqrt{\rho E} \tag{19}$$

(kp cm^{-1} sec)

Die Impedanz der mit einer Dämpfung r behafteten Säule ist also wiederum eine komplexe Grösse. Im Falle reibungsfreier Schwingung vereinfachen sich die Ausdrücke zu:

$$Z_F = -j\,F\sqrt{\rho E}\,tg\,n, \tag{20}$$

also zu einer rein imaginären Grösse. Damit ist die Identität zum einfachen Schwinger hergestellt. Die dimensionslosen Beiwerte ξ_F und μ_F der mit 19 identischen Formel $Z_F = (\xi_F + j\mu_F)F\sqrt{\rho E}$ sind in Bild 9 in Abhängigkeit vom Frequenzverhältnis n und Dämpfungsverhältnis δ dargestellt. Als Kennwerte des Materials der schwingenden Säule gehen in die Rechnung der auf die Volumseinheit bezogene Dämpfungsbeiwert r (kp cm^{-4} sec), die Dichte des Materials ρ (kp cm^{-4} sec^2) und sein Elastizitätsmodul E (kp cm^{-2}) ein.

DIE UNENDLICH LANGE SÄULE

Für die unendlich lange Säule ergibt sich mit $v(y) = 0$ und $p(y) = 0$ für die Randbedingung $y = \infty$ aus den Gleichungen 17 mit

$$\cot\,(1 + y) = \tan\,(1 + y) = 1$$

und mit $n = \infty$ die Impedanz zu:

$$Z = p_0/v_0 = -Z_t = F\sqrt{\rho E}.$$

GRUNDLAGEN ELASTIZITÄTSTHEORETISCHER BEHANDLUNG

In eine elastizitätstheoretische Behandlung des Problems eines Schwingungserregers auf einem elastischen Medium gehen die sogenannten Lamé'schen Elastizitätskonstanten λ und μ ein, die mit dem Elastizitätsmodul E und der Poisson-Zahl γ des isotropen Materials in folgendem Zusammenhang stehen:

$$E = [\mu(3\lambda + 2\mu)/(\lambda + \mu)]\gamma = \lambda/[2(\lambda + \mu)], \tag{21a}$$

oder

$$\lambda = [E\gamma/(1 - 2\gamma)(1 + \gamma)]\mu = E/2(1 + \gamma). \tag{21b}$$

Dabei ist μ mit dem Schubmodul G identisch.

An einem quaderförmigen Volumenelement mit Flächen senkrecht zu einem rechtwinkeligen Koordinatentripel x, y, z treten sechs Spannungskomponenten, davon drei Normalspannungskomponenten σ_x, σ_y, σ_z und drei Schubspannungskomponenten τ_{xy}, τ_{yz}, τ_{zx} sowie sechs Verschiebungskomponenten, und zwar drei Dehnungskomponenten ϵ_x, ϵ_y,

ϵ_z und drei Winkeländerungen γ_{xy}, γ_{yz}, γ_{zx} auf. Sie sind durch das allgemeine Hooke'sche Gesetz verbunden:

$$\begin{aligned}
\sigma_x &= c_{11}\epsilon_x + c_{12}\epsilon_y + c_{13}\epsilon_z + c_{14}\gamma_{yz} + c_{15}\gamma_{zx} + c_{16}\gamma_{xy},\\
\sigma_y &= c_{21}\epsilon_x + c_{22}\epsilon_y + c_{23}\epsilon_z + c_{24}\gamma_{yz} + c_{25}\gamma_{zx} + c_{26}\gamma_{xy},\\
\sigma_z &= c_{31}\epsilon_x + c_{32}\epsilon_y + c_{33}\epsilon_z + c_{34}\gamma_{yz} + c_{35}\gamma_{zx} + c_{36}\gamma_{xy},\\
\tau_{yz} &= c_{41}\epsilon_x + c_{42}\epsilon_y + c_{43}\epsilon_z + c_{44}\gamma_{yz} + c_{45}\gamma_{zx} + c_{46}\gamma_{xy},\\
\tau_{zx} &= c_{51}\epsilon_x + c_{52}\epsilon_y + c_{53}\epsilon_z + c_{54}\gamma_{yz} + c_{55}\gamma_{zx} + c_{56}\gamma_{xy},\\
\tau_{xy} &= c_{61}\epsilon_x + c_{62}\epsilon_y + c_{63}\epsilon_z + c_{64}\gamma_{yz} + c_{65}\gamma_{zx} + c_{66}\gamma_{xy}.
\end{aligned} \tag{22}$$

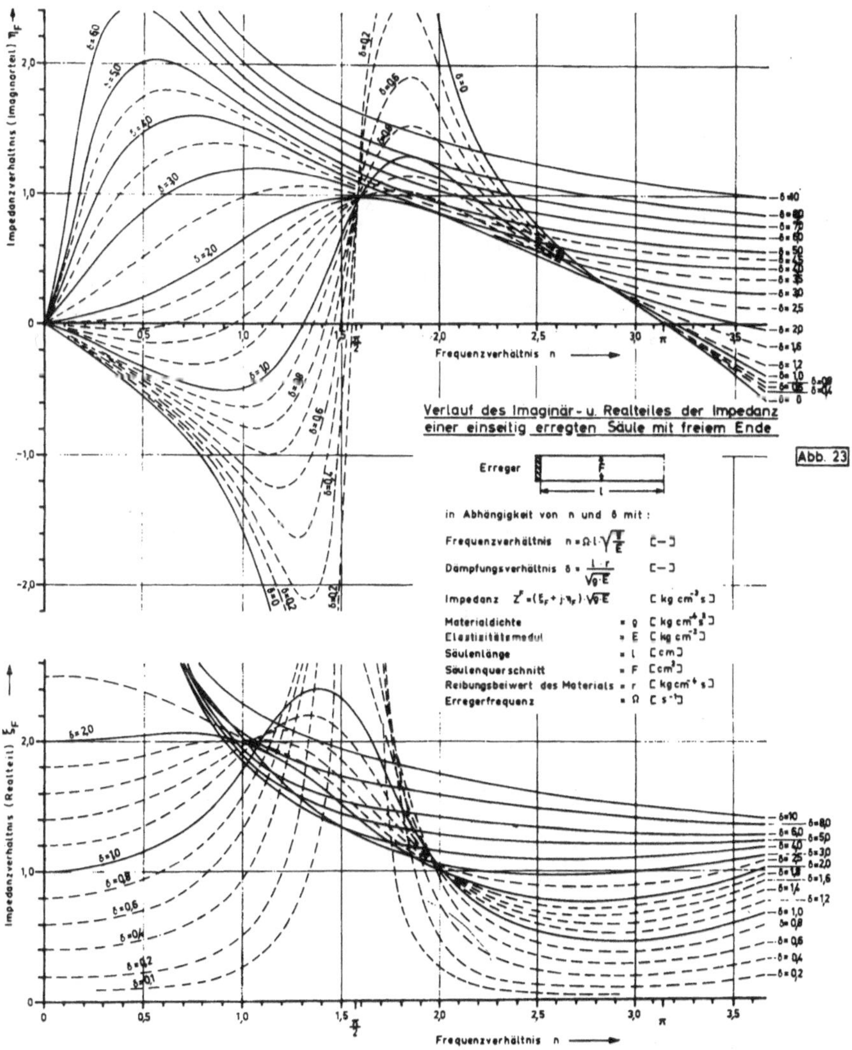

BILD 9

Von den 36 Konstanten dieser Matrix sind 15 paarweise gleich, und zwar aus Symmetriegründen ($c_{mn} = c_{nm}$), so dass sich ihre Anzahl auf 21 reduziert. In einem System mit drei zueinander senkrechten Achsen reduziert sich die Koeffizientenzahl auf 3 und bei Isotropie verbleiben wegen der Unabhängigkeit der Achsen nur 2. Damit wird:

$$\begin{aligned}
&c_{12} = c_{13} = c_{21} = c_{23} = c_{31} = c_{32} = \lambda,\\
&c_{44} = c_{55} = c_{66} = \mu,\\
&c_{11} = c_{22} = c_{33} = \lambda + 2\mu,\\
&c_{14} = c_{24} = c_{34} = c_{15} = c_{25} = c_{35} = c_{16} = c_{26} = c_{36} = 0;
\end{aligned} \tag{23}$$

und die Spannungen ergeben sich zu:

$$\sigma_x = \lambda\Delta + 2\mu\epsilon_x; \quad \sigma_y = \lambda\Delta + 2\mu\epsilon_y; \quad \sigma_z = \lambda\Delta + 2\mu\epsilon_z. \tag{24}$$

$$\tau_{yz} = \mu\gamma_{yz}; \qquad \tau_{zx} = \mu\gamma_{zx}; \qquad \tau_{xy} = \mu\gamma_{xy}; \tag{25}$$

wobei $\Delta = \epsilon_x + \epsilon_y + \epsilon_z = \partial u/\partial x + \partial v/\partial y + \partial w/\partial z$ die kubische Dilatation und u, v, und w die Verschiebungen parallel zu einer Koordinatenrichtung darstellen. Setzt man die Massenkraft parallel zu einer Koordinatenrichtung mit den Spannungskomponenten in gleicher Richtung ins Gleichgewicht (Bild 10), so erhält man mit 26:

$$\rho\frac{\partial^2 u}{\partial t^2} = (\lambda + \mu)\frac{\partial\Delta}{\partial x} + \mu\nabla^2 u\ ; \quad \rho\frac{\partial^2 v}{\partial t^2} = (\lambda + \mu)\frac{\partial\Delta}{\partial y} + \mu\nabla^2 v\ ;$$

$$\rho\frac{\partial^2 w}{\partial t^2} = (\lambda + \mu)\frac{\partial\Delta}{\partial z} + \mu\nabla^2 w \tag{26}$$

mit dem Laplace'schen Operator: $\nabla^2 = \partial^2/\partial x^2 + \partial^2/\partial y^2 + \partial^2/\partial z^2$.

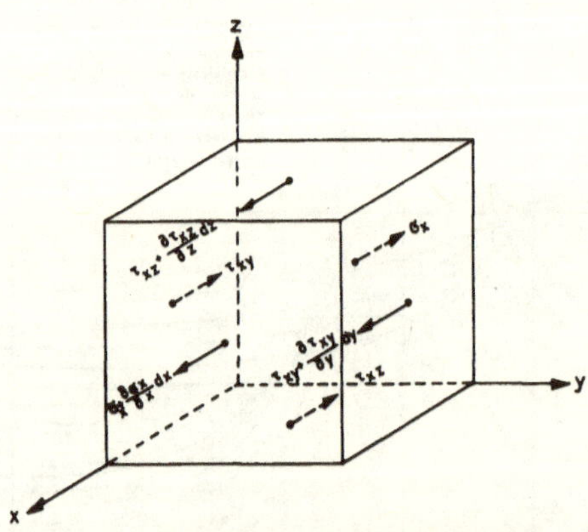

BILD 10. Spannungen am Raumelement in x-Richtung

DER VERTIKAL SCHWINGENDE ERREGER AUF DEM ISOTROPEN ELASTISCHEN HALBRAUM [5] [6]

Mit den Zylinderkoordinaten r, φ, z, den Verschiebungen u, v, w und den Verdrehungen ω_r, ω_y und ω_z müssen aus Symmetriegründen $v = 0$, $\omega_r = 0$ und $\omega_z = 0$ werden (Bild 11). Die Differentialgleichungen 26

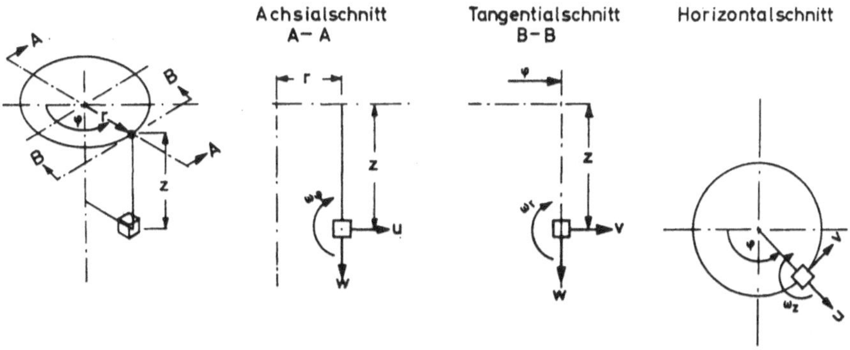

BILD 11. Volumenelement des Halbraumes, Koordinaten der Lage, der Verschiebung, und Verdrehung, $v = \omega_r = \omega_z = 0$

ergeben daher nach einigen Umformungen das System der Schwingungsgleichungen:

$$\rho \frac{\partial^2 u}{\partial t^2} = (\lambda + 2\mu) \frac{\partial \Delta}{\partial_r} + 2\mu \frac{\partial \omega_\varphi}{\partial z} \qquad [\text{kp cm}^{-3}]$$

$$\theta = (\lambda + 2\mu) \frac{1}{r} \frac{\partial \Delta}{\partial \phi}$$

$$\rho \frac{\partial^2 w}{\partial t^2} = (\lambda + 2\mu) \frac{\partial \Delta}{\partial z} - 2\mu \frac{u}{r} \frac{\partial}{\partial r} (r\omega_\varphi) \tag{27}$$

mit der kubischen Dilatation

$$\Delta = (1/r)\partial(ur)/\partial r + \partial w/\partial z = u/r + \partial u/\partial r + \partial w/\partial z$$

und mit $2\omega_\varphi = \partial u/\partial z - \partial w/\partial r$.

Den Lösungsansatz für dieses System von Differentialgleichungen hat Sezawa (7) wie folgt angegeben mit:

$$u = u_1 + u_2 + u_3;\ v = v_1 + v_2 + v_3;\ w = w_1 + w_2 + w_3,$$

und für den Fall der Achsensymmetrie:

$$
\begin{aligned}
u_1 &= -A_0(1/h^2)(\partial/\partial r)H_0^{(2)}(xr)e^{-\alpha z+j\Omega t} \\
u_2 &= 0 \\
u_3 &= C_0(\beta/k^2)(\partial/\partial r)H_0^{(2)}(xr)e^{-\beta z+j\Omega t} \\
v_1 &= v_2 = v_2 = 0 \\
w_1 &= A_0(\alpha/h^2)H_0^{(2)}(xr)e^{-\alpha z+j\Omega t} \\
w_2 &= 0 \\
w_3 &= -C_0(x^2/k^2)H_0^{(2)}(xr)e^{-\beta z+j\Omega t}
\end{aligned}
\qquad (28)
$$

mit der Hankelfunktion $H_0{}^{(2)}(xr) = J_0(xr) - jN_0(xr)$.

Ferner ist definiert:

$\alpha^2 = x^2 - h^2 = x^2 - \rho\Omega^2/(\lambda + 2\mu)$ mit $h^2 = \rho\Omega^2/(\lambda + 2\mu)$ [cm^{-2}];

$\beta^2 = x^2 - k^2 = x^2 - \rho\Omega^2/\mu$ mit $k^2 = \rho\Omega^2/\mu$ [cm^{-2}].

Nach Ausführung der Differentiationen, die in u_1 und u_3 angedeutet sind, ergibt sich für $u = u_1 + u_3$ oder für:

$$
\begin{aligned}
u &= \left[\frac{A_0x}{h^2}e^{-\alpha z} - \frac{C_0\beta x}{k^2}e^{-\beta z}\right][J_1(xr) - jN_i(xr)]e^{j\Omega t}\,; \\
w &= \left[\frac{A_0\alpha}{h^2}e^{-\alpha z} - \frac{C_0x^2}{k^2}e^{-\beta z}\right][J_0(xr) - jN_0(xr)]e^{j\Omega t}.
\end{aligned}
\qquad (29)
$$

Werden beide Ausdrücke partiell nach r und z differenziert und die Ergebnisse in 27 eingesetzt, so kann man sich von der Richtigkeit der Ansätze überzeugen. Mit diesen Ableitungen lassen sich nunmehr auch die Spannungen anschreiben, von denen ebenfalls aus Symmetriegründen $\sigma_\varphi = 0$; $\tau_{\varphi z} = 0$; $\tau_{r\varphi} = 0$ sein müssen, und zwar:

$$
\begin{aligned}
\sigma_z &= \mu e^{j\Omega t}\left[\frac{A_0(k^2 - 2x^2)}{h^2}e^{-\alpha z} + \frac{C_0 2\beta x^2}{k^2}e^{-\beta z}\right][J_0(xr) - jN_0(xr)]\,; \\
\tau_{zr} &= \mu e^{j\Omega t}\left[-\frac{A_0}{h^2}2\alpha x e^{-\alpha z} - \frac{C_0}{k^2}x(k^2 - 2x^2)e^{-\beta z}\right][J_1(xr) - jN_1(xr)].
\end{aligned}
\qquad (30)
$$

Die dritte noch verbleibende Spannung σ_r wird für die weitere Rechnung nicht benötigt. In den Lösungen stellen A_0 und C_0 die Integrationskonstanten dar, die den Randbedingungen anzupassen sind. Ferner stellen die Hankelfunktionen die Wellenausbreitung in der r-Richtung, die e-Potenzen $e^{-\alpha z}$ und $e^{-\beta z}$ die Wellenausbreitung in z-Richtung dar. Sie sind daher Ortsfunktionen, zeigen das Abklingen der Werte von u, w, σ_z, und τ_{zr} und ihre Phasenverschiebungen an, während der Faktor $e^{j\Omega t}$ anzeigt, dass die Schwingungen überall harmonisch verlaufen.

FREIE SCHWINGUNGEN

An der Oberfläche ($z = 0$) des isotropen Halbraumes muss $\sigma_z = \tau_{zr} = 0$ sein. Damit folgt aus Formel 30:

$$(k^2 - 2x^2)^2 - 4\alpha\beta x^2 = 0. \tag{31}$$

Die Gleichung hat folgende Wurzeln:

für $r = 1/2$ ($\lambda = \infty$; $\mu = G = E/3$) $\quad x_1 = 1.04678k$

$r = 1/4$ ($\lambda = \mu = 2E/5$) $\quad x_1 = 1/2\sqrt{3 + \sqrt{3}}\,k$

$r = 0$ ($\lambda = 0$; $\mu = G = E/2$) $\quad x_1 = 1/2\sqrt{3 + \sqrt{5}}\,k$

Sie stellen die Oberflächen (Rayleigh-) Wellen dar.

Mit $\alpha_1^2 = x_1^2 - h^2$ und $\beta_1^2 = x_1^2 - k^2$ und einer willkürlichen Konstanten D (bzw. D_1) folgt dann:

$$\begin{aligned} u_0 &= De^{j\Omega t}[-x_1(x_1^2 - \tfrac{1}{2}k^2)e^{-\alpha_1 z} + \alpha_1\beta_1 x_1 e^{-\beta_1 z}][J_1(x_1 r) - jN_1(x_1 r)] \\ w_0 &= De^{j\Omega t}[-\alpha_1(x_1^2 - \tfrac{1}{2}k^2)e^{-\alpha_1 z} + \alpha_1 x_1^2 e^{-\beta_1 z}][J_0(x_1 r) - jN_0(x_1 r)] \end{aligned} \tag{32}$$

oder für $z = 0$ an der Oberfläche des isotropen Halbraumes:

$$\begin{aligned} u_0 &= -D_1 x_1(2x_1^2 - k^2 - 2\alpha_1\beta_1)[J_1(x_1 r) - jN_1(x_1 r)]\,e^{j\Omega t}; \\ w_0 &= D_1 k^2\alpha_1[J_0(x_1 r) - jN_0(x_1 r)]\,e^{j\Omega t}. \end{aligned} \tag{33}$$

ERZWUNGENE SCHWINGUNGEN AM ISOTROPEN HALBRAUM

Bei erzwungenen Schwingungen geht das System homogener linearer Differentialgleichungen 2. Ordnung 27 in ein inhomogenes System über. Die partikulären Lösungen, die wiederum den eingeschwungenen Zustand darstellen, entsprechen den Formelgruppen 29 und 30, wenn die Integrationskonstanten A_0 und C_0 durch Integrationsfunktionen $A(x)$ und $C(x)$ ersetzt und die von x abhängigen Teile von 0 bis ∞ integriert werden.

Man erhält dann:

$$\begin{aligned} u &= e^{j\Omega t}\int_0^\infty\left[\frac{A(x)x}{h^2}e^{-\alpha z} - \frac{C(x)\beta x}{k^2}e^{-\beta z}\right][J_1(xr) - jN_1(xr)]dx \\ w &= e^{j\Omega t}\int_0^\infty\left[\frac{A(x)\alpha}{h^2}e^{-\alpha z} - \frac{C(x)x^2}{k^2}e^{-\beta z}\right][J_0(xr) - jN_0(xr)]dx \\ \sigma_z &= \mu e^{j\Omega t}\int_0^\infty\left[\frac{A(x)(k^2 - 2x^2)}{h^2}e^{-\alpha z} + \frac{C(x)2\beta x^2}{k^2}e^{-\beta z}\right] \\ &\qquad \times [J_0(xr) - jN_0(xr)]dx \\ \tau_{zr} &= \mu e^{j\Omega t}\int_0^\infty\left[\frac{-A(x)2\alpha x}{h^2}e^{-\alpha z} - \frac{C(x)x(k^2 - 2x^2)}{k^2}e^{-\beta z}\right] \\ &\qquad \times [J_1(xr) - jN_1(xr)]dx \end{aligned} \tag{34}$$

In erster Annäherung wird vorausgesetzt, dass zwischen der erregten Platte und dem isotropen Halbraum keine Schubspannungen durch Reibung übertragen werden. Das heisst, es muss auch hier für $z = 0$ die Schubspannung $\tau_{zr} = 0$ sein. (Nicht aber die Normalspannung σ_z.) Daraus folgt:

$$C(x) = -2\alpha k^2 A(x)/[h^2(k^2 - 2x^2)] \tag{35}$$

Damit folgt für die Oberfläche des isotropen Halbraumes (d.h. für $z = 0$):

$$u(r, 0) = e^{j\Omega t} \int_0^\infty \frac{x(k^2 - 2x^2) + 2\alpha\beta x}{h^2(k^2 - 2x^2)} A(x)[J_1(xr) - jN_1(xr)]dx \tag{36a}$$

$$w(r, 0) = e^{j\Omega t} \int_0^\infty \frac{\alpha k^2}{h^2(k^2 - 2x^2)} A(x)[J_0(xr) - jN_0(xr)]dx \tag{36b}$$

$$\sigma_z = (r, 0) = \mu e^{j\Omega t} \int_0^\infty \frac{(k^2 - 2x^2)^2 - 4\alpha\beta x^2}{h^2(k^2 - 2x^2)} A(x)[J_0(xr) - jN_0(xr)]dx) \tag{36c}$$

Die Integrationsfunktion $A(x)$ muss so bestimmt werden, dass sich:

(a) für $r \leqq r_0$, also innerhalb der Platte der Wert von $w(r, 0) = d$ also die statische Einsenkung der Platte ergibt;

(b) für $r > r_0$, also ausserhalb der Platte der Wert von $\sigma_z(r, 0) = 0$ also die Oberfläche spannungsfrei wird.

Die doppelte Integralgleichung (36b) und (36c) ist aber unlösbar und als Näherungslösung wird vorausgesetzt, dass die Einsenkung d der Einsenkung im statischen Fall unter der Belastung P entspricht. Das ist aber der bekannte Fall von Boussinesq, für den die Lösung mit:

$$d = P_\phi(1 - \gamma^2)/2r_0E = P_\phi/[8\mu(\gamma^2 - 1)r_0];$$

und der Spannungsverteilung

$$\sigma_z = P_\phi/(2\pi r_0\sqrt{r_0^2 - r^2})$$

bekannt ist. Dabei wird allerdings für $r = r_0$, also am Plattenrand $\sigma_z = \infty$. Damit kann für:

$$A(x) = \frac{P_\phi \sin(xr_0)h^2(k^2 - 2x^2)}{8\mu\pi r_0[(x^2 - \frac{1}{2}k^2)^2 - \alpha\beta x^2]} \quad \text{[cm]}, \tag{37}$$

angesetzt werden und man erhält für:

$$u(r, 0) = -\frac{P_\phi e^{j\Omega t}}{8\mu\pi r_0} \int_0^\infty \frac{x(2x^2 - k^2 - 2\alpha\beta)}{(x^2 - \frac{1}{2}k^2)^2 - \alpha\beta x^2} \times \sin(xr_0)[J_1(xr) - jN_1(xr)]dx \tag{38a}$$

$$w(r, 0) = \frac{P_\phi e^{j\Omega t}}{8\mu\pi r_0} \int_0^\infty \frac{\alpha k^2}{(x^2 - \frac{1}{2}k^2)^2 - \alpha\beta x^2} \times \sin(xr_0)[J_0(xr) - jN_1(xr)]dx \tag{38b}$$

$$\sigma_z(r,0) = \frac{P_\phi e^{j\Omega t}}{2\pi r_0}\int_0^\infty \sin(xr_0)[J_0(xr) - jN_0(xr)]dx \tag{38c}$$

Bei der Integration, die im einzelnen nicht ausgeführt wird, ist zu beachten, dass der Ausdruck $(x^2 - \frac{1}{2}k^2 - \alpha\beta x^2$ eine reelle Wurzel im Integrationsbereich hat, weshalb der Cauchy'sche Hauptwert des Integrals zu nehmen ist. Die Integrale enthalten ferner neben Ausdrücken, die divergierende Wellen darstellen, weil sie die Hankelfunktionen $H_0^{(2)}$ bzw. $H_1^{(2)}$ uerfassen, auch noch Ausdrücke, die stehende Wellen darstellen. Sie erfüllen daher noch nicht die Bedingung der Abstrahlung im Unendlichen. Erst die additive Hinzufügung adäquater freier, stehender Wellen, die immer Verträglichkeit mit den Randbedingungen besitzen, lassen sich die Randbedingungen im Unendlichen erfüllen.

Diese freien Wellen nach Formel 33 mit

$$D_1 = [-P_\phi 2\pi j \sin(x_1 r_0)]/[16\mu\pi r_0(4x^3 - 2x(k^2 + \alpha\beta)]$$

bringen die Lösung schliesslich zusammen mit 38 auf:

$$u(r,0) = \frac{P_\phi e^{j\Omega t}}{8\mu\pi r_0}\int_0^\infty \frac{x(2x^2 - k^2 - 2\alpha\beta)}{(x^2 - \frac{1}{2}k^2)^2 - \alpha\beta x^2}\sin(xr_0)\cdot[J_1(xr) - jN_1(xr)]dx$$

$$+\frac{jP_\phi e^{j\Omega t}\sin(x_1r_0)x_1(2x_1^2 - k^2 - 2\alpha_1\beta_1)}{8\mu r_0[4x^3 - 2x(k^2 + \alpha\beta)]}\cdot[J_1(x_1r) - jN_1(xr)]; \tag{39a}$$

$$w(r,0) = \frac{P_\phi e^{j\Omega t}}{8\mu\pi r_0}\int_0^\infty \frac{\alpha k^2}{(x^2 - \frac{1}{2}k^2) - \alpha\beta x^2}\sin(xr_0)[J_0(xr)$$

$$- jN_0(xr)]dx - \frac{jP_\phi^{j\Omega t}\sin(x_1r_0)k^2\alpha_1}{8\mu r_0[4x^3 - 2x(k^2 + \alpha\beta)]}$$

$$\cdot[J_0(x_1r) - jN_0(x_1r)] = \frac{P_\phi e^{j\Omega t}}{\mu r_0}[f_1 + jf_2]. \tag{39b}$$

Damit ergibt sich die Schwingungsamplitude (mit $P = P_0e^{j\Omega t}$)

$$x = w(r,0) = (P/\mu r_0)(f_1 + jf_2),$$

bw. ihr Absolutwert

$$|x| = (P/\mu r_0)\sqrt{f_1^2 + f_2^2}$$

und nach der Impedanzdefinition:

$$x = P/j\Omega Z = P/[j\Omega(X + jY)]$$
$$X = (\mu r_0/\Omega)(f_2/f_1^2 + f_2^2); \quad Y = (\mu r_0/\Omega)(f_1/f_1^2 + f_2^2) \tag{40}$$

Den Verlauf der beiden Funktionen f_1 und f_2 in Abhängigkeit von der Poissonzahl γ und dem Frequenzverhältnis $n = \Omega r_0 \sqrt{\rho}/\mu$ zeigt Abbildung 12.

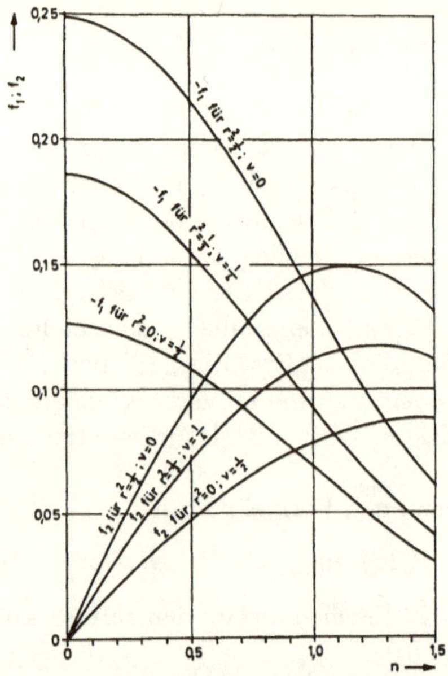

BILD 12. Verlauf der funktion f_1 und f_2 in Abhängigkeit von n und v

VERGLEICH DER ERGEBNISSE

Ein Vergleich der Impedanzbeiwerte des elastisch-isotropen Halbraumes mit der elastischen Säule zeigt nach Umformung von 40

$$X = r_0^2\sqrt{\rho\mu}\, f_2/[n(f_1^2+f_2^2)]\ ; \qquad Y = r_0^{2\phi}\sqrt{\rho\mu}\, f_1/[n(f_1^2+f_2^2)]$$

(elast. isotrop. Halbraum)
und aus 21 mit $F = r_0^2\pi$ für die zylindrische Säule und mit $E = 2(1+\gamma)\mu$

$$X = r_0^2\sqrt{\rho\mu}\cdot\pi\sqrt{2(1+\gamma)} = 4.96 r_0^2\sqrt{\rho\mu}\ ;\ Y = 0$$

konstant

(unendl. elastische Säule).

Abgesehen davon, dass der dimensionsbestimmende Faktor $r_0^2\sqrt{\rho\mu}$ übereinstimmt, ist eine Aehnlichkeit der Ergebnisse nicht gegeben. Das gleiche gilt für den Fall eines Vergleiches der Impedanzbestandteile des isotropen Halbraumes mit der schwingenden, auf endlicher Länge starr begrenzten Säule einerseits und einem einfachen Schwinger als Bodenmodell andererseits (Bild 13). Man wird daher wohl die Aussage treffen können, dass sowohl der einfache Schwinger als auch die starr begrenzte Säule als Bodenmodell die tatsächlichen Bewegungsverhältnisse nicht richtig beschreiben. Es ist zwar der Nachweis der befriedigenden Uebere-

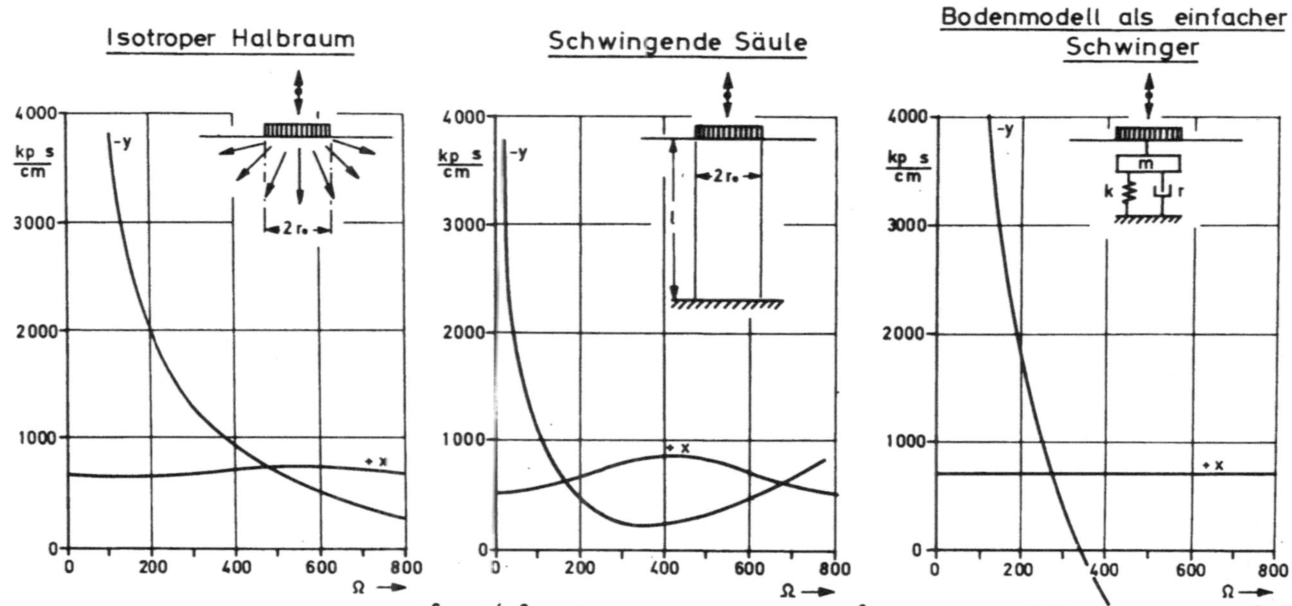

Abmessungen: $r_0 = 50\,cm$; $\varrho = 2{,}24 \cdot 10^{-6}\,kp\,cm^{-4}\,s^2$; $\mu = \frac{1}{4}$; $l = 250\,cm$; $\delta = 2{,}5$; $m = r_0^2 \pi\, l\, \varrho = 4{,}39\,kp\,cm^{-1}\,s^2$; $k = 525\,000\,kp\,cm^{-1}$

BILD 13. Verlauf des Real- und Imaginärteiles der Impedanz bei verschieden Schwingsystemen instimmung der ermittelten Ergebnisse des Schwingers am isotropen Halbraum für verschiedene Materialien, nicht aber noch für diejenigen erbracht, mit denen man es bei der Erarbei tung einer Theorie der Vibrationsverdichtung zu tun hat (7). Entsprechende Versuche, verbunden mit der Ermittlung der entsprechenden Materialkennwerte, müssen noch durchgeführt werden.

Eine Betrachtung des Bildes 14 mit dem Amplituden-Geschwindigkeits- und Beschleunigungsverhältnis zeigt für den Fall $\partial = 0$, d.h. für ein masseloses Erregersystem, das Fehlen jeder Resonanz. Die Amplituden nehmen unabhängig von der Poissonzahl mit steigendem Frequenzverhältnis kontinueirlich ab, die Geschwindigkeiten und Beschleunigungen aber zu. Die kontinuierlich von $\gamma = 0$ auf $\gamma = 1/2$ zunehmende Poissonzahl wirkt auf Amplituden, Geschwindigkeiten, und Beschleunigungen stark dämpfend, insbesondre wenn der Erreger massebehaftet ist, und zwar mit $m = \partial \rho r_0^3$.

Bild 15 zeigt, dass nur bei grosser Masse des Erregers und kleiner Poissonzahl Amplituden-, Beschleunigungs-, Geschwindigkeits- und Phasenresonanz zusammenfallen. Insoweit liegt eine gewisse Ähnlichkeit mit dem gedämpften Schwingsystem vor.

AUSBLICK

Bisher scheint die Theorie des zwangserregten, elastisch isotropen Halbraumes eine brauchbare Vorstellung über das Bewegungsverhalten einer Verdichtungsmaschine zu liefern. Die erläuterten Ergebnisse geben vorläufig nur eine Vorstellung über das Bewegungsverhalten des mit einer kreisförmigen Platte auf das Material wirkenden Erregers. Die mathematischen Untersuchungen müssen sich noch auf das Bewegungsverhalten des Materials ausserhalb der Erregerplatte an der Materialoberfläche sowie im Innern des unendlich ausgedehnten Materialkörpers sowie auf die dort auftretenden Normal- und Schubspannungen erstrecken.

Verdichtungsgeräte haben im allgemeinen keine kreisförmigen Auflager platten, jedoch kann angenommen werden, dass die quadratische oder annähernd quadratische Platte durch die Kreisplatte gut angenähert wird. Wirkungsflächen mit überwiegender Erstreckung in einer Dimension, wie sie bei Vibrationswalzen vorhanden sind, bedürfen einer besonderen mathematischen Behandlung. Ferner wäre der Fall der elastischisotropen Schicht bestimmter Dicke noch besonders zu untersuchen, die von einem starren Material oder einem Material mit anderen elastischen Eigenschaften unterlagert wird.

An die mathematische Durchleuchtung der Vorgänge wird sich eine umfangreiche Verifizierung der theoretisch ermittelten Ergebnisse im

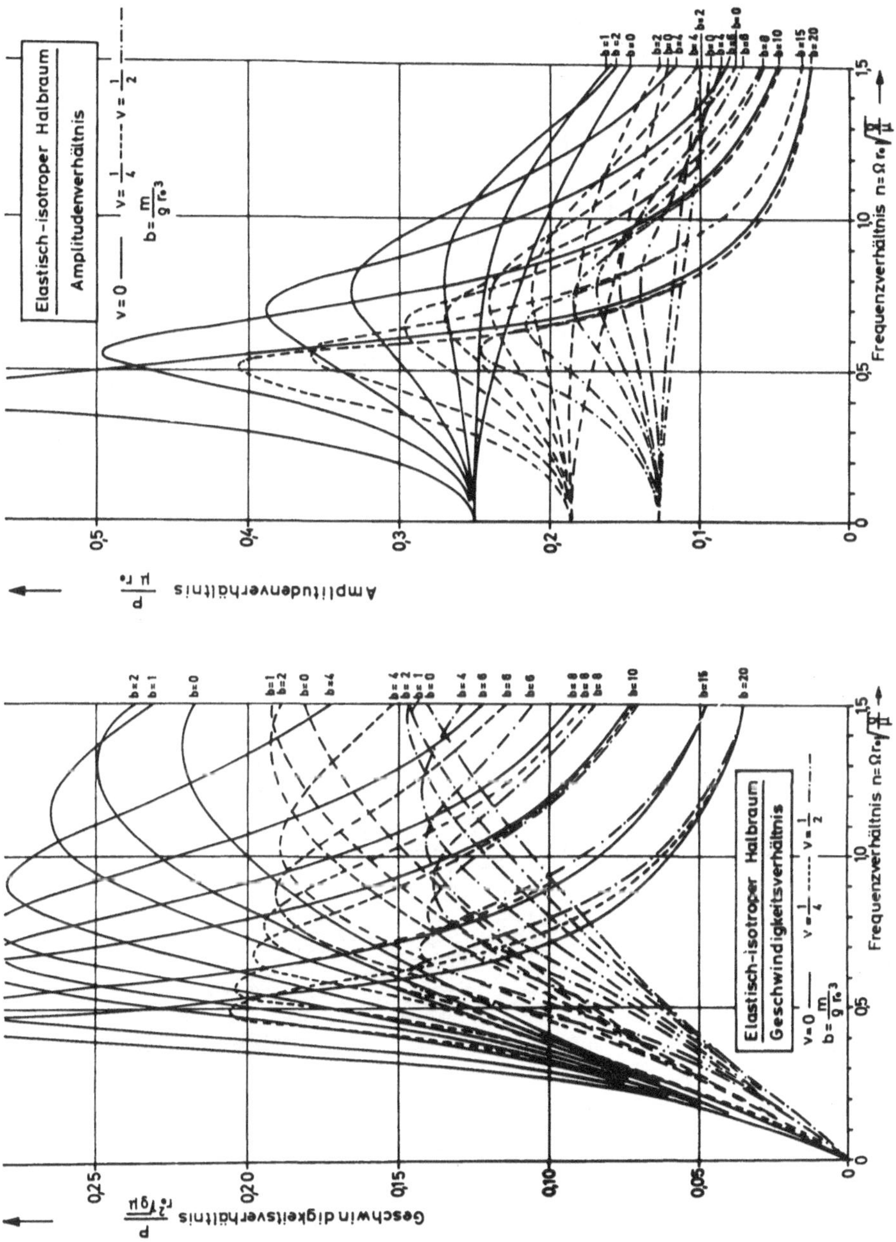
Elastisch-isotroper Halbraum
Amplitudenverhältnis
Elastisch-isotroper Halbraum
Geschwindigkeitsverhältnis
Frequenzverhältnis
Amplitudenverhältnis
Geschwindigkeitsverhältnis

BILD 14A

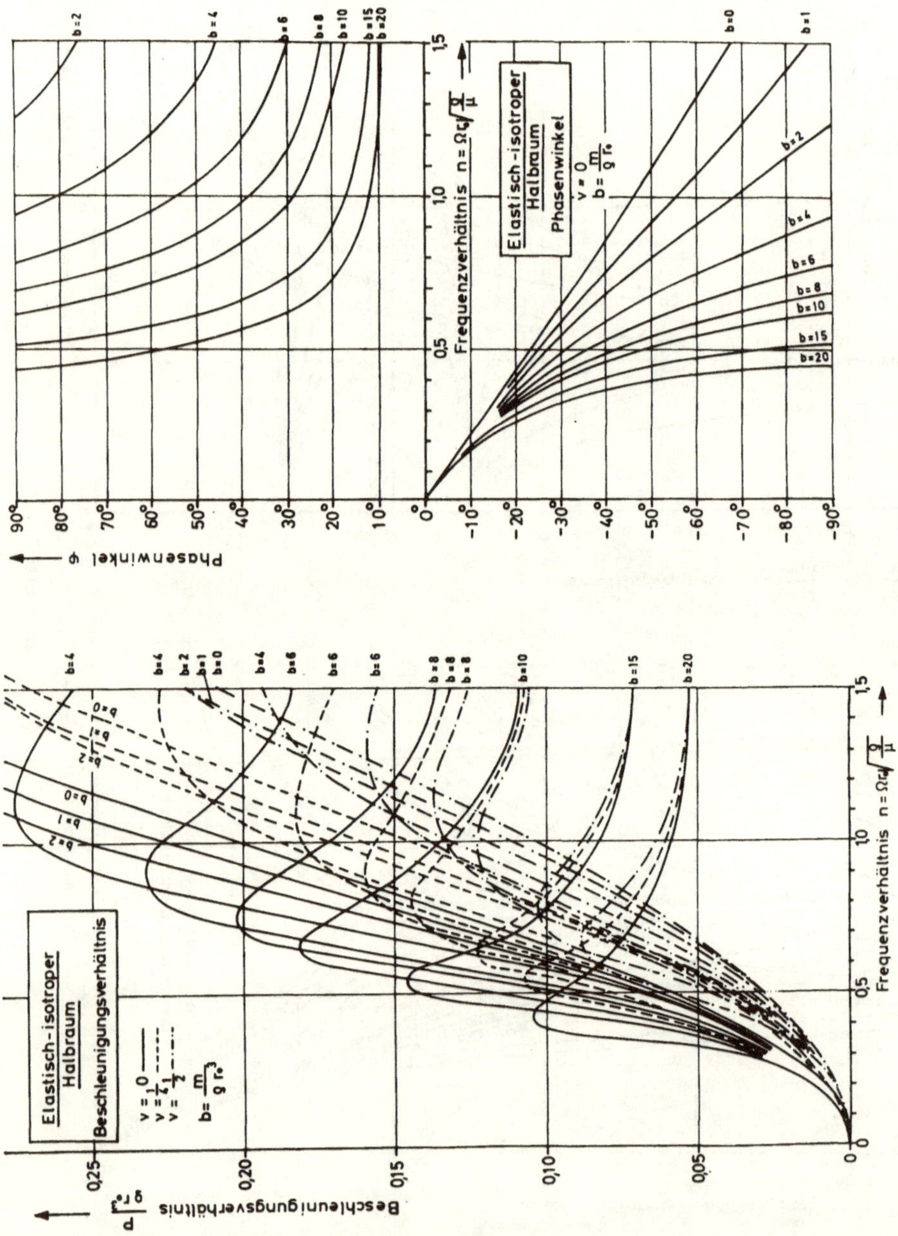

Elastisch-isotroper
Halbraum
Phasenwinkel
Phasenwinkel φ
Frequenzverhältnis
Elastisch-isotroper
Halbraum
Beschleunigungsverhältnis
Beschleunigungsverhältnis
Frequenzverhältnis

BILD 14B

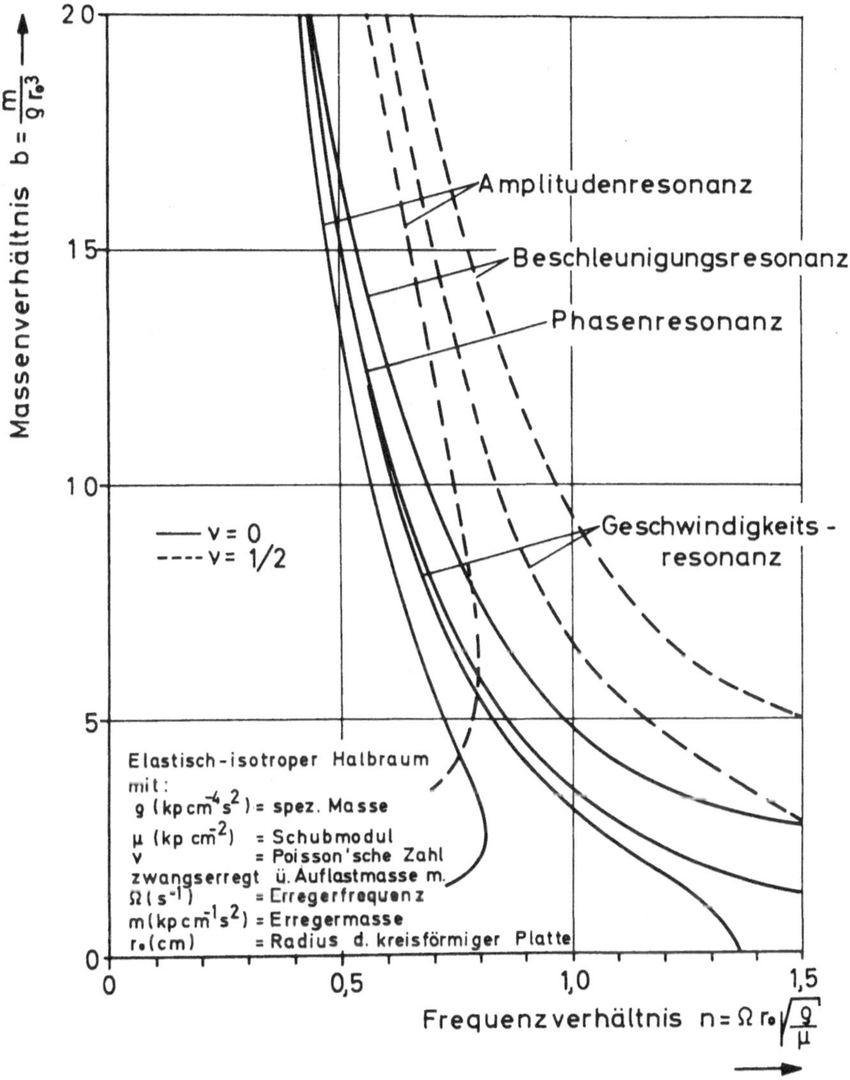

BILD 15. Resonanzverhältnisse

Versuchswege zusammen mit einer Ermittlung der Materialkennwerte anschliessen müssen. Erst dann—wenn diese Verifizierung die Richtigkeit der gewählten Ansätze nachweist, wird es möglich sein, die Theorie der Vibrationsverdichtung zur praktischen Anwendung auf die Konstruktion besserer und nutzbringenderer Verdichtungsgeräte zu verwenden.

LITERATUR

1. BATHELT, U., *Das Arbeitsverhalten des Rüttelverdichters auf plastisch-elastichem Untergrund*, Berlin: Ernst und Sohn, 1956.
2. HARTMANN, K., *Das Arbeitsverhalten des Einmassen-Rüttelverdichters*, Berlin: Ernst und Sohn, 1963.
3. KUPPERS, E., Beitrag zur Dimensionierung von Rütteleinrichtungen für die Verdichtung von Frischbeton, *Mitteilungen des Instituts für Baumaschinen und Baubetrieb*, Aachen, 1965.
4. JURECKA, W., Das Bewegungsverhalten von Vibrationsmaschinen, *Mitteilungen des Instituts für Baumaschinen und Baubetrieb*, Aachen, 1965.
5. REISSNER, E., Stationäre, achsialsymmetrische, durch eine schüttelnde Masse erregte Schwingungen eines homogenen, elastischen Halbraumes, *Ing. Archiv.*, 1936.
6. BYCROFT, G. N., Forced Vibrations of a Rigid Circular Plate on a Semi-Infinite Elastic Space and on an Elastic Stratum, *Phil. Transact. Roy. Soc.*, London, 1956.
7. SEZAWA, K., *Bul. Earthq. Res. Inst.*, Tokyo, 1929.
8. BYCROFT, G. N., *J. of Appl. Mechanics (American)*, 1955.

Evaluation of Single-Wheel Testing Techniques

A. J. GREEN*

N. R. MURPHY†

ABSTRACT

One of the current problems in mobility research at the U.S. Army Engineer Waterways Experiment Station is the application of single-wheel laboratory test data to the performance of multiwheeled vehicles that are free to develop the amount of slip appropriate to the trafficability conditions. It is important, therefore, that the laboratory testing techniques be evaluated to determine their influence on test results.

The study reported herein is a comparative evaluation of the effects of testing techniques on the performance of pneumatic tires operating in an air-dry sand and a nearly saturated fat clay. Controlled-pull tests and several types of controlled-slip tests were conducted during the programme. Also, the initial or starting conditions were varied, that is, tests were started on a firm surface, on the surface of the soil, and with the tire in a preformed depression.

It was tentatively concluded that the pull-slip and torque-slip relations developed for single wheels were valid and independent of testing techniques, and thus amenable to extrapolation to multiwheel vehicle performance.

This study was conducted under the sponsorship and guidance of the Directorate of Research and Development, U.S. Army Materiel Command.

*Engineer, Mobility Section, Army Mobility Research Branch, Mobility and Environmental Division, USAE Waterways Experiment Station, Vicksburg, Miss.

†Mathematician, Mobility Section, Army Mobility Research Branch, Mobility and Environmental Division, USAE Waterways Experiment Station, Vicksburg, Miss.

Spectral Techniques of Identification for Vehicle Systems

F. KOZIN

J. L. BOGDANOFF*

ABSTRACT

The problem of identification for the purpose of modeling complex systems is becoming paramount in its importance. Modeling such phenomena as the response of the human eye to stimuli, the response of complex chemical processes to various inputs, and the determination of the analytical model of missile and satellite mechanical systems require techniques that are presently in the process of evolution. The vehicle system, being a highly complex mechanical system, is also a system that must be identified for purposes of response characteristics, instabilities, and controllability.

In this paper, the basic spectral techniques of identification are reviewed, a few examples are given and feasibility for future practical use in the identification of real vehicles is discussed.

*Purdue University, Lafayette, Indiana.

www.ingramcontent.com/pod-product-compliance
Lightning Source LLC
LaVergne TN
LVHW100245210926
848480LV00003B/833